ADVANCED CELLULAR NETWORK PLANNING AND OPTIMISATION

ADVANCED CELLULAR NETWORK PLANNING AND OPTIMISATION
2G/2.5G/3G...EVOLUTION TO 4G

Edited by

Ajay R Mishra
Nokia Networks

John Wiley & Sons, Ltd

Copyright © 2007   John Wiley & Sons Ltd, The Atrium, Southern Gate, Chichester,
West Sussex PO19 8SQ, England

Telephone   (+44) 1243 779777

Email (for orders and customer service enquiries): cs-books@wiley.co.uk
Visit our Home Page on www.wiley.com

Reprinted February 2007

All Rights Reserved. No part of this publication may be reproduced, stored in a retrieval system or transmitted in any form or by any means, electronic, mechanical, photocopying, recording, scanning or otherwise, except under the terms of the Copyright, Designs and Patents Act 1988 or under the terms of a licence issued by the Copyright Licensing Agency Ltd, 90 Tottenham Court Road, London W1T 4LP, UK, without the permission in writing of the Publisher. Requests to the Publisher should be addressed to the Permissions Department, John Wiley & Sons Ltd, The Atrium, Southern Gate, Chichester, West Sussex PO19 8SQ, England, or emailed to permreq@wiley.co.uk, or faxed to (+44) 1243 770620.

Designations used by companies to distinguish their products are often claimed as trademarks. All brand names and product names used in this book are trade names, service marks, trademarks or registered trademarks of their respective owners. The Publisher is not associated with any product or vendor mentioned in this book.

This publication is designed to provide accurate and authoritative information in regard to the subject matter covered. It is sold on the understanding that the Publisher is not engaged in rendering professional services. If professional advice or other expert assistance is required, the services of a competent professional should be sought.

Other Wiley Editorial Offices

John Wiley & Sons Inc., 111 River Street, Hoboken, NJ 07030, USA

Jossey-Bass, 989 Market Street, San Francisco, CA 94103-1741, USA

Wiley-VCH Verlag GmbH, Boschstr. 12, D-69469 Weinheim, Germany

John Wiley & Sons Australia Ltd, 42 McDougall Street, Milton, Queensland 4064, Australia

John Wiley & Sons (Asia) Pte Ltd, 2 Clementi Loop #02-01, Jin Xing Distripark, Singapore 129809

John Wiley & Sons Canada Ltd, 6045 Freemont Blvd, Mississauga, ONT, L5R 4J3

Wiley also publishes its books in a variety of electronic formats. Some content that appears in print may not be available in electronic books.

British Library Cataloguing in Publication Data

A catalogue record for this book is available from the British Library

ISBN-13 978-0-470-01471-4 (HB)

Typeset in 9/11pt Times by TechBooks, New Delhi, India.
Printed and bound in Great Britain by Antony Rowe Ltd, Chippenham, England.
This book is printed on acid-free paper responsibly manufactured from sustainable forestry in which at least two trees are planted for each one used for paper production.

*Dedicated
to
The Lotus Feet of my Guru*

Contents

Forewords	xiii
Acknowledgements	xvii
Introduction	xix
1 Cellular Networks	**1**
Ajay R Mishra	
1.1 Introduction	1
1.2 First Generation Cellular Networks	1
1.2.1 NMT (Nordic Mobile Telephony)	1
1.2.2 AMPS (Advanced Mobile Phone System)	2
1.3 Second Generation Cellular Networks	2
1.3.1 D-AMPS (Digital Advanced Mobile Phone System)	3
1.3.2 CDMA (Code Division Multiple Access)	3
1.3.3 GSM (Global System for Mobile Communication)	3
1.3.4 GPRS (General Packet Radio Service)	9
1.3.5 EDGE (Enhanced Data Rate for GSM Evolution)	10
1.4 Third Generation Cellular Networks	10
1.4.1 CDMA2000	10
1.4.2 UMTS	10
1.4.3 HSDPA in UMTS	12
2 Radio Network Planning and Optimisation	**15**
Johanna Kähkönen, Nezha Larhrissi, Cameron Gillis, Mika Särkioja, Ajay R Mishra and Tarun Sharma	
2.1 Radio Network Planning Process	15
2.1.1 Network Planning Projects	15
2.1.2 Network Planning Project Organisation	16
2.1.3 Network Planning Criteria and Targets	17
2.1.4 Network Planning Process Steps	18
2.2 Preplanning in a GSM Radio Network	21
2.2.1 GSM Network Planning Criteria	21
2.2.2 Introducing GPRS in the GSM Network	22
2.2.3 Introducing EGPRS in the GSM Network	23
2.2.4 WCDMA in UMTS	26
2.3 Radio Network Dimensioning	29
2.3.1 Link Budget Calculations	29

	2.3.2 Dimensioning in the EGPRS Network	34
	2.3.3 Dimensioning in the WCDMA Radio Network	34
2.4	Radio Wave Propagation	40
	2.4.1 Okumura–Hata Model	41
	2.4.2 Walfish–Ikegami Model	43
	2.4.3 Ray Tracing Model	44
	2.4.4 Model Tuning	45
2.5	Coverage Planning	46
	2.5.1 Coverage Planning in GSM Networks	46
	2.5.2 Coverage Planning in EGPRS	54
	2.5.3 Coverage Planning in WCDMA Networks	57
2.6	Capacity Planning	57
	2.6.1 Capacity Planning in GSM Networks	57
	2.6.2 EGPRS Capacity Planning	62
	2.6.3 Capacity Planning in WCDMA Networks	69
2.7	Frequency Planning	71
	2.7.1 Power Control	73
	2.7.2 Discontinuous Transmission	74
	2.7.3 Frequency Hopping	74
	2.7.4 Interference Analysis	75
2.8	Parameter Planning	76
	2.8.1 Parameter Planning in the GSM Network	76
	2.8.2 Parameter Planning in the EGPRS Network	78
	2.8.3 Parameter Planning in the WCDMA Network	87
2.9	Radio Network Optimisation	106
	2.9.1 GSM Radio Network Optimisation Process	106
	2.9.2 Optimisation in the EGPRS Network	162
	2.9.3 Optimisation in the WCDMA Network	181
3	**Transmission Network Planning and Optimisation**	**197**
	Ajay R Mishra and Jussi Viero	
3.1	Access Transmission Network Planning Process	197
	3.1.1 Master Planning	197
	3.1.2 Detail Planning	198
3.2	Fundamentals of Transmission	199
	3.2.1 Modulations	199
	3.2.2 Multiple Access Schemes	199
3.3	Digital Hierarchies – PDH and SDH	201
	3.3.1 Plesiochronous Digital Hierarchy (PDH)	201
	3.3.2 Synchronous Digital Hierarchy (SDH)	203
	3.3.3 Asynchronous Transfer Mode (ATM)	217
3.4	Microwave Link Planning	234
	3.4.1 Microwave Link	236
	3.4.2 Microwave Tower	242
	3.4.3 Microwave Link Design	242
	3.4.4 LOS Check	246
	3.4.5 Link Budget Calculation	247
	3.4.6 Repeaters	251
3.5	Microwave Propagation	253
	3.5.1 Slow Fading	253

3.5.2 Fast Fading	256
3.5.3 Overcoming Fading	264
3.6 Interface Planning	268
3.6.1 A_{bis} Planning	269
3.6.2 Dynamic A_{bis}	269
3.6.3 Interface Planning in the UMTS Access Transmission Network	272
3.7 Topology Planning	280
3.8 Frequency Planning and Interference	282
3.8.1 Loop Protection	284
3.9 Equipment Planning	285
3.9.1 BSC and TCSM Planning	285
3.10 Timeslot Planning	286
3.10.1 Linear TS Allocation	286
3.10.2 Block TS Allocation	286
3.10.3 TS Grouping	286
3.10.4 TS Planning in the EDGE Network	287
3.12 Transmission Management	291
3.12.1 Element Master	292
3.12.2 Management Buses	292
3.13 Parameter Planning	293
3.13.1 BTS/AXC Parameters	293
3.13.2 RNC Parameters	294
3.14 Transmission Network Optimisation	296
3.14.1 Definition of Transmission	296
3.14.2 GSM/EDGE Transmission Network Optimisation	299
3.14.3 UMTS Transmission Network Optimisation	303
4 Core Network Planning and Optimisation	**315**
James Mungai, Sameer Mathur, Carlos Crespo and Ajay R Mishra	
Part I Circuit Switched Core Network Planning and Optimisation	315
4.1 Network Design Process	315
4.1.1 Network Assessment	315
4.1.2 Network Dimensioning	319
4.2 Detailed Network Planning	323
4.3 Network Evolution	327
4.3.1 GSM Network	327
4.3.2 3GPP Release 99 Network	328
4.3.3 3GPP Release 4 Network	329
4.3.4 3GPP Release 5 and 6 Networks	332
4.4 3GPP Release 4 Circuit Core Network	335
4.4.1 Release 4 Core Network Architecture	335
4.4.2 CS Network Dimensioning	335
4.5 CS Core Detailed Network Planning	352
4.5.1 Control Plane Detailed Planning	352
4.5.2 Control Plane Routing	358
4.6 User Plane Detailed Planning	359
4.6.1 Configuring Analyses in the MSS	359
4.6.2 Routing Components of the MSC Server	360
4.6.3 User Plane Routing	360

4.7 CS Core Network Optimisation 368
    4.7.1 Key Performance Indicators 368
    4.7.2 Network Measurements 369
    4.7.3 CS Core Network Audit 370
    4.7.4 Audit Results Analysis 375
    4.7.5 Network Optimisation Results 379

Part II Packet Switched Core Network Planning and Optimisation 379
  4.8 Introduction to the PS Core Network 379
    4.8.1 Basic MPC Concepts 380
    4.8.2 Packet Routing (PDP Context) 382
    4.8.3 Interface of the GPRS with the 2G GSM Network 384
  4.9 IP Addressing 385
    4.9.1 Types of Network 385
    4.9.2 Dotted-Decimal Notation 386
    4.9.3 Subnetting 387
  4.10 IP Routing Protocols 388
  4.11 Dimensioning 390
    4.11.1 GPRS Protocol Stacks and Overheads 390
  4.12 IP backbone Planning and Dimensioning 397
    4.12.1 Current Network Assessment 397
    4.12.2 Dimensioning of the IP Backbone 398
    4.12.3 Bandwidth Calculations 399
  4.13 Mobile Packet Core Architecture Planning 399
    4.13.1 VLAN 401
    4.13.2 Iu-PS Interface 401
    4.13.3 Gn Interface Planning 401
    4.13.4 Gi Interface Planning 402
    4.13.5 Gp Interface Planning 403
  4.14 Packet Core Network Optimisation 404
    4.14.1 Packet Core Optimisation Approaches 404
    4.14.2 Packet Core Optimisation – Main Aspects 405
    4.14.3 Key Performance Indicators 408
    4.14.4 KPI Monitoring 409
  4.15 Security 410
    4.15.1 Planning for Security 410
    4.15.2 Operational Security 411
    4.15.3 Additional Security Aspects 411
  4.16 Quality of Service 412
    4.16.1 Introduction to QoS 412
    4.16.2 QoS Environment 412
    4.16.3 QoS Process 412
    4.16.4 QoS Performance Management 414

5 Fourth Generation Mobile Networks **417**
    Ajay R Mishra
  5.1 Beyond 3G 417
  5.2 4G Network Architecture 417
  5.3 Feature Framework in a 4G Network 418
    5.3.1 Diversity in a 4G Network 418
  5.4 Planning Overview for 4G Networks 419

5.4.1 Technologies in Support of 4G ... 420
5.4.2 Network Architectures in 4G ... 421
5.4.3 Network Planning in 4G Networks ... 421
5.5 OFDM ... 423
5.5.1 What Is OFDM? ... 423
5.5.2 MIMO Systems ... 425
5.6 All-IP Network ... 427
5.6.1 Planning Model All-IP Architecture ... 428
5.6.2 Quality of Service ... 432
5.7 Challenges and Limitations of 4G Networks ... 435
5.7.1 Mobile Station ... 435
5.7.2 Wireless Network ... 435
5.7.3 Quality of Service ... 437

Appendix A: Roll-Out Network Project Management ... **439**
Joydeep Hazra
A.1 Project Execution ... 439
A.2 Network Implementation ... 440
A.2.1 Site Selection and Acquisition ... 440
A.2.2 Provision of Site Support Elements ... 444
A.2.3 Site Planning and Equipment Installation ... 444
A.2.4 Legal Formalities and Permissions ... 445
A.2.5 Statutory and Safety Requirements ... 446
A.3 Network Commissioning and Integration ... 446
A.3.1 Confirming the Checklist ... 447
A.3.2 Powering-Up and System Precheck ... 447
A.3.3 Commissioning ... 447
A.3.4 Inspection and Alarm Testing ... 448
A.3.5 Parameter Finalisation ... 448
A.3.6 Tools and Macros ... 449
A.3.7 Integration of Elements ... 449
A.3.8 System Verification and Feature Testing ... 451
A.3.9 System Acceptance ... 452
A.4 Care Phase ... 453
A.4.1 Care Agreement ... 453
A.4.2 Care Services ... 455
A.4.3 Other Optional O&M Assistance Services ... 462

Appendix B: HSDPA ... **467**
Rafael Sánchez-Mejías
B.1 Introduction ... 467
B.2 HSDPA Performance ... 468
B.3 Main Changes in HSDPA ... 468
B.3.1 HSDPA Channels ... 469
B.3.2 MAC Layer Split ... 470
B.3.3 Adaptive Modulation and Coding (AMC) Scheme ... 471
B.3.4 Error Correction (HARQ) ... 471
B.3.5 Fast Packet Scheduling ... 472
B.3.6 Code Multiplexing (Optional) ... 473
B.3.7 Impact on the Iub Interface ... 474
B.4 Handset Capabilities ... 475

B.5	HSDPA Planning and Dimensioning	476
	B.5.1 Planning Basics	476
	B.5.2 HSDPA Dimensioning	476
	B.5.3 HSDPA Planning	478
B.6	Further Evolution: Release 6 HSDPA, HSUPA and HSPA	480
	B.6.1 HSDPA Release 6 Improvements	480
	B.6.2 HSUPA	480

Appendix C: Digital Video Broadcasting — **483**
Lino Dalma

C.1	Introduction	483
C.2	Handheld Television: Viewing Issues	483
C.3	System Issues of DVB-H and Broadband Wireless	484
C.4	Mobile Broadcasting	485
C.5	Mobile Broadcasting	485
	C.5.1 DVB-H Technology Overview	486
C.6	DVB-H Introduction	486
	C.6.1 Motivation for Creating DVB-H	486
	C.6.2 Overview of Mobile TV Broadcasting Technology (DVB-H)	487
	C.6.3 Overview of DVB-T	488
	C.6.4 DVB-H Innovative Elements	490
	C.6.5 DVB-T and DVB-H Coexistence	491
	Conclusion	492

Appendix D: TETRA Network Planning — **493**
Massimiliano Mattina

D.1	TETRA Standard	493
D.2	TETRA Services	495
D.3	TETRA Network Elements	496
D.4	TETRA Main Features	499
	D.4.1 Physical Layer	500
	D.4.2 TETRA Memorandum of Understanding	502
D.5	Introduction to TETRA Network Planning	502
	D.5.1 Radio Network Planning	503
	D.5.2 Traffic Capacity Planning	505

Suggested Reading — **507**

Index — **511**

Forewords

Mobile Revolution

A global revolution is going on – and most people do not even realise that it is happening. I am talking about the Mobile Revolution.

During each of the next 5 years there will be about 350 million additional people on this earth who will be connected to the communications network or, in other words, 1 million new subscribers every day or 10 new subscribers every second. By the time you will have finished reading this page, another 1000 more people will have become members of the global communications community. For most of them it will not only be their first mobile phone, it will be their first phone, and for at least half of them it will be their first camera, their first music player and, of course, their first Internet access.

Mobile Revolution is a big word, you might say. Why not be more humble and call it a Mobile Evolution? Actually, the growth pattern of the cellular industry in the 1990s was very much evolutionary. However, in the last couple of years the momentum has attained a different quality. From making voice mobile we are moving to all aspects of our lives. Mobile music, mobile TV, mobile email and mobile office all allow us to liberate ourselves from a fixed location. People no longer need to rush back home in order not to miss the first minutes of a sports event on television and sales people update their sales catalogue and price lists over-the-air directly into the mobile device on the way to their clients. Mobile services with video capabilities have the capacity to transform the way production plants are being monitored and maintained. Complete value chains are just now being redefined. Sociologists have started academic studies on the effect of mobile communications on societies, for instance regarding community building among teenagers. Whereas tribal behaviour among teenagers used to be tied to a certain location, e.g. the local park, communities nowadays are formed through communications devices irrespective of location. This will have a significant effect on the way societies will develop.

In April 1991, I saw the first commercial GSM Base Station in the field. Most probably the last pieces of this first generation equipment have long found their way into the technical museums of this world. Since then thousands of innovative and experienced engineers have worked together in order to take technology from GSM to GPRS, EDGE, WCDMA and HSPA; soft switching is replacing traditional switching concepts and IP technology enables true technology and service convergence. The true challenge of the future, however, does not lie in bringing even higher speeds into the networks or in connecting even more people faster. In my mind the real engineering masterpiece will be to create superior quality-of-experience for the end user. Consumers are not interested at all in three- or four-letter abbreviations, let alone in understanding how they work. Consumers want ease-of-use and superior quality – anytime and anywhere. The ultimate engineering challenge is to understand the desired experience and implement the technical ecosystem from the user interface and the operating system of the mobile device through the radio network and the air interface, passing through the application middleware and connecting to the Internet, the voice network or the corporate environment.

Mobile networks enable an uninterrupted phone conversation to be enjoyed while driving on German motorways with no speed limits at speeds of 200 km/h with handovers every 50–60 s or allow businessmen to check their emails while riding elevators in some of the highest buildings in the world in Shanghai or

Hong Kong. However, do they know about the technology behind it, the radio propagation, the cell design, the handover optimisation or the latency management? Imagine what technical understanding is needed to plan, design, implement and optimise the service and the network, end-to-end. Broadband, real-time services like video calls are just being added as an additional dimension to the already complex equation. Both the service providers and operators need to have a detailed understanding of what is required to create a quality network for mobile users in order to provide them with a quality experience. This is where this book *Advanced Cellular Network Planning and Optimisation* comes in handy. Covering the aspects that are required to design and optimise various types of networks ranging from GSM to EGPRS to UMTS and across all domains, radio, transmission and core, this book will definitely help people in the cellular industry to get their networks to a level where the subscribers will have a quality experience. Also, with an introduction to fourth generation technologies, this book offers a window towards what the future has in store for us.

In this book, Ajay R Mishra and his colleagues have put in their years of experience, ranging from the R&D labs to actual network planning in the field across all six continents, on paper for the benefit of professionals in the mobile industry.

One hundred years ago, railways and roads were built in order to connect cities. The Internet is connecting computers and machines. The Mobile Revolution is now connecting the world – connecting people and their lives.

Bosco Novak
Senior Vice President and General Manager
Nokia Networks
Dusseldorf, Germany

On the Crossroads of History

The mobile telecommunications industry has reached an important crossroads in its development: one road leads to a maturing 2G voice-centric end user industry, while the other offers incredible possibilities for end users to enjoy a variety of data-centric 3G services on top of the conventional voice services.

The profitability of the industry is at reasonably high levels, although the industry still suffers from infrastructure overcapacity, which arguably leads to aggressive price reductions as mobile operators fight for market share. The recent strong growth in 2G mobile subscribers, in emerging markets in particular, resulted in the expected passing of the 2 billion benchmark in mobile subscriptions during the year 2005. The long-awaited issuance of 3G licenses in China and the start of 3G roll-outs in the US should take place during 2006. In Western Europe, where 3G services have been launched commercially, subscriber adoption has remained below previous forecasts, despite the coverage build-outs in urban areas.

It is believed that data offers – not only to offset declining voice ARPUs (annual revenues per user) – substantial additional revenue opportunities for mobile operators. The current data revenues account for 16 per cent of ARPU. Moving to higher data speed technologies, such as WCDMA/HSDPA and CDMA 1X/EV-DO, which are specifically designed for higher data throughput and capacity, can greatly boost operators' revenues.

The competitive outlook of the telecommunications industry calls for a new approach in managing costs: one has to go beyond cost cutting and find the underlying ways to improve efficiency, while delivering value-added services for mobile end users. One solution lies in the planning and optimisation of networks, through which improved asset utilisation and fine-tuned add-on network investments can be achieved, which in turn enable enhanced network performance.

The challenges in the planning and optimisation of networks are exhaustively covered by Ajay R Mishra and his colleagues in this book. Notwithstanding their several publications on the subject area, they have been able to maintain a direct involvement with the actual network planning and optimisation discipline. This rare combination is evident throughout the text. In addition to academic soundness, this book offers hands-on guidance to solve a variety of practical network planning and optimisation issues.

Timo S. Hanninen
Vice President
Nokia Networks
Helsinki, Finland

Acknowledgements

I would like to thank, on behalf of all the contributors, various people within the organisation who encouraged each of us to write this book. Firstly, I would like to thank Kari Suneli and Antti Rahikainen for their immense encouragement to take on this project.

I would like to also thank Veli-Pekka Somila, Reema Malhotra and Juha Sarkioja for their support in the difficult times in the course of this project.

Special thanks are due to the following colleagues and friends for taking time to contribute and read the manuscript and give their valuable comments: Johanna Kähkönen, Nezha Larhrissi, Tarun Sharma, Cameron Gillis, Mika Särkioja, Jussi Viero, Sameer Mathur, James Mungai, Carlos Crespo, Pauli Aikio, Tomi Nurmi, Olli Nousia, Manuel Blasco, Christophe Landemaine, and Irina Nicolescu.

Many thanks are due to Bosco Novak and Timo Hanninen for donating their precious time to write the visionary forewords for this book.

I would like to also thank students of University of Delhi, Rajat Budhiraja and Sandeep Makker, whose contributions during the course of writing this book were immense.

Thanks are also due to Azizah Aziz for helping me during the last phases of writing this book. I would also like to thank Joydeep Hazra, Rafael Sanchez, Massimiliano Mattina and Lino Dalma who provided the material for the appendices.

A big thanks goes to the team at John Wiley & Sons for their guidance and legendary patience during the course of writing this book.

Finally, I would like to thank my parents Mrs Sarojini Devi Mishra and Mr Bhumitra Mishra who gave me the inspiration to undertake this project and deliver it to the best of my capability.

Ajay R Mishra

Introduction

With each passing day, the maturity level of the mobile user and the complexity level of the cellular network reaches a new high. The networks are no more 'traditional' GSM networks, but a complex mixture of 2G, 2.5G and 3G technologies. Not only this, but new technologies beyond 3G are being utilised in these cellular networks. The very existence of all these technologies in one cellular network has brought the work of designing and optimisation of the networks to be viewed from a different perspective. Gone are the days of planning GSM, EGPRS or WCDMA networks individually. Now the cellular network business is about dimensioning for new and advanced technologies, planning and optimising 3G networks, while upgrading 2G/2.5G networks. This is not to mention the hard work required to maintain these networks in the phases after designing, implementing and commissioning.

This book has been written keeping the present day in mind. There are many instances where a practical approach has been followed in writing this book; e.g. problems faced by design and optimisation engineers have been provided with the solutions in an easy-to-follow approach. The project management related issues that have an impact on network planning have been covered as it gives the reader an insight into the real life situation. For this purpose an appendix on project management in the roll-out project has been included, giving the reader an end-to-end view of a roll-out project process. Though the fundamentals for most of the concepts are covered in the book *Fundamentals of Cellular Network Planning and Optimisation* (published by John Wiley & Sons, 2004), some places in this book do have some basic concepts for the benefit of the reader. However, *Fundamentals* is highly recommended.

The book has been divided into five chapters. The first chapter deals with the introduction to the cellular networks. This chapter takes us on a journey from the first generation to the third generation networks.

Chapter 2 describes radio network planning and optimisation. The chapter deals with the issues of planning and optimisation for GSM, EGPRS and WCDMA networks. Every time a concept is explained, e.g. planning in GSM, it is followed by planning in EGPRS networks followed by planning in WCDMA networks. This is done while keeping in mind the philosophy behind this book. It will give the engineer working in the field quick information on how he/she should handle a particular technology network.

Chapter 3 discusses transmission network planning and optimisation. Microwave planning has been covered in much more detail for the benefit of the transmission/microwave planning engineers. PDH, SDH and ATM have been covered in much more detail as compared to *Fundamentals*. Again the writing methodology is similar to Chapter 2.

Chapter 4 is about core network planning and optimisation. This chapter is divided into two parts: circuit switched core and packet switched core network planning and optimisation. As core planning engineers are aware, the planning of core networks these days is based on releases, e.g. release 99, release 4, etc. Therefore the presentation is based on the release rather than the technologies (GSM, EGPRS, etc.).

Chapter 5 discusses technologies beyond 3G. However, as the standardisation for 4G is not yet there, we have just tried to touch this field from the perspective of the design engineers in the field, so that they have some idea of what is to come.

A few appendices are also given. These appendices contributed by the experts in the respective fields deal with aspects such as Cellular Network Roll-Out Project Management, High Speed Packet Switched

Data, Digital Video Broadcasting and TETRA Network Planning. I hope that engineers will find these extremely useful for their work.

In the end, there is a list of carefully chosen books and papers that I am sure the reader will find useful.

I would be very grateful if readers would send in feedback to fcnp@hotmail.com, making any comments/suggestions that might improve the book.

Ajay R Mishra
(Editor)

1

Cellular Networks

Ajay R Mishra

1.1 Introduction

The cellular technology evolution has been going on since the late 1950s, though the first commercial systems came into being in the late 1970s and early 1980s. Here is a brief overview of the cellular technologies and the networks that made an impact on the development and the fast evolution of the mobile communications.

1.2 First Generation Cellular Networks

Since the late 1970s when the cellular era started, mobile communication has gone through an evolutionary change every decade in terms of technology and usage. Japan took the lead in the development of cellular technology, which resulted in the deployment of the first cellular networks in Tokyo. Within a couple of years Nordic Mobile Telephony (NMT) started cellular operations in Europe. Along with it, systems such as AMPS (Advanced Mobile Phone Service) started in the USA, while TACS (Total Access Communication System) started in the UK. These formed a part of what was called 'First Generation Mobile Systems', which catered for speech services and were based on analogue transmission techniques. The geographical area was divided into small sectors, each called a cell. Hence, the technology came to be known as cellular technology while the phones were called cell phones. All the systems that were initially developed were quite incompatible with each other. Each of these networks implemented their own standards. Facilities such as roaming within the continent were impossible and most countries had only one operator. The penetration was also low; e.g. penetration in Sweden was just 7 %, while countries like Portugal had a penetration of only 0.7 %. Handsets were also expensive, the minimum being more than $1000. Apart from higher costs and incompatibility with other cellular networks, first generation technology also had an inherent limitation in terms of channels, etc.

1.2.1 NMT (Nordic Mobile Telephony)

The NMT mobile phone system was created in 1981 as a response to the increasing congestion and heavy requirements of the ARP (auto radio puhelin, or car radio phone) mobile phone network. The technical principles of NMT were ready by 1973 and specifications for base stations were ready in 1977. It is based

---

Advanced Cellular Network Planning and Optimisation   Edited by Ajay R Mishra
© 2007 John Wiley & Sons, Ltd

on analogue technology (first generation or 1G) and two variants exist: NMT 450 and NMT 900. The numbers indicate the frequency bands used. NMT 900 was introduced in 1986 because it carries more channels than the previous NMT 450 network. The NMT network has mainly been used in the Nordic countries, Baltic countries and Russia, but also in the Middle East and in Asia. NMT had automatic switching built into the standard from the beginning. Additionally, the NMT standard specified billing and roaming. The NMT specifications were free and open, allowing many companies to produce NMT hardware and pushing prices down. A disadvantage of the original NMT specification is that traffic was not encrypted. Thus, anyone willing to listen in would just have to buy a scanner and tune it to the correct frequency. As a result, some scanners have had the NMT bands 'deleted' so they could not be accessed. This is not particularly effective as it is not very difficult to obtain a scanner that does not have these restrictions; it is also possible to re-program a scanner so that the 'deleted' bands can be accessed. Later versions of the NMT specifications defined optional analogue encryption, which was based on two-band audio frequency inversion. If both the base station and the mobile station supported encryption, they could agree upon using it when initiating a phone call. Also, if two users had mobile stations supporting encryption, they could turn it on during conversation, even if the base stations did not support it. In this case audio would be encrypted all the way between the two mobile stations. While the encryption method was not at all as strong as encryption in newer digital phones, it did prevent casual listening with scanners. The cell sizes in an NMT network range from 2 km to 30 km. With smaller ranges the network can service more simultaneous callers; e.g. in a city the range can be kept short for better service. NMT used full duplex transmission, allowing for simultaneous receiving and transmission of voice. Car phone versions of NMT used transmission power of up to 6 watts and handsets up to 1 watt. Signalling between the base station and the mobile station was implemented using the same RF channel that was used for audio, and using the 1200 bps (bits per second) FFSK modem. This caused the periodic short noise bursts that were uniquely characteristic of NMT sound.

1.2.2 AMPS (Advanced Mobile Phone System)

The first cellular licences in the US were awarded in 1981, and the cellular services started in 1983 in Chicago and The Baltimore–Washington area using the AMPS. The AMPS was based on the FDMA (frequency division multiple access) technology, which allowed multiple users in a cell or cell sector. Initially, cell size was not fixed and an eight mile radius was used in urban areas and a twenty-five mile radius in rural areas. However, as the number of users began to increase, new cells were added. With the addition of every new cell, the frequency plan was to be re-done to be able to avoid interference related problems. This system not only had capacity related problems, but the security system was also poor. If you are able to get hold of another person's serial code, it would be possible to make illegal calls. Although efforts were made to address these problems, especially the ones related to capacity, the results were not sufficient and the industry started to look into other options, such as the next generation digital systems. The TACS was similar to the AMPS and operated in the 900 MHz frequency range.

1.3 Second Generation Cellular Networks

Due to the incompatibility of the various systems in place, the European commission started a series of discussions that tried to change the then existing telecommunication regulatory framework, leading to a more harmonised environment which resulted in the development of a common market for the telecommunication services and equipment. In the early 1990s, digital transmission technology came into force, bringing with it the next generation system, called the 'Second Generation Mobile System'. Digitisation means that the sound of the speaker's voice was processed in a way that imitated a human ear through techniques such as sampling and filtering. This made it possible for many more mobile users to be accommodated in the radio spectrum. Key 2G systems in this generation included GSM (Global

Systems for Mobile Communications), TDMA IS-136, CDMA IS-95, PDC (Personal Digital Cellular) and PHS (Personal Handy Phone System).

1.3.1 D-AMPS (Digital Advanced Mobile Phone System)

IS 54 and IS 136 (where IS stands for Interim Standard) are the second generation mobile systems that constitute the D-AMPS. This was the digital advancement of the then existing AMPS in America. TDMA (Time Division Multiple Access) was used as the air interface protocol. The D-AMPS used existing AMPS channels and allows for smooth transition between digital and analogue systems in the same area. Capacity was increased over the preceding analogue design by dividing each 30 kHz channel pair into three time slots and digitally compressing the voice data, yielding three times the call capacity in a single cell. A digital system also made calls more secure because analogue scanners could not access digital signals.

IS-136 added a number of features to the original IS-54 specification, including text messaging, circuit switched data (CSD) and an improved compression protocol. The short message service (SMS) and CSD were both available as part of the GSM protocol, and IS-136 implemented them in a nearly identical fashion. D-AMPS used the 800 and 900 MHz frequency bands – as does the AMPS – but each 30 kHz channel (created by FDMA) is further subdivided into three TDMA, which triples the channels available and the number of calls.

1.3.2 CDMA (Code Division Multiple Access)

CDMA has many variants in the cellular market. N-CDMA, i.e. Narrowband CDMA (or just CDMA), was developed by Qualcomm, known in the US as IS-95, and was a first generation technology. Its typical characteristic was high capacity and small cell radius. CDMAone (IS-95) is a second generation system, offering advantages such as an increase in capacity (almost 10 times that of the AMPS), improved quality and coverage, improved security system, etc. Enhancement of the CDMAone was IS-95B, also called 2.5G of CDMA technology, which combined the standards IS-95A, ANSI-J-STD-008 and TSB-74. Major advantages of this system include frequency diversity (i.e. frequency dependent transmission impairments have less effect on the signal), increased privacy as the spread spectrum is obtained by noise like signals, an interference limited system, etc., while some disadvantages of this system include the air interface, which is the most complicated, soft hand-off, which is more complicated than the ones used in the TDMA/ FDMA system, signals near to the receiver, which are received with less attenuation than the ones further from it, etc.

1.3.3 GSM (Global System for Mobile Communication)

GSM was first developed in the 1980s. From 1982 to 1985, in the GSM group (originally hosted by CEPT) discussions were held to decide between building an analogue or a digital system. After multiple field tests, etc., it was decided to build a digital system and a narrowband TDMA solution was chosen. The modulation scheme chosen was Gaussian minimum shift keying (GMSK). The technical fundamentals were ready by 1987 and by 1990 the first specification was produced. By 1991, GSM was the first commercially operated digital cellular system with Radiolinja in Finland. GSM is by far the most popular and widely implemented cellular system with more than a billion people using the system (by 2005). Features such as prepaid calling, international roaming, etc., enhanced the popularity of the system. Of course, this also led to the development of smaller and lighter handsets with many more features. The system became more user friendly with many services also provided apart from just making calls. These services included voice mail, SMS, call waiting, etc. SMS was a phenomenal success, with almost 15 billion SMS sent every month by the year 2000. The key advantage of GSM systems has been higher digital voice quality and low cost alternatives to making calls, such as text messaging. The advantage for network operators has been the ability to deploy equipment from different vendors because the open standard allows easy interoperability.

The GSM system operates at various radio frequencies, with most them operating at 900 MHz and/or 1800 MHz. In the US and Canada, the operation is at 850 MHz and/or 1900 MHz. The uplink frequency band in the 900 MHz band is 935–960 MHz and the downlink frequency is 890–915 MHz. Thus, in both the uplink and downlink the band is 25MHz, which is subdivided into 124 carriers, each being 200 kHz apart. Each radio frequency channel contains eight speech channels. The cell radius in the GSM network varies depending upon the antenna height, antenna gains, propagation conditions, etc. These factors vary the cell size from a couple of hundred metres to a few kilometres. Due to this cell sizes are classified into four kinds in GSM networks; macro, micro, pico and umbrella, with macro cells being the biggest and pico and umbrella cells being the smallest.

System Architecture

A network mobile system has two major components: the fixed installed infrastructure (network) and the mobile subscribers, who use the services of the network. The fixed installed network can again be subdivided into three subnetworks: radio networks, mobile switching network and management network. These subnetworks are called subsystems. The respective three subsystems are:

- Base Station Subsystems (BSS);
- Switching and Management Subsystem (SMSS);
- Operation and Management Subsystems (OMSS).

Radio Network – Base Station Subsystem (BSS)
This comprises the Base Station Controller (BSC) and the Base Transceiver Station/Base Station (BTS/BS). The counterpart to a Mobile Station (MS) within a cellular network is the Base Transceiver Station, which is the mobile's interface to the network. A BTS is usually located in the centre of a cell. The BTS provides the radio channels for signalling and user data traffic in the cells. Besides the high frequency part (the transmitter and receiver component) it contains only a few components for signal and protocol processing. A BS has between 1 and 16 transceivers, each of which represents a separate radio frequency channel.

The main tasks of the BSC include:

- frequency administration;
- control of the BTS;
- exchange functions.

The hardware of the BSC may be located at the same site as the BTS, at its own stand-alone site, or at the site of the Mobile Switching Centre (MSC).

Mobile Switching Network
The Mobile Switching Subsystem (MSS) consists of Mobile Switching Centres and databases, which store the data required for routing and service provisions. The switching node of a mobile network is called the Mobile Switching Centre (MSC). It performs all the switching functions of a fixed network switching node, e.g. routing path search and signal routing. A public land mobile network can have several Mobile Switching Centres with each one being responsible for a part of the service area. The BSCs of a base subsystem are subordinated to a single MSC.

Dedicated Gateway MSC (GMSC)

This passes voice traffic between fixed networks and mobile networks. If the fixed network is unable to connect an incoming call to the local MSC, it routes the connection to the GMSC. This GMSC requests the routing information from the Home Location Register (HLR) and routes the connection to the local MSC in whose area the mobile station is currently staying. Connections to other mobile international networks are mostly routed over the International Switching Centre (ISC) of the respective country.

Home and Visitor Location Registers (HLR and VLR)

A given mobile network has several databases. Two functional units are defined for the synchronisation of registration of subscribers and their current location: a home location register (HLR) and the visitor location register (VLR). In general, there is one central HLR per public land mobile network (PLMN) and one VLR for each MSC.

Home Location Register (HLR)

The HLR stores the identity and user data of all the subscribers belonging to the area of the related GMSC. These are permanent data such as the International Mobile Subscriber Identity (IMSI) of an individual user, the user's phone number from the public network (not the same as IMSI), the authentication key, the subscribers permitted supplementary service and some temporary data. The temporary data on the Subscriber Identity Module (SIM) may include entries such as:

- the address of the current VLR;
- the number to which the calls may be forwarded;
- some transit parameters for authentication and ciphering.

Visitor Location Register (VLR)

The VLR stores the data of all mobile stations that are currently staying in the administrative area of the associated MSC. A VLR can be responsible for the areas of one or more MSCs. Mobile Stations are roaming freely and therefore, depending on their current location, they may be registered in one of the VLRs of their home network or in the VLR of a foreign network.

Operation and Maintenance Subsystem (OMSS)

The network operation is controlled and maintained by the Operation and Maintenance Subsystem (OMSS). Network control functions are monitored and initiated from an Operation and Maintenance Centre (OMC). The OMC has access to both the GMSC and BSC. Some of its functions are:

- administration and commercial operations (subscribers, end terminals, charging, statistics);
- security management;
- network configuration, operation, performance management;
- maintenance tasks.

The OMC configures the BTS via the BSC and allows the operator to check the attached components of the system.

User Authentication and Equipment Registration

Two additional databases are responsible for the various aspects of system security. They are based primarily on the verification of the equipment and subscriber identity; therefore, the databases serve for user authentication, identification and registration. Confidential data and keys are stored or generated in the Authentication Centre (AUC). The Equipment Identity Register (EIR) stores the serial numbers (supplied by the manufacturer) of the terminals (IMEI), which makes it possible to block service access for mobile stations reported as stolen.

Addresses and Identifiers
Mobile Station (MS)
These are pieces of equipment used by mobile subscribers for accessing the services. They consist of two major components: the Mobile Equipment (ME) and the Subscriber Identity Module (SIM). In addition to the equipment identifier the International Mobile Station Equipment Identity (IMEI) the mobile station has subscriber identification (IMSI and MSISDN, or the Mobile Subscriber ISDN Number) as subscriber dependent data.

Subscriber Identity Module (SIM)
The Subscriber Identity Module (SIM) provides mobile equipment with an identity. Certain subscriber parameters are stored on the SIM card, together with personal data used by the subscriber. The SIM card identifies the subscriber to the network. To protect the SIM card from improper use, the subscribers have to enter a 4-bit Personal Identification Number (PIN) before using the mobile. The PIN is stored on the card. If the wrong PIN is entered three times in a row, the card blocks itself and may only be unblocked with an 8-bit personal blocking key (PUK), also stored in the card.

International Mobile Station Equipment Identity (IMEI)
This serial number uniquely identifies mobile stations internationally. It is allocated by the equipment manufacturer and registered by the network operators who store them in the Equipment Identity Register (EIR). IMEI is a hierarchical address, containing the following parts:

- Type Approval Code (TAC): 6 decimal places, centrally assigned;
- Find Assembly Code (FAC): 6 decimal places, assigned by the manufacturer;
- Serial Number (SNR): 6 decimal places, assigned by the manufacturer;
- Spare (SP): 1 decimal place.

Hence, IMEI = TAC + FAC + SNR + SP.

International Mobile Subscriber Identity (IMSI)
While registering for service with a network operator, each subscriber receives a unique identifier, the International Mobile Subscriber Identity (IMSI), which is stored in the SIM. A mobile station can be operated if a SIM with a valid IMSI is inserted into equipment with a valid IMEI. The IMSI also consists of the following parts:

- Mobile Country Code (MCC): 3 decimal places, internationally standardised;
- Mobile Network Code (MNC): 2 decimal places, for unique identification of mobile networks across the country;
- Mobile Subscriber Identification Number (MSIN): maximum 10 places, identification number of the subscriber in his/her mobile home network.

Thus IMSI = MCC + MNC + MSIN and a maximum of 15 digits is used.

Mobile Subscriber ISDN Number (MSISDN)
The real telephone number of the MS is the Mobile Subscriber ISDN Number. It is assigned to the subscriber, such that an MS can have several MSISDNs depending on the SIM. The subscriber identity cannot be derived from the MSISDN unless the association of IMSI and MSISDN as stored in the HLR is known.

In addition to this a subscriber can hold several MSISDNs for selection of different services. Each MSISDN of a subscriber is reserved for a specific service (voice, data, fax, etc.). In order to realise this service, service specific resources need to be activated, which are done automatically during the setup of a connection. The MSISDN categories have the following structure:

- Country Code (CC): up to 3 decimal places;
- National Destination Code (NDC): typically 2–3 decimal places;
- Subscriber Number (SN): maximum 10 decimal places.

Thus, MSISDN = CC + NDC + SN.

Mobile Station Roaming Number (MSRN)
The Mobile Station Roaming Number (MSRN) is a temporary location dependent ISDN number. It is assigned by a locally responsible VLR to each MS in its area. Calls are routed by the MS using the MSRN. On request, the MSRN is passed from the HLR to the GMSC. The MSRN has the same structure as the MSISDN:

- Country Code (CC) of the visited network;
- National Destination Code (NDC) of the visited network;
- Subscriber Number (SN) in the current mobile network.

The components CC and NDC are determined by the visited network and depend on the current location. The SN is assigned by the current VLR and is unique within the mobile network. The assignment of an MSRN is done in such a way that the currently responsible switching node MSC in the visited network (CC + NDC) can be determined from the subscriber number, which allows routing decisions to be made.

The MSRN can be assigned in two ways by the VLR: either at each registration when the MS enters a new Location Area (LA) or each time when the HLR requests it for setting up a connection for the incoming calls to the mobile station. In the first case, the MSRN is also passed on from the VLR to the HLR, where it is stored for routing. In the case of an incoming call, the MSRN is first requested from the HLR of the MS. In this way the currently responsible MSC can be determined, and the call can be routed to this switching node. In the second case, the MSRN cannot be stored in the HLR, since it is only assigned at the time of the call setup. Therefore the address of the current VLR must be stored in the tables of the HLR. Once routing information is requested from the HLR, the HLR itself goes to the current VLR and uses a unique subscriber identification (IMSI and MSISDN) to request a valid MSRN. This allows further routing of the call.

Location Area Identity (LAI)
Each LA has its own identifier. The Location Area Identifier (LAI) is also structured hierarchically and is internationally unique. It consists of the following parts:

- Country Code (CC): 3 decimal digits;
- Mobile Country Code (MNC): 2 decimal places;
- Location Area Code (LAC): maximum 5 decimal places, or maximum twice 8 bits.

The LAI is broadcast regularly by the base station on the Broadcast Control Channel (BCCH). Thus, each cell is identified uniquely on the radio channel and each MS can determine its location through the LAI. If the LAI that is heard by the MS notices this LA change it requests the updating of its location information in the VLR and HLR (location update). The LAI is requested from the VLR if the connection for an incoming call has been routed to the current MSC using the MSRN. This determines the precise location of the MS where the mobile can be currently paged.

Temporary Mobile Subscriber Identity (TMSI)
The VLR being responsible for the current location of a subscriber can assign a Temporary Mobile Subscriber Identity (TMSI), which only has significance in the area handled by the VLR. It is used in place of the IMSI for the definite identification and addressing of the MS. Therefore nobody can

determine the identity of the subscriber by listening to the radio channel, since the TMSI is only assigned during the MS's presence in the area of one VLR, and can even be changed during this period (ID (identity) hopping). The MS stores the TMSI on the network side only in the VLR and it is not passed to the HLR.

Local Mobile Subscriber Identity (LMSI)
The VLR can assign an additional searching key to each MS within its area to accelerate database access, called the Local Mobile Subscriber Identity (LMSI). Each time messages are sent to the VLR concerning an MS, the LMSI is added, so the VLR can use the short searching key for transactions concerning this MS.

Mobile Call Origination and Termination
The case is considered where a person makes a call from a telephone connected to a public switched telephone network (PSTN) or ISDN, to a mobile subscriber going from city A to city B. The call will take place only if the subscriber's mobile is switched on. Assuming the mobile to be switched on, the MS searches for the cellular network by scanning the relevant frequency band for some control channel transmitted by a nearby MS. After location updating the MS accesses the network and acquires a unique serial number. Once an MS has successfully registered its location with the network it enters the idle mode, whereby it listens to the paging channels from the selected BS.

Since the subscriber is presently in city A the MS will have identified a BS in this area. The MS will notice that the signal begins to fall as it is moving from city A to city B and it will now look for a more appropriate BS to take over. When the MS identifies a more appropriate BS it examines its control channels to determine the location area to which it belongs. If it belongs to the same location area as the previous BS, the MS simply re-tunes to a paging channel on the new BS and continues to monitor this new channel for incoming paging calls. If the MS has moved between BSs in different location areas, then it performs a location update and informs the network of its new position. This process of transition between BSs while in the idle mode is termed the idle mode handover.

The entire process of the call is initiated by the person lifting the handset and dialing the number of the mobile subscriber. On receiving a number with the area code, the PSTN/ISDN network will route the call to the gateway switch of the mobile network and will also provide the telephone number of the mobile subscriber. The gateway switch then interrogates the mobile network's HLR to recover the subscriber's records.

Once the call arrives at the MSC, the MS is paged to alert it to the presence of an incoming call. A paging call is then issued from each BS in the location area in which the subscriber is registered. On receiving a paging call the MS responds by initiating the access procedure. The access procedure commences with the MS sending a message to the BS requesting a channel. The BS replies by sending the MS details of a dedicated channel and the MS re-tunes to this channel. A certain degree of handshaking occurs to ensure that the identity of the subscriber is correct.

Once the dedicated signalling channel is established, security procedures such as subscriber authentication, take place over this channel. Following this, the network allocates a dedicated speech channel and both the BS and MS re-tune to this channel and establish a connection. It can be seen that until this point is reached all processes are carried out autonomously by the MS and no interaction is required from the subscriber. It is only once all these processes are completed that the MS begins to ring.

The subscriber can now talk and the handover can take place between different BSs. Once the call has ended the call clear process initiates, which consists of a small exchange of signalling information ensuring that both the network and the MS know that the call has ended. The MS again returns to the idle mode and monitors the paging channel of its current cell.

Several cryptographic algorithms are used for GSM security, which include the features link user authentication, over-the-air voice privacy, etc. The security model of GSM, however, lacked some features such as authentication of the user to the network and not vice versa (a feature that came in the Universal Mobile Telecommunications System, or UMTS).

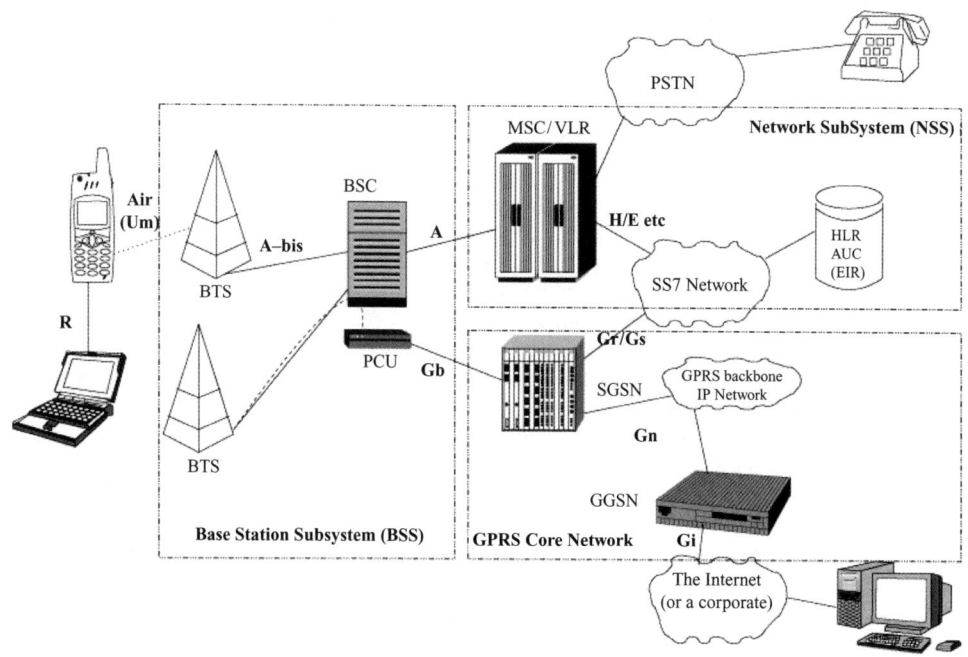

Figure 1.1 GSM and GPRS System

Limitations of 2G Networks
- *Low transfer rates.* The 2G networks are primarily designed to offer voice services to the subscribers. Thus the transfer rates offered by these networks are low. Though the rates vary across technologies, the average rate is of the order of tens of kilobits per second.
- *Low efficiency for packet switched services.* There is a demand for Internet access, not just at home or the office but also while roaming. Wireless Internet access with the 2G networks is not efficiently implemented.
- *Multiple standards.* With a multitude of competing standards in place, a user can roam in only those networks that support the same standard. This allows the user only limited roaming. Therefore the 2G network technology was semi-global in this respect.

1.3.4 GPRS (General Packet Radio Service)

GPRS is a nonvoice, i.e. data, value added service to the GSM network. This is done by overlaying a packet based air interface on the existing circuit switched GSM network (see Figure 1.1). In infrastructure terms, the operator just needs to add a couple of nodes and some software changes to upgrade the existing voice GSM system to voice plus data GPRS system. The voice traffic is circuit switched while data traffic is packet switched. Packet switching enables the resources to be used only when the subscriber is actually sending and receiving the data. This enables the radio resources to be used concurrently while being shared between multiple users. The amount of data that can be transferred is dependent upon the number of users. Theoretical maximum speeds of up to 171.2 kilobits per second (kbps) are achievable with GPRS using all eight timeslots at the same time. GPRS allows the interconnection between the network and the Internet. As there are the same protocols, the GPRS network can be viewed as a subnetwork of the Internet, with GPRS capable mobile phones being viewed as mobile hosts.

However, there are some limitations in the GPRS network, such as low speed (practical speed is much lower than theoretical speeds).

1.3.5 EDGE (Enhanced Data Rate for GSM Evolution)

The limitation of the GPRS network was eliminated to a certain extent by the introduction of the EDGE technology. EDGE works on TDMA and GSM systems. It is considered to be a subset of the GPRS as it can be installed on any system that has GPRS deployed on it. It is not an alternative to UMTS but a complimentary technology for it. In EDGE, 3G services can be given at a lower but similar data rate as UMTS, with the data rates going up to 500 kbps (theoretically). This is done by introducing a new modulation scheme 8-PSK (phase-shift keying) and will coexist with the GMSK that is used in GPRS. However, the major advantage is that existing GSM networks can be upgraded for the same, thus preventing huge costs needed to roll-out the 3G networks and at the same time giving services like 3G. General features of EDGE include enhanced throughput per timeslot (8.8–59.2 kbps/timeslot), modulation changes from GMSK to 8-PSK, decreased sensitivity of the 8-PSK signal and higher capacity and coverage. Though, not many changes in the hardware are required by EDGE, except for some hardware upgrades in the BTS and some software upgraded in the network.

However, the second generation system lacked capacity, global roaming and quality, not to mention the amount of data that could be sent. This all led to the industry working on a system that had more global reach (e.g. the user did not need to change phones when going to Japan or the US from SE Asia or Europe). This was the beginning of the evolution of third generation systems.

1.4 Third Generation Cellular Networks

The third generation cellular networks were developed with the aim of offering high speed data and multimedia connectivity to subscribers. The International Telecommunication Union (ITU) under the initiative IMT-2000 has defined 3G systems as being capable of supporting high speed data ranges of 144 kbps to greater than 2 Mbps. A few technologies are able to fulfil the International Mobile Telecommunications (IMT) standards, such as CDMA, UMTS and some variation of GSM such as EDGE.

1.4.1 CDMA2000

CDMA2000 has variants such as 1X, 1XEV-DO, 1XEV-DV and 3X. The 1XEV specification was developed by the Third Generation Partnership Project 2 (3GPP2), a partnership consisting of five telecommunications standards bodies: CWTS in China, ARIB and TTC in Japan, TTA in Korea and TIA in North America. It is also known as the High Rate Packet Data Air Interface Specification. It delivers 3G like services up to 140 kbps peak rate while occupying a very small amount of spectrum (1.25 MHz per carrier). 1XEV-DO, also called 1XEV Phase One, is an enhancement that puts voice and data on separate channels in order to provide data delivery at 2.4 Mbps. EV-DV, or 1XEV Phase Two promises data speeds ranging from 3 Mbps to 5 Mbps. However, CDMA2000 3 × is an ITU-approved, IMT-2000 (3G) standard. It is part of what the ITU has termed IMT-2000 CDMA MC. It uses a 5 MHz spectrum (3 × 1.25 MHz channels) to give speeds of around 2–4 Mbps.

1.4.2 UMTS

The Universal Mobile Telecommunications System (UMTS) is one of the third generation (3G) mobile phone technologies. It uses W-CDMA as the underlying standard. W-CDMA was developed by NTT DoCoMo as the air interface for their 3G network FOMA. Later it submitted the specification to the

International Telecommunication Union (ITU) as a candidate for the international 3G standard known as IMT-2000. The ITU eventually accepted W-CDMA as part of the IMT-2000 family of 3G standards. Later, W-CDMA was selected as the air interface for UMTS, the 3G successor to GSM. Some of the key features include the support to two basic modes FDD and TDD, variable transmission rates, intercell asynchronous operation, adaptive power control, increased coverage and capacity, etc. W-CDMA also uses the CDMA multiplexing technique, due to its advantages over other multiple access techniques such as TDMA. W-CDMA is merely the air interface as per the definition of IMT-2000, while UMTS is a complete stack of communication protocols designated for 3G global mobile telecommunications. UMTS uses a pair of 5 MHz channels, one in the 1900 MHz range for uplink and one in the 2100 MHz range for downlink. The specific frequency bands originally defined by the UMTS standard are 1885–2025 MHz for uplink and 2110–2200 MHz for downlink.

UMTS System Architecture

A UMTS network consists of three interacting domains: Core Network (CN), UMTS Terrestrial Radio Access Network (UTRAN) and User Equipment (UE). The UE or ME contains the mobile phone and the SIM (Subscriber Identity Module) card called USIM (Universal SIM). USIM contains member specific data and enables the authenticated entry of the subscriber into the network. This UMTS UE is capable of working in three modes: CS (circuit switched) mode, PS (packet switched) mode and CS/PS mode. In the CS mode the UE is connected only to the core network. In the PS mode, the UE is connected only to the PS domain (though CS services like VoIP (Voice over Internet Protocol) can still be offered), while in the CS/PS mode, the mobile is capable of working simultaneously to offer both CS and PS services.

The components of the Radio Access Network (RAN) are the Base Stations (BS) or Node B and Radio Network Controllers (RNCs). The major functions of the BS are closed loop power control, physical channel coding, modulation/demodulation, air interface transmissions/reception, error handling, etc., while major functions of the RNC are radio resource control/management, power control, channel allocation, admission control, ciphering, segmentation/reassembly, etc.

The main function of the Core Network (CN) is to provide switching, routing and transit for user traffic. The CN also contains the databases and network management functions. The basic CN architecture for UMTS is based on the GSM network with GPRS. All equipment has to be modified for UMTS operation and services. The CN is divided into the CS and PS domains.

Circuit switched elements are the Mobile Services Switching Centre (MSC), Visitor Location Register (VLR) and Gateway MSC. Packet switched elements are the Serving GPRS Support Node (SGSN) and the Gateway GPRS Support Node (GGSN). Network elements like EIR, HLR, VLR and AUC are shared by both domains. The Asynchronous Transfer Mode (ATM) is defined for UMTS core transmission. The ATM Adaptation Layer type 2 (AAL2) handles the circuit switched connection and the packet connection protocol AAL5 is designed for data delivery. A typical 3G network is shown in Figure 1.2.

UMTS QoS Classes

UMTS network services have different quality of service (QoS) classes for four types of traffic:

- conversational class (e.g. voice, video telephony, video gaming);
- streaming class (e.g. multimedia, video on demand);
- interactive class (e.g. web browsing, network gaming, database access);
- background class (e.g. email, SMS, downloading).

Conversational Class

The best examples of this class are voice traffic and real time data traffic such as video telephony, video gaming, etc. This traffic runs over CS bearers. The quality of this class is dependent totally on subscriber

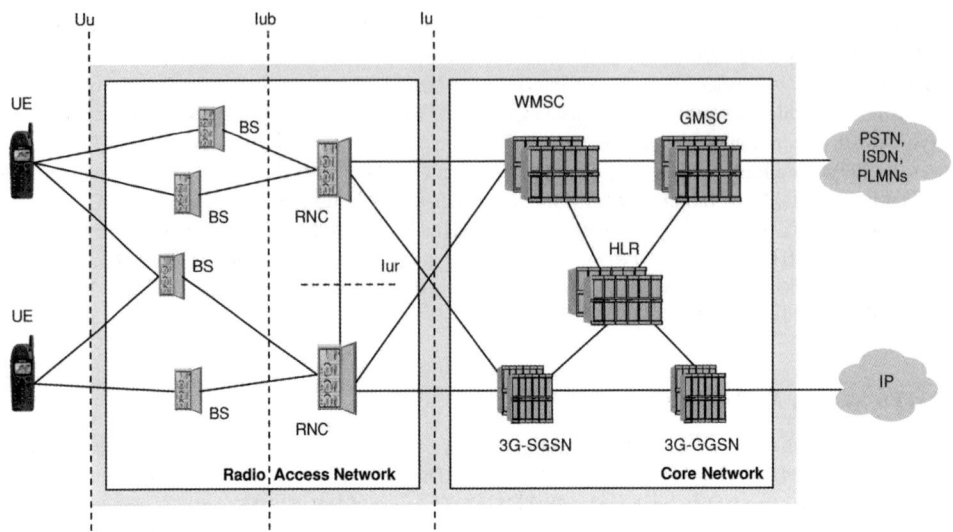

Figure 1.2 UMTS network

perception. The key aspect of this class is low end-to-end delays (e.g. less than 400 ms). For speech coding/decoding the adaptive multirate (AMR) technique will be used. Upon request, AMR coders can switch the bit rates every 20 ms of speech frame. Thus, during busy hours, bit rates can be lowered to offer higher capacity by sacrificing quality. Also, the coverage area of the cell can be increased by decreasing bit rates. Thus, this technique helps in balancing coverage, capacity and quality of the network.

Streaming Class
Multimedia, video on demand, etc., are examples of the streaming class. The data are transferred in a steady and continuous stream. How does this work? On the Internet, the display starts even when the entire file has not been downloaded. The delay in this class is higher than the conversational class.

Interactive Class
The Internet is a classical example of the interactive class. The subscriber requests the information from the server and waits for the information to arrive. Thus, delay is not the ster minimum under this class as the time to download also depends upon the number of subscribers logged on to the system and the system capacity itself. Another aspect of this service is that the transfer of data is transparent. Location based services and computer games are other examples of this class of service.

Background Class
Other applications such as SMS, fax, emails, etc., fall under the background class. Delay is the highest in this class of service. Also, the data transfer is not transparent as in the interactive class.

1.4.3 HSDPA in UMTS

High Speed Downlink Packet Access (HSDPA) is a packet based data service in the downlink having a transmission rate up to 8–10 Mbps over the 5 MHz bandwidth. This means that implementation of this technique will allow data speeds to increase to almost five times that of the most advanced Wideband Code Division Multiple Access (WCDMA) networks. Also, the base station capacity increased by double. The system capacity and the user data rates are increased by implementation of HSDPA, which includes MIMO (Multiple Input Multiple Output), cell search, advanced receiver design, HARQ (Hybrid Automatic Request) and AMC (Adaptive Modulation and Coding). HSDPA is mainly intended for non-real-time traffic, but can also be used for traffic with tighter delay requirements (for more details refer to Appendix B).

2

Radio Network Planning and Optimisation

Johanna Kähkönen, Nezha Larhrissi, Cameron Gillis, Mika Särkioja,
Ajay R Mishra and Tarun Sharma

2.1 Radio Network Planning Process

The network planning process itself is not standard. Though some of the steps may be common, the process is determined by the type of projects, criteria and targets. The process has to be applied case by case.

2.1.1 Network Planning Projects

Network planning projects can be divided into three main categories based on how much external planning services the operator is using. No services means simply that the operator is responsible for the network planning from the very beginning until the end. This type of comprehensive responsibility for the network planning is more suitable for traditional network operators, who have extensive knowledge of their existing network and previous network planning experience than newcomers in this technology field. There is risk, however, that if the operator is the only person responsible for network planning there might be a difficulty in maintaining knowledge of the latest equipment and features.

The opposite network planning solution is when the network operator buys the new network with a turnkey agreement (Greenfield case). In this case, the operator is involved only in defining the network planning criteria. After the network roll-out has been finished and enters the care phase an agreement about the future has to be made. The care services can be outsourced as well, but the operator might also be interested to take some portion of the network operations and start to learn the process. An operator taking all the responsibility after the outsourced planning phase includes some risk. A better solution is to learn the network operation at a pace agreed with the network vendor.

The network operator can also buy network planning consultancy services. In this, the operator performs majority of the planning function and outsource selected aspects of the job. In this way some special know-how can be bought to supplement the knowledge of the network planning group. This is generally used in cases where new technologies need to be introduced in mature networks.

Advanced Cellular Network Planning and Optimisation   Edited by Ajay R Mishra
© 2007 John Wiley & Sons, Ltd

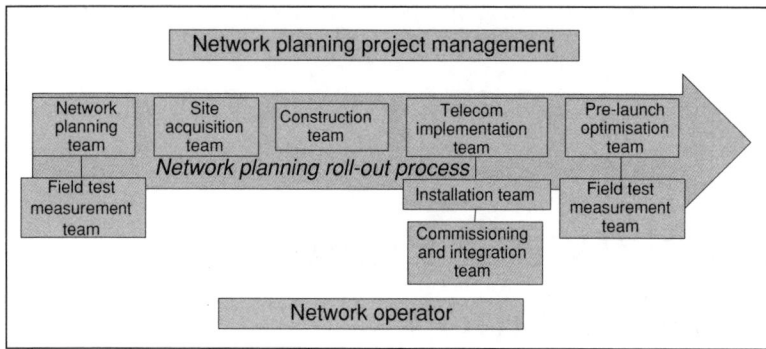

Figure 2.1 Network planning project organisation

The background of companies offering network planning services for operators is diverse. One group are the equipment vendors with the newest technical information about the equipment and technology. Another group are consultancy companies, who offer network planning services. These companies are independent from the vendors, which is on the one hand an advantage but on the other hand a disadvantage. When there needs to be selection between different vendors, the fair choice for cooperation is an independent consulting company. Network planning services are in some cases also offered by basic infrastructure firms, who are also involved in building the network (refer to Appendix A for more information on the roll-out projects).

2.1.2 Network Planning Project Organisation

The network planning project organisation is based on the network planning roll-out process steps. The final target of the network planning roll-out process is to deliver a new network for the operator according to the agreed requirements. The process steps, inputs and outputs will be discussed in more detail later, as well as network planning tasks and deliverables. Here the general frame of the roll-out process will be introduced.

The network planning project organisation is pictured according to network planning roll-out process flow in Figure 2.1. The roll-out process applies both for individual base stations as well as for the whole network. Due to the limited time in the project, bases stations need to be managed in groups, but not all base stations can be guided through, for example, the construction phase at once. The process steps need to be phased and overlapped in order to keep the whole process inside a reasonable time limit. The network planning project management takes care of the whole project organisation. Some support functions, e.g. marketing, selling, logistics and technical support, are also project organisation wide and are not specifically connected to any of the project teams.

The network planning team is responsible for both network preplanning and actual network planning, giving site proposals as the output. The network planning team has the assistance of the field measurement team. The site proposals are an input for the site acquisition team, which is responsible for finding the actual site locations. The site acquisition team makes technical site surveys ending up with site lease agreements for the best possible site locations – a decision that is always the sum of several factors. The construction works are carried out by the construction team and the target is to prepare the site ready for telecom implementation. The site location can vary from an existing building to a mast, which has to be built purposely. Therefore the construction work varies a lot from one site to another. Telecom implementation covers installation, commissioning and integration. Installation is the setting up of the base station equipment, antennas and feeders. Commissioning stands for functional testing of stand-alone network entities. In the commissioning phase it is also verified that the site data depend on the network plan and, for example, the billing and routing data meet the operator requirements. The integration phase verifies that the site is operational as a part of the network. After this it is ready for commercial use. A separate optimisation team or the network planning team is responsible for the prelaunch optimisation phase. Here the field test measurement team is giving support and the aim of this phase is to verify the functionality of the network. It should be shown that the parameter settings in the network are correct and that the planning targets can be met.

2.1.3 Network Planning Criteria and Targets

Network planning is a complicated process consisting of several phases. The final target for the network planning process is to define the network design, which is then built as a cellular network. The network design can be an extension of the existing GSM network or a new network to be launched. The difficulty in network planning is to combine all of the requirements in an optimal way and to design a cost-effective network.

Before the actual planning is started for a new network the current market situation is analysed. The market analysis covers all the competitors and the key information from them: market share, network coverage areas, services, tariffs, etc. Based on the market situation it is possible to create a future deployment strategy for the new operator. Questions arise about the nature of the targeted user group, how large is the coverage provided in the beginning and how it will grow in the future. It is also decided in the beginning what kind of services will be offered and which is connected to the target user group. This leads to estimations of market share in the beginning and objectives for the future. More detailed estimations are needed on how much each user of a certain type is using the services provided. The needed capacity for each service and onwards for the whole network can be calculated from the estimated average usage.

The basic requirements for the cellular network are to meet coverage and quality targets. These requirements are also related to how the end user experiences the network. Coverage targets firstly mean the geographic area the network is covering with an agreed location probability, i.e. the probability to get service. The requirements also specify the signal strength values that need to be met inside different area types. The quality targets are related to factors such as the success of the call, the drop call ratio, which should not exceed the agreed value, and the success ratio for the call setup and for handovers.

Environmental factors also greatly affect network planning. The propagation of radio waves varies depending on the area morthography. The attenuation varies, for instance, when comparing rural, suburban and urban tactors and also indoor and outdoor differences caused by buildings. Most importantly, the frequency range has an impact on propagation. The topography of the planned area, the location of cities, roads and other hotspots are obviously factors having an impact on planning. As the frequency band is a limited resource the available bandwidth partly determines the tactics for network planning.

All previously mentioned factors – data based on market analysis, operator requirements, environmental factors and other boundary conditions – help to define planning parameters and frames for the network plan. Due to various design parameters the network planning process requires optimisation and compromises in order to end up with a functional cellular network. The network planning target is to build as high a quality network as possible. On the other hand, there is the cost-efficiency – how much money the operator can spend for the investments so that the business is financially profitable. The two factors – network quality and investments – are connected to profit. To simplify, the better the end users can be served and the more traffic the network can handle, the more impact there is on the profits. This explains the complexity of network planning, where sufficient cellular network coverage and capacity need to be created with as low investments as possible.

A summary of the main factors affecting network planning are listed below:

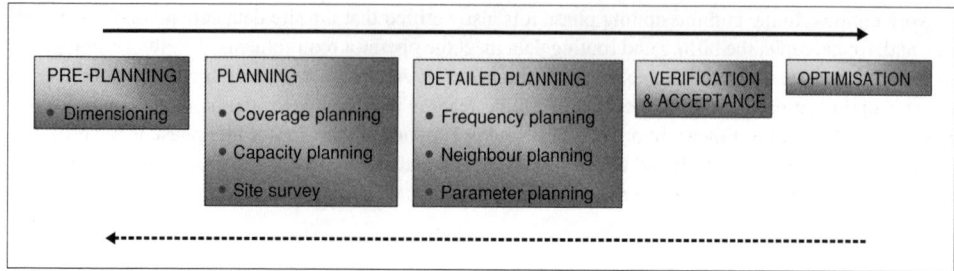

Figure 2.2 Network planning process steps

Market analysis

- Competitor analysis
- Potential customers
- User profiles: services required and usage

Customer requirements

- Coverage requirements
- Capacity requirements
- Quality targets: call setup success, drop call rate, etc.
- Financial limitations
- Future deployment plans

Environment factors and other boundary conditions

- Area morthography
- Area topography
- Hotspot locations
- Available frequency band
- Recommended base transceiver station (BTS) locations

The radio network planning deliverables are final BTS configurations and site locations. The final coverage predictions including dominance and composite maps are delivered. Power budgets are calculated for all the configurations. Related to the frequency plan, the allocated frequencies are documented and an interference analysis is also presented. Deliverables also include the adjacency plan and allocated parameters, either default or optimized ones.

2.1.4 Network Planning Process Steps

The network planning process consists of several phases, which can be combined at a higher level to main phases that differ depending on the logics. The radio network planning process is divided into five main steps, from which four are before the network launch and the last one after the network has been launched. The five main phases then include subphases, which are introduced in this section and then each is explained in detail in later sections. The flowchart for the network planning process is shown in Figure 2.2. After detailed planning the network is ready for commercial launch, but the postplanning phase continues the process and targets the most optimal network configuration. Actually the network planning process is a never ending cycle due to changes in the design parameters.

The five main steps in the network planning process are: preplanning, planning, detailed planning, acceptance and optimisation. The input for the preplanning phase is the network planning criteria. The main activity is dimensioning, which gives the initial network configuration as a result. The first step in the planning phase is nominal planning; it provides the first site locations in the map based on input from the dimensioning phase. The process continues with more detailed coverage planning after site hunting and transmission planning. Detailed Capacity planning is also included in the planning phase. Detailed planning covers frequency, neighbour and parameter planning. After detailed planning the network is ready for verification and acceptance, which finishes the prelaunch activities. After the launch the activities continue with optimisation.

Preplanning

The preplanning phase covers the assignments and preparation before the actual network planning is started. As in any other business it is an advantage to be aware of the current market situation and competitors. The network planning criteria is agreed with the customer. As specified earlier, the requirements depend on many factors, the main criteria being the coverage and quality targets. Also several limitations exist, like the limited frequency band and the budget for the investments. The priority for the planning parameters comes from the customer. Due to the fact that the network plan can not be optimized with regard to all the parameters the priorities need to be agreed with the customer throughout the whole process.

The network planning criteria is used as an input for network dimensioning. In the following are listed basic inputs for dimensioning:

- coverage requirements, the signal level for outdoor, in-car and indoor with the coverage probabilities;
- quality requirements, drop call rate, call blocking;
- frequency spectrum, number of channels, including information about possible needed guard bands;
- subscriber information, number of users and growth figures;
- traffic per user, busy hour value;
- services.

The dimensioning gives a preliminary network plan as an output, which is then supplemented in coverage and parameter planning phases to create a more detailed plan. The preliminary plan includes the number of network elements that are needed to fulfil the quality of service requirements set by the operator, e.g. in GSM the number of BTSs and TRXs (transceivers). It also needs to be noted that dimensioning is repeated in the case of network extension.

The result of dimensioning has two aspects; it tells the minimum number of base stations due to coverage or capacity reasons. Both of these aspects need to be analysed against the original planning targets. It is also important to understand the forecasts for the subscriber growth and also the services that are going to be deployed.

The dimensioning result is an average capacity requirement per area type like urban, suburban, etc. More detailed capacity planning, capacity allocation for individual cells, can be done using a planning tool having digital maps and traffic information. The dimensioning results are an input for coverage planning, which is the next step in the network planning process.

The radio network configuration plan also provides information for preplanning of the transmission network. The topology can be sketched based on the initial configuration and network design criteria.

Planning

The planning phase takes input from the dimensioning, initial network configuration. This is the basis for nominal planning, which means radio network coverage and capacity planning with a planning tool.

The nominal plan does not commit certain site locations but gives an initial idea about the locations and also distances between the sites.

The nominal plan is a starting point for the site survey, finding the real site locations. The nominal plan is then supplemented when it has information about the selected site locations; as the process proceeds coverage planning becomes completed. The acquisition can also be other inputs, existing site locations or proposals from the operator. The final site locations are agreed together with the radio frequency (RF) team, transmission team and acquisition team.

The target for the coverage planning phase is to find optimal locations for BSs to build continuous coverage according to the planning requirements. Coverage planning is performed with a planning tool including a digital map with topography and morthography information and a tuned model for propagation. The propagation model tuning measurements have been performed with good accuracy.

In the capacity planning phase the final coverage plan including composite and dominance information is combined with the user density information; in this way the capacity can be allocated. Boundary conditions for capacity allocation are agreed with the customer earlier, i.e. the maximum TRX number per base station. The known capacity hot spots are treated with extra care and special methods can be used to fulfil the estimated need.

The output of the planning phase is the final and detailed coverage and capacity plans. Coverage maps are made for the planned area and final site locations and configurations.

Detailed Planning

After the planning phase has finished and the site location and configurations are known detailed planning can be started. The detailed planning phase includes frequency, adjacency and parameter planning.

Planning tools have frequency planning algorithms for automatic frequency planning. These require parameter setting and prioritization for the parameters as an input for the iteration. The planning tool can also be utilised in manual frequency planning. The tool uses interference calculation algorithms and the target is to minimise firstly the co-channel interference and also to find as low an adjacent channel interference as possible. Frequency planning is a critical phase in network planning. The number of frequencies that can be used is always limited and therefore the task here is to find the best possible solution.

Neighbour planning is normally done with the coverage planning tool using the frequency plan information. The basic rule is to take the neighbouring cells from the first two tiers of the surrounding BTSs: all cells from the first circle and cells pointing to the target cell from the second circle.

In the parameter planning phase a recommended parameter setting is allocated for each network element. For radio planning the responsibility is to allocate parameters such as handover control and power control and define the location areas and set the parameters accordingly. In case advanced system features and services are in use care must be taken with parameter planning. The output of the detailed planning phase is the frequency plan, adjacencies and the parameter plan.

Verification and Acceptance

After the planning phase has been finished the aim of the prelaunch optimisation phase is to ensure optimal operation of the network. In addition to fine-tuning a search is made for possible mistakes that might have occurred during the installation. Prelaunch optimisation is high level optimisation but does not go into detail. Network optimisation continues after the launch and goes into a more detailed level. At that point the detailed level is easier to reach due to growing traffic.

The quality of service requirements for the cellular network, i.e. coverage, capacity and quality requirements, are the basis for dimensioning. The targets are specified with key performance indicators (KPIs), which show the target to meet before network acceptance. Drive testing is used as the testing

Table 2.1 Example coverage thresholds

Area type	Coverage threshold
Urban (indoor)	> −75 dB m
Suburban (indoor)	> −85 dB m
In-car	> −90 dB m
Rural	> −95 dB m

method for the network functionality verification. During the verification the functionality of different services agreed with the operator has to be tested.

Optimisation

After the network has been launched the planning and optimisation related activities do not end because network optimisation is a continuous process. For the optimisation the needed input is all available information about the network and its status. The network statistic figures, alarms and traffic itself are monitored carefully. Customer complaints are also a source of input to the network optimisation team. The optimisation process includes both network level measurements and also field test measurements in order to analyse problem locations and also to indicate potential problems.

2.2 Preplanning in a GSM Radio Network

Preplanning precedes actual network planning. The main part of the preplanning consists of dimensioning, but before proceeding with that the network planning criteria has to be agreed with the operator.

2.2.1 GSM Network Planning Criteria

The definition of the radio network planning criteria is done at the beginning of the network planning process. The customer requirements form the basis of the negotiations and the final radio network planning criteria is agreed between the customer and the radio network planner. The network operator has performance quality targets for the cellular network and these quality requirements are also related to how the end user experiences the network. The network planner's main target is to build as high a quality network as possible. On the other hand, there must be cost-efficiency – how much money the operator can spend for the investments so that the business is financially profitable. The two factors – network quality and investments – are connected to profit. The link is not straighforward but, is one factor, if the network end user quality perception is good it has an impact on the profits. This explains the complexity of network planning; sufficient cellular network coverage and capacity needs to be created with as low an investment as possible.

The coverage targets include the geographical coverage, coverage thresholds for different areas and coverage probability. Examples of coverage thresholds are presented in Table 2.1. The range for a typical coverage probability is 90–95 %. The geographical coverage is case-specific and can be defined in steps according to network roll-out phases.

The quality targets are those agreed in association with the customer and network planning. The main quality parameters are call success or drop call rate, handover success, congestion or call attempt success and customer observed downlink (DL) quality. The DL quality is measured according to BER as defined in GSM specifications and mapped to RXQUAL values. Normally downlink RXQUAL classes 0 to 5 are considered as a sufficient call quality for the end user. The classes 6 and 7 represent poor performance

Table 2.2 Typical network quality targets

Quality parameter	Target value
Drop call rate	< 5 %
Handover success rate	> 95 %
Call attempt success rate	> 98 %
DL quality	$\geq$ RXQUAL 5

and thus need to be avoided. The target value for RXQUAL can be, for example, 95 % or the time equal or better than 5. Example values for network quality targets are shown in Table 2.2.

The coverage and quality targets need to be considered in connection with the network evolution strategy. The subscriber forecast predicts the need and pace of network enlargement. Due to this obvious connection it is important to verify the subscriber forecast from time to time and keep it up to date. The coverage and quality targets need to be adjusted for the different network evolution phases. Interference probability becomes more important as the network capacity enlarges and has to be added as part of the quality targets.

The network features that are used have an effect on the dimensioning phase. The capacity and quality requirements need to be adjusted according to the features in use.

Some parameters that affect network planning cannot be controlled and therefore it is important to know what they are and then to take them into account in the planning. The topology and morphology is always area specific and due to this the area related planning parameters are case specific. An accurate digital map is needed in network planning. The map is used together with the propagation model to calculate the coverage areas in the planning phase. The propagation model is customised for the planning area with propagation measurements.

Population data are needed when estimating subscriber numbers. Population data as a layer on the map are useful in the planning tasks in order to cover a dense population area and allocate needed capacity.

The available bandwidth is a critical network planning parameter. Some basic decisions are dependent on the bandwidth, BTS configuration and frequency planning.

The quality of the cellular network is highly dependent on the quality of the network plan. The network performance will be measured and analysed to prove that it is working according to the planning requirements.

2.2.2 Introducing GPRS in the GSM Network

The GSM networks did not previously feature any efficient method for dealing with larger data quantities (packets). Excessive connecting times and signalling traffic did not allow the effective transfer of large quantities of data. The GPRS system, however, fulfills the requirements of future markets for mobile data communication in a far more innovative and effective manner. In the GSM system, connections are made on a circuit switched basis; i.e. the network establishes an air interface connection. A radio channel is then assigned to a mobile station and the data are transferred via this channel. The MS occupies this radio channel for the entire duration of the transfer process and the user must therefore pay for the full duration of the connection. With the new packet switching transfer system, a data packet is only transferred when there is actually data ready for transfer. This means that several MSs can use a single radio channel simultaneously. The MS is able to use a maximum of eight radio timeslots simultaneously. Whenever the MS sends a data packet, the network forwards the packet to the recipient through the first free radio channel. Since data transfer frequently consists of transfer 'bursts', this system makes efficient use of the radio channels.

Coding Schemes in GPRS

ETSI has specified four coding schemes in the GPRS network: CS-1, CS-2, CS-3 and CS-4. These coding schemes have been developed based on a compromise between the amount of user data carried and error protection. Coding scheme CS-1 has the lowest data rate and the highest protection while CS-4 has the highest data rate but the lowest protection.

When introducing GPRS to the network the operator has certain requirements for the coverage and capacity for the GPRS service and at the same time it is also required that the existing GSM service is not degraded by the introduction of the GPRS service. The planning aspects of introducing GPRS can be considered as two separate approaches, a high level plan and a detailed plan. The high level plan considers the strategy of the GPRS service, whereas the detailed plan places the strategy requirements in the network on a cell-by-cell basis.

To meet these requirements the operator needs to consider the following tasks:

- Select the area in which GPRS will be offered.
- Identify the cell groups that will offer the GPRS service:
  - macro cells, micro cells, pico cells, dual band.
- Define the GPRS parameters that are operator-definable on a cell group basis (i.e. micro cell 2 carrier, macro cell 3 carrier) understanding the impact to network performance of both the existing GSM service and GPRS service:
  - dedicated timeslots;
  - default GPRS capacity;
  - territory occupancy;
  - prefer BCCH frequencies;
  - uplink power control parameters – gamma channel and alpha;
  - define routing area boundaries.
- Revisit the existing GSM parameter setting to ensure optimum GPRS operation:
  - C1/C2 parameters for mobility management;
  - traffic management features – directed retry.
- Analyse GPRS service areas to check that the GPRS service covers the wanted area to the required throughput levels.
- Analyse the capacity of GPRS and GSM services to ensure that the GPRS service has the required capacity and the GSM capacity is not degraded.

In practice this is an iterative process and all the tasks can be done several times to meet all the requirements. Adding new cells or carriers may also be necessary. The flowchart of the process of introducing GPRS to an existing network is presented in Figure 2.3.

2.2.3 Introducing EGPRS in the GSM Network

The most fundamental difference between EGPRS (enhanced GPRS) and GSM/GPRS is the capability to offer higher throughput using a new 8-PSK modulation to transmit 3 bits per symbol and offer a full range of nine modulation and coding schemes (MCSs). Despite the new 8-PSK modulation, EGPRS timeslots still fit into the standard burst structure of GSM and GPRS.

Figure 2.4 outlines how flow EGPRS Radio Link Control (RLC) blocks are sent over the air interface. Each RLC block is split into four bursts and fits on to the GSM TDMA frame structure (in this case, timeslot 7 in the 52 Packet Data Channel (PDCH) multiframe). The figure also includes the concept of how to calculate effective timeslot throughput (radio bearer channel rate) for any EGPRS data connection. Since 12 RLC/MAC (media access control) blocks fit into a standard 52 PDCH multiframe and the block

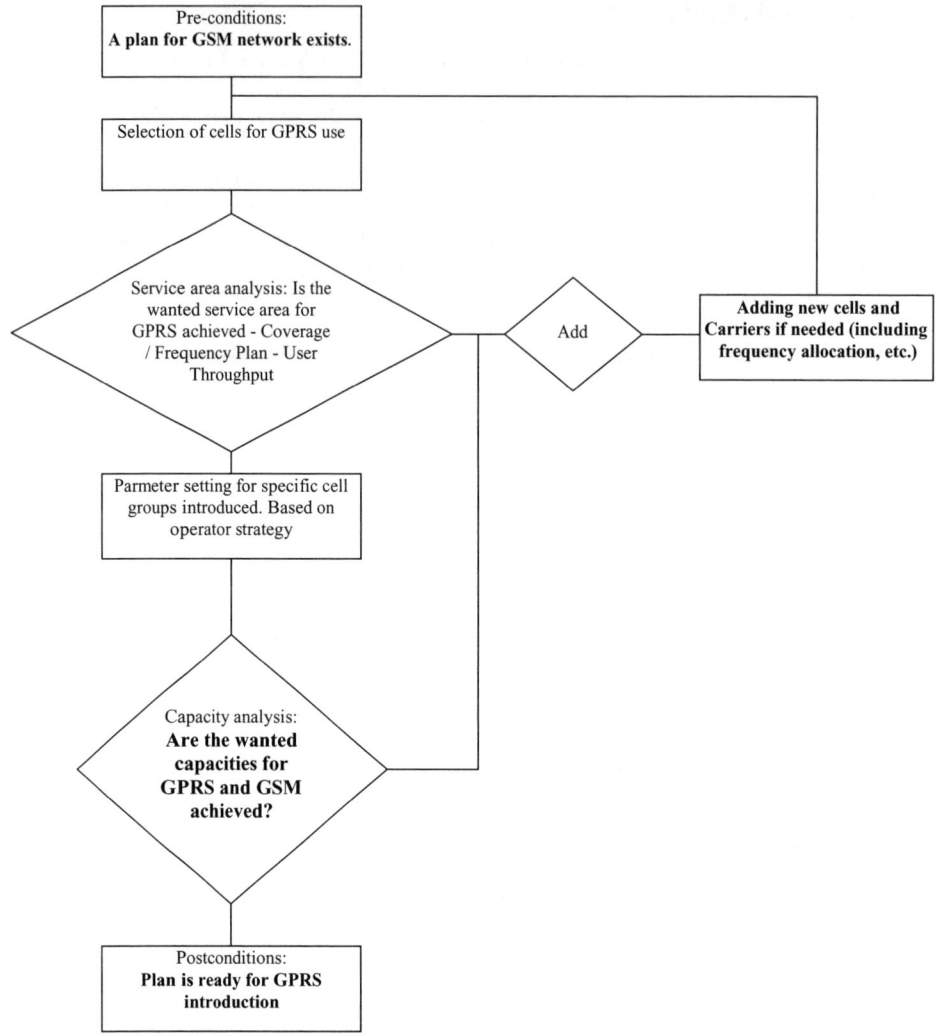

Figure 2.3 Flowchart of the planning process when GPRS is introduced to network

transfer rate is 50 RLC/MAC blocks/second, this can be multiplied by the payload (in bits)/RLC/MAC block (depending on the RLC Coding Scheme utilized) to calculate the effective timeslot throughput.

Since most EGPRS devices are capable of two to four simultaneous timeslots, the additional radio timeslots assigned to the mobile will increase the throughput accordingly. Multiple EGPRS timeslots will be assigned to the mobile based on the availability of idle timeslots. To balance the load, congestion and blocking of GSM voice and EGPRS data calls, various algorithms and parameters must be considered.

To maintain link quality, link adaptation and incremental redundancy techniques are introduced in the EGPRS system. Link adaptation dynamically adjusts the modulation and coding schemes based on estimated channel conditions. Incremental redundancy adjusts the code rate to actual channel conditions. This is done by incrementally transmitting redundancy until decoding is successful.

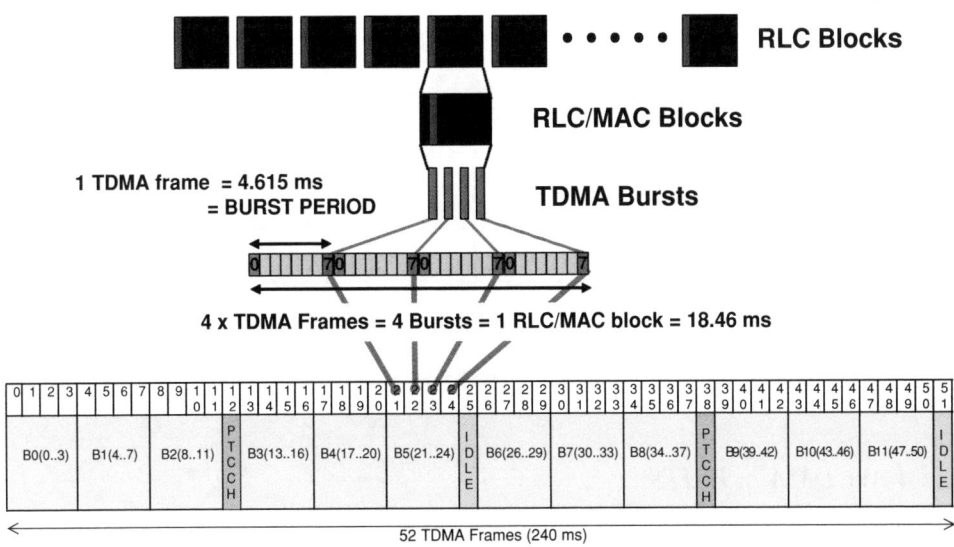

Figure 2.4 RLC block transmission

A_{bis}

Standard GSM and GPRS require only 16 kbps A_{bis} channels since all radio link codings fit within this transmission channel structure. However, EGPRS radio transmission requires additional A_{bis} channel capacity to support the higher coding schemes. This is a key factor when planning and optimising EGPRS networks since inadequate transmission planning will lead directly to lower radio and end user throughput in the specific cell served by the allocated transmission (Chapter 3 contains further discussion on this concept).

PCU

While Packet Control Unit (PCU) architecture varies between different vendors, it must be dimensioned and balanced adequately to support the additional demand that EGPRS places on the processing load and logical A_{bis} channel load. Inadequate PCU capacity leads to lower end user throughput due to less processing allocated to requested transmission, MCS, or EGPRS capable radio timeslots. In general, each PCU serves a large portion of the BSS coverage area so it is very important to ensure that PCU capacity is dimensioned accordingly.

Gb Link

The transmission links between the BSS PCU and the Packet Core SGSN must be expanded to support the additional bandwidth demand that EGPRS places on the packet data network. Running out of Gb link capacity severely reduces the throughput of all cells connected to the Gb link. While adequate PCU, A_{bis} and radio resources may be available in the BSS, users will experience degraded or no throughput for the times that Gb link congestion is present.

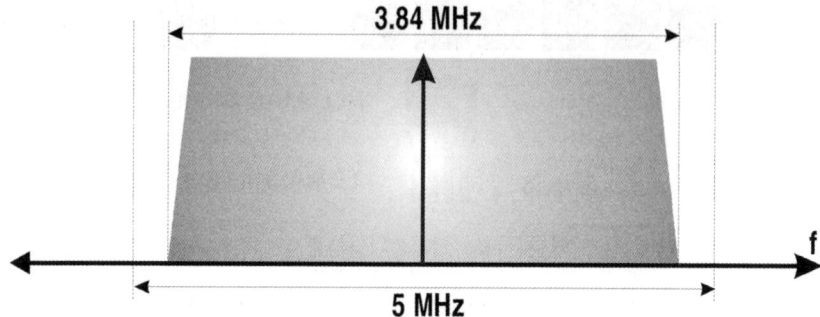

Figure 2.5 WCDMA carrier

2.2.4 WCDMA in UMTS

WCDMA – Air Interface in UMTS

WCDMA is the air interface chosen for the UMTS. It is quite different from the air interface used in GSM networks. The quality (and delay) requirement is much higher in UMTS networks as compared to GSM networks. The bit rates are higher (up to 2 Mbps with Release 99 and up to 10 Mbps with HSDPA), which means that a larger bandwidth of 5 MHz is required to support these higher bit rates. The possibility of offering subscriber variable bit rates and bandwidth – on demand – is an attractive feature in UMTS networks. For this, ATM technology (described in detail in Chapter 3) becomes handy. Asymmetric traffic is supported. Transmit diversity can be used to improve downlink capacity in WCDMA networks. Also, as only one frequency is used, frequency planning is therefore not such a tedious task as it is in the GSM networks. Packet data scheduling is load based as compared to timeslot based as in the GSM network, thus making the system more efficient. The algorithms that are used for Radio Resource Management (RRM) functionality are more advanced as compared to GSM networks.

WCDMA Carrier

WCDMA is a DS-CDMA, or Direct Sequence CDMA, system. This technology enables multiple accesses that are based on a spread spectrum. This means that user information bits are spread over a wide bandwidth by multiplying user information with the quasi-random bits called chips, derived from CDMA spreading codes. The rate at which the data spreads is called the chip rate. The ratio of the chip rate to the symbol rate is called the spreading Factor (SF). The destination mobile phone uses the same spreading code as the one used at the transmission point and performs correlation detection. Each user is identified by a unique spreading code assigned to it. The chip rate of 3.84 Mcps (megachips per second) is used, which leads to a carrier bandwidth of 5 MHz, as shown in Figure 2.5.

The narrowband signal R has different bit rates according to the amount of information to be transmitted. The SF tells how many baseband chips are used to transmit one narrowband information bit:

$$R\, SF = \text{constant} = W = 3.84\ \text{Mcps} \qquad (2.1)$$

All the users in DS-CDMA can be allocated the same frequency band and timeframe for communication. As a result, the signal-to-interference ratio at the receiver can easily be less than unity as the power of the signal of a user is typically lower than the aggregate power of signals of other users.

Preplanning in a GSM Radio Network

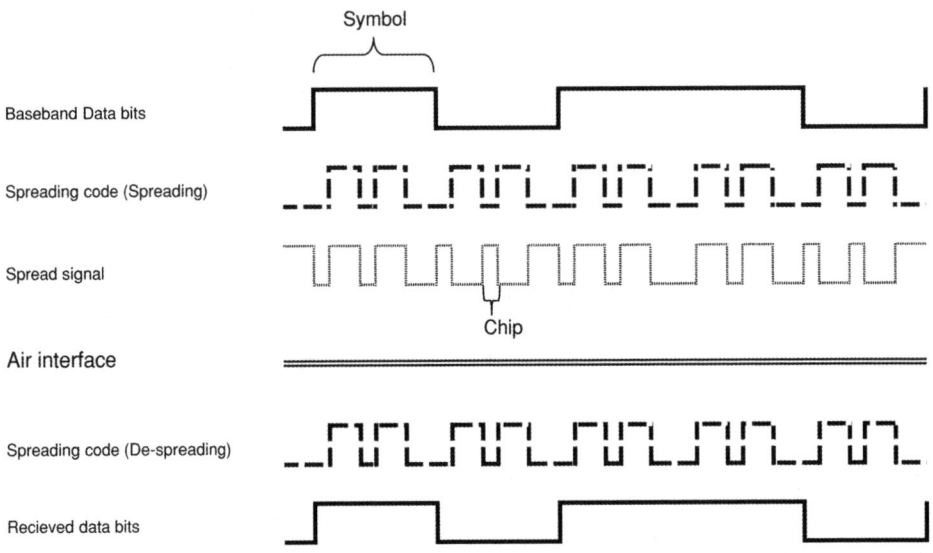

Figure 2.6 Spreading and de-spreading

Spreading and De-spreading

The spreading codes are Walsh–Hadamard codes: orthogonal variable spreading factor (OVSF) codes. The Walsh–Hadamard codes are selected because the code property is called orthogonality. By definition, orthogonality is a property of two codes where their cross-correlation equals zero. In a more practical example, orthogonality implies that information coded with orthogonal codes can be separated without incurring any errors.

Multiplying a baseband chip sequence with one spreading code and multiplying the same baseband chip sequence with another spreading code, both change the order of bits in the baseband chip sequence. However, the spreading code is also applied again when information is to be extracted from the baseband chips, after the signal has been received on another end of the air interface. This is shown in Figure 2.6.

De-spreading the baseband bit sequence with one spreading code gives the first coded information bit sequence; de-spreading the same baseband bit sequence with another spreading code gives another coded bit sequence. No errors are due to coding and decoding. The same baseland bit sequence can carry several coded information sequences.

Spreading codes are also called channellisation codes. This is because spreading codes are used in uplink to separate user and control data (one user) and in downlink to separate common and dedicated channels within a cell (multiple users). The length of the spreading code varies according to the symbol rate (shown in Figure 2.7).

Spreading codes must be orthogonal. When selecting a scrambling code to be used, other spreading codes belonging to the same code branch cannot be used. This kind of hierarchical selection of scrambling codes forms a 'code tree'. If the branch is in use spreading codes belonging to that same code branch cannot be used. Running out of code resources would create code blocking. The resource manager functionality has functions to optimise the code tree.

Another step after spreading is scrambling. The baseband bit sequence is scrambled using a scrambling code. This code is used to separate UEs and BSs. This operation does not change the signal bandwidth but only separates the signals from different sources. The chip rate is already achieved in spreading by channellisation codes and the symbol rate is not affected by scrambling.

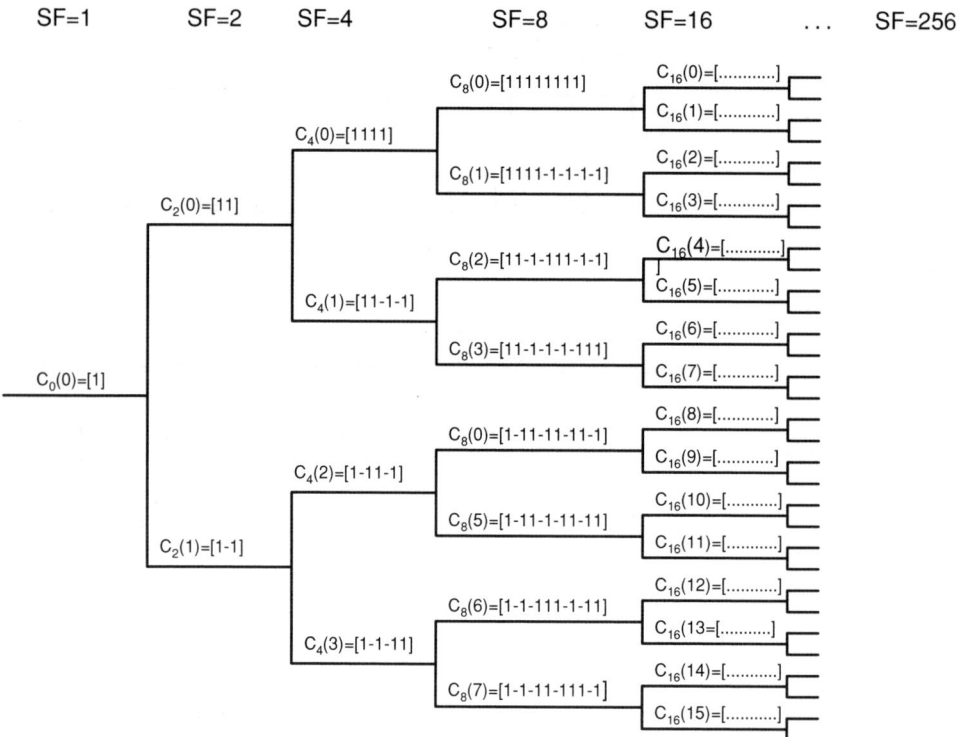

Figure 2.7 Code tree

Processing Gain

Due to spectral spreading of the user signal, the D-CDMA system is more resistant to interference. This is because, with respect to the interference signal, the multiplication code used at the receiver is similar to the one used at the transmitter. In an effort to understand this, assume that the user signal during the course of spreading is multiplied by a factor of, say, eight. This means that the signal increases by a factor of eight with respect to the interfering signal from another user. This is called the processing gain.

Processing gain is a power density factor that tells how much less energy in W/Hz is required to transmit information on a wideband signal in comparison with transmitting the same information on a narrowband signal. According to information theory the energy × time × frequency band required to transmit an information bit is constant. Spreading the narrowband signal into a wideband signal gives the processing gain. The greater the difference between the narrowband bit rate and the wideband chip rate the greater the processing gain:

$$G_p[\text{dB}] = \frac{B_{U_u}}{B_{\text{Bearer}}} = \frac{W}{R} = \text{SF}$$

Processing gain gives the WCDMA system its robustness against self-interference, making it possible to reuse the 5 MHz carrier frequencies in close geographical distances.

Reception in WCDMA Networks

The signal received by the receiving antenna consists of direct and indirect signals. The indirect signals can be reflected from natural obstacles such as buildings, hills, towers, etc., or from the ground. Thus the signal that reaches the receiving antenna after reflection will be delayed by certain time intervals (the time interval may vary depending upon the terrain it was travelling on, e.g. less delay in urban areas and more delay in hilly terrains). Also, the amplitude of these reflected signals will be different from that of the direct signal. In WCDMA, efficient reception takes place by using RAKE receivers. Any difference between two different multipath components that is more than one chip period can be resolved by this kind of receiver. The compensation of delay is accounted for by the difference in arrival times of the RAKE fingers while the RAKE combiner sums up the channel compensated symbols. Also, the shorter the duration of one chip period, the better the network is able to combat against interference due to multipath reception. RAKE receivers are different in the UE and BS, but the fundamental principle remains the same.

Radio Resource Management (RRM)

The Radio Resource Management (RRM) functionality controls system resources. The physical resources that limit the number of users and the services users have available are transmission power, number of hardware channel units installed, transmission capacity, code resources and number of frequencies available. The RRM allocates these resources and controls the use of them. The basic principle is to save resources and guarantee the quality of the connection for existing users, in a case of resource limitation at the expense of new incoming users.

The RRM consists of power control (PC), admission control (AC), load control (LC), packet scheduler (PS) and resource manager (RM) functionality. These are explained in detail in the planning sections.

2.3 Radio Network Dimensioning

Dimensioning is the main part of the preplanning phase. In addition to the dimensioning parameters the priority of the parameters also needs to be agreed. The radio network dimensioning parameters have an impact on each other and therefore it is important to decide the emphasis in order to get an optimal dimensioning result within the agreed parameter ranges. It is imperative to agree the network layout, usage of three sector sites or a combination of three sector and omni sites with the operator. One important planning issue is also whether only macro cells are used in the beginning or a combination of macro, micro and pico cells. The macro, micro and pico cells vary with the size of the cell and the antenna placement. The macro cells are used in rural and suburban areas to cover large areas. Micro cells are used in city areas to cover areas close by and antennas are on the walls. The pico cells are used to cover either very specific hot spot in an outdoor area or to give indoor coverage. The optional network features like frequency hopping also have an impact on dimensioning. All of these need to be discussed and agreed between the operator and vendor.

The macro cell has a cell range of 1–35 km and is characterised by an outdoor antenna, which covers a large area. Antennas are higher that rooftops. The micro cell has a cell range of less than 1 km and is for outdoor coverage. The antennas are typically mounted on walls and below the average rooftop level. The pico cell has a cell range less than 500 m and is characterised by antennas mounted low on the walls, clearly below the rooftop level. They are used for both indoor and outdoor coverage.

2.3.1 Link Budget Calculations

The radio link budget aims to calculate the cell coverage area. One of the required parameters is radio wave propagation to estimate the propagation loss between the transmitter and the receiver. The other required parameters are the transmission power, antenna gain, cable losses, receiver sensitivity and margins, as shown in Figure 2.8.

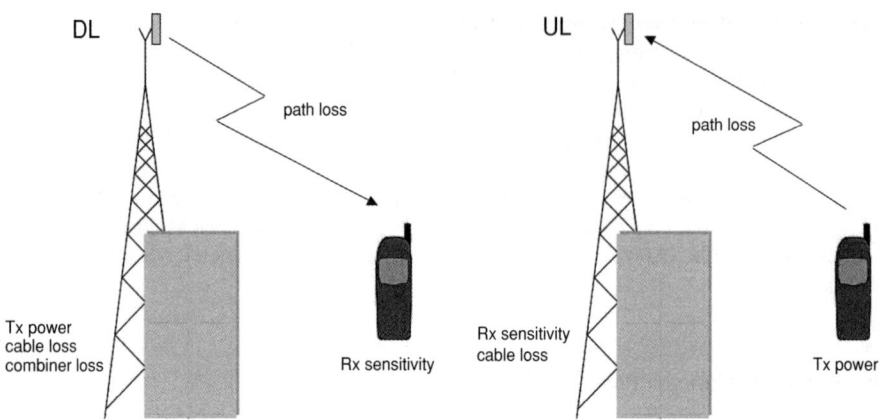

Figure 2.8 Link budget parameters

When defining the cell coverage area, the aim is to balance the uplink and downlink powers. The links are calculated separately and are different from the transmission powers. The BTS transmission power is higher than the MS transmission power and therefore the reception of the BTS needs to have high sensitivity. The other differences in calculations between uplink and downlink are explained later. The parameter that is the same for both downlink and uplink calculations is the propagation loss. The radio signal experiences the same path loss when travelling from the BTS to the MS as from the MS to the BTS.

The GSM link budget parameters are explained in the following before proceeding to the calculations in more detail:

- *BTS sensitivity* is specified on the ETSI GSM recommendation 05.05 and the recommended value is -106 dBm. This is a general recommendation and therefore when preparing a link budget with a certain manufacturer's equipment this vendor's recommendations can be used.
- *MS sensitivity* is also specified in the ETSI recommendation 05.05, where the receiver sensitivity value is separate for each MS class. For MS class 4, which means GSM 900, the recommended value is -102 dBm. Correspondingly for class 1, GSM 1800, the value is -100 dBm. The MS sensitivity can also be calculated using the information of receiver noise F and minimum E_b/N_0. The value for the noise is 10 dB and the minimum E_b/N_0 is 8 dB, as defined in the ETSI recommendation 03.30. The receiver sensitivity S_i is solved from the following equation, where the input noise power N_i is the product of three parameters: the Boltzman constant k, temperature $T_0 = 290$ K and bandwidth $W = 271$ kHz (54 dB):

$$F = \frac{S_i/N_i}{S_0/N_0} = \frac{S_i/(kT_0W)}{E_b/N_0}, \qquad N_i = kT_0W \tag{2.2}$$

where

$$\begin{aligned} S_i &= \frac{E_b}{N_0} F k T_0 W \\ &= 8\,\text{dB} + 10\,\text{dB} + (290\,\text{K} \times 1.38 \times 10^{-23}\,\text{J/K}) + 54\,\text{dB} \\ &= 8\,\text{dB} + 10\,\text{dB} - 174\,\text{dBm} + 54\,\text{dB} = -102\,\text{dBm} \end{aligned} \tag{2.3}$$

BTS sensitivity can also be calculated using this equation with the noise value specified for the BTS.

Table 2.3 Approximate values for cable losses

Cable type	Loss/100 m, 900 MHz	Loss/100 m, 1800 MHz
$\frac{1}{2}$ inch	7 dB	10 dB
$\frac{7}{8}$ inch	4 dB	6 dB
$1\frac{1}{4}$ inch	3 dB	4.5 dB

- *MS and BTS powers* are important along with the sensitivities. The MS TX (transmission) power is defined by the MS class in ETSI specifications. For MS class 4 (GSM 900) the maximum TX power is 2 W and for class 1 (GSM 1800) 1 W. BTS TX power depends on the BTS type and vendor. The TX power is adjustable, which enables the link budget to be balanced.
- *Antenna gains* can be found for both BTS and MS antennas. The BTS antenna gain is dependent on the antenna type and whether the antenna is omnidirectional or directional. The first mentioned BTS antenna type has a gain of approximately 10 dBi. The gain of a directional BTS antenna is dependent on the horizontal and vertical half power beam widths. It is also dependent on the physical size of the antenna which in turn has an impact on the frequency range. Also frequency range is inversely proportional to the size of the antenna, which is then connected to the radiating aperture of the antenna. The antenna gain is around 16–20 dBi when there is a widely used antenna with 60–65° horizontal half power beam width and 5–10 vertical half power beam width. In the link budget calculations for the MS antenna the gain is generally 0 dBi. The actual MS antenna gain is complicated to estimate, because the gain is highly dependent on the mobile user's relative location towards the base station when the amount of body loss varies.
- *Diversity gain* can be used for correcting unbalance between the uplink and downlink. The typical way to arrange diversity is to have it in the BTS reception. One basic method is to separate receiver antennas vertically or horizontally; the method is called space diversity. The diversity decreased fading effect and gain achieved can be around 5 dB.
- *Cable and connector losses* are case specific and need to be measured or calculated separately. Thicker cable causes less loss, but it is more difficult to install due to weight and a larger bending radius. Thicker cables are also more expensive and therefore the usage needs to be considered. Obviously a longer cable gives a higher loss and because of this the shortest possible route for the cable should be used. Some idea of the cable losses can be seen from Table 2.3. An individual connector gives a loss of around 0.1 dB, but depending on the cable installations there can be several in one antenna line.
- *Other equipment loss factors* consist of *isolator*, *combiner* and *filter* losses. The isolator isolates the transmitted signal from the transmitter. The combiner combines TX (transmission) signals to one antenna. The filter combines transmitted and received signals to one feeder cable as well as to a single antenna and antenna cable. The isolator, combiner and filter together give around 2–3 dB losses.
- Two other gain factors, which need to be considered in the link budget if used, are the *mast head amplifier (MHA)* and *booster*. The MHA, which is located close to the antenna in BTS reception, is used to amplify the received signal. This decreases the unbalance between the uplink and downlink by giving extra gain in the uplink the direction. The booster can be used to amplify the BTS transmission power.
- *The interference degradation margin* describes the loss due to frequency reuse. Therefore the frequency reuse rate corresponds to the degradation margin value. The suggested value for the interference degradation margin in an average suburban and rural area is 3 dB according to the ETSI recommendation 03.30. In the case of high frequency reuse in urban areas the degradation margin can have a value of 4–5 dB.

An example of power budget calculations is presented in Table 2.4. This example is for GSM 1800 and the more calculations are described in more detail in the following using the values of this example. The

Table 2.4 Power budget for GSM 1800

RADIO LINK POWER BUDGET		MS CLASS	1	
General information				
Frequency [MHz]: 1800		System: GSM		

Receiving end	Unit	BS	MS	
RX RF input sensitivity	dBm	−106.00	−100.00	A
Interference degradation margin	dB	3.00	3.00	B
Cable loss + connector	dB	2.00	0.00	C
RX antenna gain	dBi	18.00	0.00	D
Diversity gain	dB	5.00	0.00	E
Isotropic power	dBm	−124.00	−97.00	$F = A + B + C - D - E$
Field strength	dB μV/m	18.31	45.31	$G = F + Z^a$

Transmitting end		MS	BS	
TX RF output peak power	W	1.00	25.12	
Mean power over RF cycle	dBm	30.00	44.00	K
Isolator + combiner + filter	dB	0.00	3.00	L
RF peak power, combiner output	dBm	30.00	41.00	$M = K - L$
Cable loss + connector	dB	0.00	2.00	N
TX antenna gain	dBi	0.00	18.00	O
Peak EIRP	W	1.00	501.19	
EIRP = ERP + 2 dB	dBm	30.00	57.00	$P = M - N + O$
Isotropic path loss	dB	154.00	154.00	$Q = P - F$

$^a Z = 77.2 + 20 \log (\text{frequency [MHz]})$.

power budget calculations start with the receiving end by calculating isotropic power (F) both for the BTS and the MS. The isotropic power is the minimum received power, which enables the link balance situation. The receiver sensitivity (A) is the basis for the isotropic power with the power budget gain parameters subtracted from it and the loss parameters added to it. The isotropic power for the BTS is

$$F = A + B + C - D - E = -106\,\text{dBm} + 3\,\text{dB} + 2\,\text{dB} - 18\,\text{dBi} - 5\,\text{dB} = -124\,\text{dBm} \qquad (2.4)$$

and for the MS is

$$F = -100\,\text{dBm} + 3\,\text{dB} + 0\,\text{dB} - 0\,\text{dBi} - 0\,\text{dB} = -97\,\text{dBm} \qquad (2.5)$$

The isotropic power is next converted to dB μV/m and the result is field strength (G). For the BTS:

$$G = F + Z^* = -124\,\text{dBm} + 77.2 + 20\,\log(1800\,\text{MHz}) = 18.31\,\text{dB}\mu\text{V/m} \qquad (2.6)$$

and field strength for the MS:

$$G = -97\,\text{dBm} + 77.2 + 20\,\log(1800\,\text{MHz}) = 45.31\,\text{dB}\mu\text{V/m} \qquad (2.7)$$

The power budget calculation continues with the MS peak EIRP (effective isotropic radiated power) (P) calculation. The MS radiation power is presented in this way as the antenna gain uses decibels over an isotropic antenna (dBi) format. If the antenna gain format is in decibels over a dipole (dBd) the radiated power format is ERP (effective radiated power). The transformation from EIRP to ERP is given by

EIRP = ERP + 2 dB. As written earlier, the MS antenna gain is case dependent due to the body loss of the MS user and therefore a 0 dBi gain is normally used in the calculations. The losses are also zero and therefore the EIRP of the MS is the TX output power (K) and the MS power is fixed according to the MS class:

$$P = K - L - N + O = 30\,\text{dBm} - 0\,\text{dB} - 0\,\text{dB} + 0\,\text{dBi} = 30\,\text{dBm} \qquad (2.8)$$

The isotropic path loss (Q) for the uplink is the EIRP minus the isotropic power:

$$Q = P - F = 30\,\text{dBm} - (-124\,\text{dBm}) = 154\,\text{dB} \qquad (2.9)$$

The path loss for the downlink direction equals the path loss for the uplink direction. With this the BTS TX output peak power calculation can be started. As mentioned before, the BTS TX power is critical for the link balance. Firstly the BTS peak EIRP is

$$P = Q + F = 154\,\text{dB} + (-97\,\text{dBm}) = 57\,\text{dBm} \qquad (2.10)$$

Then continuing to the BTS TX power, the downlink EIRP is subtracted by the BTS antenna gain (O) and added to the cable loss (N) and isolator, combiner and filter loss (L):

$$K = M + L = P + N - O + L = 57\,\text{dBm} + 2\,\text{dB} - 18\,\text{dBi} + 3\,\text{dB} = 44\,\text{dBm} \qquad (2.11)$$

A typical value for dimensioning the traffic created by one subscriber is 15–30 mErl (milli-Erlang). What does, for example, 25 mErl mean in practice? The user talks during a busy hour for $60 \times 60 \times 25$ mErl = 90 seconds. This much capacity needs to be reserved for one user to meet the planned requirements.

Network dimensioning is performed for the busy hour traffic, which ensures that the availability for users is as good as possible. The blocking probability describes the probability that the MS cannot initiate a call because all of the traffic channels have been reserved at that time moment. The blocking probability and estimated traffic give the required number of channels using the Erlang formulas, where formula B for the non-queing situation and formula C for the situation where queing is possible can be used for dimensioning. In traffic calculations the first assumption is that the subscribers are evenly distributed inside the planned area. The number of required channels, traffic and signalling channels specify the numbers of TRXs and BTSs. From these the minimum number of BSCs and MSCs can be dimensioned. The limitation of the capacity comes from the availability of frequency bands, the number of frequencies and the frequency re-use rate.

Dimensioning Tools

Several commercial tools are available for dimensioning purposes. Those are provided by software companies, but many vendors and operators also have their own tools for dimensioning. Complexity of the tool depends on the functionalities. The dimensioning tool can in the simplest case consist of self-made spreadsheets. Those are adequate for small dimensioning cases and supplementing capacity calculations. More sophisticated tools provide dimensioning for several features and are suitable for large dimensioning cases. Some additional software might be needed when dimensioning the capacity needed for advanced features.

The dimensioning tool provides the capacity calculations to meet the planning requirements: capacity, coverage and quality targets. The link budgets are an essential part of the calculations and the dimensioning results are given per area type – the required number of network elements for each area type.

The network evolution can also be sketched using dimensioning tools . Roll-out occurs in different phases, with network dimensioning following the phases structure. As the plan to activate new features has a capacity effect, dimensioning can be done according to time schedules. The output the dimensioning tool gives, the number of network elements, is used as an input for the coverage planning phase and coverage planning tool.

2.3.2 Dimensioning in the EGPRS Network

Probably the single most important knowledge to have when planning any EGPRS network is to understand where the heavy traffic is located. Considering this, some provision must be made to support initial optimisation around these dense data carrying cells to ensure a desired quality of service for the new EGPRS subscribers. After the EGPRS network is deployed and operational, periodic optimisation of the key parameters will ensure a smooth running network.

Table 2.5 summarises the key radio and transmission parameters that need to be considered when planning and optimising an EGPRS network. (Note that the parameters are listed with a conceptual explanation in order to keep a generic description regardless of the equipment supplier.) An example of the link budget for the EGPRS network is shown in Figure 2.9.

2.3.3 Dimensioning in the WCDMA Radio Network

In WCDMA the coverage and capacity are interrelated. This means that in dimensioning the coverage has to be planned for a service having a load. The used services and their load are input parameters for dimensioning. If the network load changes the coverage has to be planned for a new load situation and if new services are introduced they must be evaluated so as not to sacrifice coverage.

The dimensioning is always a compromise between coverage and capacity. By reducing the maximum load the coverage can be extended. If more capacity is required the coverage area of an individual cell shrinks, and this increases the number of required network elements.

Coverage Limited Scenario

In this case the load on the system is lower than the initially used value for computation of the cell range. The link budget calculation is repeated with a lower value of system loading to match the actual value of the one used in the link budget. This process leads to a bigger cell range and a lower interference floor, thereby permitting more users in the cell. This reduction in system loading used in the link budget calculation is continued until it matches the actual system loading computed by the traffic profile, resulting in the final cell range.

Coverage Related Input
- Area type information as accurate as possible
- Coverage area for each roll-out phase
- Percentage of the area for each morphoclass (DU, U, SU, R)
- Building penetration loss and fading margin
- Propagation models for path loss calculation
- Correction factors for the propagation model

Service scenarios should be defined, such as which kind of service is to be offered and where (with a big impact on the number of sites).

Capacity Limited Scenario

In this case the load on the system is greater than the initially used value for computation of the cell range, which means that either the cell size decreases or the capacity increases. If the system load remains more

Table 2.5 Summary of the key EGPRS parameters

	Parameter(s)	Conceptual explanation
Radio	Timeslot allocation	Affects how the packet switched allocation of timeslots is performed. This includes prioritisation between different bands (850/1900 or 900/1800) and also prioritisation between GSM voice and multislot EGPRS data calls
	Link adaptation thresholds and modulation coding scheme allocation	Adjusts the algorithm for changing between different MCSs to maximise throughput and latency while ensuring adequate forward error correction in changing channel conditions
	BTS power back-off	Reduces the maximum transmitted power of the GSM transmitter. Can be used to balance an uplink limited network and reduce interference at the cost of reducing the coverage footprint. Also can be used to minimise the effect of link imbalance caused by changing between 8-PSK and GMSK modulation
	One-phase access	Sets the radio network to support an 11-bit RACH (random access channel) and directly assign EGPRS radio timeslots based on the mobile capability. This effectively reduces the latency of initial access to the network
	TBF release delay	Adjusts the time (in seconds) in which the network will release the TBF (temporary block flow) after idle transmission. Optimum settings will be based on the predominant application since long release delays significantly reduce latency of successive access to the network at the cost of reserving TBF radio resources
Transmission	EGPRS A_{bis} transmission	Allocates the appropriate number of 16 kbps A_{bis} subtimeslots and 64 kbps DS0 T1/E1 transmission between the BSC and BTS. This ensures that multislot EGPRS data traffic demand for higher 8-PSK MCS radio timeslots can be matched by the supplied transmission to the site
	LLC/GTP buffers and BVC flow control	Used to match the PCU segmentation processing speed of LLC (logical link control) frames into RLC blocks over the downlink GPRS Tunnelling Protocol (GTP) flow through each base station virtual circuit (BVC)
	Gb NSVC CIR	Sets the transmission bandwidth between the SGSN and PCU/BSS. An adequate bandwidth or CIR committed information rate must be allocated on the Gb link for each Network Service Virtual Circuit (NSVC) to support the instantaneous demand from all BSs supporting EGPRS data transmission connected to a given NSVC

DOWNLINK

BTS Transmitter:	Voice (>=Q3)	MCS-1	MCS-2	MCS-3	MCS-4	MCS-5	MCS-6	MCS-7	MCS-8	MCS-9	Units
Peak TX Power	37.0	37.0	37.0	37.0	37.0	37.0	37.0	37.0	37.0	37.0	dBm
Average TX Power (2 dB backoff for 8-PSK)	37.0	37.0	37.0	37.0	37.0	35.0	35.0	35.0	35.0	35.0	dBm
Isolator + Combiner + Filter Losses	3.5	3.5	3.5	3.5	3.5	3.5	3.5	3.5	3.5	3.5	dB
Cable and connector losses	3.0	3.0	3.0	3.0	3.0	3.0	3.0	3.0	3.0	3.0	dB
Antenna Gain	11.0	11.0	11.0	11.0	11.0	11.0	11.0	11.0	11.0	11.0	dBi
EIRP:	**41.5**	**41.5**	**41.5**	**41.5**	**41.5**	**39.5**	**39.5**	**39.5**	**39.5**	**39.5**	**dBm**

MS Receiver:	Voice (>=Q3)	MCS-1	MCS-2	MCS-3	MCS-4	MCS-5	MCS-6	MCS-7	MCS-8	MCS-9	Units
Receiver Sensitivity (Static 1900) *	-107.0	-108.0	-106.9	-105.0	-102.6	-101.8	-99.8	-96.7	-93.7	-91.7	dBm
Receiver Sensitivity (TU50iFH 1900) *	-105.0	-103.4	-101.4	-97.3	-91.5	-97.5	-95.4	-90.4	-82.5	-84.5	dBm
Incremental Redundancy Gain *	0.0	0.0	0.0	0.0	0.0	0.0	0.0	0.0	1.0	2.0	dB
* At the following QoS Service Levels: Voice (>=Q3): 1% FER; MCS-1 to MCS-9: 10% BLER; **MCS-9: 30% BLER at TU50iFH 1900											
Antenna Gain	0.0	0.0	0.0	0.0	0.0	0.0	0.0	0.0	0.0	0.0	dBi
Body Loss	3.0	0.0	0.0	0.0	0.0	0.0	0.0	0.0	0.0	0.0	dB
Isotropic Power Required at Antenna (Static 1900)	-104.0	-108.0	-106.9	-105.0	-102.6	-101.8	-99.8	-96.7	-94.7	-93.7	dBm
Isotropic Power Required at Antenna (TU50iFH 1900)	-102.0	-103.4	-101.4	-97.3	-91.5	-97.5	-95.4	-90.4	-83.5	-86.5	dBm

Downlink Path Loss	Voice (>=Q3)	MCS-1	MCS-2	MCS-3	MCS-4	MCS-5	MCS-6	MCS-7	MCS-8	MCS-9	Units
Downlink Path Loss (Static 1900)	145.5	149.5	148.4	146.5	144.1	141.3	139.3	136.2	134.2	133.2	dB
Downlink Path Loss (TU50iFH 1900)	143.5	144.9	142.9	138.8	133.0	137.0	134.9	129.9	123.0	126.0	dB

UPLINK

MS Transmitter:	Voice (>=Q3)	MCS-1	MCS-2	MCS-3	MCS-4	MCS-5	MCS-6	MCS-7	MCS-8	MCS-9	Units
Peak TX Power	30.0	30.0	30.0	30.0	30.0	30.0	30.0	30.0	30.0	30.0	dBm
Average TX Power (4 dB backoff for 8-PSK)	30.0	30.0	30.0	30.0	30.0	26.0	26.0	26.0	26.0	26.0	dBm
Antenna Gain	0.0	0.0	0.0	0.0	0.0	0.0	0.0	0.0	0.0	0.0	dBi
Body Loss	3.0	0.0	0.0	0.0	0.0	0.0	0.0	0.0	0.0	0.0	dB
EIRP:	**27.0**	**30.0**	**30.0**	**30.0**	**30.0**	**26.0**	**26.0**	**26.0**	**26.0**	**26.0**	**dBm**

BTS Receiver:	Voice (>=Q3)	MCS-1	MCS-2	MCS-3	MCS-4	MCS-5	MCS-6	MCS-7	MCS-8	MCS-9	Units
Receiver Sensitivity (Static 1900) *	-111.0	-112.0	-110.9	-109.0	-106.6	-105.8	-103.8	-100.7	-97.7	-95.7	dBm
Receiver Sensitivity (TU50iFH 1900) *	-109.0	-107.4	-105.4	-101.3	-95.5	-101.5	-99.4	-94.4	-86.5	-88.5	dBm
Incremental Redundancy Gain *	0.0	0.0	0.0	0.0	0.0	0.0	0.0	0.0	1.0	2.0	dB
* At the following QoS Service Levels: Voice (>=Q3): 1% FER; MCS-1 to MCS-9: 10% BLER; **MCS-9: 30% BLER at TU50iFH 1900											
Cable and connector losses	3.0	3.0	3.0	3.0	3.0	3.0	3.0	3.0	3.0	3.0	dB
NET MHA Gain	0.0	0.0	0.0	0.0	0.0	0.0	0.0	0.0	0.0	0.0	dB
Antenna Gain	11.0	11.0	11.0	11.0	11.0	11.0	11.0	11.0	11.0	11.0	dBi
Isotropic Power Required at Antenna (Static 1900)	-119.0	-120.0	-118.9	-117.0	-114.6	-113.8	-111.8	-108.7	-106.7	-105.7	dBm
Isotropic Power Required at Antenna (TU50iFH 1900)	-117.0	-115.4	-113.4	-109.3	-103.5	-109.5	-107.4	-102.4	-95.5	-98.5	dBm

Uplink Path Loss	Voice (>=Q3)	MCS-1	MCS-2	MCS-3	MCS-4	MCS-5	MCS-6	MCS-7	MCS-8	MCS-9	Units
Uplink Path Loss (Static 1900)	146.0	150.0	148.9	147.0	144.6	139.8	137.8	134.7	132.7	131.7	dB
Uplink Path Loss (TU50iFH 1900)	144.0	145.4	143.4	139.3	133.5	135.5	133.4	128.4	121.5	124.5	dB

Figure 2.9 Dimensioning in the EGPRS network for the EGPRS link budget

than the permitted value, the range of cell should be decreased. This is followed by performing system loading calculations iteratively until the actual system load matches the permissible one, thereby defining cell range.

Capacity Related Inputs

The main requirements for capacity dimensioning are:

- The number of subscribers
- User profile
- Spectrum available

A traffic forecast should be done by analysing the offered busy hour traffic per subscriber for different service bit rates in each roll-out phase.

Traffic Data
Voice

- Erlang per subscriber during the busy hour of the network
- Codec bit rate
- Voice activity

RT (real-time) data

- Erlang per subscriber during the busy hour of the network
- Service bit rates

NRT (non-real-time) data

- Average throughput (kbps) of the subscriber during the busy hour of the network
- Target bit rates
- Asymmetry between UL (uplink) and DL (downlink) traffic for NRT services (downloading 1/10) should be taken into consideration

A network and subscribers evolution forecast is also needed.

Quality Related Input
Quality is as perceived by the subscriber. A couple of important ones that are used during the dimensioning phase are the location probability and the blocking probability. The former one is related to the fact that if UE is able to detect more than one cell, the location probability increases. The range is usually between 90 and 99 %. Factors such as customer requirements, environment, etc., have an impact on this. The latter factor of blocking probability should be lower as the number of calls blocked due to unavailability of resources should be less (1–2 %).

RNC Dimensioning
The purpose of RNC dimensioning is to provide the number of RNCs needed to support the estimated traffic. The main limitations that should be taken into account are:

- maximum number of cells;
- maximum number of base stations;
- maximum I_{ub} throughput;
- amount and type of interfaces;
- input data from the customer or RF planning;
- RNC dimensioning example;
- RNC areas.

Some Parameters Used in WCDMA Dimensioning
E_b/N_0 is the ratio of the average bit energy and noise spectral density. The value of E_b/N_0 is dependent on the type of service, UE speed and radio channel:

$$\frac{E_b}{N_0} = \frac{P_{rx}}{I}\frac{W}{R} \quad \text{dB} \tag{2.12}$$

$$I_{UL} = I_{own} + I_{oth} + P_N \tag{2.13}$$

$$I_{DL} = I_{own}(1-\alpha) + I_{oth} + P_N \tag{2.14}$$

where

P_{rx} = received power
I_{own} = total power received from the serving cell (excluding own signal)
I_{oth} = total power received from other cells
α = orthogonality factor
R = bit rate
W = bandwidth
P_N = noise power

Thus, the required RF C/I (channel-to-interference) ratio needed to meet the baseband E_b/N_0 criteria, also known as the ratio of energy per chip to total spectral density, can be given as

$$\frac{E_c}{I_0} = \frac{E_b}{N_0} \frac{R}{W} \quad \text{dB} \tag{2.15}$$

In WCDMA systems, the E_b/N_0 is larger than the S/N (signal-to-noise) ratio by an amount of the spreading factor. Thus, communication is possible in the WCDMA system even when S/N for a connection at the receiver is typically much smaller than unity.

Macro Diversity Gain (MDC)

The concept of a soft handover (described later) is used for this. It is calculated using the average overall connections by taking into account the difference of received signal branched and the UE speed. In the UL direction it is about 0 dB and in the downlink direction it is 1 dB.

Interference Margin

This is calculated from UL/DL loading values (η). UL indicates the loss in link budget calculations due to load while in DL it shows the loss of sensitivity of the BS due to load:

$$\text{NR} = -10 \log_{10}(1 - \eta) \tag{2.16}$$

Power Control Headroom

At the cell edge the UE (usually a slow moving one) does not have enough power to follow the fast fading dips. The margin against fast fading is described by the power control headroom. This can be calculated as:

$$\begin{aligned} \text{power control headroom} &= (E_b/I_0)\text{average received without fast PC} \\ &\quad - (E_b/I_0)\text{average received with fast PC} \end{aligned}$$

Thus, the required signal power can be calculated as:

$$\begin{aligned} \text{Required signal power} &= \text{receiver noise power} + \text{required } E_c/I_0 \\ &\quad - \text{interference margin} + \text{MDC gain} \end{aligned}$$

Body loss in usually used only for speech services and can be safely assumed to be zero for data services. Based on the inputs, an example of a link budget in 3G is shown in Figure 2.10 (where LCD is long constrained delay data and UDD is unconstrained delay data).

Link Budgets:		Voice		LCD				UDD				UDD				UDD			
Data rate (kbps):		12.2	12.2	64	64	64	64	144	144	384	384
Load:		50%	50%	50%	50%	50%	50%	50%	50%	50%	50%
		Uplink	Downlink	Uplink	Downlink	Uplink	Downlink	Uplink	Downlink	Uplink	Downlink
RECEIVING END		Node B	UE	Node B	UE	Node B	UE	Node B	UE	Node B	UE
Thermal Noise Density	dBm/Hz	-174	-174	-174	-174	-174	-174	-174	-174	-174	-174
BTS Receiver Noise Figure	dB	3.00	8.00	3.00	8.00	3.00	8.00	3.00	8.00	3.00	8.00
BTS Receiver Noise Density	dBm/Hz	-171.00	-166.00	-171.00	-166.00	-171.00	-166.00	-171.00	-166.00	-171.00	-166.00
BTS Noise Power [NoW]	dBm	-105.16	-100.16	-105.16	-100.16	-105.16	-100.16	-105.16	-100.16	-105.16	-100.16
Required Eb/No	dB	1.50	5.50	-0.50	4.50	-0.50	4.50	-1.00	4.00	-1.50	3.50
Soft handover MDC gain	dB	0.00	0.50	0.00	0.50	0.00	0.50	0.00	0.50	0.00	0.50
Processing gain	dB	24.98	24.98	17.78	17.78	17.78	17.78	14.26	14.26	10.00	10.00
Interference margin (NR)	dB	3.01	3.01	3.01	3.01	3.01	3.01	3.01	3.01	3.01	3.01
Required BTS Ec/Io [q]	dB	-20.47	-16.97	-15.27	-10.77	-15.27	-10.77	-12.25	-7.75	-8.49	-3.99
Required Signal Power [S]	dBm	-125.63	-117.13	-120.43	-110.93	-120.43	-110.93	-117.41	-107.91	-113.65	-104.15
Cable loss	dB	3.00	0.00	3.00	0.00	3.00	0.00	3.00	0.00	3.00	0.00
Body loss	dB	0.00	5.00	0.00	0.00	0.00	0.00	0.00	0.00	0.00	0.00
Antenna gain RX	dBi	18.00	0.00	18.00	0.00	18.00	0.00	18.00	0.00	18.00	0.00
Soft handover gain	dB	2.00	2.00	2.00	2.00	2.00	2.00	2.00	2.00	2.00	2.00
Power control headroom	dB	3.00	0.00	3.00	0.00	3.00	0.00	3.00	0.00	3.00	0.00
Sensitivity	dBm	-139.63	-114.13	-134.43	-112.93	-134.43	-112.93	-131.41	-109.91	-127.65	-106.15
TRANSMITTING END		UE	Node B	UE	Node B	UE	Node B	UE	Node B	UE	Node B
Power per connection	dBm	21.00	26.50	21.00	27.50	21.00	27.50	26.00	32.50	26.00	32.50
Maximum Power per connection	dBm	21.00	40.00	21.00	40.00	21.00	40.00	26.00	40.00	26.00	40.00
Cable loss	dB	0.00	3.00	0.00	3.00	0.00	3.00	0.00	3.00	0.00	3.00
Body loss	dB	5.00	0.00	0.00	0.00	0.00	0.00	0.00	0.00	0.00	0.00
Antenna gain TX	dBi	0.00	18.00	0.00	18.00	0.00	18.00	0.00	18.00	0.00	18.00
Peak EIRP	dBm	16.00	41.50	21.00	42.50	21.00	42.50	26.00	47.50	26.00	47.50
Maximum Isotropic path loss	dB	155.63	169.13	155.43	167.93	155.43	167.93	157.41	164.91	153.65	161.15
Isotropic path loss to the cell border			155.63		155.43		155.43		157.41		153.65

Figure 2.10 Dimensioning in the WCDMA radio network

2.4 Radio Wave Propagation

Propagation models have been developed to be able to estimate the radio wave propagation as accurately as possible. Models have been created for different environments to predict the path loss between the transmitter and receiver. How much power needs to be transmitted using the BTS to be able to receive a certain power level from the MS? The complexity of the model affects the applicability as well as the accuracy. Two well-known models are those of Okumura–Hata and Walfish–Ikegami. The first mentioned is created for large cells, i.e. for rural and suburban areas, while the Walfish–Ikegami model is used for small cells, i.e. for urban areas.

The basic electromagnetic wave propagation mechanisms are free space loss, reflection, diffraction and scattering. Free space loss describes the ideal situation, where the transmitter and receiver have line-of-sight and no obstacles are around to create reflection, diffraction or scattering. In this ideal case the attenuation of the radio wave signal is equivalent to the square of the distance from the transmitter.

When the signal has been transmitted in the free space towards the receiver antenna, the power density S at the distance from the transmitter d can be written as

$$S = \frac{P_t G_t}{4\pi d^2} \qquad (2.17)$$

where P_t is the transmitted power and G_t is the gain of the transmission antenna. The effective area A of the receiver antenna, which affects the received power, can be expressed as

$$A = \frac{\lambda^2 G_r}{4\pi} \qquad (2.18)$$

where λ is the wavelength and G_r is the gain of the receiver (RX) antenna. The received power density can also be written as

$$S = \frac{P_r}{A} \qquad (2.19)$$

Combining these equations, previous the format for the received power is

$$P_r = P_t G_t G_r \left(\frac{\lambda}{4\pi d}\right)^2 \qquad (2.20)$$

The free space path loss is the ratio of transmitted and received power. Here is the equation in simplified format, when the antenna gains are excluded:

$$L = \left(\frac{4\pi d}{\lambda}\right)^2 \qquad (2.21)$$

and the free space loss converted in decibels

$$L = 32.4 + 20 \log_{10}(f) + 20 \log_{10}(d) \qquad (2.22)$$

where f is the frequency in megahertz and d is the distance in kilometres.

In reality the radio wave propagation path is normally a non-line-of-sight situation with surrounding obstacles like buildings and trees. Therefore the applicability of the free space propagation loss is limited. The received signal actually consists of several components, which have been travelling through different paths facing reflection, diffraction and scattering. This effect is called multipath and one component represents one propagation path. The different components, signal vectors, are summarised as one signal considering the vector phases and amplitudes.

The attenuation of the radio wave signal power depends on the frequency band and terrain types between the transmitting and receiving antenna. When estimating the total path loss of the radio signal, the travelled path can be split into sections according to terrain types. As the propagation varies according to the area type, this has to be taken into account in the propagation model. The difference can be explained using the measured correction factor for each terrain type.

One more phenomenon of the mobile environment is the different fading types. Slow fading happens when the radio wave signal is diffracted due to buildings or other big obstacles in the signal path. The receiver, the mobile phone, is in a way in the shadow of these obstacles. Slow fading is log-normal fading and therefore modelled with a Gaussian distribution.

The previously mentioned multipath propagation causes short term fades, which can be relatively deep, in the received signal due to the summarised signal vectors, which are having different phases and amplitudes. This fading is known as fast fading or Rayleigh fading. As the second name implies, fast fading can be modelled using the Rayleigh distribution.

The third fading type is a combination of the previous two and is called Rician fading. When speaking about fast fading only the scattered components are taken into account, but in this case a line-of-sight component also exists. Supposedly this fading can be modelled using the Rician distribution.

2.4.1 Okumura–Hata Model

The Okumura–Hata model is a well-known propagation model, which can be applied for a macro cell environment to predict median radio signal attenuation. Having one component the model uses free space loss. The Okumura–Hata model is an empirical model, which means that it is based on field measurements. Okumura performed the field measurements in Tokyo and published results in graphical format. Hata applied the measurement results into equations. The model can be applied without correction factors for quasi-smooth terrain in an urban area but in case of other terrain types correction factors are needed. The weakness of the Okumura–Hata model is that it does not consider reflections and shadowing. The parameter restrictions for this model are:

Frequency f: 150–1500 MHz, extension 1500–2000 MHz
Distance between MS and BTS d: 1–20 km
Transmitter antenna height H_b: 3–200 m
Receiver antenna height H_m: 1–10 m

The Okumura–Hata model for path loss prediction can by written as

$$L = A + B \ \log_{10}(f) - 13.82 \ \log_{10}(H_b) - a(H_m) \\ + \left[44.9 - 6.55 \ \log_{10}(H_b)\right] \log_{10}(d) + L_{\text{other}} \quad (2.23)$$

where f is the frequency (MHz), H_b is the base station antenna height (m), $a(H_m)$ is the mobile antenna correction factor, d is the distance between the BTS and MS (km) and L_{other} is an additional correction factor for area type correction. The correction factor for the MS antenna height is represented as follows for a small or medium sized city:

$$a(H_m) = \left[1.1 \ \log_{10}(f) - 0.7\right] H_m - \left[1.56 \ \log_{10}(f) - 0.8\right] \quad (2.24)$$

and for a large city:

$$a(H_m) = \begin{cases} 8.29 \left[\log_{10}(1.54 \ H_m)\right]^2 - 1.1 : & f \leq 200 \text{ MHz} \\ 3.2 \left[\log_{10}(11.75 \ H_m)\right]^2 - 4.97 : & f \geq 400 \text{ MHz} \end{cases} \quad (2.25)$$

where H_m is the MS antenna height:

$$1 \leq H_m \leq 10 \quad (H_m \text{ in metres}) \quad (2.26)$$

Table 2.6 Parameter set for Okumura–Hata calculations

	Parameter set 1	Parameter set 2
City type	Small/medium city	Large city
Frequency f	900 MHz	1800 MHz
BTS height	30 m	30 m
MS height H_m	1.5 m	1.5 m
A	69.55	46.30
B	26.16	33.90

The parameters A and B are dependent on the frequency as follows:

$$A = \begin{cases} 69.55, & f = 150 - 1500 \text{ MHz} \\ 46.30, & f = 1500 - 2000 \text{ MHz} \end{cases}$$
$$B = \begin{cases} 26.16, & f = 150 - 1500 \text{ MHz} \\ 33.90, & f = 1500 - 2000 \text{ MHz} \end{cases} \quad (2.27)$$

In an example of the path loss calculation using the Okumura–Hata model the calculations are done with two parameter sets and compared to the free space loss (see the graphs in Figure 2.11). The parameter sets are presented in Table 2.6. The distance between the BTS and MS is from 1 to 20 km in these calculations, which is also the validity range for the Okumura–Hata model. The calculation shown below is for the parameter set 2 when the distance from the BTS is 2 km:

$$\begin{aligned} L &= A + B \log_{10} f - 13.82 \log_{10}(H_b) - a(H_m) + \left[44.9 - 6.55 \log_{10}(H_b)\right] \log_{10}(d) \\ &= 46.30 + 33.90 \log_{10}(1800) - 13.82 \log_{10}(30) - \left\{3.2\left[\log_{10}(11.75 \times 1.5)\right]^2\right\} \\ &\quad + \left[44.9 - 6.55 \log_{10}(30)\right] \log_{10}(2) \\ &= 146.8246 \end{aligned} \quad (2.28)$$

The Okumura–Hata model is valid for the frequency ranges 150–1500 MHz and 1500–2000 MHz. The range for the base station antenna height is from 30 to 200 metres, the mobile antenna height from 1 to 10 meters and the cell range, i.e. the distance between the BTS and MS, from 1 to 20 km. With

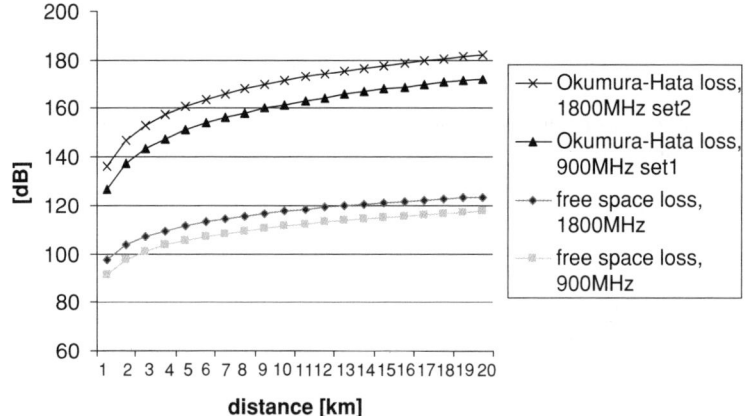

Figure 2.11 Free space loss and Okumura–Hata propagation loss comparison

Table 2.7 Propagation loss for different area types

Area type	Propagation loss [dB/decade]
Free space	20
Open area (ground reflection)	25
Suburban	35
Urban	40

the additional correction factor (L_other) the Okumura–Hata model can be applied for all terrain types, meaning different morphological areas. The correction factors for each area are received as a result of model tuning including field measurements in the particular areas. Some examples of propagation loss values can be seen in Table 2.7.

2.4.2 Walfish–Ikegami Model

The Walfish–Ikegami model is an empirical propagation model for an urban area, which is especially applicable for micro cells but can also be used for macro cells. The parameters related to the Walfish–Ikegami model are illustrated in Figure 2.12. The mean value for street widths (w) is given in metres and the road orientation angle (φ) in degrees. The mean value for building heights (h_roof) is an average over the calculation area and is given in metres. The mean value for building separation (b) is calculated from the centre of one building to the centre of another building and is also given in metres.

The Walfish–Ikegami model separates into two cases, which are the line-of-sight (LOS) and non-line-of-sight situations. The formula for path loss prediction in the LOS condition can be written as

$$L = 42.6 + 26 \log d + 20 \log f \qquad (2.29)$$

where d is distance (km) and f frequency (MHz). When no line-of-sight condition exists, the path loss formula can be written as

$$L = 32.4 + 20 \log_{10} d + 20 \log_{10} f + L_\text{rts} + L_\text{msd} \qquad (2.30)$$

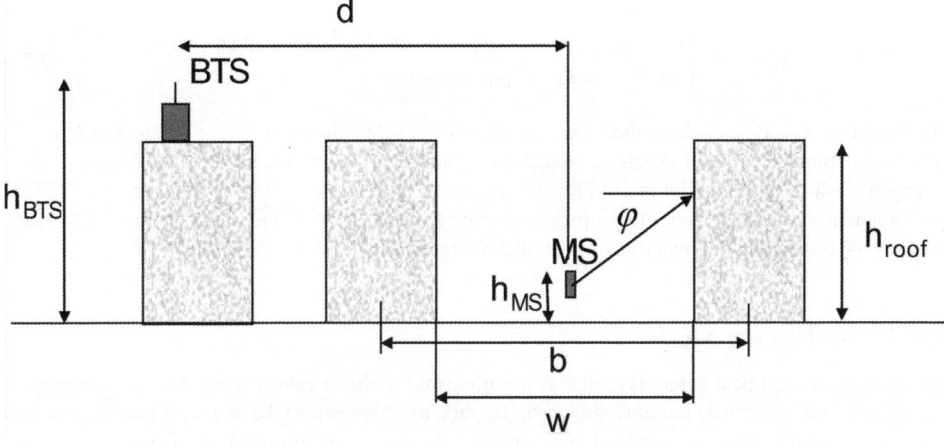

Figure 2.12 Parameters in the Walfish–Ikegami model

where L_{rts} is the rooftop–street diffraction and scatter loss while L_{msd} is the multiscreen diffraction loss. The path loss in the non-line-of-sight situation actually consists of three components: rooftop–street diffraction and scatter loss, multiscreen diffraction loss and free space loss:

$$L_0 = 32.4 + 20 \log d + 20 \log f \tag{2.31}$$

The rule follows:

$$L = \begin{cases} L_0 + L_{rts} + L_{msd} : & L_{rts} + L_{msd} > 0 \\ L_0 : & L_{rts} + L_{msd} \leq 0 \end{cases} \tag{2.32}$$

The rooftop–street diffraction and scatter loss is the loss occurring when the radio wave propagates from the closest rooftop to the receiver:

$$L_{rts} = -16.9 - 10 \log_{10} w - 10 \log_{10} f - 20 \log_{10}(h_{roof} - h_{RX}) - L_{Ori} \tag{2.33}$$

where L_{Ori} is the street orientation loss:

$$L_{Ori}(\varphi) = \begin{cases} -10 + 0.354\varphi : & \text{for } 0 \leq \varphi < 35° \\ 2.5 + 0.075(\varphi - 35) : & \text{for } 35 \leq \varphi < 55° \\ 4.0 - 0.114(\varphi - 55) : & \text{for } 55 \leq \varphi < 90° \end{cases} \tag{2.34}$$

The multiscreen diffraction loss is caused by propagation from the BTS to the rooftop, which is closest to the MS:

$$L_{msd} = L_{bsh} + k_a + k_d \log_{10} d + k_f 10 \log_{10} f - 9 \log b \tag{2.35}$$

where

$$L_{bsh} = \begin{cases} -18(1 + (h_{BTS} - h_{roof})) : & h_{BTS} > h_{roof} \\ 0 : & h_{BTS} < h_{roof} \end{cases} \tag{2.36}$$

$$k_a = \begin{cases} 54 : & h_{BTS} > h_{roof} \\ 54 - 0.8(h_{BTS} - h_{roof}) : & d \geq 0.5 \text{ km and } h_{BTS} \leq h_{roof} \\ 54 - 0.8(h_{BTS} - h_{roof}) \frac{d}{0.5} : & d < 0.5 \text{ km and } h_{BTS} \leq h_{roof} \end{cases} \tag{2.37}$$

$$k_d = \begin{cases} 8 : & h_{BTS} > h_{roof} \\ 18 - 15 \frac{h_{BTS} - h_{roof}}{h_{roof} - h_{MS}} : & h_{BTS} < h_{roof} \end{cases} \tag{2.38}$$

$$k_f = -4 \begin{cases} 0.7 \left(\frac{f}{925} - 1\right) : & \text{medium sized city and suburban centres} \\ 1.5 \left(\frac{f}{925} - 1\right) : & \text{urban centres} \end{cases} \tag{2.39}$$

The parameter k_a increases the path loss in case the BTS is below the rooftop. The parameters k_d and k_f are for adjusting the correlation between the distance and frequency with multiscreen diffraction.

The Walfish–Ikegami model is valid for the frequency range 800–200 MHz. The range for the base station antenna height is from 4 to 50 metres, mobile antenna height from 1 to 3 metres and distance between the transmitter and receiver from 20 to 5000 metres.

2.4.3 Ray Tracing Model

The accuracy of empirical modes is limited in the urban microcellular environment, because of multipath propagation. The previously introduced Walfish–Ikegami model does not take into account reflections and diffractions and therefore the propagation estimation using this model is complicated in an environment where this is typical. Ray tracing meets this claim.

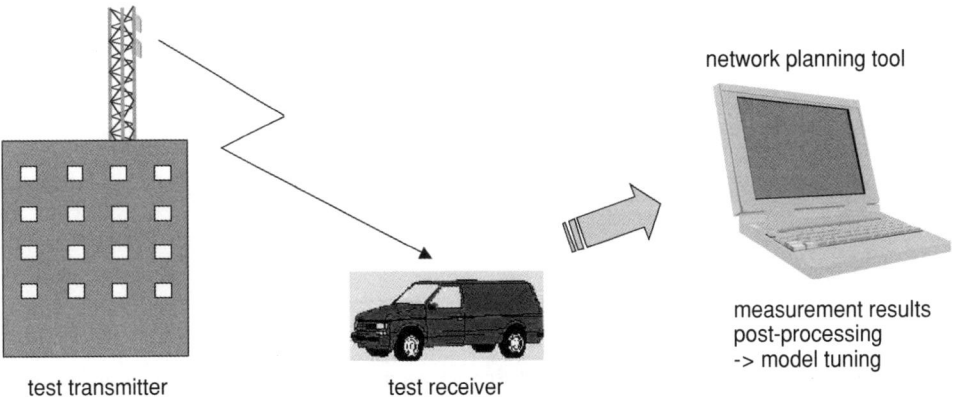

Figure 2.13 Model tuning process

The principle of ray tracing is that a selected number of rays is used to make the propagation estimation. How many reflections and diffractions are taken into account is dependent on the network planning tool's algorithm. When talking about a complicated algorithm it is clear that the more accurate the propagation estimation that is required the more capabilities from the planning tool and especially time are needed.

2.4.4 Model Tuning

The aim of model tuning is to investigate the propagation conditions in the planning area, in a way to customise the used propagation model for the area (Figure 2.13). A carefully tuned model is especially important in the coverage planning phase, because the coverage estimation is based on the propagation.

Model tuning basically consists of three phases: preparations, model tuning field measurements and measurement analysis with a planning tool. Before the field measurements can be started some preparations are needed. Firstly, a proper and as accurate a digital map as possible is required. The model tuning is based on the map information and therefore the validity and resolution are critical; the model cannot be more accurate than the map. One part of the preparations, obviously when talking about field measurements, is to collect and test the measurement equipment. A test transmitter is needed as well as a receiver for the measurement car. The measurement locations for the test transmitter have to be considered carefully and therefore it is suggested that a visit is made to the candidate locations. As they are related to the test transmitter locations the test measurement routes also have to be well planned.

The following are instructions for selecting the test transmitter locations and planning the measurement routes for model tuning:

- The locations and routes need to represent the area well, with typical places inside the area.
- All the clutter types are covered by the measurement locations.
- Several locations are required in the same area (e.g. one city) in order to get a good general view.
- Extensive measurement routes are needed to get statistically valid data of all the clutter types, so that all types are covered equally.
- Restrictions of the tuned propagation model have to be considered when planning the measurements.

The model tuning measurements start by setting up the test transmitter in the first measurement location. The drive test is run according to the preplanned measurement route. The tests are continued until enough data are collected for the planned analysis. According to good measurement practices, the

Table 2.8 Correction factors

Clutter type	Correction factor [dB]
Water	−20
Rural	−16
Forest	−10
Suburban	−7
Industrial	−5
Urban	0
Dense urban	+2

field measurements need to be well documented. Special parameters related to model tuning are required in the analysis phase:

- coordinates for the test transmitter location;
- test transmitter antenna type, height and direction;
- power of the test transmitter;
- cable types and length, or if possible measured cable losses.

After the field measurements are finished it is time to analyse the data, i.e. tune the model. Different planning tools provide help for model tuning. The planning tool requires the measurement data, which is exported from the receiver and model tuning related parameters documented during the tests. The location of the test transmitter is first specified on the planning tool's digital map and then all the other data related to it is added. The data need to be verified carefully and in the case of measurement errors the incorrect parts need to be filtered. As an output the tool gives correction factors for each measured clutter type. Example correction factors for some clutter types using the Okumura–Hata model are presented in Table 2.8. A small value in the correction factor means easy propagation conditions.

2.5 Coverage Planning

2.5.1 Coverage Planning in GSM Networks

The target for coverage planning is to find optimal locations for base stations to build continuous coverage according to the planning requirements. Especially in the case of a coverage limited network the BTS location is critical. With a capacity limited network the capacity requirements also need to be considered.

Coverage planning is performed with a planning tool including a digital map with topography and morthography information. The propagation model is selected and customised with model tuning measurements before the coverage planning phase has been started. The model selection is done according to the planning parameters, e.g. frequency, macro/micro cell environment, BTS antenna height. The coverage prediction is based on the map and the model and therefore the accuracy is dependent on those as well.

As the link budget calculations are created for all the network configurations, which means different combinations of the planning parameters, the cell size determination can be started. The number of different combinations is targeted to be kept as small as possible, but some different BTS profiles are normally needed. Sometimes there can be restrictions in the antenna usage, e.g. only small ones can be used in the city area, and this creates different link budgets for urban and rural areas. The link budget defines the maximum allowed path loss with certain configurations. Firstly, the theoretical maximum for the cell size is calculated using the selected basic propagation model.

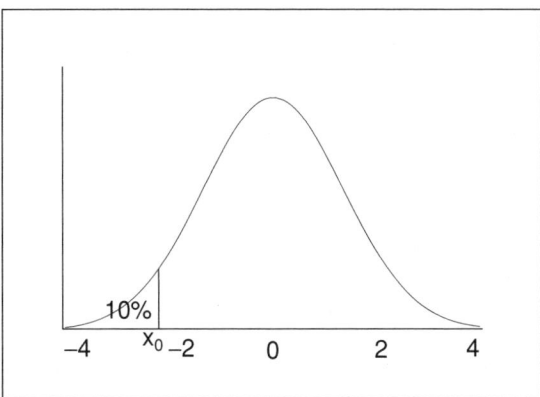

Figure 2.14 Lower tail of normal distribution for the location probability calculation

A theoretical maximum for the cell size is impossible to achieve in practice. Slow fading causes variations of the received signal level, due to obstacles in the signal propagation path. Therefore the term location probability is introduced, which describes the probability of the receiver being able to capture the signal; i.e. the signal level is higher than the receiver sensitivity. In reality there can never be 100 % location probability because it is impractical with only a reasonable amount of resources.

To determine the location probability a distribution for the received signal has to be defined. The slow fading variations in the average received signal level are normally distributed, which is presented in Figure 2.14. The distribution function for slow fading is

$$p(\bar{r}) = \frac{1}{\sqrt{2\pi\sigma^2}} e^{-(\bar{r}-\bar{r}_m)^2/(2\sigma)^2} \qquad (2.40)$$

where $\bar{r}$ is the random variable and $\bar{r}_m$ the mean value for it, and σ is the standard deviation, which is measured in dB. The standard deviation σ depends on the area type and is normally 5–10 dB. The value generally rises in dense areas. The slow fading is described by the normal random variable $\bar{r}$.

The location probability can be expressed by an equation, which is upper tail probability of Equation (2.40). The probability p_{x_0} gives the location probability at a certain point when the random variable $\bar{r}$ exceeds some threshold x_0:

$$p_{x_0} = \int_{x_0}^{\infty} \frac{1}{\sqrt{2\pi\sigma^2}} e^{-(\bar{r}-\bar{r}_m)^2/(2\sigma^2)} d\bar{r} = \frac{1}{2}\left[1 - \operatorname{erf}\left(\frac{x_0 - \bar{r}_m}{\sigma\sqrt{2}}\right)\right] \qquad (2.41)$$

The location probability can be expressed as well as the lower tail in Equation (2.41), and therefore the probability can be calculated below a certian margin. A practical example of the lower tail probability would be to calculate certain location probabilities at the cell edge; e.g. a 70 % location probability at the cell edge can be calculated by finding the x_0 value where below it the signal can be received with a 30 % probability. The planning target for the location probability is normally 90–95 % over the whole cell area.

The location probability, slow fading margin $(x_0 - \bar{r}_m)$, maximum path loss and cell range are all connected. If the location probability is, for example, 80 % on the cell edge with a certain slow fading margin, to get a higher 95 % location probability the slow fading margin has to be increased, which also has an effect on the maximum allowed path loss. The cell range is dependent on the maximum allowed path loss and therefore improvement in the location probability causes a decrease in the cell range.

The cell range leads to calculation of the coverage threshold, which is the minimum allowed downlink signal strength at the cell edge with a certain location probability. For the coverage threshold calculations are needed for the MS isotropic power, propagation model with calculation parameters, standard deviation, area type correction factor and building penetration loss. Using the standard deviation and location probability the value of the slow fading margin is first calculated. Area type correction factors come from the propagation model tuning measurements. Building penetration loss is needed in the case where indoor coverage thresholds are calculated. ETSI recommendation 03.30 suggests values for the average building penetration loss (BPL); in urban areas it is approximately 15 dB for 1800 MHz and 18 dB for 900 MHz and in rural areas around 10 dB, due to the smaller size of buildings. For an indoor coverage threshold calculation the deviation of building penetration loss indoor is also needed, which is used as the standard deviation indoor. The approximate value for the BPL deviation is 10 dB.

Table 2.9 shows an example of cell range and coverage threshold calculations for three different area types: urban, suburban and rural. The network basic parameters are frequency GSM 900 MHz and location probability 90 % for indoor and 95 % for outdoor. The link budget, which is used as the basis for these cell size and coverage threshold calculations, gives -100 dBm for the isotropic power. The MS RX sensitivity is -102 dBm (for class 4 mobile) and 2 dB is used for the interference degradation margin. The area type correction factors are the same as those given in the correction factor table in the network dimensioning section: 0 dB for urban, -7 dB for suburban and -16 dB for rural. The ETSI recommended values for GSM 900 are used for the building penetration loss, which are 18 dB for urban+suburban and 10 dB for rural. The standard deviation is 7 dB and the BPL deviation is 10 dB.

The outdoor coverage thresholds with 95 % location probability are the same in urban, suburban and rural areas: -95.5 dBm. The indoor coverage thresholds with 90 % location probability in urban, suburban and rural areas are accordingly -74.2 dBm, -74.2 dBm and 82.2 dBm, the maximum cell ranges for outdoor coverage in urban, suburban and rural are 5 km, 7.89 km and 14.22 km and for indoor coverage are 1.24 km, 1.96 km and 5.96 km respectively.

As shown in the equations above, the point location probability is normally calculated for the cell edge. From this an equation can be derived for the area location probability, which can be used for calculation of the cell coverage area probability. Parameter F_u defines the part of the useful service area when R is the radius for the whole are a with at least a certain threshold x_0. Parameter $\bar{r}$ is again the received signal and p_{x_0} the probability that the received signal exceeds the threshold x_0 inside area dA. The equation for the useful service area is

$$F_u = \frac{1}{\pi R^2} \int p_{x_0} \, dA \qquad (2.42)$$

The mean value of the received signal strength r can be expressed as

$$r = P_0 - 10\gamma \log \frac{d}{R} \qquad (2.43)$$

for d in connection with R, as shown in Figure 2.15. The average received carrier-to-interference ratio (CIR) is

$$\bar{r}_m = \alpha - 10\gamma \log \frac{d}{R} \qquad (2.44)$$

where α represents the average received CIR at the distance R.

The equation for the cell area location probability is

$$F_u = \frac{1}{2}\left[1 + \operatorname{erf}(a) + e^{(2ab+1)/b^2}\left(1 - \operatorname{erf}\frac{ab+1}{b}\right)\right] \qquad (2.45)$$

Table 2.9 Cell size and coverage threshold calculations

RADIO LINK POWER BUDGET GENERAL INFORMATION			MS CLASS:	4
Frequency [MHz]: 900			System:	GSM

Receiving end	Unit	BS	MS	
RX RF input sensitivity (TU50, R)	dBm	−106.00	−102.00	A
(Additional) fast fading margin	dB	3.00	2.00	B
Cable loss + connector	dB	2.00	0.00	C
RX antenna gain	dBi	18.00	0.00	D
Diversity gain	dB	3.50	0.00	E
Isotropic power	dBm	−122.50	−100.00	$F = A + B + C - D - E$
Field strength	dB μV/m	13.78	36.28	$G = F + Z^*$

Transmitting end		MS	BS	
TX RF output peak power	W	2.00	28.25	
Mean power over RF cycle	dBm	33.01	44.51	K
Isolator + combiner + filter	dB	0.00	4.00	L
RF peak power, combiner output	dBm	33.01	40.51	$M = K - L$
Cable loss + connector	dB	0.00	3.00	N
TX antenna gain	dBi	0.00	18.00	O
Peak EIRP	W	2.00	355.66	
(EIRP = ERP + 2 dB)	dBm	33.01	55.51	$P = M - N + O$
Isotropic path loss	dB	155.51	155.51	$Q = P - F$

*$Z = 77.2 + 20 \log(\text{freqeuency [MHz]})$

Common information	Urban	Suburban	Rural
MS antenna height [m]	1.5	1.5	1.5
BS antenna height [m]	30.0	30.0	30.0
Standard deviation [dB]	7.0	7.0	7.0
BPL average [dB]	18.0	18.0	10.0
Standard deviation indoors [dB]	10.0	10.0	10.0
Okumura–Hata [OH]			
Area type correction [db]	0.0	−7.0	−16.0
Walfish–Ikegami [WI]			
Roads width [m]	30.0	30.0	30.0
Road orientation angle [degrees]	90.0	90.0	90.0
Building separation [m]	40.0	40.0	40.0
Building average height [m]	30.0	30.0	30.0
Indoor coverage			
Propagation model	OH	OH	OH
Slow fading margin [dB]	7.8	7.8	7.8
SFM + BPL [dB]	25.8	25.8	17.8
Coverage threshold [dB μV/m]	62.1	62.1	54.1
Coverage threshold [dBm]	−74.2	−74.2	−82.2
Location probability over cell area [%]	90.0	90.0	90.0
Cell range [km]	1.24	1.96	5.96
Outdoor coverage			
Propagation model	OH	OH	OH
Slow fading margin [dB]	4.5	4.5	4.5
Coverage threshold [dB μV/m]	40.8	40.8	40.8
Coverage threshold [dBm]	−95.5	−95.5	−95.5
Location probability over cell area [%]	95.0	95.0	95.0
Cell range (km)	5.00	7.89	14.22

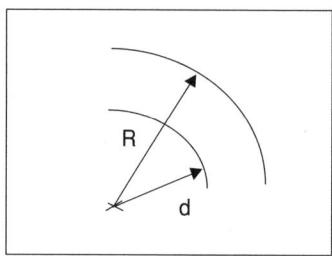

Figure 2.15 Coverage at the cell edge

where

$$a = \frac{x_0 - \alpha}{\sigma\sqrt{2}} \quad \text{and} \quad b = \frac{10\gamma \log(e)}{\sigma\sqrt{2}} \tag{2.46}$$

For the normal case of urban propagation with a standard deviation of 7 dB and a distance exponential of 3.5, a 90 % area coverage corresponds to about a 75 % location probability at the cell edge.

The actual network coverage planning is done with a network planning tool using a digital map and a propagation model verified using model tuning measurements. Some of the vendors have their own coverage planning tools, but planning tools are also provided by some specialised tools vendors. One such tool, the Nokia NetAct Planner, is used extensively in this book. The coverage plot shown in Figure 2.16 has been done using a NetAct Planner.

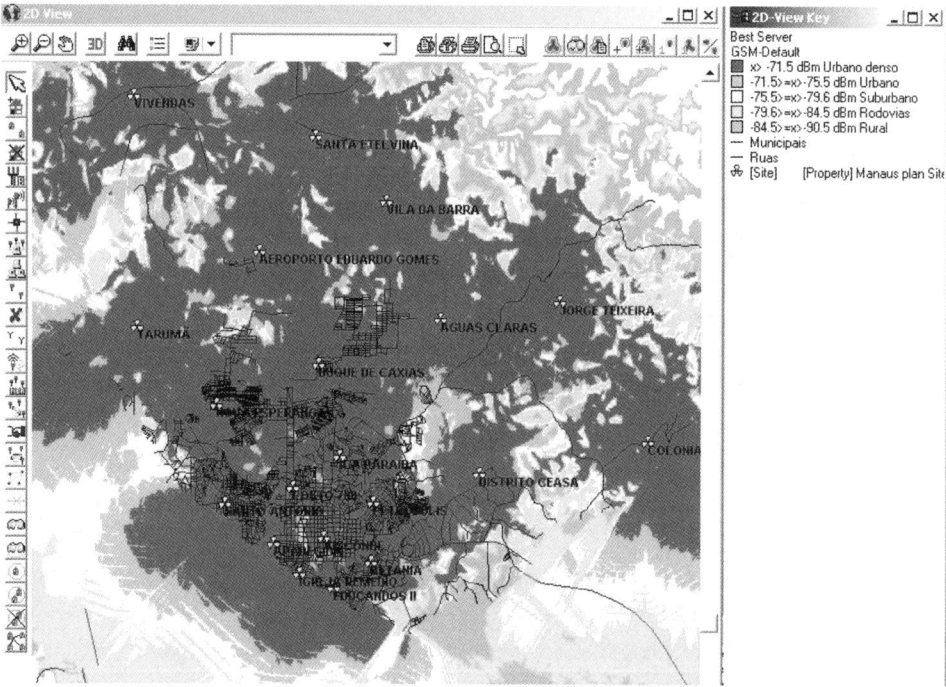

Figure 2.16 Example of the coverage plot using planning tools

Coverage Planning

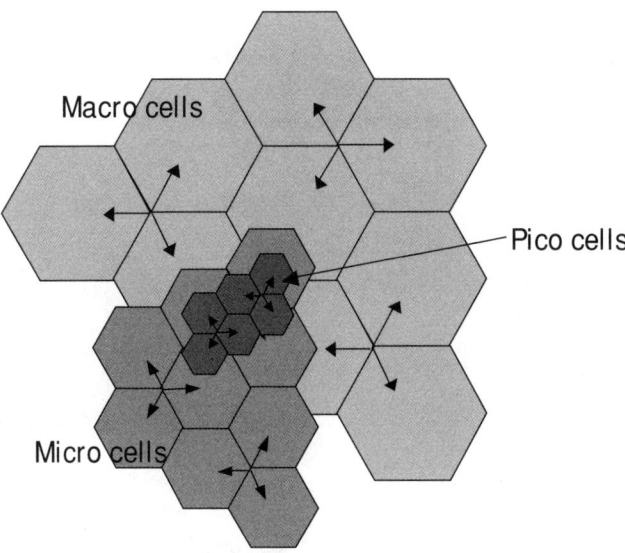

Figure 2.17 Macro, micro and pico cells drawn as hexagons

The coverage estimates given in the dimensioning phase produced only very rough figures, because they were based on the hexagonal model. In practice the cells of a GSM network are completely different from theoretical hexagons, the shape and size being dependent on the surrounding area and also the BTS parameters. Example hexagons are shown in Figure 2.17 and the terms macro, micro and pico cells are explained. In more dense areas smaller cells are used, because the capacity need limits the cells sizes. In this example all the cells are triple sectorised.

As a practical example in Figure 2.18 is shows the coverage area of a real GSM cell. The coverage is for a single cell with a sectorised antenna.

The first step in coverage planning is to create a preliminary plan based on the calculated number of base stations from the dimensioning phase, which is agreed with the operator. The BTS locations are theoretical in this phase, because they have not been verified in the field. The usage of omni and/or sectorised antennas is part of the planning strategy. Omni cells can be used in rural or other sparsely inhabited areas, where there are not high capacity requirements. As stated before for base stations having sectorised antennas, it is easier to give coverage to precise locations.

The next step is to start to find actual base station locations, which is a task of the site acquisition team. To find an optimal BTS location is an essential but complicated task. What makes it complicated are the many practical requirements and especially the transmission that need to be arranged. When the actual BTS location has been found the preliminary location changes and the plan is updated and the cell coverage areas are calculated again using the new parameters. All the preliminary BTS locations are gone through in the order the actual BTS locations are found and the coverage areas are always recalculated.

Normally the plan is divided into smaller segments, each consisting of some base stations located close by. The segment can be drawn on the map as a geographically logical area, with the particular base stations inside the area. The aim is to find actual BTS locations segment by segment and to finalise the preliminary coverage plan one area at a time. When finding the actual base station locations it is important to reach the planning requirements. The coverage and capacity requirement needs to be reached with the actual BTS locations.

When calculating the cell coverage with a planning tool the calculation range has to be wide, especially when the later interference estimation needs to be accurate. The common way to show the calculated

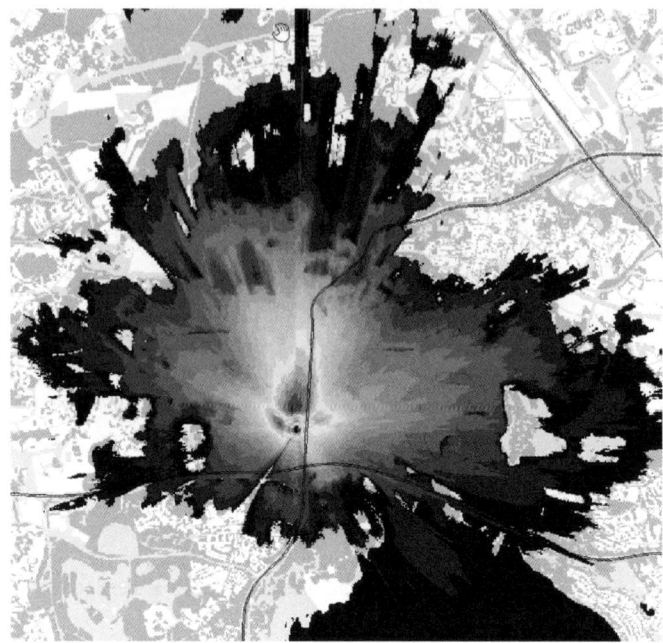

Figure 2.18 Coverage area of a real cell

coverage for a certain area is to view a composite plot with specified thresholds. In this type of presentation it is easy to see possible gaps in the coverage area. The contrary situation, when the overlapping areas between the cells are too large, can also be investigated. By fine tuning the planning parameters inside the given ranges can be made to fill the coverage gaps and reduce the unneeded overlapping; e.g. the BTS output power or antenna height, direction or down tilt can be changed.

Another useful way to view the coverage is a dominance map. In this presentation each cell has a particular colour, which is used to show where this cell gives the dominant coverage. The coverage thresholds are therefore not used in this presentation. In practice the coverage borders are not this strict, because of the handover margin. When leaving the dominance area of one cell, another cell starts be to dominant – the handover is done only when the second cell's coverage is better than the handover margin. Therefore the mobile is still for a short time connected to the first cell while the dominant cell is already the second one.

The theoretical calculation of coverage thresholds was explained earlier. The coverage thresholds are a way to analyse the coverage area in the planning phase and compare it to the planning requirements. Therefore it is useful to set the coverage thresholds used in the planning tool in line with the planning targets. Sometimes thresholds used in the printouts are separately agreed with the operator in order to make it easier to compare the different versions. There can also be several threshold sets, with the first one giving the thresholds for all the morphological types in the area. Separately there can be a threshold for different types of receivers, according to defined MS classes. One threshold is set to illustrate indoor coverage, so that building and car, etc., penetration losses are taken into account.

Careful coverage planning is important to encourage carefulness throughout the network planning process. The accuracy of the coverage plan is a sum of several factors. Map resolution and accuracy affect the propagation model as well as the accuracy of the model tuning measurements. Important factors during the coverage planning phase are accuracy of the link budget parameters, BTS coordinates and other coverage planning parameters. The measurable link budget parameters, cable loss and BTS

Table 2.10 Measured signal level mapping to RXLEV

	Signal level minimum	Signal level maximum
RXLEV 0		-110
RXLEV 1	-110	-109
RXLEV 2	-109	-108
RXLEV 3	-108	-107
...	...	...
RXLEV 61	-50	-49
RXLEV 62	-49	-48

power need to be verified with measurements when possible. The parameters need to be verified so that the same numbers are used from the planning through to the actual BTS. The coordinates measured for the actual BTS during site hunting are essential because they are used for coverage planning in the planning tool. Construction parameters related to the antenna are remarkable when talking about accuracy issues, actual antenna height and direction.

In complicated planning situations there are some ways to enhance the coverage. These tricks are best to use during the planning phase, because it is more complicated and costly to do it later. These tricks do not reduce the importance of careful planning, as there is no way to find a better location of incorrectly located base stations except to change it to a better position. One method to enhance the coverage is to optimise the link budget parameters within the given range, using as high BTS powers as possible and having high gain antennas. Another method is to use supplementary hardware for coverage enhancement. A booster amplifies the BTS transmission and in this way strengthens downlink. A mast head amplifier strengthens uplink by giving more sensitivity to BTS and enables the reception of weaker signals. The coverage and capacity requirements are important to meet but at the same time planning has to target clear cell dominance areas. BTS powers and antenna heights need to be considered and downtilts (both electrical and mechanical) used when needed. As already mentioned, in the planning phase the coverage calculation range has to be wide in order to detect interfering reflections.

It is specified that the BTS and MS must measure the received signal level within the range of -110 dBm to -48 dBm. The measured signal levels are averaged while mapped to RXLEV (received level) values, which are between 0 and 63. The signal levels are reported as RXLEV values. The mapping is presented in Table 2.10.

The radio link measurements also include received signal quality measurements. In a similar way to signal level measurements the measured signal quality is averaged and mapped to the RXQUAL (received quality) value. There are eight RXQUAL categories and the mapping table is gevin in Table 2.11.

Table 2.11 BER mapping to RXQUAL

RXQUAL	BER range
RXQUAL 0	$BER < 0.2\%$
RXQUAL 1	$0.2\% < BER < 0.4\%$
RXQUAL 2	$0.4\% < BER < 0.8\%$
RXQUAL 3	$0.8\% < BER < 1.6\%$
RXQUAL 4	$1.6\% < BER < 3.2\%$
RXQUAL 5	$3.2\% < BER < 6.4\%$
RXQUAL 6	$6.4\% < BER < 12.8\%$
RXQUAL 7	$12.8\% < BER$

Table 2.12 C/N versus coding scheme

Coding scheme	QoS	C/N Nonhopping [dB]	C/N frequency hopping [dB]
CS-1	BLER < 10%	9.0	6.2
CS-2	BLER < 10%	11.3	9.8
CS-3	BLER < 10%	12.7	12
CS-4	BLER < 10%	17	19.3

The radio link measurements, i.e. the reported RXLEV and RXQUAL values, are needed when performing handovers and power control. The radio link measurements are performed over a slow associated control channel (SACCH) multiframe, which consists of 104 TDMA frames in the case of full rate traffic channel (TCH). The 104 TDMA frames last for 480 ms.

2.5.2 Coverage Planning in EGPRS

The coverage planning aspects of EGPRS implementation is centered around ensuring sufficient carrier-to-noise (C/N) ratios on both uplink and downlink to allow for successful data transmission across the coverage area. Each coding scheme defined for GPRS (and MCS for EGPRS) is suited to a particular range of C/N for a given block error rate (BLER). Generally, it is found that the higher the level of error protection, the lower the required C/N. Table 2.12 shows some sample simulation results for GPRS coding schemes.

The objective of coverage planning is to provide a coverage area for both the uplink and downlink in a balanced way. This is achieved by calculating the link budget. The link budget calculations allow a comparison to be made between achievable cell ranges of EGPRS using different coding schemes and those on current GSM networks.

As an input to the link budget, a number of key elements need to be established, some of which are:

- Transmitter output power (MS/BS);
- Receiver performance for different coding schemes (MS/BS);
- Antenna configurations (diversity).

Relative Coverage Areas of GPRS Coding Schemes

Due to the differing C/N requirements of the modulation coding schemes, the relative coverage area of each will obviously be different. It is useful to compare the relative predicted coverage areas of the coding scheme, in addition to the existing GSM voice service.

Figure 2.19 illustrates coverage over a nominal cell pattern (three sectors, each of a hexagonal shape). In this GPRS example, the difference in coverage of the four coding schemes can clearly be seen, with the covered area decreasing progressively from CS-1 to CS-4. Figure 2.19 assumes the use of frequency hopping. Figure 2.20 illustrates the equivalent coverage areas for the nonhopping case. With the exception of CS-4, all coverage areas are reduced compared to the hopping case. The difference with CS-4 is the absence of error protection so this coding scheme has increased vulnerability to fading (and hence increased block errors) when hopping. EGPRS and GPRS share the same concepts for relative coverage areas where EGPRS varies over the full range of modulation coding schemes 1 to 9.

These results refer to a convenient cell size of 4000 m. Continuing with the GPRS example, coding scheme CS-1 does not quite cover the whole cell at this distance (with frequency hopping), and hence the overall coverage for CS-1 is a little less that 100 %. This fact explains the values presented in the next figure. Figure 2.21 illustrates the perceived cell radius of the different coding schemes, as a percentage of the 4000 m cell radius, both with and without frequency hopping.

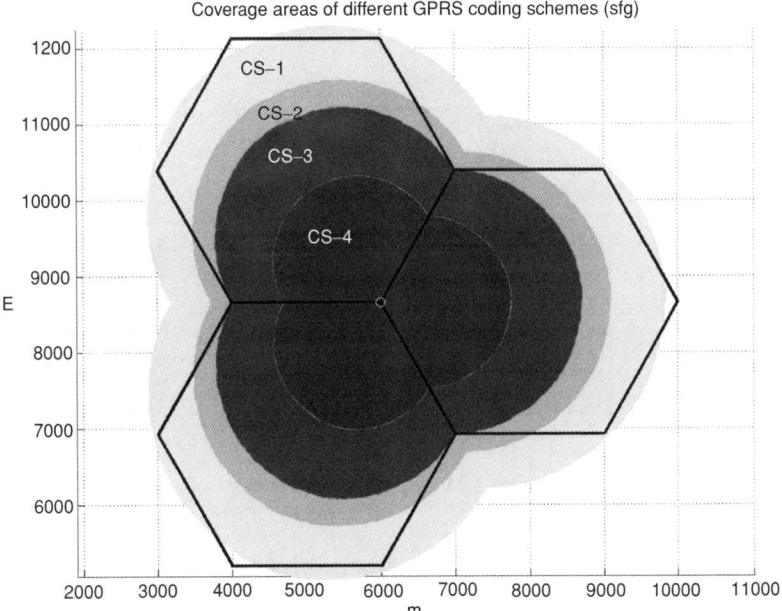

Figure 2.19 Coverage with frequency hopping

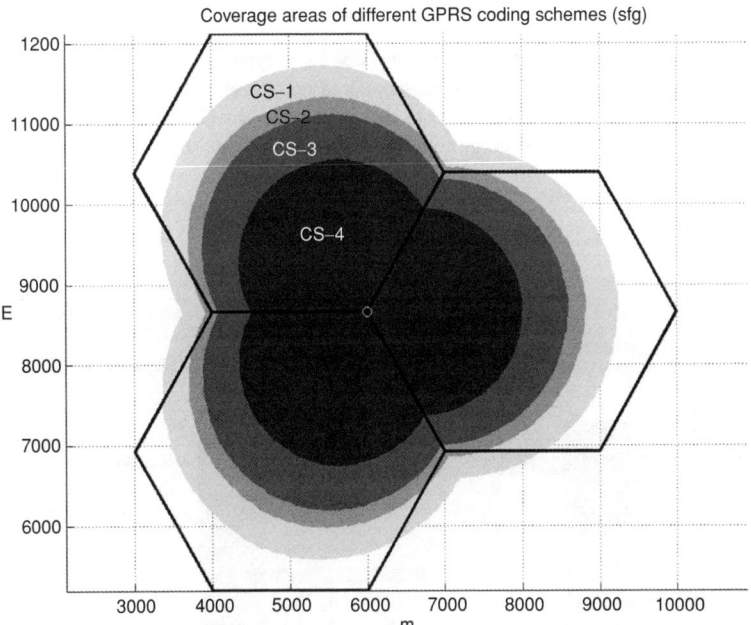

Figure 2.20 Coverage without frequency hopping

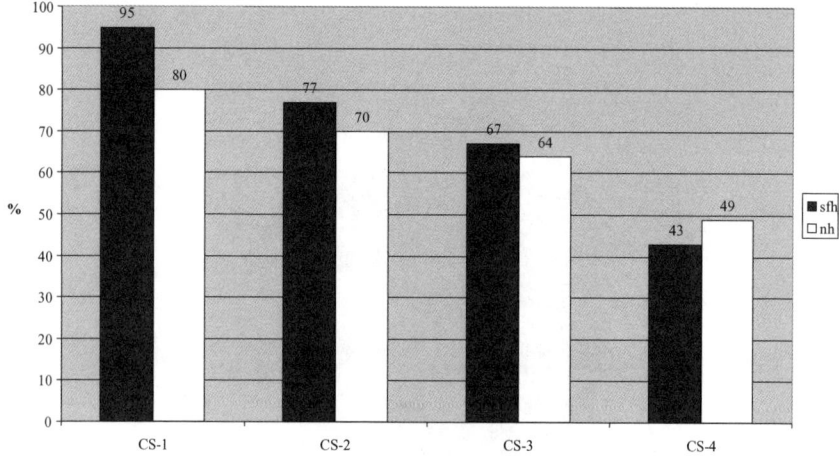

Figure 2.21 Perceived cell radius

Figure 2.22 shows the relative coverage areas (within a cell) of the four schemes (relative to CS-1) in the two hopping scenarios.

Both figures are based on simulations employing the path loss model used in UMTS 30.03 v3.2.0. Assuming 900 MHz with a base station antenna height of 15m above average rooftop level, the path loss L is given by the following formula:

$$L = 120.9 + 37.6 \log 10(R)\,\text{dB} \tag{2.47}$$

It has been shown that the service area for CS-1 and CS-2 will typically cover the same area covered by a GSM voice network. For CS-2, this is due to the lack of body loss in the data environment. For CS-1, this is due to the coding scheme being similar to that already employed in GSM (stand-alone dedicated control channel, or SDCCH). The coverage areas for CS-3 and CS-4 are somewhat reduced, however, due

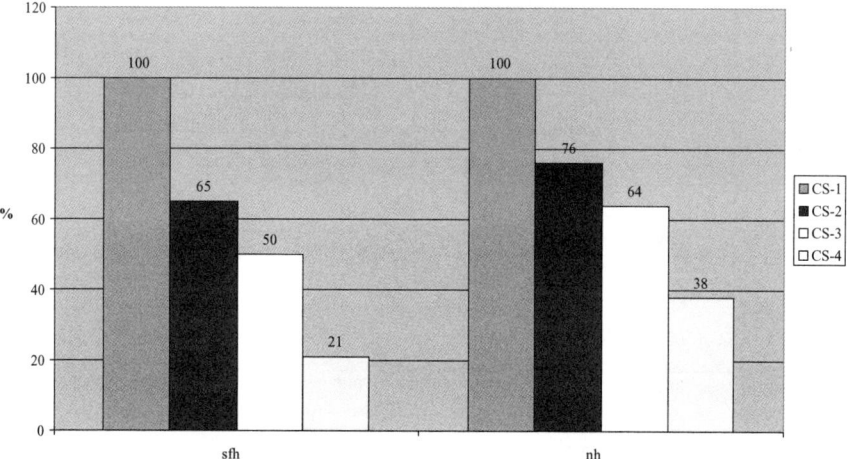

Figure 2.22 Coverage areas of coding schemes relative to CS-1

to their reduced error protection and hence higher required C/I. These services will therefore typically be available closer to the cell centre. Similarly, EGPRS Modulation Coding Schemes 5 through 9 will be further reduced based on their C/I requirements.

The C/I distribution (associated with frequency re-use) will determine the coverage area for the different coding schemes. In addition, new sites may be required to ensure contiguous high bit rate coverage. In many cases, EGPRS performance will be dictated more by the cell interference than by the absolute sensitivity; i.e. the network coverage is more often interference-limited than noise-limited. It is possible to estimate typical C/I distributions across a cell for a given re-use pattern. In this way, it is possible to predict the user throughput available across the cell. Under normal circumstances, the higher the re-use factor, the higher the cell C/I. Therefore, a system with more spectrum will be likely to offer higher average data rates than a system with limited spectrum.

2.5.3 Coverage Planning in WCDMA Networks

The link budget calculation for the WCDMA system and coverage planning for the GSM system have already been discussed. The fundamental process for coverage planning in the WCDMA system is quite similar to that of the GSM system. However, propagation models need to be adjusted to take into consideration the WCDMA technology. The cell range R can be calculated using the Okumara–Hata or the Walfish–Ikegami models. After this, the site are a can be calculated, which is $2.6R^2$. However, in the WCDMA networks, some additional measurements and adjustments were done in the framework of European Cooperation in the Field of Scientific and Technical Research, also called COST. The validity for this extended Okumara–Hata model is

- Frequency f: 150–2000 MHz
- Distance R: 1–20 km
- UE height : 10–200 m
- MS height: 1–10 m

The correction factor c is given as

$$\text{Correction factor} = \begin{cases} 2 \log_{10}^2\left(\frac{f}{28}\right) + 5.4 & \text{for suburban areas} \\ 4.78 \log_{10}^2(f) - 18.33 \log_{10}(f) + 44.94 & \text{for rural areas} \end{cases} \quad (2.48)$$

This correction factor is added to obtain the actual losses in the WCDMA environment. Similarly for the Walfish–Ikegami model, the COST model is applied, which is based on the typical antenna placements and has the validity range

- Frequency f: 800–2000 MHz
- BS height h_{bs}: 4–50 m
- UE height h_{ms}: 1–3 m
- Distance d: 0.02–5 km

2.6 Capacity Planning

2.6.1 Capacity Planning in GSM Networks

The preliminary capacity planning has already been done in the dimensioning phase of the network planning process. In that phase only rough figures in the area type level were estimated without going into the capacities of individual base stations or cells. The planning parameters came from customer requirements and the estimated number of users. The user estimates are for every network roll-out phase. The capacity requirement per user is different depending on the user profile. The dimensioning can be simplified by having one user profile per area type; i.e. all the users inside the same area type have a similar user profile. The user profiles define the average usage as well as the busy hours. The capacity has to be planned based on the maximum simultaneous usage.

In the capacity planning phase a detailed capacity per cell level is estimated. The prior task was to select the base station locations and calculate the coverage area using actual BTS parameters. The capacity allocation is based on these coverage maps and traffic estimates, which can be a separate layer on the map of the planning tool. The coverage dominance map provides the information for the cell borders.

As mentioned, the maximum simultaneous usage is the main planning target for the network capacity. The capacity peaks are momentary and therefore define a blocking probability, which is the accepted level for unsuccessful call attempts due to lack of resources. This parameter has already been defined by the customer at the beginning of the planning process.

The amount of traffic is expressed in Erlangs, which is the magnitude of telecommunications traffic. An Erlang describes the amount of traffic in one hour. The definition for Erlang is the following:

$$\text{Erlang} = \frac{\text{(number of calls in hour) (average call length)}}{3600\,\text{s}} \tag{2.49}$$

as an example of an Erlang calculation, 25 users make a phone call for each an average of three minutes in an hour. How much traffic are the users creating in Erlangs?

$$\text{Traffic in Erlangs} = \frac{25\,(3 \times 60\,\text{s})}{3600\,\text{s}} = 1.25\,\text{Erl} \tag{2.50}$$

A quite normal dimensioning value for traffic per user is 15–30 mErl during the network busy hour, which means 54–108 s call times. This value varies a lot depending on the user, but one average value is needed in the network dimensioning phase. The value of traffic per user can be said to be linked to the culture. In detailed capacity planning different user profiles can be used to simulate different types of usages. The capacity can be calculated using Erlang formulas, where the B formula is for the case without queuing and the C formula for the case which takes queuing into account. The Erlang formulas are presented below.

The Erlang formula B calculates the call blocking probability P_B for traffic T with the amount of channels C. The traffic is given in Erlangs. The amount of traffic to be served with an accepted blocking probability level is $T' = T/(1 - B)$. According to this formula the blocked calls are terminated without queuing:

$$P_B = \frac{T^C/C!}{\sum_{i=0}^{C} T^i/i!} \tag{2.51}$$

The Erlang C formula calculates the probability for queuing, i.e. the probability that the user has to wait to be served. The calculation parameters traffic T and number of traffic channels C are the same as for the Erlang B formula:

$$P_C = \frac{T^C C/[C!(C - T)]}{\sum_{i=0}^{C-1} T^i/i! + T^C C/[C!(C - T)]} \tag{2.52}$$

Tables and automatic calculators based on Erlang formulas can be found from different sources. These tables and calculators are useful in the capacity planning work and can be used for solving any of the three parameters when the other two are known. The calculations are done using the busy hour figures, i.e. when there is the maximum amount of users. Therefore the service is more predictable and the capacity is sufficient. Table 2.13 presents the short Erlang B table showing how much traffic can be served, e.g. with 10 channels, when the blocking probability is 1, 2, 3, 5 or 10 %. For example, with a blocking probability of 2 % the 10 channels can serve 5.09 Erlangs of traffic.

Capacity Planning

Table 2.13 Erlang B for blocking probabilities of 1, 2, 3, 5 and 10 %.

Channels	1 %	2 %	3 %	5 %	10 %
1	0.01	0.02	0.03	0.05	0.11
2	0.15	0.23	0.28	0.38	0.60
3	0.46	0.60	0.71	0.90	1.27
4	0.87	1.09	1.26	1.53	2.05
5	1.36	1.66	1.87	2.22	2.88
6	1.91	2.28	2.54	2.96	3.76
7	2.50	2.93	3.25	3.74	4.67
8	3.13	3.63	3.99	4.54	5.60
9	3.78	4.34	4.75	5.37	6.55
10	4.46	5.09	5.53	6.21	7.51
15	8.11	9.01	9.65	10.63	12.48
20	12.03	13.18	14	15.25	17.61
25	16.12	17.5	18.48	19.98	22.83
30	20.34	21.93	23.06	24.80	28.11

As an example of a trunking gain calculation, the difference is to be found in trunking gains between the cases of 7 and 30 traffic channels with a 2 % blocking probability. The first, seven traffic channels, corresponds to one TRX case when one of the eight time slots is reserved for signalling. The second one corresponds to three TRX cases, when two time slots of 32 are reserved for signalling. The trunking gains can be calculated with the Erlang B formula and in this case the required traffic figures can be found from Table 2.13:

$$\text{Trunking gain with 7 channels} = 2.93 \, \text{Erl}/7 \, \text{Erl} = 0.42$$
$$\text{Trunking gain with 30 channels} = 21.93 \, \text{Erl}/32 \, \text{Erl} = 0.69$$

As another example, calculate the number of required channels when the traffic per user is 20 mErl and the number of subscribers is 500. The blocking probability is less than 2 %. Firstly, the 500 users create traffic of $500 \times 20 \, \text{mErl} = 10 \, \text{Erl}$. The Erlang B table is used to calculate the number of required channels. The column is selected based on the blocking probability. Table 2.14 is a snapshot of the Erlang B table for a 2 % blocking probability. The required number of channels is 17 to be able to serve 10 Erl of traffic. The second step would be to calculate the total number of needed TRXs. The general dimensioning rule with a noncombined SDCCH channel is: 7 timeslots, 1 TRX; 14 timeslots, 2 TRX; and 22 timeslots, 3 TRX. Thus the required number of TRXs is 3.

Channel Structure

Due to the fact that the GSM frequency band is limited, the system needs to be able to carry out intelligent frequency band division. The GSM system combines time division multiple access (TDMA) and frequency division multiple access (FDMA) for effective use of the frequency band. Frequency

Table 2.14 Erlang B table for a 2 % blocking probability

Traffic [Erl]	7.40	8.20	9.01	9.28	10.67	11.49
Channels	13	14	15	16	17	18

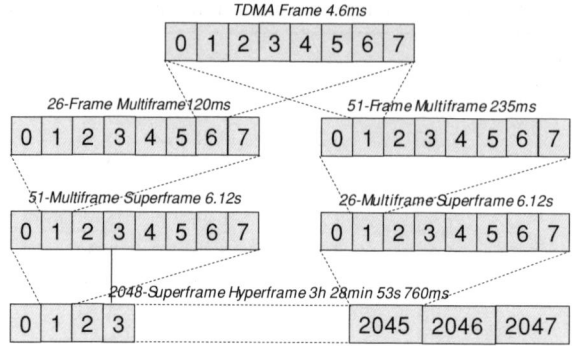

Figure 2.23 GSM channel structure

division means the division of the bandwidth to carrier frequencies, which are spaced 200 kHz apart. The frequency ranges for the GSM frequency bands are specified in ETSI specifications. The GSM 900 frequency ranges are 890–915 MHz for uplink and 935–960 MHz for downlink. The GSM 1800 frequency ranges are 1710–1785 MHz for uplink and 1805–1880 MHz for downlink. The GSM 850 frequency ranges are 824–849 MHz for uplink and 869–894 MHz for downlink. The GSM 1900 frequency ranges are 1850–1910 MHz for uplink and 1930–1990 MHz for downlink. Finally, The GSM 400 frequency ranges are 450.4–457.6 MHz for uplink and 460.4–467.6 MHz for downlink or 478.8–486 MHz for uplink and 488.8–496 MHz for downlink. For example, the GSM 900 sub-bands are divided into 124 carrier frequencies. The time division divides each carrier frequency into timeslots numbered from 0 to 7 (see Figure 2.23). The length of one timeslot is 0.577 ms.

As stated above, the GSM system divides the carrier frequencies into timeslots. The TDMA frame consists of eight timeslots and therefore the length is 4.6 ms. To identify the TDMA frames, each has a 22 bit part used for identification information. The structure of a traffic channel (TCH) is the following: one hyperframe, which is consists of 2048 superframes each having 51 multiframes and 26 TDMA frames. The structure of a control channel (CCH) is the following: 2048 superframes each consisting of 26 multiframes and 51 TDMA frames. One burst in the TDMA frame, one timeslot, is equal to a physical channel.

Two concepts are good to separate at this point, physical and logical channels. The first are the actual channels for carrying the information between the BSS and MS. The timeslots are in other words the physical channels. The logical channels structure the transferred information, data and signalling and which specific type. Traffic channels are for user data and control channels are for signalling. The user data can be either voice or data.

In the GSM system the control channels can be of three basic types: broadcast control channels, common control channels and dedicated control channels. The usage is different as common channels are used in the idle mode and dedicated channels are allocated to the mobile when it is in the active mode. As indicated by their name, the broadcast channels broadcast the information to the mobile phones, whereas the dedicated and common control channels are of a point-to-multipoint type. In the following the logical channels of the GSM system are described in more detail.

Traffic Channel (TCH)

A traffic channel can be either full, half rate or enhanced full rate. In the case of a full rate channel the bit rate of the codec is 13 kbps with a half rate of 5.6 kbps. The enhanced full rate (EFR) works at 12.2 kbps. The TCH is a dedicated channel for one user. Another speech coding type is the adaptive multirate (AMR) specified by the 3GPP (3G partnership project). The idea of the AMR is to adapt dynamically the source and channel coding according to the network conditions. The better the channel, the less channel coding is needed. Overall this improves the quality and saves capacity. AMR allows eight bit rates, which are 12.2, 10.2, 7.95, 7.40, 6.70, 5.90, 5.15 and 4.75 kbps.

Broadcast Control Channels

Broadcast Control Channel (BCCH)
The BCCH broadcasts the system information continuously in the downlink direction, e.g. frequencies of the neighbouring cells and paging group information.

Cell Broadcast Channel (CBCH)
This broadcasting contains user information with no signalling.

Frequency Correction Channel (FCCH)
This is for the MS to find the FCCH burst, which contains the frequency information.

Synchronisation Channel (SCH)
This is used for synchronisation of the MS. The broadcast contains information of the BS identity code and TDMA frame number.

Common Control Channels

Access Grant Channel (AGCH)
The AGCH channel is used to allocate the SDCCH for the MS and acknowledge the MS channel request.

Notification Channel (NCH)
The purpose of this channel is to inform the mobile about voice broadcast calls; it operates only to the downlink.

Paging Channel (PCH)
The PCH channel is used to inform the MS about terminating a call.

Random Access Channel (RACH)
The RACH channel is for MS access requests and location updates for instance; it operates to the uplink direction only.

Dedicated Control Channels

Slow Associated Control Channel (SACCH)
This is used for monitoring the link, e.g. transmitting the measurements supporting the handover (HO) decision making. The SACCH is always connected to the TCH. The SDCCH, as do all the dedicated control channels, operates in both directions (uplink and downlink) and is a point-to-point channel.

Stand-alone Dedicated Control Channel (SDCCH)
This channel is used for signalling, e.g. in the call setup phase, for assignment of the traffic channel and in location updates.

Fast Associated Control Channel (FACCH)
This channel is used in critical signalling cases; it takes the burst from the TCH to transmit signalling. The critical situation is, for example, the handover.

Capacity Enhancement Techniques
Capacity enhancement techniques can be used if the required capacity is higher than can be provided by macro cells. One example of enhancement techniques is frequency hopping. The theory will be explained in the frequency planning section (Section 2.7.3). Micro and pico cells are another example of how to offer extra capacity.

2.6.2 GPRS and EGPRS Capacity Planning

The following sections consider the dimensioning for a mixed traffic profile of GPRS / EGPRS and GSM voice. Note that GPRS and EGPRS data is able to perform in a discontinuous mode with resources being allocated as and when available (capacity on demand) while GSM voice requires a continuous connection for extended periods (tens of seconds). For the purposes of this discussion, all GPRS examples that are provided can also be directly applied to EGPRS capacity planning.

Another important concept in GPRS and EGPRS capacity planning relates to the resource assignment priorities of the different services. Due to the relative delay insensitivity of many packet based services, it would not be unreasonable for a packet data transfer to be interrupted if a channel is required for a voice call. In this way, the capacity allocated to GPRS traffic will be time variable, allowing a degree of freedom in the overall system planning. Note, however, that it is possible to dimension additional timeslots per carrier for GPRS should a minimum level of GPRS capacity be required.

Circuit Switched Traffic Dimensioning

Since it is assumed that circuit switched traffic will take priority over packet traffic, and be likely to constitute the majority of the system load, it is appropriate to consider this first.

Circuit Switched Traffic and the Erlang B Equation

Circuit switched traffic design usually involves the application of the Erlang B formula to give a figure for the required number of channels given a required traffic load and blocking probability:

$$B(N) = \frac{A^N/N!}{\sum_{i=0}^{N} A^i/i!} \quad (2.53)$$

where B is the blocking probability (%), A is the traffic load (Erl) and N is the number of channels (timeslots). Fortunately, direct evaluation of Equation (2.53) is rarely required, as tables containing the results are widely available.

As an example, the Erlang B calculation indicates that 22 traffic channels would support 14.9 Erlangs of traffic with <2 % blocking. An area requiring, at an even distribution, an overall load of around 100 Erlangs could therefore be built up from seven cells each carrying 22 traffic channels, maintaining the <2 % blocking criterion.

A design based on the above would also incorporate a design C/I threshold, thus ensuring that, across the area, the C/I is sufficient for reliable transmission. This C/I is chosen to suit the transmission format in question, e.g. speech and circuit switched data.

Available GPRS Resource within the Circuit Switched Design

A system designed for circuit switched traffic will usually allow for a basic GPRS throughput: since the system has been designed for a sufficient margin to permit a low blocking level, some of the spare instantaneous capacity can be utilised for packet data transmission. As long as the packet traffic can be temporarily interrupted to accommodate the peaks in circuit switched traffic, no degradation in the circuit switched services will result.

From the example above, a cell offering a circuit switched load of 14.9 Erlangs with 22 circuits will, on average, have 7.1 spare circuits. These could carry packet data on an on-demand basis, relinquishing the channels for circuit switched traffic when required. In this way, the blocking probability of the circuit switched facility is not degraded, even though the traffic channels are subject to a higher utilisation. However, there is a certain overhead associated with the division of the circuit switched area and the

Capacity Planning

Table 2.15 Mean number of timeslots available

Number of (TCH) carrier	GSM traffic at 1 % blocking (Erl)	GSM traffic at 2 % blocking (Erl)	Mean free TCH for GPRS (1 % blocking)	Mean free TCH for GPRS (2 % blocking)
1 (7)	0.5	2.9	5.5	3.1
2 (14)	7.4	8.2	5.1	4.3
3 (22)	13.7	14.9	6.8	5.6
4 (30)	20.3	21.9	7.2	5.6
5 (38)	27.3	29.2	7.7	5.8
6 (46)	34.3	36.5	8.7	6.5
7 (54)	41.5	43.9	9.5	7.1
8 (62)	48.7	51.5	9.3	6.5

GPRS area. This is described in the following sections, and results in a reduction in the number of channels available to GPRS, in this case from 7.1 channels to 5.6 channels.

Table 2.15 shows the mean number of timeslots (TCHs) available for GPRS for different numbers of carriers per cell and for circuit switched blocking probabilities of 1 and 2 %. It takes account of the channel overhead (see Table 2.16 and the associated description later). It would be generally expected that the number of timeslots available to the GPRS would increase as the number of carriers increases, but there are several examples in Table 2.15 where this does not occur. This is a result of the channel overhead taking on increasing values with increasing numbers of carriers, in a step like manner.

Theoretically, 5.6 timeslots are, on average, available for the GPRS in the previous example. It is possible to use all of these and to load each with a number of users (a maximum of 9 in the downlink direction and 7 in the uplink). However, under such circumstances the data rate offered to each user will be low in comparison with that theoretically offered by the coding scheme, and the overall end-to-end message delay will tend to be high where user message sizes are large. Since the distribution of required data transmissions is not expected to be steady (it may be likely to be Poisson distributed, for example) some overhead should be allowed in system resources to accommodate peaks in the load, in order to preserve a satisfactory user data rate.

Increasing the GPRS Capacity

As shown, a typical network will, at most times, have some limited available capacity for GPRS traffic. To increase the available capacity, further resources will be required in the form of traffic channels (and, hence, carriers). Adding an extra carrier to a cell will create a further seven or eight channels, depending on whether the first timeslot is defined as a control channel or a traffic channel. In all of the cases shown in Table 2.15, the addition of seven or eight TCHs adds significant capacity to the GPRS; in the case of three carriers and 2 % circuit switched blocking, the typical resource available for the GPRS is effectively more than doubled by the addition of a single carrier (for the same circuit switched traffic level).

Circuit Switched and GPRS Load Division – The Territory Method *

The primary technique for dividing resources between circuit and packet traffic is known as the Territory Method. Here, the timeslots within a cell are dynamically divided into GSM and GPRS territories. This means that a certain number of consecutive traffic timeslots are reserved for circuit switched GSM calls, with the remainder being available for GPRS traffic. A dynamic variation of the territory boundary (and

---

* Note that this type of method may be vendor dependent.

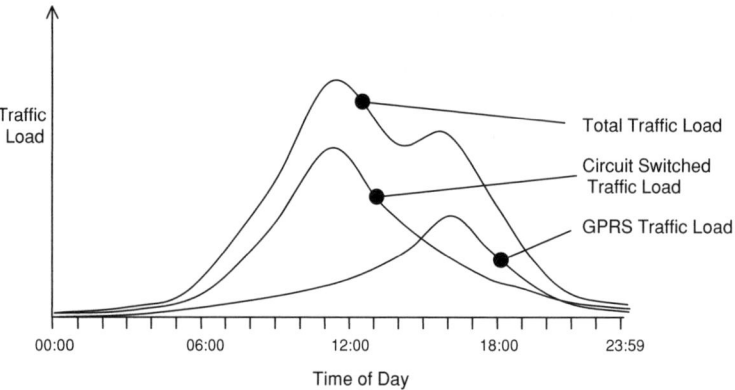

Figure 2.24 Example traffic profiles

hence the number of timeslots in each territory) permits the system to adapt to different load levels and traffic proportions, thus offering optimised performance under a variety of load conditions.

Traffic Profile

An optimal assignment of the various territory parameters depends on the availability of realistic system traffic profiles. Example traffic profiles for a possible system are shown in Figure 2.24. In this case, distinct peaks in the packet traffic and circuit traffic loads are readily observed, suggesting that long-term territory boundary changes would be likely. These changes would be in addition to short-term changes taking account of fluctuations in the load profile (Figure 2.24).

The main parameters of concern in the overall network dimensioning are:

- peak circuit switched traffic load (Erlangs);
- peak packet switched load (kbps);
- circuit switched traffic load during the overall traffic peak hour (Erlangs);
- packet switched load during the overall traffic peak hour (kbps).

The first two parameters give an indication of the limiting optimal resource allocations at the respective peak times (or gross-scale territory boundaries). The latter two will aid the overall dimensioning under peak load conditions, indicating overall carrier requirements. The fact that the respective peaks in traffic load do not necessarily coincide is an important one from the overall dimensioning point of view.

Dedicated GPRS Territory

It is possible to dedicate traffic channels to the GPRS service if a minimum level of service is required when circuit switched traffic reaches peak levels. If this is done on a system that has been dimensioned for circuit switched traffic at 2 % blocking, then it is clear that the resulting blocking level will increase due to the reduction in the number of channels. Figure 2.25 shows the effect of dedicating GPRS timeslots on circuit switched blocking performance. As would be expected, the increase in blocking is more pronounced where there are few carriers per cell. The graph does show, however, that even dedicating just one or two timeslots to the GPRS in the case where there are relatively high numbers of carriers per cell can significantly affect the call blocking (e.g. 5 carriers, 38 TCHs: 2 % blocking increases to 3.5 % when 2 TCHs are dedicated to the GPRS).

There is, therefore, a compromise between providing a minimum GPRS service level and increasing the blocking probability for circuit switched services. Whether resources should be dedicated is dependent on the priorities of the network operator and the existing system performance. Unless there is an absolute

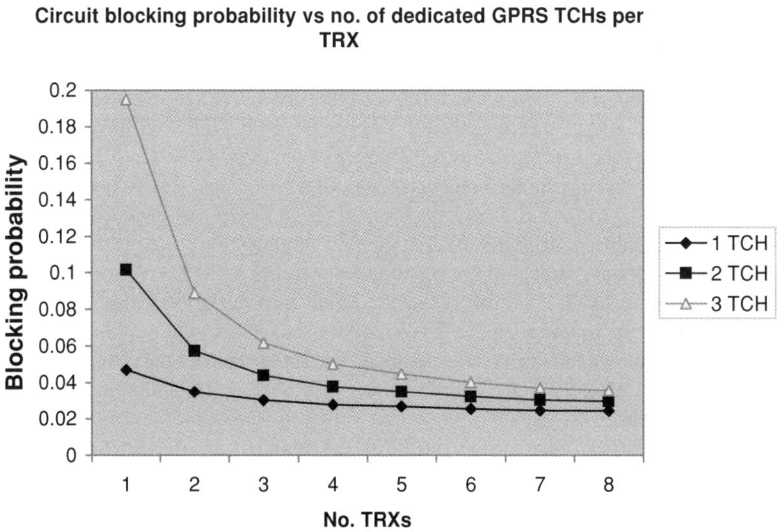

Capacity Planning

Circuit blocking probability vs no. of dedicated GPRS TCHs per TRX

Figure 2.25 Effect of dedicating traffic channels to the GPRS on the circuit switched blocking probability

need to guarantee a minimum GPRS service (albeit at the expense of circuit switched blocking), where this might otherwise be unlikely (such as under very high circuit switched traffic levels), it would probably be better not to dedicate slots to the GPRS.

Default GPRS Territory

The default GPRS territory is an area that will always be included in the instantaneous GPRS territory should the circuit switched traffic level permit. If circuit switched traffic is decreasing, having occupied some of the GPRS default territory, then re-allocation of timeslots to the GPRS will occur automatically, independent of the actual GPRS load. Such timeslots will therefore be available immediately to GPRS users about to set up a call, or for a re-allocation of the existing users with the aim of increasing their data rates.

Where circuit switched traffic levels are falling, but are outside the GPRS territory, then automatic re-allocation does not take place. The free timeslots made available in this case will only be requested should the GPRS load reach a predefined level (at which time the users may already be experiencing slightly degraded throughputs, depending on the predetermined configuration). Thus, setting the GPRS default territory to a higher level will tend to increase the QoS experienced by the GPRS users.

There is a downside to increasing the default GPRS territory, however. One issue is that there will be a tendency for increasing numbers of intracell handovers of circuit switched users to occur, with the aim of keeping the GPRS default territory free for the GPRS. As stated above, this will happen despite the fact that the GPRS load may be low. The second issue relates to the number of timeslots available per PCU. This number is fixed, and the overall system GPRS capacity may be affected if large numbers of GPRS timeslots are taking up PCU connections when not actually carrying GPRS traffic. There is, therefore, a compromise between the effects. Overall, it is initially recommended that the default territory is set to a level a little below the anticipated load level, as a compromise between the above factors.

Territory Upgrade/Downgrade

The GPRS territory upgrade and downgrade procedures are triggered under the following circumstances.

GPRS Downgrade

There are two dominant scenarios under which GPRS downgrade occurs:

- If the circuit switched (CSW) territory were to become fully occupied and it was necessary to accommodate a further CSW connection, then a timeslot from the GPRS territory would need to be re-allocated. This is a process that involves signalling between the CS RRM and the PCU, and also on air to the GPRS user(s) occupying the appropriate timeslot, and is therefore subject to delay. In order not to pass on this delay to the CSW user, the system tries to keep one or more timeslots (depending on the number of carriers per cell) free for such CSW allocations. Where the CSW traffic load steadily increases, therefore, there will be a continuous increase in the CSW territory, with timeslots being re-allocated from the GPRS territory, ideally in advance of the incoming CSW call requests. When the number of free timeslots in the CSW territory falls below that required to be free, then a GPRS downgrade is initiated to correct the situation. The (predefined) number of free slots allocated is based on the probability of 95 % that a further downgrade will not be required while there is already a downgrade in progress.
- If the GPRS territory has been enlarged beyond the default boundary and the GPRS traffic requirements are decreasing, a downgrade will occur to re-allocate the freed timeslot(s) back to the circuit switched territory.

GPRS Upgrade

There are two corresponding scenarios that result in GPRS upgrade:

- Where the CSW territory has moved into the default GPRS territory and the CSW traffic is decreasing, the CS RRM will initiate a GPRS upgrade to re-allocated timeslots back to the GPRS. Again, the number of free slots required to trigger the upgrade is dependent on the number of carriers per cell. In the upgrade case, however, the number is that which would result in there being no need (at 95 % probability) for a downgrade within four seconds of an upgrade having occurred, thus allowing for a degree of hysteresis.
- Where the GPRS service requirements are not being met by the number of timeslots in the GPRS territory (which is already at the boundary of, or into, the default circuit switched territory), and where the traffic level in the CSW territory permits, a GPRS upgrade is performed to increase the size of the GPRS territory. The rules governing this are based on the number of temporary block flows (TBFs) per timeslot; once the average number of users sharing the GPRS timeslots becomes too great (as defined by a nonconfigurable system parameter), the upgrade will be attempted.

Preliminary values for the number of free timeslots in the CSW territory are given in Table 2.16. The mean number of free timeslots in the CSW territory is also given, on the assumption that there are, on average, equal numbers of upgrades and downgrades.

Data Throughput

The overall EGPRS data throughput depends on a number of factors. GPRS has four coding schemes while EGPRS has nine modulation coding schemes, each offering different theoretical data rates. The actual performance of each of these is dependent on the C/I conditions, as shown in Figures 2.26(a) and (b).

In the GPRS example, an ideal link adaptation would ensure that the coding scheme changes from CS-1 to CS-2 as the C/I increases from approximately 6 dB to 7 dB, thus maximising the user data rate. It should be realized, however, that at the crossover point high retransmission rates could be expected with the higher rate scheme. Allthough this is taken into account in the user bit rate calculation, it does result in inefficiency on the channel. Therefore, a system designed for a C/I of 10 dB will offer higher CS-2 throughput than one designed for 7 dB. In practice, many systems will already be designed for C/I levels

Table 2.16 Number of free timeslots in the circuit switched territory

Number of carrier	Free TSLs (after downgrade)	Free TSLs (after upgrade)	Mean free TSL in CSW
1	1	1	1
2	1	2	1.5
3	1	2	1.5
4	2	3	2.5
5	2	4	3
6	2	4	3
7	2	4	3
8	3	5	4
9	3	5	4
10	3	6	4.5
11	3	6	4.5
12	3	6	4.5

significantly higher than this, and hence the regions of interest would be more likely to be the CS-2 and CS-3 boundary and the CS-3 and CS-4 boundary.

The C/I for a given user will depend on the location within the cell. The link adaptation will aim to ensure that the correct coding scheme is used to allow the highest data throughput per user, typically based on measured C/I or BLER, as stated above. In a given cell, therefore, there will be a number of users achieving differing data rates and using different coding schemes. In addition, there will be different types of users – some being capable of single-slot only operation, others capable of multislot operation. The overall data throughput will be a function of these, and other, factors.

Load Effects on Types of Active Users

Since a limited number of timeslots are available, the average chance of a user having access to the desired number of traffic timeslots will decrease as the number of users increases. With busy conditions, several users may share a single timeslot, and hence the net data rate per user will be reduced proportionally, while the overall message transmission time will increase. Although the system throughput may not be compromised under busy conditions, the grade of service experienced by the users will be a result of the reduced data rates/increased transmission delay times.

Efficient multislot operation is obviously very dependent upon the number of available timeslots. Under high load conditions, it becomes increasingly unlikely that a user will have full access to, for example, three consecutive timeslots. Therefore, it will be found that the mean data rate for a three-timeslot user will rarely meet those that are theoretically available (i.e. three times the single slot rate), except on systems operating at low loading levels.

Based on the above, it might be anticipated that the data rates achieved with multislot operation will often fall short of those theoretically offered. The high rates will be available during low load conditions, but often reduced at medium–high load levels.

Quality of Service and Rate Reduction

In terms of dimensioning a network, it is necessary to consider the quality of service (QoS) that is offered to EGPRS users. For applications involving the transfer of relatively large files (email, web browsing, etc.), the overall data rate and message delay will be the best measure of QoS. While dimensioning for

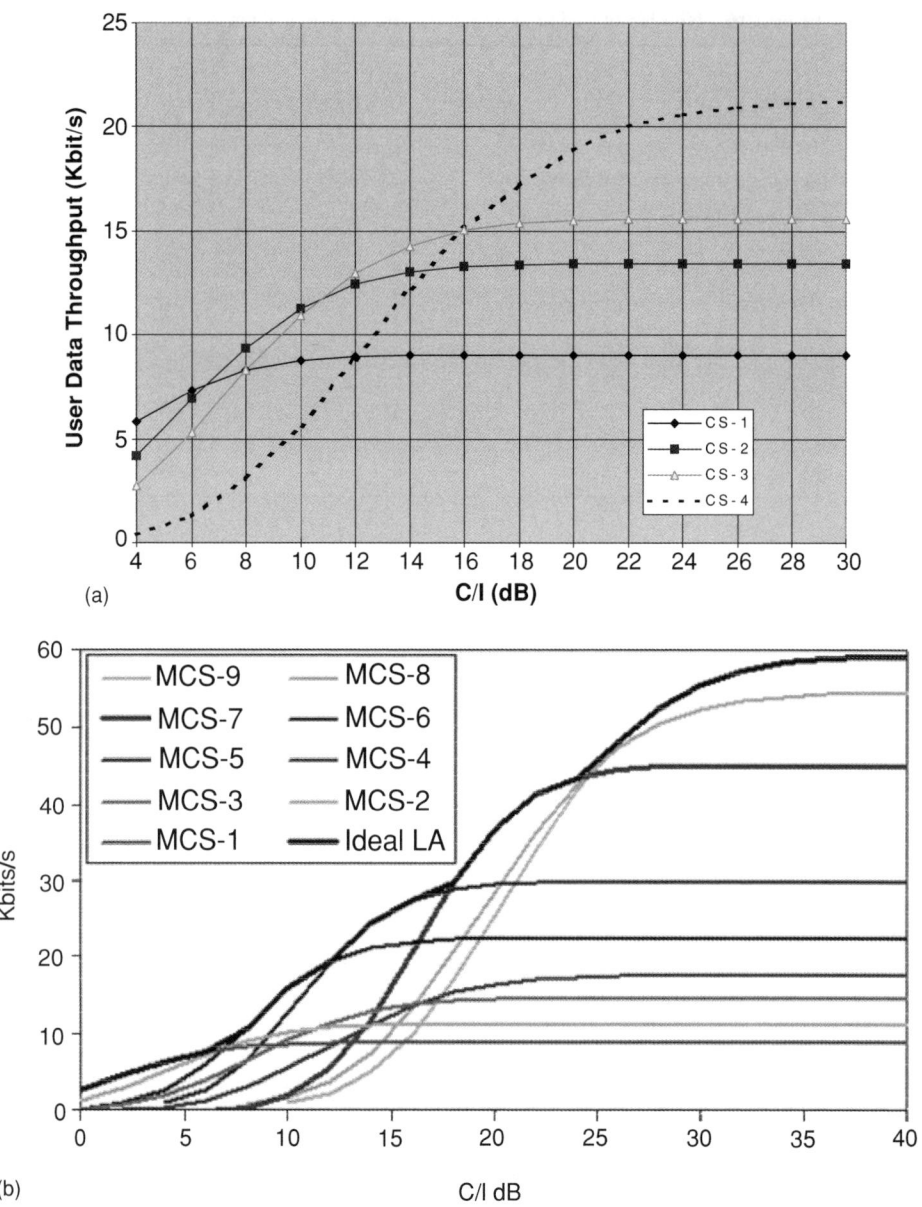

Figure 2.26 (a) Data throughput versus C/I for GPRS coding schemes. (b) EGPRS timeslot throughput versus C/I

a given set of QoS requirements, the rate reduction factor can be applied since this provides an estimate of how heavily the EGPRS timeslots can be loaded before the offered QoS falls to unacceptable levels (The user data rates decrease as the EGPRS load increases). High occupancy (caused by multiple simultaneous EGPRS users) increases the rate reduction factor and reduces the effective throughput per user or QoS.

Summary of GPRS and EGPRS Traffic Dimensioning

Basic guidelines have been outlined for dimensioning a mixed circuit switched and GPRS/EGPRS network. These should be treated as a 'rule of thumb' to enable initial designs to be performed. As the data network is deployed and carrying live traffic, two key "assumptions" require re-visiting and quantification for calibrating and re-dimensioning: data and traffic profiles:

- *Data profiles.* Initially, it is assumed that the majority of users will be using EGPRS for web browsing and email applications. This implies relatively large message sizes (typically upwards of several kilobits). The QoS requirements for these applications may differ somewhat from those associated with other applications, such as those involving short messaging. Should the application profile change significantly from that assumed in the simulation work performed, then, for example, it may be possible to relax the EGPRS loading constraints, allowing for higher channel loading and higher EGPRS throughput.
- *Traffic profiles.* Dimension a data network on the assumption that the circuit switched and EGPRS traffic peaks simultaneously may result in a degree of over dimensioning. Dimensioning should be adjusted based on revisiting this assumption through live traffic profiling as the heavy data cells start loading EGPRS traffic. For the busiest cells, the overall peak hour of primary concern so dimensioning should reflect this.

2.6.3 Capacity Planning in WCDMA Networks

Capacity planning in WCDMA networks is much more complicated than in GSM/EGPRS. Factors that affect the coverage calculations are load, interference, traffic behaviour, speed of subscribers, etc.

Uplink

WCDMA is an interference system limited by the air interface. Hence, capacity planning would need to calculate the interference and the cell capacity, i.e. the amount of traffic that is supported by a base station. The amount of uplink interference has a great impact on the cell capacity and radius. The interference margin (η) indicates the total amount of interference (including thermal noise power) in comparison to the thermal noise:

$$\eta_u = \frac{E_b RN}{W N_0}(1+i)v_j \qquad (2.54)$$

where

E_b/N_0 = signal energy per bit/noise spectral density
N = total number of users/cell
R = bit rate
W = chip rate
i = other cell-to-own cell interference
v_j = activity factor of user j

Downlink

In the downlink, the power transmitted by the BS is shared between all users. The capacity is determined by the power transmitted by the BS, locations of UE and interference. Thus, the parameters needed for downlink calculations include the power transmitted by BS and power allocation to the Common Control

Channel (CCCH). Thus, in downlink the capacity is determined by the power transmitted by the BS, locations of UE and interference. This makes the calculations in downlink more complicated than the uplink directions, for in the uplink each user has its own amplifier to transmit the power. Thus, coverage becomes a function of the number of users. In DL the own cell interference is reduced by the factor $(1 - \alpha)$. This is due to the synchronised orthogonal channellisation codes, which are used in DL. The downlink load factor can be calculated as

$$\eta_{DL} = [(1 - \alpha_j) + i] \sum_{j}^{N} \text{load}_j \quad (2.55)$$

where

$$\text{load}_j = \frac{1}{1 + (W/R_j)/(E_b/N_0)_j 1/v_j} \quad (2.56)$$

and the orthogonality factor α_j is between 0.4 and 0.9 (the ITU vehicular subscriber for the macro cell is 0.6 and the ITU pedestrian subscriber for the micro cell is 0.9).

In the WCDMA system, the traffic can be asymmetric in the uplink and downlink directions and thus the load can also be different in either direction. The DL load is, however, higher than the UL load. The link performance also differs in either direction (the noise figure is higher for the UE than the BS). Soft handover heads are only in the DL direction.

The load factor for different services has to be calculated separately. The total load is then the sum of different services in the cell area.

Soft Capacity

The principle of soft capacity is that a cell can be more loaded when surrounding cells are unloaded. The less interference there is coming from all the neighbouring cells, the more users can be admitted before the load (interference or transmitted power) of a cell reaches the load target. If the average loading is low, there is extra capacity available in the neighbouring cells. As this capacity can be borrowed from the neighbouring cells, the interference sharing gives soft capacity. The soft capacity has more inpact on high bit rate real time users because of a larger relative change for higher bit rates.

The soft capacity can be approximated based on the total interference at the BTS. The total interference includes that of both the own cells and other cells. Therefore, the total channel pool can be obtained by taking the number of channels per cell in the equally loaded case and multiplying that by $1 + i$, which then gives the single isolated cell capacity.

The basic Erlang B formula is then applied to this larger channel pool. This obtained Erlang capacity is then equally shared between neighbouring (interfering) cells by dividing the maximum offered traffic by $1 + i$ (in UL the power rise is also taken into account).

From the planned load, the number of channels available in the resource pool (an average condition) can be calculated:

$$N_{UL} = \frac{\eta_{UL}}{(1+i)} \left(1 + \frac{W/R}{E_b/N_0} \frac{1}{v}\right) \quad (2.57)$$

The soft blocking capacity (in Erlang) for RT services can be calculated using the Erlang B table and the equation

$$\text{Soft capacity/cell} = \frac{\text{Erlang B }[N(1+i)\text{ blocking \%}]}{1+i} \quad [\text{Erl}] \quad (2.58)$$

The DL soft capacity is calculated using a similar method.

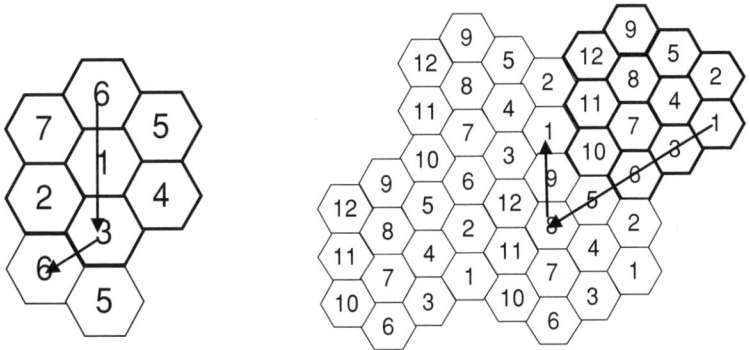

Figure 2.27 Frequency re-use patterns of 7 and 12 cells

2.7 Frequency Planning

The dilemma behind frequency planning is to provide needed capacity and coverage within a given frequency band. The frequency channels therefore need to be re-used, but it is wise not to increase the interference level. Interference is caused when two network cells use the same channel too close to each other; more precisely this is a co-channel interference situation. When the interfering channels are consecutive there is some neighbour channel interference, but this is less serious. The interference level cannot be high when building a functional network. The interference level increases with high transmission power in a close location.

The frequency re-use rate is simplest to explain using a hexagonal model. Frequency re-use patterns are not used in practice because the cells are not hexagons, as already explained in the coverage planning section. The cell shapes are different and cells do not have equal sizes. Therefore the frequency re-use rate is not a constant throughout the network, but varies from one place to another and can also vary between BCCH and TCH layers. The available frequency band and the capacity plan give boundary conditions for the largest possible frequency re-use rate. Figure 2.27 shows two examples of the frequency re-use rate for hexagons. The first describes the re-use pattern for 7 cells and the second one the pattern for 12 cells.

The following is an example of the connection between frequency and capacity planning. If the operator bandwidth is 6 MHz, the number of channels is 30. In general, frequency re-use affects the number of carriers that can be used for one sector. The number of carriers per sector can be calculated by dividing the available bandwidth by the product of the re-use rate and bandwidth for a single carrier, i.e. dividing the number of channels by the frequency re-use rate. Using a re-use rate of 12, the number of TRXs, carriers per sector, is calculated below:

$$\frac{6\,\text{MHz}}{0.2\,\text{MHz} \times 12} = \frac{30}{12} = 2.5(\text{TRX})$$

Continuing, the capacity of the network with 100 BTSs and 600 cells can be calculated as

$$600 \times 2.5\,\text{TRX} = 1500\,\text{TRX}$$

An interesting factor in frequency re-use is the distance between base stations using the same frequency. This distance is known as D and the cell (hexagon) radius as R. The re-use distance is explicit for the hexagonal model, cells with equal sizes and frequency re-use patterns repeating the frequencies in the

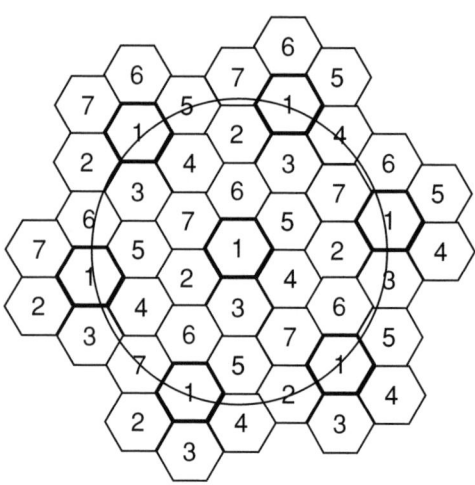

Figure 2.28 Co-channel interference with hexagonal cells

same order. Parameters D and R are illustrated in Figure 2.28. The re-use pattern factor K can be calculated geometrically:

$$K = D_x^2 + D_x D_y + D^2 \qquad (2.59)$$

From Figure 2.27, K can be calculated as

$$K = 1^2 + 1 \times 2 + 2^2 = 7 \qquad (2.60)$$

Using the re-use distance gives

$$D = \sqrt{3 \times 7}\, R = \sqrt{21}\, R \approx 4.58R \qquad (2.61)$$

To enable maximum capacity, the parameter K should be optimised to be as small as possible when the system is operational and fulfilling the planning requirements. Parameter q is the co-channel interference reduction factor, where the higher the value of q the smaller is the co-channel interference (see Fegure 2.28):

$$q = \frac{D}{R} \qquad (2.62)$$

The co-channel interference can be calculated as the ratio of the carrier (C) to the sum of the interferers (I_n):

$$\frac{C}{I} = \frac{C}{\sum_{n=1}^{N} I_n} \qquad (2.63)$$

where N is the number of interferers and the received carrier is

$$C = \alpha d^{-\gamma} \qquad (2.64)$$

where d describes the distance between the transmitter and the receiver, α is a constant and γ is the propagation path loss slope. Interference for the first tier is

$$\frac{C}{I} = \frac{R^{-\gamma}}{\sum_{n=1}^{6} D_n^{-\gamma}} = \frac{1}{\sum_{n=1}^{6} q^{-\gamma}} \qquad (2.65)$$

and is simplified for the first tier, as all the interferers in the same tier are equally strong:

$$\frac{C}{I_{\text{first tier}}} = \frac{1}{6q^{-\gamma}} \qquad (2.66)$$

Interference for the second tier is

$$\frac{C}{I_{\text{second tier}}} \approx \frac{1}{6(2q)^{-\gamma}} \qquad (2.67)$$

The C/I relation can be to improve the network using the following methods: decrease transmission power, fine tuning of the antenna azimuth and antenna down-tilting. All the methods have an impact on the cell coverage area and therefore they need to be used carefully, keeping in mind the coverage targets. Also the overall situation needs to be verified, preferably with field test measurements, because the changes made in one area can have other types of impact in the adjacent area.

The spectrum efficiency techniques, as indicated by their name, provide ways to utilise the frequencies more effectively. This is critical in network planning due to the limited resources, i.e. limited frequencies. More effective usage of the frequency band is possible without increasing the interference level. In this way the network capacity increases.

2.7.1 Power Control

Power control can be used for both the downlink and uplink directions to decrease co-channel interference. Power control increases the spectrum efficiency and at the same time saves the mobile station battery lifetime, because less power needs to be transmitted. The transmission power can be reduced in case the received power level is high and decreased until the received level still remains adequate. The power ranges for the base station and the mobile station as well as the power control steps are specified in the GSM specifications.

The mobile station transmission power is defined by the MS classes, according to the specifications. For example, the MS output power with a class 4 mobile (GSM 900) is 2 W. The MS power can be decreased with steps of 2 dB, the range for the MS power control being 15 steps, or in other words 30 dB. Similarly, the base station power control also works in 2 dB steps and the range is 30 dB. Another parameter besides the power ranges and the steps defined in the GSM specifications is time – how fast the power can be decreased or increased. One 2 dB change can be done once during 13 TDMA frames, which is 13×4.615 ms $= 60$ ms. The GSM channel structure is specified in Section 2.6.1 (Capacity Planning in GSM Networks).

When the connection is initialised the MS first transmits using a fixed power. The transmission power level is defined for the whole cell, which is always used when initialising a connection. The information is sent to the MS via the BCCH. After a while the power control can be used to find a more suitable MS power. The initial power is selected to be adequate in all situations for that particular cell.

The power control is an optional network feature, so can be activated when desired. The base station subsystem (BSS) controls the power control both in the downlink and uplink directions. The MS measures the received power level and sends the results via the BTS to the BSS. The BTS also measurs the level of the received signal. Based on these measurement results reported to the BSS the needed power level can be calculated for both the MS and BTS transmissions. The measurements will ensure that the transmission power is kept at a certain level that the quality of the connection does not fall below required level.

2.7.2 Discontinuous Transmission

Discontinuous transmission (DTX) is another method used to decrease co-channel interference and therefore improve spectrum efficiency. When co-channel interference decreases the quality of those particular cells improves. DTX works for both downlink and uplink. In the uplink direction, discontinuous transmission also saves the battery lifetime.

The discontinuous transmission feature is based on recognition of silent moments. During any telephone conversation silent moments occur, because in an average conversation one person normally speaks for half of the time. The mobile is able to separate the silent and the conversation moments using voice activity detection (VAD). During silent moments DTX minimizes the transmission.

2.7.3 Frequency Hopping

Frequency hopping is a feature of the GSM system used to decrease the simultaneous usage of the same frequencies and in this way averages the interference level. Compared to normal frequency re-use, which is static, frequency hopping provides a benefit by allowing the dynamic frequency to change. In practice, by using the frequency hopping algorithm the carrier frequency changes either cyclically or randomly. By frequency hopping the interference is averaged more effectively, which increases the overall perceived quality. Frequency diversity and interference averaging can be gained using frequency hopping.

The GSM channel structure was referred to in Section 2.7.1 (Power Control) and has been specified in Section 2.6.1 (Capacity Planning in GSM Networks). As discribed, the carrier frequencies are divided timewise into eight timeslots, according to time division multiple access (TDMA). One timeslot is allocated to serve one call, i.e. one subscriber. The length of one timeslot is 0.577 ms and the length of the TDMA frame consisting of eight timeslots is 4.6 ms. The frequency hopping method implemented in the GSM system is slow frequency hopping; after every TDMA frame the frequency can be changed. By changing the frequency after each 4.615 ms, in one second the TDMA frame makes 217 changes.

The frequency hopping types of the GSM system are base band (BB) frequency hopping and synthesised frequency (RF) hopping. The broadcast control channel (BCCH) in the first timeslot of the TRX and the traffic channels (TCHs) are treated differently. The BCCH frequency transmits continuously whereas the TCH slots are based on traffic.

In BB hopping the TRXs actually have fixed frequencies and the frequency hopping operates so that the bursts are shifted from one TRX to another according to the hopping sequence. The number of hopping frequencies therefore comes from the number of TRXs. The length of the mobile allocation (MA) list is the number of TRXs. The BCCH TRX does not hop and therefore is excluded from the hopping groups. With BB hopping there are two actual hopping groups, one for the first timeslots of each TRX (in one cell) excluding the BCCH TRX and another for all the other timeslots. The first hopping group has TRX-1 frequencies, due to the dedicated BCCH frequency. See Figure 2.29 for an illustration of base band hopping.

In RF hopping the frequency of the TRX changes but a single call uses only one TRX. This does not have the same limitation to the number of frequencies as in BB hopping. The RF hopping pattern can be either random or cyclic. In random frequency hopping the frequencies are randomly selected from the available frequencies. Cyclic hopping has a defined cycle for the frequencies in the MA list. Figure 2.30 shows the basic functionality of RF hopping.

Frequency Planning

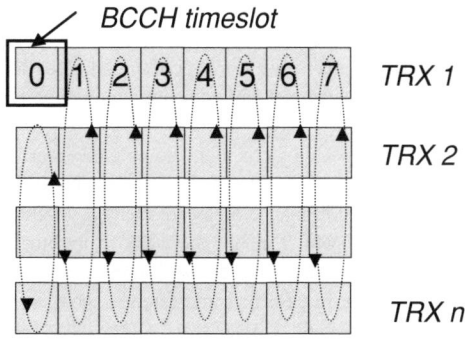

Figure 2.29 BB hopping principle

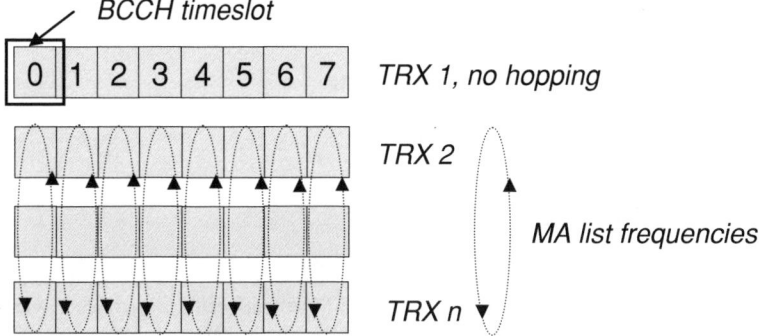

Figure 2.30 RF hopping principle

In the following section the parameters controlling frequency hopping in the GSM are introduced. The Hopping Sequence Number (HSN) can be selected from 64 options and defines the sequence showing how the frequencies change. In case the HSN is set to 0 cyclic hopping is selected. The HSN values from 1 to 63 are used to select different random sequences. In RF hopping all the timeslots have the same HSN number. With BB hopping the story is different, as the first timeslot (timeslot 0) uses one HSN and the other timeslots another HSN. To avoid different TRXs using the same HSN to transmit simultaneously at the same frequencies a parameter called the Mobile Allocation Index Offset (MAIO) is developed. Setting the MAIO differently to the TRXs in the same cell ensures that those TRXs start to transmit at different frequencies; i.e. they start the sequence from different locations. The maximum length of the MA list is 64 frequencies, which also includes the BCCH frequency, and therefore the maximum number of hopping frequencies on the list is 63.

The key advantages of the frequency hopping feature are that it does not require changes in the network configuration and it is automatically applicable to all MS types.

2.7.4 Interference Analysis

The target of frequency planning is to allocate the frequencies so that the network is operational and the quality of service requirements are fulfilled. Co-channel and adjacent channel interference is analysed at the network level and also on a cell-by-cell basis. Frequency planning is often iterative and some allocated frequencies need to be changed in order to decrease the interference level. Naturally later during the network live time the frequency plan is renewed either partly or fully when adding more TRXs or new base stations. The GSM specifications define values for co-channel and adjacent channel interference. In addition, the customer may give extra recommendations for the interference analysis. In the case of an option between the co-channel and adjacent channel interference the system can bear adjacent channel interference better.

Interference is normally analysed channel by channel, showing the interference layer on the planning tool. The frequency channels are reallocated to make the interference layer, both co-channel and adjacent channel, as clean as possible. The cell dominance areas are cleared. Some planning tools also provide numerical statistics of the interference, which makes the work of finding the interfered channels and comparing them after changing the plan easier. Co-channel interference clearance of the BCCH layer is preferred over that of the TCH layer. In case bad interference spots are found in this phase a short re-check of the dominance areas can be made. Unnecessarily large coverage makes frequency planning complicated. Earlier mentioned techniques can be used to clear the dominance areas and in this way create a better basis for successful frequency planning:

- For co-channel interference : $C/I_c = 9$ dB
- For adjacent (200 kHz) interference : $C/I_{a1} = -9$ dB
- For adjacent (400 kHz) interference : $C/I_{a2} = -41$ dB
- For adjacent (600 kHz) interference : $C/I_{a3} = -49$ dB

2.8 Parameter Planning

2.8.1 Parameter Planning in the GSM Network

The Cell Identity (CI) Code, the unique naming of the GSM network cells, is guaranteed with CI coding. This identification is used in the idle mode. Each code is unique inside one location area. The length of the CI code is 16 bits.

The Location Area Code (LAC) defines the identification of a location area. The code consists of 2 bits. The location area combines nearby cells to one logical area. In practice, the location areas are physically large. Two types of location updates exist: the MS crosses the border of two location areas or a timer doing the update expires. The location update involves both the Home Location Register (HLR) and Visitor Location Register (VLR). The critical issue in planning the location areas is to define the borders so that unnecessary location updates can be avoided and no extra signalling is created.

The Base Station Identity Code (BSIC) is used for cell identification in the dedicated mode and is broadcast on the SCH. The BSIC is a six-digit code consisting of two parts, each of three digits, a Network Colour Code (NCC) and a Base Station Colour Code (BCC). The NCC part is reserved for the separation of different networks and the BCC separates the base stations. One critical issue in BSIC planning is not to allocate same BCCs to cells with the same frequency in the same area. The NCC in practice separates different operators using the same frequencies, which might be the case in the border areas of two countries. Both NCC and BCC parameters can have values between 0 and 7.

The Training Sequence Code (TSC) is transmitted in the middle of each burst and is used to keep the timing. The training sequence (TS) is a known code, 26 bits in length, which can be identified by the receiver and transmitter. In addition, the RXQUAL is calculated based on the bit error ratio from the training sequence.

The carrier frequency lists of own and neighbouring cells are informed to the MS differently depending on whether it is connected or in idle mode. In the case where the MS is in idle mode, the list of the neighbouring cell BCCH frequencies is transmitted with its own BCCH frequency and the MS listens. Therefore the MS can measure only neighbouring cell frequencies. In addition the list may also include some other frequencies, which have been defined separately. When the MS is in dedicated mode the content of the list can be basically similar. The information is given instead of BCCH to SACCH. The

Parameter Planning

GSM specifications allow 255 frequency lists per BSC. The number of frequencies in one list can be up to 32, which also defines the maximum number of neighbouring cells.

Cell selection in idle mode is basically controlled by two parameter based criteria: C1 criterion is mainly for initial cell selection while C2 criterion, if it has been activated, is for cell reselection. The C1 criterion can also be used for cell reselection if the C2 criterion is not a possible option. The C1 criterion is called the criteria for path loss. The C1 value determines the selection of the cell in the idle mode. The following equation is the C1 criterion, with the parameters explained below:

$$C1 = (RxLev - RxLevAccessMin - MAX\,[(MSTxPower - MSTxPowerMax), 0] \qquad (2.68)$$

where RxLev is the level of the signal received on the BCCH. This signal level is compared to the RxLevAccessMin and if lower the MS continues searching another BCCH. This test ensures that the downlink is strong enough for a proper connection. RxLevAccessMin is the minimum requirement for the received signal level in order for the MS to access the cell. In other words, the idle mode parameter RxLevAccessMin limits the cell size. Normally a signal level close to $-100\,\text{dBm}$ is required. MSTxPower is the maximum power the MS can use based on the network parameter settings. As the RxLevAccessMin limits the cell in the downlink direction the MSTxPower limits the uplink direction and cell size. MSTxPowerMax is the maximum transmission power of the MS based on its type and specified class.

The C2 criterion is called the cell reselection criteria. The following is the equation for C2 and the parameters are defined below:

$$C2 = C1 + CellReselectionOffset - TemporaryOffset^* H(PenaltyTime - T) \qquad (2.69)$$

Basically the selection for a new cell can take place after 5 seconds if the C2 value has been higher for another cell than the current one. For the C2 criterion the current cell and six strongest neighbours are compared. The neighbours are added to a comparison list and PenaltyTime is applied. In the beginning the timer T is zero until T is smaller than the PenaltyTime parameter value. TemporaryOffset is used which negatively affects selection of the cell. In order to prevent cell reselections that are too fast as well as the possible ping-pong effect after previous cell selection additional criteria are applied for 15 seconds. The C2 value for the new cell has to be 5 dBm or better compared to the C2 value of the current cell. Another exception rule is the situation where the new cell and old cell belong to different location areas. In this case the C2 of the new cell has to be higher with CellReselectHysteresis for a time of 5 seconds time or more. The CellReselectionOffset parameter can be used for prioritization, with certain cells having a higher offset, making selection easier. TemporaryOffset is applied for the cells added to the comparison list. PenaltyTime is the time controlling the TemporaryOffset usage.

The timing advance (TA) parameter is used to indicate how far the MS is from the BTS. The actual measure is time – how much time it takes for the signal to travel the distance from the BTS to the MS. The speed of the RF signal is equal to the speed of light, and in this way the distance can be calculated. The maximum cell radius of a normal GSM cell is 35 km. This radius is divided into 64 equal TA steps. Each step is then approximately 550 m. For example, the TA value for a mobile that is 1500 metres distance from the BTS is 2. For other examples see Table 2.17.

Location update is a procedure that makes the network aware of the MS location. This is a prerequisite for mobility where the MS movement can be tracked and its position known in the case of incoming calls, short message services (SMSs), etc. The network is always sending parameter information via broadcasting to the MS, as one part there is also LAC. The MS keeps this information and incase of a change in location area or PLMN compared to the previous information the MS makes a request for a location update. On the network side the location update changes the MS to be assigned to the new VLR, which informs the HLR about the update. The old VLR re-sets the information related to the MS. The parameter IMSIAttach defines the location update procedure to be automatic. If IMSIAttach is activated when the mobile is powered it informs the network to be active. A very similar parameter is IMSIDetach,

Table 2.17 Timing advance (TA) parameter values

Distance (MS-BTS) [m]	TA	Distance (MS-BTS) [m]	TA
<550	0	3300–3850	6
550–1100	1	3850–4400	7
1100–1650	2	4950–5500	8
1650–2200	3	...	...
2200–2750	4	...	...
2750–3300	5	35 000	63

which defines the automatic procedure when the MS is powered down. The MS informs the network that it is not available for communication. The advantage of attach and detach parameters is decreased signalling – more precisely the automatic procedure prevents unnecessary paging.

Parameter Consistency Tool

Tools maintaining the network parameter information helps to keep the consistency. The initial parameter planning is done with the network planning tool, e.g. the frequency plan. These parameters can be first exported from the planning tool and then imported to the configuration tool in the network management system. In the configuration for management tools first a plan is created that can be sent to the network during a low traffic period so that the end users are not disturbed.

The network planning tool does not coverall the parameters that need to be set in the network and therefore more planning is required. The parameter consistency tool gives good support to maintain the information, where the data are coming from and the different versions. The parameter consistency tool can maintain data from both the network and the planning tool. When testing new parameter settings, it is worthwhile to keep the old setting as a backup.

2.8.2 Parameter Planning in the EGPRS Network

The introduction of EGPRS brings a new set of BSS parameters to control the EGPRS traffic and make the most efficient use of the resources available. This section covers some of the generic parameters. Parameter names are vendor specific and therefore only a generic description of the parameter functionality is provided here. All references are made to GPRS only since the same concepts also apply to EGPRS parameter planning.

Parameters

GPRS Enabled
This parameter defines whether the cell/radio is allowed to handle GPRS traffic.

Default GPRS Capacity
Timeslots are always allocated to the GPRS territory unless pre-empted by CSW traffic (Figure 2.31).

Dedicated GPRS Capacity
A number of TCHs can be dedicatedly reserved for GPRS use and are thereby removed for circuit switched traffic.

Additional GPRS Capacity
Additional timeslots, over and above the default territory, may be used by GPRS traffic if circuit switched traffic permits.

Parameter Planning 79

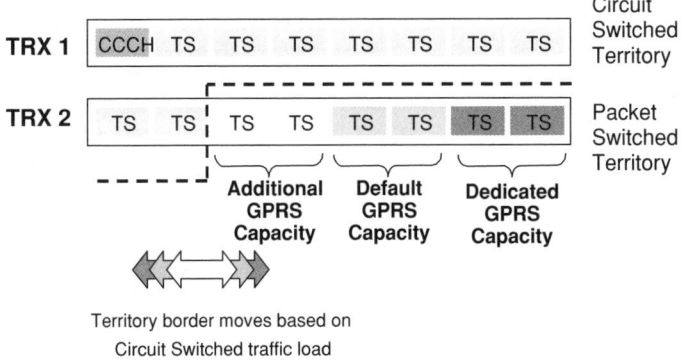

Figure 2.31 EGPRS territory

Maximum GPRS Capacity
This is the maximum number of timeslots in a given cell that can be used up for GPRS traffic.

GPRS Traffic Preference
This decides if GPRS traffic should be assigned to the BCCH carrier prior to other TCH carriers within the same cell.

Territory Update Guard Time
This determines a timer value, i.e. the guard time between two subsequent territory updates.

Intracell Handover for a GPRS Territory Upgrade
During the GPRS territory upgrade procedure, CSW calls in timeslots to be included in the territory are handed over to other available timeslots in the same cell (Figure 2.32).

Signalling Capacity in the GPRS Network

The negotiation between the terminal and the base station has to be considered in order to understand the extra load placed on the existing GSM signalling channels, unless separate channels for the GPRS are used. Figure 2.33 introduces this new set of logical GPRS radio interface channels.

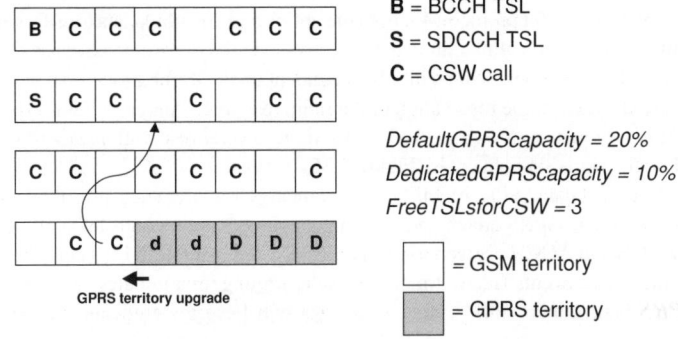

Figure 2.32 EGPRS territory upgrade

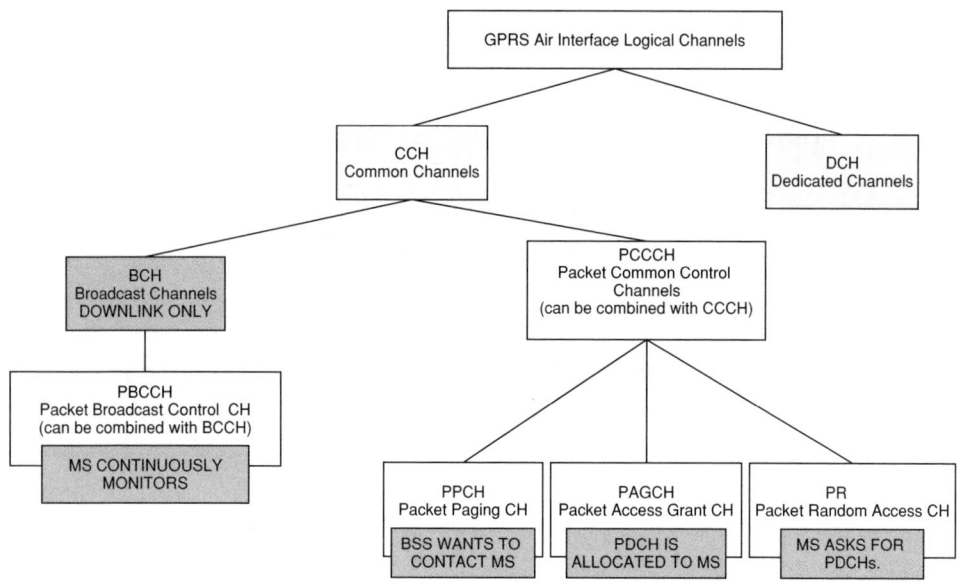

Figure 2.33 GPRS radio interface channels

In Release 1, the GPRS common control channels are combined with the corresponding logical channels. Therefore, with the introduction of the GPRS extra load will be placed on the Random Access Channel (RACH), the Access Grant Channel (AGCH) and the Paging Channel (PCH).

Data Transfer

In the downlink direction, the BS uses the packet PPCH and PAGCH for the call assignment, with the PPCH being used for the paging of a terminal and the PAGCH providing the information of the assigned traffic channel to the terminal. In the uplink direction, a terminal sends information to the BS by using just one broadcasting channel, which is called PAGCH. On this channel, the mobile sends a request for service to the BS (or to the network) in both mobile originating and mobile terminating cases.

The GPRS packet data transfer protocols for both the uplink and downlink data transfers are presented in Figure 2.34. Both uplink and downlink initiated data transfers will require extensive use of the RACH and AGCH. The load on these channels will be highly dependent on the traffic model. An email message can usually be transferred with a single reservation of the radio resource, whereas a www type of application may produce a bursty sequence of data packets that require several reservations, therefore generating a lot of extra load on the signalling channels (Figure 2.35).

In addition to the packet data traffic, mobility management procedures such as routing area updates and cell re-selections as well as GPRS attach and detach generate common control signalling. The resource required on the RACH and AGCH is given for every GPRS event in Table 2.18. Therefore the parameters associated with the above events (ready timers, periodic routing area updates) and implementation of the network (micro cells/macro cells/routing area size) will have a significant impact on the CCCH load.

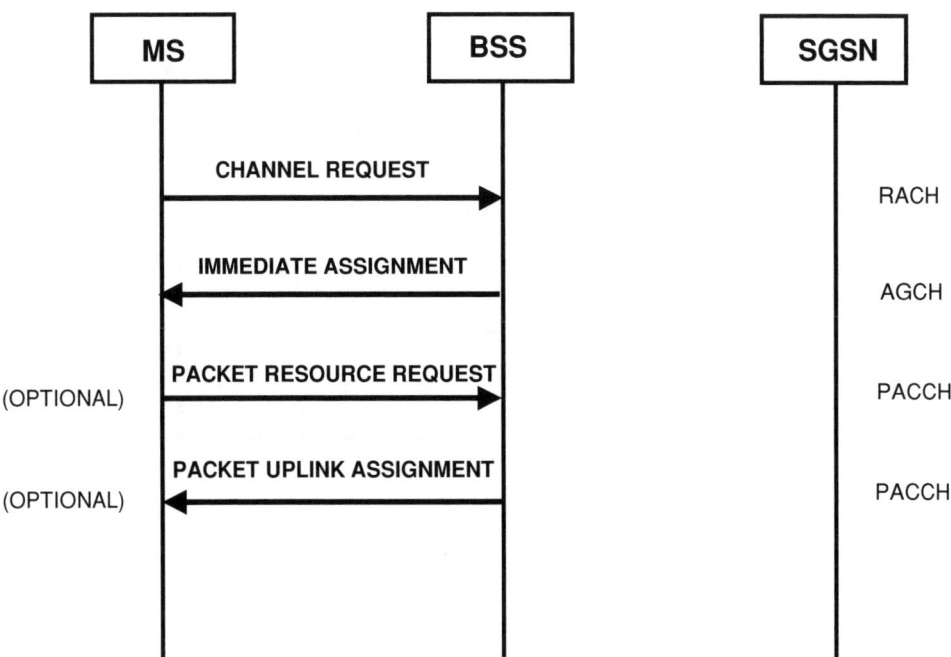

Figure 2.34 Packet data transfer: uplink initiated

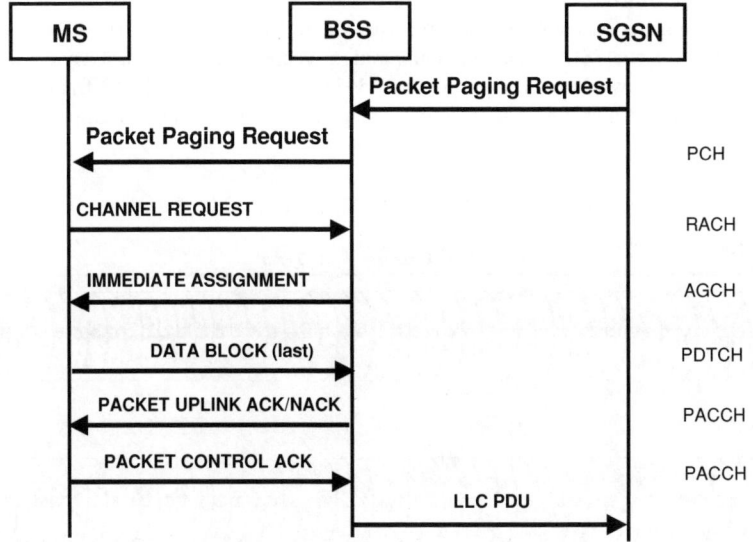

Figure 2.35 Packet data transfer: downlink initiated

Table 2.18 Resource requirement for an EGPRS event

	RACH	AGCH
GPRS attach (no authentication)	2	3
GPRS attach (authentication)	2	5
GPRS detach (MO, or mobile originated)	1	2
PDP (Packer Data Protocol) context activation (MO)	1	2
Paging	1	2
Cell update	1	1
Routing area update	1	2
SMS (MT, or mobile terminated) (with paging)	2	5
SMS (MT) (without paging)	1	3
SMS (MO) (with stats report)	3–4	6–8
SMS (MO) (without stats report)	2	3

Channel Combinations

In the existing GSM infrastructure, timeslots 0 and 1 in the first carrier are usually needed for the use of all of the above-mentioned channels. Due to capacity constraints, there are two main configurations of these channels:

- combined channel structure;
- uncombined channel structure.

Combined Configuration

The combined configuration combines the Broadcast Control Channel (BCCH) with the Stand-alone Dedicated Control Channel (SDCCH). This configuration occupies only one timeslot and is usually used with cells having up to two carriers requiring limited signalling capacity. The 51-frame multiframe is shown in Figure 2.36 and indicates the location of the signalling channels for both the uplink and downlink

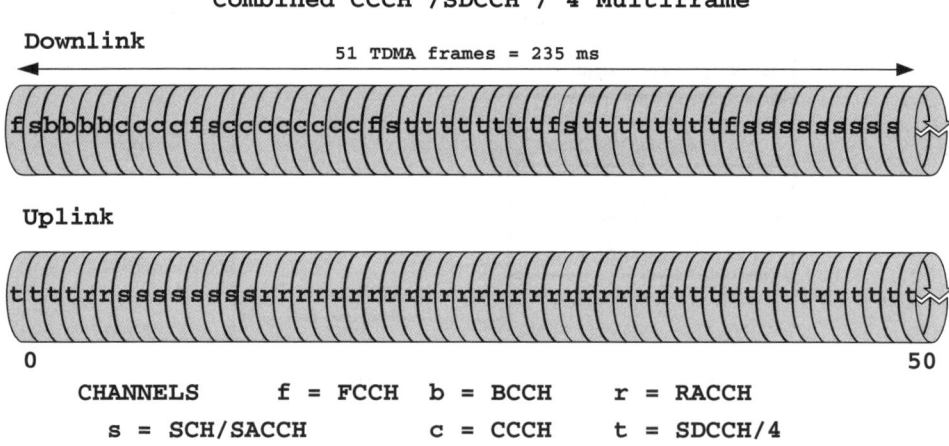

Figure 2.36 Combined BCCH/SDCCH structure

BCCH/CCCH Multiframe

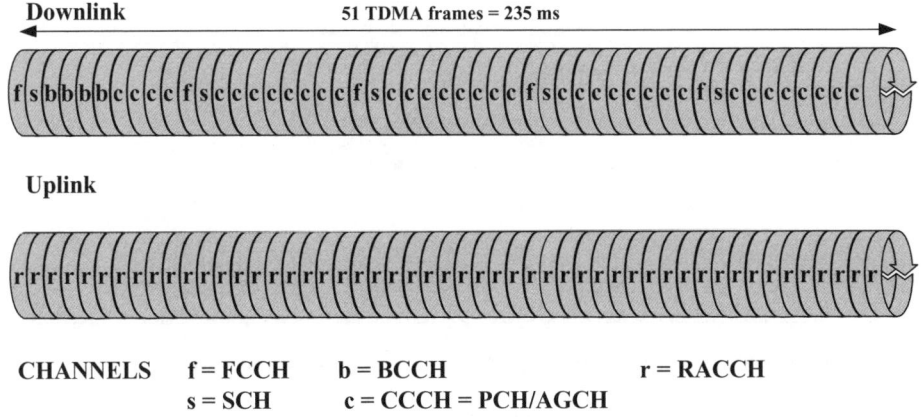

Figure 2.37 BCCH multiframe

directions. The GPRS will place an extra load on both the RACH and AGCH and therefore the main focus will be aimed at these channels.

In the combined configuration the RACH has allocated 27 TDMA frames and three blocks of 4 TDMA frames to the CCCH. The CCCH is shared by both the AGCH and the PCH. The capacity allocated to the AGCH can be specified by the operator with the parameter NumberOfBlocksForAccessGrant. In a combined configuration this is usually defined as:

- 1 block used for AGCH → 2 blocks for paging

Therefore the RACH and AGCH capacity in a combined configuration is:

- RACH cpacity = 27 TDMA frames/235 ms = 115 frames/second
- AGCH capacity = 1 block/235 ms = 4.25 blocks/second

With the immediate assignment extended, the access grant can support two assignment messages. However, this is dependent on the message type and therefore the expected number of messages per block is 1.8.

$$\text{Immediate assignment extended} = 4.25 \times 1.8 = 7.65 \text{ messages/second}$$

Uncombined Configuration

In the uncombined configuration a separated channel structure is adopted with the BCCH remaining on timeslot 0 and the SDCCH allocated to a second timeslot (Figures 2.37 and 2.38). This allows extra signalling capacity to both the SDCCH and the CCCH. A different configuration is used in cells with more than three carriers.

The uncombined configuration allows the RACH the full 51 frames and the CCCH 9 blocks, which are split between the AGCH and PCH:

- 3 blocks used for AGCH → 6 blocks for paging

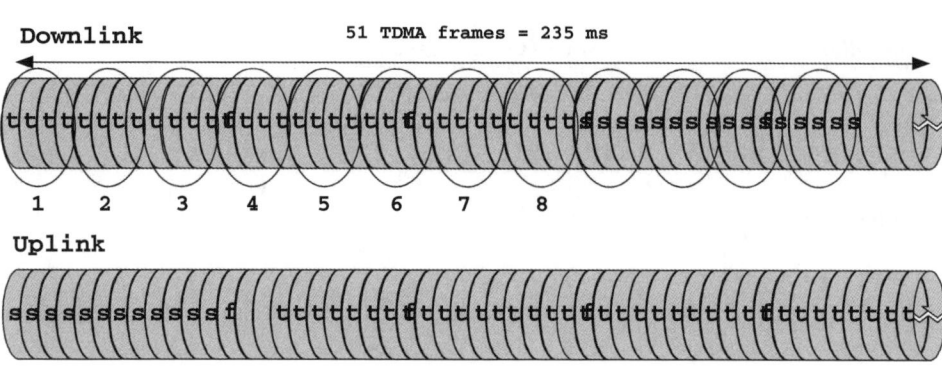

Figure 2.38 SDCCH multiframe

With the above assumption, the RACH and AGCH capacity available is:

- RACH capacity 51 TDMA frames = 217 frames/second
- AGCH capacity 3 blocks = 12.8 blocks/second

$$\text{Immediate assignment extended} = 12.8 \times 1.8 = 23 \text{ messages/second}$$

Capacity (PCH and AGCH)

Signalling capacity in existing voice networks depends mostly on the paging channel (PCH) capacity and on the SDCCH capacity. However, GPRS functionality puts a far greater requirement on both the RACH and the AGCH. This will impact on the signalling configuration and care will have to be taken when dimensioning the size of the Location Area (LA)/Routing Area (RA).

It is possible to calculate the RACH capacity available for circuit switched users by subtracting the GPRS associated RACH throughput for different offered load levels from the total RACH capacity (combined configuration).

PCH Load and Routing Area Dimensioning

The mobility management function in the GPRS network is handled in a similar way to that done for the GSM system. One or more cells form a routing area, which is a subset of one location area. Every routing area is served by one SGSN. The tracking of the location of a terminal depends on the mobility management (MM) state.

Mobility Management States

IDLE (GPRS) State
In the GPRS IDLE state, the subscriber is not attached to the GPRS mobility management. The MS and SGSN context hold no valid location or routing information for the subscriber. The subscriber related mobility management procedures are not performed.

STANDBY State

In the STANDBY state, the subscriber is attached to the GPRS mobility management. The MS and SGSN have established MM contexts for the subscriber's IMSI. Pages for data or signalling information transfers may be received. It is also possible to receive pages for the CS services via the SGSN.

The MS performs GPRS RA and GPRS cell selection and re-selection locally. The MS executes mobility management procedures to inform the SGSN when it has entered a new RA. The MS does not inform the SGSN of a change of cell in the same RA. Therefore, the location information in the SGSN MM context contains only the GPRS RA identity for MSs in the STANDBY state. The MS may initiate activation or deactivation of PDP contexts while in the STANDBY state. A PDP context will be activated before data can be transmitted or received for this PDP context.

The SGSN may have to send data or signalling information to an MS in the STANDBY state. The SGSN then sends a Paging Request in the routing area where the MS is located. The MM state in the MS is changed to READY when the MS responds to the page and in the SGSN when the page response is received. Also, the MM state in the MS is changed to READY when data or signalling information is sent from the MS and, accordingly, the MM state in the SGSN is changed to READY when data or signalling information is received from the MS.

READY State

In the READY state, the SGSN MM context corresponds to the STANDBY MM context extended by location information for the subscriber on a cell level. The MS performs mobility management procedures to provide the network with the actual selected cell. GPRS cell selection and re-selection is done locally by the MS.

The MS may send and receive point-to-point packet data units (PDUs) in this state. The network initiates no GPRS pages for an MS in the READY state, but pages for other services may be done via the SGSN. The SGSN transfers downlink data to the BSS responsible for the subscriber's actual GPRS cell.

The MS may activate or deactivate PDP contexts while in the READY state. Regardless of whether a radio resource is allocated to the subscriber or not, the MM context remains in the READY state even when there is no data being communicated. A timer supervises the READY state. The MM context moves from the READY state to the STANDBY state when the READY timer expires. In order to move from the READY state to the IDLE state, the MS initiates the GPRS detach procedure.

Paging for GPRS Downlink Transfer

An MS in the STANDBY state is paged by the SGSN before a downlink transfer to that MS. The paging procedure will move the MM state to READY to allow the SGSN to forward downlink data to the radio resource. Therefore, any uplink data from the MS that moves the MM context at the SGSN to the READY state is a valid response to paging.

The SGSN supervises the paging procedure with a timer. If the SGSN receives no response from the MS to the paging request message, it will repeat the paging. The repetition strategy is implemented using the parameter RA paging repetition (RPR), which is set to 3 seconds as a default. The number of repetitions is dependent on the parameter RA paging area (RPA) (1–5). If the MS has still not responded, the page is sent to the whole SGSN paging area. The parameter SGSN paging area defines the maximum number of SGSN wide pages.

The MS will accept pages also in the READY state if no radio resource is assigned. This supports recovery from inconsistent MM states in MS and SGSN. The paging procedure is illustrated in Figure 2.39:

- The SGSN receives a downlink PDP PDU for an MS in the STANDBY state. Downlink signalling to a STANDBY state MS initiates paging as well.
- The SGSN sends a BSSGP (Base Station System GPRS Protocol) Paging Request (IMSI, P-TMSI, Area, Channel Needed, QoS, DRX Parameters) message to the BSS serving the MS. IMSI is needed

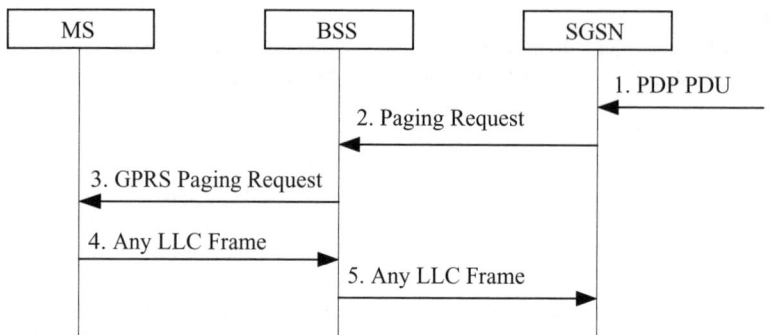

Figure 2.39 Paging procedure

by the BSS in order to calculate the MS paging group. P-TMSI is the identifier by which the MS is paged. Area indicates the routing area in which the MS is paged. Channel Needed indicates GPRS paging. DRX Parameters indicates whether the MS uses discontinuous reception or not. If the MS uses discontinuous reception, then DRX Parameters also indicates when the MS is in a nonsleep mode able to receive paging requests.
- The BSS pages the MS with one Paging Request (P-TMSI, Channel Needed) message in each cell belonging to the addressed routing area.
  Upon receipt of a GPRS Paging Request message, the MS will respond with any single valid LLC frame (e.g. a Receive Ready or Information frame) that implicitly is interpreted as a page response message by the SGSN. When responding, the MS changes the MM state to READY. The response is preceded by the Packet Channel Request and Packet Immediate Assignment procedures as described in GSM 03.64.
- Upon reception of the LLC frame, the BSS adds the Cell Global Identity including the RAC (routing area code) and LAC of the cell and sends the LLC frame to the SGSN. The SGSN will then consider the LLC frame to be an implicit paging response message and stop the paging response timer.

Routing Area Identity

Routing Area Identity (RAI) is a parameter defined by the operator and identifies one or several cells. RAI is broadcast as system information and is used by the MS to determine, when changing a cell, whether an RA border was crossed. If that was the case, the MS initiates the RA update procedure.

The location of an MS in the STANDBY state is known in the SGSN on an RA level. This means that the MS is paged in the RA where the MS is located when mobile-terminated traffic arrives in the SGSN. An RA is a subset of one, and only one, LA, meaning that an RA cannot span more than one LA. An RA is served by only one SGSN. An optimal RA is balanced between the PCH load and RA updates. If the RA size is too large, paging channels and capacity will be saturated due to a limited LAPD (Link Access Procedure on the D channel) A_{bis} or radio interface CCCH paging capacity. On the other hand, with a small RA there will be a larger number of RA updates performed. Based on traffic behaviour of subscribers and the performance of the network (in terms of paging success), it is possible to derive guidelines regarding the maximum number of subscribers per LA/RA.

The routing area dimensioning approach assumes the same rule sets as applied to dimensioning the LA of the existing GSM service. This approach is devised by balancing paging traffic from subscribers and the paging capacity offered by a given paging channel configuration.

An example of the paging capacity of a combined channel configuration with three pages per block is shown in Figure 2.40.

CAPACITY of PAGING CHANNEL

Paging Capacity in Air Interface			
Combined BCCH	[yes/no]	Yes	
Blocks for Access Grant	[0-2]	1	
Number of Pages/Block	[2-4]	3	
Max. Capacity:		25	pages / sec
		91915	Pages per Hour

Figure 2.40 Capacity of the paging channel

The number of pages that are sent by the BTS within an LA/RA indicates the number of Mobile Terminating Calls (MTC) that are being sent to subscribers in the LA/RA. This paging demand is a function of three factors:

- number of MTC;
- number of subscribers in the LA/RA;
- the paging parameters within the SGSN defined by the Operator, RA Paging Repetition (RPR), RA Paging Area (RPA), SGSN Paging Area (SPA).

The higher the number of MT sessions for subscribers in the RA, the higher the number of pages that have to be sent by the BTS in the RA. The number of subscribers also has a strong influence on the number of pages being sent, since with an increased number of subscribers the number of MT sessions will inevitably rise. The success of paging, i.e. the number of times that a paging message has to be re-sent before it is answered, also has a profound effect. It can be seen that with one block reserved for AGCH and assuming three paging numbers per paging block, 25 pages can be sent per second.

The paging traffic observed in a typical network can be found by means of:

- the number of pages per second per user;
- the number of subscribers;
- the paging success ratio.

The signalling load is highly dependent on the parameters. In the same LA/RA, the paging load should be monitored. Note that if there is only one small cell in a given LA/RA, where a combined channel structure is in use, this will be the bottleneck if paging blocking criteria are strictly followed. In other words, the smallest cell in the LA/RA will set the PCH limit.

2.8.3 Parameter Planning in the WCDMA Network

Physical layer procedures

Logical Channels

Logical channels were created to transmit a specific content. There are, for instance, logical channels to transmit the cell system information paging information or user data. Logical channels are offered as a data transfer service by the Medium Access Control (MAC) layer to the next higher layer. Consequently, logical channels are in use between the mobile phone and the RNC. Logical channels are characterised by the specific content to be transmitted: user data (DTCH), control messages (DCCH, CCCH), broadcast data (CTCH) and cell system information (BCCH).

Control Channels (CCH)

Broadcast Control Channel (BCCH)
System information is made available on this channel. The system information informs the UE about the serving PLMN, the serving cell, neighbourhood lists, measurement parameters, etc.
This information is permanently broadcast in the downlink.

Paging Control Channel (PCCH)
Given the BCCH information, the UE can determine at what times it may be paged. Paging is required when the RNC has no dedicated connection to the UE. The PCCH is a downlink channel.

Common Control Channel (CCCH)
Control information is transmitted on this channel. It is in use when there is no dedicated connection between the UE and the network. It is a bidirectional channel, i.e. it exists both in the uplink and downlink directions.

Dedicated Control Channel (DCCH)
Dedicated resources were allocated to a UE. These resources require radio link management, and the control information is transmitted both uplink and downlink on DCCHs.

Traffic Channels (TCH)

Dedicated Traffic Channel (DTCH)
User data has to be transferred between the UE and the network. Therefore dedicated resources can be allocated to the UE for the uplink and downlink user data transmission.

Common Traffic Channel (CTCH)
Dedicated user data can be transmitted point-to-multipoint to a group of UEs.

Common Packet Channel (CPCH)
Similar to the RACH, it is a contention based uplink channel. In contrast to the RACH, it can be used to transmit larger amounts of (bursty) traffic.

Transport Channels (TrCH)
The transport channels determine how the content is organised to be transmitted. The MAC layer uses the transport channels as service for the lower physical layer. The MAC layer is responsible for organising the logical channel data on transport channels. This process is called mapping. The MAC layer determines the used transport format, the used cyclic redundancy check (CRC) length, channel coding (convolutional/turbo, coding rate), etc.

User Dedicated Channel (DCH), common (FACH/RACH)
The MAC layer uses the transport service of the lower layer, the physical layer. The MAC layer is responsible for organising the logical channel data on transport channels. This process is called mapping. In this context, the MAC layer is also responsible for determining the used transport format. The transport of logical channel data takes place between the UE and the RNC.

Dedicated Transport Channels

Dedicated Channel (DCH)
Dedicated resources can be allocated both uplink and downlink to a UE. Dedicated resources are exclusively in use for the subscriber.

Physical Channels (PhyCH)

The physical layer offers the transport of data to the higher layer. The characteristics of the physical transport need to be described. When information is transmitted between the RNC and the UE, the physical medium changes. Between the RNC and Node B, where the interface Iub is discussed, the transport of information is physically organised in so-called frames.

Between Node B and the UE, where the WCDMA radio interface Uu is found, the physical transmission is described by physical channels. A physical channel is defined by the carrier frequency number (UARFCN) and the spreading code (SC) in the FDD mode.

Primary Synchronisation Channel (P-SCH)

The P-SCH uses the first 256 chips of every timeslot. In a P-SCH a primary synchronisation code is transmitted. This is done in every UMTS cell in every timeslot. If the UE detects the P-SCH it has performed chip synchronisation. The P-SCH (as well as the secondary (S)-SCH) is not transmitted under the cell scrambling code, but uses its own predefined code, which is the same for all cells in the UMTS network.

Secondary Synchronisation Channel (S-SCH)

The S-SCH also uses only the first 256 chips of a timeslot. In an S-SCH the secondary synchronisation code is transmitted. There are 16 different secondary synchronisation codes that are organised into 64 different combinations. The 64 combinations are grouped with 64 scrambling code groups, each consisting of 8 scrambling codes.

Common Pilot Indication Channel (CPICH)

CPICH carries a predefined bit/symbol sequence at a fixed rate (15 kbps, SF = 256). It is used for channel estimation and for measurement of the neighbour cells. It is also used in an initial cell search to find the correct scrambling code of a cell. It is also used as the phase reference for most physical channels.

Primary Common Control Physical Channel (PCCPCH)

PCCPCH is the physical channel that carries broadcast channel (BCH) information. It is a fixed rate channel without power control because it must be decoded by all the mobiles in the cell. The channelisation code is fixed by specification and has $SF = 256$. The channel bit rate is 30 kbps but in order to reduce the total interference it is sent alternatives with the SCH giving a 'net' bit rate of 27 kbps. The PCCPCH does not have any pilot bits in the frame because the channel estimation is done using the CPICH.

Secondary Common Control Physical Channel (SCCPCH)

SCCPCH carries two different common transport channels, the FACH (Forward Access Channel) and the PCH (Paging Channel). It is on air only when it has something to transmit. There can be up to three secondary CCPCHs configured. FACH and PCH can be mapped in two different physical channels. In addition, if the Service Area Broadcast (SAB) service is implemented it requires an additional SCCPCH.

Paging Indicator Channel (PICH)

The Paging Indicator Channel (PICH) operates together with the Paging Channel (the transport channel is sent on the physical channel: SCCPCH). The paging indicator is sent on PICH, and the corresponding paging message is sent on the associated SCCPCH. Having one channel for indicators and one for messages provides terminals for an efficient sleep mode operation.

The paging indicator (PI) uses a channelisation with SF = 256. Depending on the paging indicator ratio there can be 18, 36, 72 or 144 paging indicators per PICH frame. To each terminal registered to the network is allocated a paging group that corresponds to a PI. When the mobile detects the PI it decodes the next PCH frame transmitted on the secondary CCPCH. If the PICH is received with low reliability

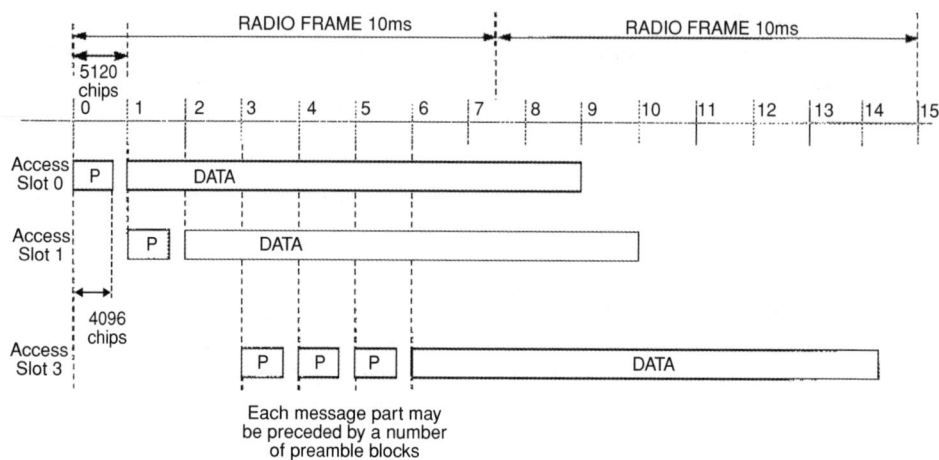

Figure 2.41 PRACH radio frame

then the PCH is decoded. The less the mobile needs to listen to the PICH, the longer the battery life. The drawback is a longer response time for a mobile terminated call.

Physical Random Access Channel (PRACH)
In a PRACH two radio frames are joined to form a '15 TS RACH frame' in a PRACH transmission consisting of transmitting the preamble and a message part (see Figure 2.41). The preamble PRACH is a repetition of a specific signature which maps the channelisation code used for the PRACH data part. There are 16 different signatures and the mobile randomly picks up one of those. In case no acquisition Indication Channel (AICH) is detected it is re-transmitted with a different signature and with higher power. The preamble is 4096 chips long and is made by 256 repetitions of the signature, which is 16 chips long. The preamble PRACH is then scrambled with a PRACH scrambling code related to the scrambling code of the serving cell. For each primary scrambling code (downlink) there are 16 RACH scrambling codes (uplink) associated with it.

The scrambling code use for the message part of the RACH is the same as the one for the preamble. The channelisation code is related to the signature used for the preamble while the SF is decided by the mobile. The signature specifies one of the 16 nodes in the code tree, which corresponds to channelisation with SF 16. The message part can be one or at a maximum two frames long. Because the message is short there is no power control.

Acquisition Indication Channel (AICH)
The AICH is a downlink physical channel with SF 256 in which an echo of the preamble RACH is sent from the WBTS (WCDMA BTS). The WBTS knows that there will be a message part coming and starts to listen to the channellisation code indicated by the signature. At this point the WBTS does not have any information regarding the user.

Dedicated Physical Control Channel (DPCCH)
The DPCCH has a constant bit rate and carries all information needed to keep a physical connection running. On the DPCCH the reference symbols (pilots) are sent for a channel estimation in coherent detection and for signal-to-interference ratio (SIR) estimation in fast power control. The power control signalling bits (transmission power control, or TPC) are also sent. The information of how data are coded is carried on the DPCCH and the Transport Format Combination Information (TFCI) is also sent as it contains information, for example, about bit rate and interleaving.

Parameter Planning

Dedicated Physical Data Channel (DPDCH)
The DPDCH is a variable bit rate channel. The DPDCH bit rate is indicated by TFCI bits on the DPCCH. User data, higher layer signalling, e.g. mobile measurements, active set updates and packet allocations are all sent on the DPDCH.

Downlink Dedicated Physical Channel (DDPCH)
In the downlink direction the DPCCH and DPDCH are time multiplexed. This causes discontinuous transmission.

Uplink Dedicated Physical Channel (UDPCH)
In the uplink direction the transmission is continuous. The DPCCH and DPDCH are I-Q/code multiplexed into the UDPCH.

Planning Idle Mode

Common Channel Power Settings
The transmit power of the common channels is controlled by common channel power setting parameters. The common pilot indication channel (CPICH) carries the pilot bit sequence and is used as the quality estimation channel for handover and cell reselection procedures.

PLMN Selection
The PLMN selection procedure is used to find the network that the mobile is allowed to use its resources. The selected PLMN can be the home PLMN, the network that the subscriber has a subscription with, or any other network that allows the mobile to use its resources. The PLMN search procedure starts in two cases:

1. After power-on. The UE looks for the home PLMN. If the UE has been on another PLMN before switch-off or the previously used PLMN was not the home PLMN, the UE starts the PLMN selection procedure.
2. Periodically. If the UE is roaming (camping in other than home PLMN) it searches for the home PLMN.

In a PLMN search procedure the UE scans through all the carrier frequencies. In the manual mode, the UE gives the user an option to select from a list of found PLMNs where the signal level is considered good enough (RSCP, or received signal code power > -95 dBm). In the automatic mode, the UE automatically selects the PLMN in the following priority order:

1. Home PLMN
2. User controlled PLMN list (SIM)
3. Operator controlled PLMN list (SIM)
4. Randomly one of the PLMNs (RSCP > -95 dBm)
5. The best of the other found PLMNs (highest RSCP)

The frequency with which the mobile searches its home PLMN can be set by a parameter: HomePLMNSeachPeriodTime.

Cell Search
During the cell search, the UE searches for a cell and determines the downlink scrambling code and common channel frame synchronisation of that cell. The cell search is performed in idle mode when the UE makes an initial cell selection or when the mobile makes a cell re-selection. The search procedure consists of three phases: scanning for the primary synchronisation channel, detecting the secondary

synchronisation channel and detecting the cell primary scrambling code using CPICH. The cell search uses primary (P-SCH) and secondary (S-SCH) synchronisation channels, which are transmitted in all cells in the network with the same code sequence. The P-SCH and S-SCH are the only channels not scrambled with the cell primary scrambling code. In addition to synchronisation channels the CPICH is used to find the correct scrambling code of a particular cell.

Scanning for the Primary Synchronisation Channel (P-SCH)
The primary synchronisation channel carries the P-SCH code (PSC). The same code is repeated at the beginning of every slot. Every cell sends an identical code. The receiver correlates the received signal on the selected carrier frequency with the P-SCH code. Each correlation peak corresponds to one cell and each cell transmits the same P-SCH code. The UE selects the strongest peak, which gives the slot synchronisation to the transmitting cell.

Detecting the Secondary Synchronisation Channel (S-SCH) Sequence
At the same time as the P-SCH is transmitting the P-SCH code the S-SCH transmits one of 16 different S-SCH codes. The S-SCH codes are repeated in a predefined sequence. The UE detects the S-SCH codes and correlates the detected S-SCH sequence to possible combinations of S-SCH sequences. From the detected S-SCH sequence the UE detects the starting point of the sequence, which gives the frame synchronisation. The S-SCH sequence of the codes is also used to identify the primary scrambling code group. The UE checks from a predefined table which primary scrambling code group corresponds to the detected sequence of the codes. There are 64 possible S-SCH code sequence combinations that correspond to 64 primary scrambling code groups. After detecting the S-SCH sequence the mobile knows to which scrambling code group the scrambling code of a cell belongs.

Detecting the Cell Primary Scrambling Code from the PCPICH
The PCPICH contains the predefined pilot bit sequence, which is repeated in every frame (10 ms). The primary scrambling codes are divided into 64 scrambling code groups. Each scrambling code group consists of 8 codes, which give 512 primary scrambling codes. After the UE has detected the scrambling code group, it applies the trial-and-error method to detect which one of the 8 codes is the primary scrambling code of the cell. It applies each primary scrambling code from a scrambling code group until it finds which one gives the correct pilot bit sequence. When the UE is able to decode the PCPICH it can also decode the PCCPCH that carries the BCCH. It is then able to read the system information, which enables it to transmit or receive other common channels as well. The PCPICH pilot bit sequence is measured constantly to evaluate the quality of the cell.

The exact implementation of a cell search depends on available preinformation (e.g. the cell primary scrambling code) and is UE vendor specific. The cell search procedure is based on usage of the P-SCH, S-SCH and PCPICH in the cell. The parameters related to the configuration of these channels are the transmit power parameters and the timing offset (T-offset), which configures the timing offset for each cell within a site. The offset is given to make sure that synchronization channels are transmitted at different times in each cell of the same site. Each cell belonging to the same WBTS (WCDMA BTS) should transmit the synchronisation channels with a given offset (T-offset) from each other. If the T-offset is given to all cells and the scrambling codes of a base station belong to the same scrambling code group, there is a gain when detecting P-SCH, but detecting S-SCH in soft handover areas is likely to be unsuccessful.

Camping on a Cell
When the mobile is on idle mode it is camping on a cell and the network knows the mobile location on a location area/routing area basis. The mobile monitors the PCPICH channel to measure cell quality; the PCCPCH is monitored for BCCH information and PICH for possible paging requests. The target for the mobile in idle mode is to camp on the best possible cell. In addition to measured cell quality, cells are also ranked according to suitability.

Parameter Planning

Suitable Cell
A suitable cell provides a normal service to the UE/user. It belongs to a selected PLMN and allows the UE to camp normally on the cell. The UE can receive system and paging information from the PLMN and can initiate a call setup for outgoing calls or other actions from the UE. A suitable cell meets the cell selection criteria (S-criteria).

Acceptable Cell
This provides a limited service (emergency calls). An acceptable cell also meets the cell selection criteria.

Barred Cell
The UE is not allowed to camp on. If a cell is barred the indication is sent in the system information.

Reserved Cell
The UE is not allowed to camp on. If a cell is reserved the indication is sent in the system information.

Cell Reservations and Access Restrictions

There are two mechanisms that allow an operator to impose cell reservations or access restrictions: cell status indication and access control.

The cell status indication and special reservations are sent on the BCCH system information. The cell status controls cell selection and re-selection; if cell selection is not allowed for a user, not even emergency calls are possible. There are three states in system information block 3 (SIB3) reserved for cell reservations: cell barred, cell reserved for operator use and cell reserved for future extension.

Access control controls the connection setup. If the call has already been set up and the mobile is moving in the network the access control has no effect on the connected mode users making handovers from one cell to another. The access control is used for load control purposes to prevent certain services being established in certain cells. Access class specific restrictions are sent in system information block 3. The access control does not affect emergency calls, if not specifically prevented.

Cell Re-selection

The cell (re-)selection procedure is used in idle mode and in connected mode states of Cell_FACH, Cell_PCH and URA(UTRAN routing area)_PCH. The target is to find the most suitable cell to camp on. If the VE does not have stored information on the target cell must use the 'initial cell selection' procedure, or otherwise the 'stored information cell selection' procedure.

Cell Re-selection in WCDMA

The mobile obtains neighbour cell information from the BCCH, which consists of possible cell individual offsets, scrambling code, carrier frequency and radio access technology (RAT). In the cell re-selection procedure the UE tries to find a better cell to camp on. The neighbour cell measurements are started only when the measurement criteria are met. The mobile measures the serving cell all the time, but it measures neighbouring cells only when the serving cell E_c/N_0 falls below the measurement criteria. The measurement criteria can be set differently for different neighbour types:

Same carrier neighbours (intrafrequency)	Sintrasearch
Different carrier neighbours (interfrequency)	Sintersearch
GSM neighbours (intersystem)	SsearchRAT

When the neighbour cells are evaluated it must first be verified that they fulfil the minimum quality criteria, also called the S-criteria. The S-criteria (Squal) consist of a minimum acceptable quality and/or a minimum acceptable level for the neighbouring cell. If a neighbour cell passes the S-criteria the cell re-selection is performed when a neighbour cell is ranked better than a serving cell. The ranking is based on the R-criteria. There are two parameters that affect the ranking: for a serving cell it is possible to define a hysteresis (Qhyst2) that applies on ranking to all neighbour cells and also individual offsets (Qoffset2) for each neighbour cell, which affects only one neighbour cell. Cell re-selection is performed when a neighbour cell is ranked over a current serving cell (Rn > Rs).

The cell re-selection procedure is different when GSM cells are involved in the ranking. The GSM measurements are activated when inter-RAT measurement criteria are met and the cell has GSM neighbours defined. There are no quality (E_c/N_0) measurements and the S-criteria only includes the level criteria (Qrxlevmin). Also, if even one GSM cells the ranking, the R-criteria is always evaluated using signal level measurements; for the WCDMA cells RSCP and for the GSM cells RX. The quality based hysteresis and individual offset values cannot be used; instead level based hysteresis (Qhyst1) and cell individual offset (Qoffset1) values are always used for signal level evaluations.

Measurement Requirements

The UE evaluates the S-criteria for the serving cell at least every DRX cycle. If the UE has not found any suitable cell in the neighbour list, it will start the PLMN selection procedure 12 seconds after the serving cell did not meet the S-criteria. If the measurement criteria (Sintrasearch, Sintersearch, SsearchRAT) are met the neighbour cells are measured based on TmeasureFDD and TmeasureGSM parameters. The measurement frequency is different for different types of neighbours:

Intrafrequency neighbours: every TmeasureFDD
Interfrequency neighbours: every (Ncarrier-1)*TmeasureFDD
GSM neighbours: every TmeasureGSM

The UE evaluates the cell re-selection criteria for the cells, which have new measurement results available.

System Information

After the mobile has decoded PCPICH it can also decode PCCPCH, which is the physical channel that carries BCCH. In BCCH the network information is sent on system information blocks (SIBs). Different system information blocks carry different information elements. For example, system information block 7 (SIB7) carries information about the uplink interference situation, which is needed to estimate transmission power in open loop power control. Due to the fact that the radio environment changes constantly SIB7 needs to be transmitted often. As another example, SIB11 carries information about neighbours needed for cell re-selection and handovers. The mobile needs to read SIB11 after cell re-selection, but after that it does not need to update neighbour information.

The master information block (MIB) carries reference and scheduling information to all SIBs; different SIBs have different characteristics regarding their repetition rate and requirements. The MIB also indicates whether the information in one of the SIBs has changed and the mobile needs to read that SIB again in order to update an information element.

Connection Setup

Random Access Procedure

The random access procedure is used to establish the RRC connection setup when the mobile wants to move from idle mode to connected mode or if the mobile is in connected mode (Cell_FACH, Cell_PCH, URA_PCH) and it is to transmit information on uplink. The PRACH and AICH channels are involved in the PRACH procedure.

In the random access procedure the UE sends a trial transmission burst called preamble (see Figure 2.42). The transmission power of a preamble is estimated using open loop power control. If the mobile does not receive an indication from the BTS that it has received the preamble, the mobile will ramp up the transmission power and send another preamble. The power ramp-up step size is configured by a parameter. Only after receiving confirmation on the AICH that the BTS has received the preamble does the mobile send the RACH message part. These is also an offset given in between the last preamble

Parameter Planning

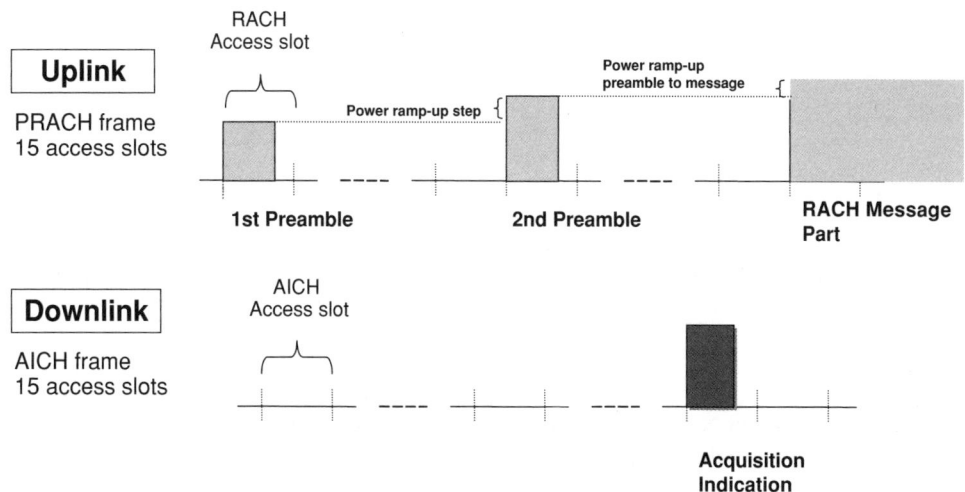

Figure 2.42 Power ramp-up process

sent and the message part to guarantee reception of the message. As the RACH message is short no power control is used. The mobile sends a predefined number of preambles, ramping up power in between every step. If the mobile has ramped up the power a predefined number of times and receives no confirmation from the base station, if starts the sequence from the begining and repeats it, but the sequence of ramping up the TX power is repeated only a predefined number of times. The UE will exit the random access procedure if it does not receive a response from the base station before it runs out the maximum number of times it is allowed to repeat the power ramp-up procedure.

PRACH and AICH channels are structured to a 20 ms frame with 15 access slots. The PRACH frame is divided into two access slot sets: access slot set 1 (access slots 0 to 7) and access slot set 2 (access slots 8 to 14).

Random Access Procedure: Preamble and Message
The preamble is 4096 chips long. It consists of 256 × 16 chip signatures. There are 16 possible signatures in each cell. The uplink scrambling code of a cell is selected from 16 dedicated UL scrambling codes.

The RACH message consists of one or two radio frames (10–20 ms). The scrambling code number is the same as that used for the preamble. The message part can have different spreading factors. The spreading factor is based on a selected signature and spreading factors between 256 and 32 can be used.

Random Access Procedure: AICH
Each acquisition indication channel (AICH) slot carries a separate acquisition indication for each signature (total 16). The acquisition indication has three different values:

- 0, no indication → the BTS has not received this signature;
- 1, positive indication → the UE is allowed to transmit a message;
- −1, negative indication → the UE must exit the RACH procedure.

Each slot contains a 32 symbol waveform, which is scrambled with the cell primary scrambling code.

Random Access Procedure: UE Required Information
The preamble scrambling code is transmitted on the BCCH (SIB5), based on the parameter. The allowed signatures are based on the BCCH (SIB5) and the UE Access Service Class (ASC). For each preamble one

of the allowed signatures is randomly selected. The UE is allowed to transmit the preamble on PRACH access slots based on the BCCH (SIB5) and the UE ASC. For the first preamble one of the allowed access slots in the next full access slot set (1 or 2) is randomly selected. For re-transmission the first allowed access slot is selected.

Paging Procedures

The paging procedure uses two physical channels. The paging indication (PI) is sent over the paging indication channel (PICH). The paging indication shows that the UE should read the corresponding message on the secondary common control physical channel (SCCPCH).

Overview

The paging procedure is required to enable the network to contact the UE in idle mode or when in connected mode the Cell_PCH and URA_PCH states. In idle mode the UE is paged over the whole location area (LA) in the case of CS paging or the routing area (RA) in the case of PS paging. This means that in idle mode the paging indication and message is sent on all cells belonging to the LA or RA. In connected mode the UE is paged over one individual cell if the UE is in the Cell_PCH state or over the UTRAN registration area if the UE is in the URA_PCH state.

There are three sources that might start a paging procedure:

- the circuit switched core, if the UE is registered but not connected;
- the packet switched core, if the UE is registered but not connected;
- the RNC in the case where the UE is in the CELL_PCH and URA_PCH states.

The paging procedure has two phases: the UE detects an indication on the PICH and the UE decodes the paging message from the SCCPCH and checks whether is for that UE. Cells can have one or more SCCPCHs configured. There can be one SCCPCH dedicated to carry paging.

PICH

The PICH is a fixed rate physical channel, with a spreading factor of 256. It is used to carry the paging indicators. One PICH radio frame of length 10 ms consists of 300 bits. Of these 300 bits 288 bits are used to carry paging indicators. Discontinuous transmission is used and the remaining 12 bits are not transmitted. In each PICH frame N_p paging indicators are transmitted, where $N_p = 18, 36, 72$ or 144; N_p is a radio network parameter.

The UE has to check the PI from the PICH on every discontinuous reception (DRX) cycle length. The DRX cycle length is $2k$ radio frames, where k is the smallest of the defined DRX cycle length coefficients. There can be k defined for the CS core (kCS), for the PS core (kPS) and also if in the connected mode for UTRAN (kRAN).

The UE has to check the PI from the PICH on every paging occasion. The paging occasion is calculated using the following formula:
(IMSI div K) mod (DRX cycle length) where IMSI is the subscriber identity and K is the number of SCCPCHs configured for paging. The paging indication is calculated for one particular UE: (IMSI mod 8192) mod N_p.

SCCPCH

The PICH and SCCPCH configurations are sent on SIB5. The paging procedure has two phases. The UE detects an indication on the PICH and then decodes the paging message from SCCPCH and checks whether it is for this UE. Cells can have one or more SCCPCHs configured. There can be one SCCPCH dedicated to carry paging only.

Parameter Planning

Handover Control

WCDMA networks have large numbers of different types of handovers taking place compared to GSM networks. A distinctive feature of a WCDMA system is the introduction of a soft handover (SHO) – a situation where the UE is connected to two or more sectors simultaneously. A special case of a soft handover is a softer handover, where the UE is connected to two or more sectors from the same Node B.

An intrafrequency hard handover is also possible in the case where an SHO is not possible. Usually an intrafrequency handover is used as the 'last chance' in the case where an SHO is not possible. If neither an SHO nor an intrasystem hard handover is possible, but there are measurement reports confirming that a neighbour cell E_c/I_0 is stronger than a serving cell, a forced RRC connection release takes place.

In the case of a lack of WCDMA coverage, poor E_c/I_0, quality deterioration reports from the outer loop power control function or high transmit power from either the UE or WBTS, interfrequency or intersystem handover is triggered. The interfrequency and intersystem handovers require measurements on another frequency band. If the UE is not designed with a dual receiver the needed performing measurements require compressed mode. In compressed mode the continuous transmission of the WCDMA is interrupted and transmission gaps are used to perform measurements on another frequency band. If a neighbouring cell is found and decision criteria are fulfilled the interfrequency or intersystem handover is performed.

Also FDD (frequency division duplex)/TDD (time division duplex), TDD/FDD and TDD/TDD are specified (TR 25.922), but as they are not implemented in existing networks they are not covered in this chapter.

Different Types of WCDMA Handovers
- FDD soft/softer handover (intrasystem intrafrequency HO)
- FDD interfrequency (intrasystem interfrequency HO)
- FDD/TDD handover (intersystem HO)
- TDD/FDD handover
- TDD/TDD handover
- Handover WCDMA–GSM (intersystem HO)
- Handover GSM–WCDMA (intersystem HO)

A specific (soft) handover situation is a crossing of an RNC border, when serving RNC resources need to be relocated.

Soft Handover
Soft handover (SHO) is a situation where the UE is connected to two or more sectors simultaneously. A definition to separate soft handover from softer handover can be that soft handover is a situation where the UE is simultaneously connected to two or more sectors from different NodeBs. The NodeBs participating in soft handover do not necessary need to be under the same RNC.

The SHO is a necessary and interference reducing feature of the WCDMA system. However, if the SHO area is more than 40 % of the dominance area it is consuming a large amount of I_{ub}, code and NodeB resources, and so should be limited. On the other hand, if the cell separation does not leave much overlap and the SHO area is less than 30 % of the cell dominance area the risk of having coverage holes increases. The lack of cell overlap increases the transmit powers when the UE is close to the cell edge, compared to the situation where the mobile is connected to two cells where SHO gain allows transmit powers to be reduced. The third problem relates to small SHO areas where there is a risk taht the SHO will fail. If the serving cell E_c/I_0 reduces very rapidly when the new cell enters the reporting range, there is a risk that the SHO will fail due to the serving link failing so fast that the UE measurement reports are not received by the WBTS and the active set update command is not received by the UE. As a general guideline the SHO area should be around 30–40 % of cell dominance area.

Soft handover takes place when the radio conditions are such taht the mobile moves in the network and the radio conditions change. The mobile is measures and evaluates the situation according to the measurement control information. In the measurement control message the mobile is informed of the measurements that need to be performed and the conditions that require the mobile to report the situation to the network. If the triggering conditions are fulfilled, the mobile sends a measurement report to the network.

In SIB11/12 or in the measurement control message the network sends information needed to measure the neighbouring cells. In this case triggering conditions are fulfilled and the mobile sends a measurement report requesting an active set update. The measurement control message contains two different types of information. Firstly, the network can modify the measurements and triggering criteria of the UE. Secondly, a neighbour list is sent on a measurement control message. In the case of SHO the neighbour list needs to be combined from individual cell neighbour lists participating in the SHO.

The UE evaluates network quality (E_c/I_0) to determine whether handover should be triggered. If the radio conditions change and triggering conditions are fulfilled the UE sends an event to the network. The event tells what conditions have triggered the handover. When the triggering conditions are valid the UE sends events in regular intervals; this is called periodical measurement reporting. If the UE sends only information that a defined measurement triggering condition is true, this is called event based measurement reporting. As the conditions to trigger the SHO are mobile evaluated, and it is the mobile requesting the SHO, the SHO is a mobile evaluated handover (MEHO).

The UE measurement reports are sent to the RNC and the handover decision is made in the RNC. If the admission control or resource manager functions of the RNC confirm that there are resources to add another radio link, the RNC allocates a new radio link from a requested neighbour to the UE. The availability of channel cards is also needed due to extra transmissions across I_{ub} (BTS-RNC). If resources are available, the RNC sends an active set update command to the UE using the existing radio link(s). The UE confirms that it has received the command and starts receiving using the new radio link. As there is now a new serving cell the network sends a new combined neighbour list to the UE to monitor. The neighbour list defines the possible cells that the mobile can try to make an SHO.

The objectives of an SHO are:

- optimum fast closed loop PC as the terminal is always connected to the; best cell;
- seamless HO with no disconnections of the RAB (radio access bearer);
- Diversity gain by combining the received signals from different sectors;
- Better coverage;
- Less transmission power needed (better capacity);
- Both RT (real time) and NRT (nonreal time) RABs are supported;
- Intrafrequency HO is divided in to soft handover (SHO), and softer handover (SfHO), which can be made simultaneously.

Softer Handover (SfHO)

A special case of soft handover is softer handover, where the UE is connected to two or more sectors from the same WBTS. There are different scrambling codes used for each DL radio link transmission. In the UL the RAKE receiver is used to combine signals from two sectors; thus additional RAKE receiver channels need to be allocated. The WBTS combines the signals and no extra transmission across I_{ub} (WBTS-RNC) is needed. However, in downlink two or more sectors are transmitting and the mobile needs to allocate resources for each radio link.

The power control function has only one power control loop active in the case of softer handover. The same power control command is sent in all softer handover radio links. This is different from SHO where each DL radio link (DPCCH) transmits power control commands independently. The SfHO requires additional channelisation codes as well as additional DL transmission power.

Softer handover occurs in 5–15 % of the network area. As a planning tool it is easy to spot softer handover areas and there is clear-cut dominance in areas between sectors within the same Node B. If SHO areas are visualised the areas between the sectors are where SfHO occurs.

SHO Process

The soft handover process controls triggering conditions. A triggering condition is the measured E_c/I_0. If it becomes lower than the defined threshold, the triggering conditions are true. There is also a mechanism called 'time to trigger', which is a time that triggering conditions need to be true before the UE sends a reporting event. A reporting event tells which cell has triggered an event. In the same measurement message the UE also sends measurement results from other measured cells to the RNC. When the RNC receives a reporting event, it allocates/removes an SHO connection if there are available resources. The RNC allocates them and sends an active set update message to the UE. The active set update message is followed by an active set complete message from the UE.

The reporting range is the measured reference value based on E_c/I_0 from which the UE evaluates the triggering thresholds. The reporting range can be calculated from the best cell in the active set or from the sum of all cells that are included in the active set. Which option is used is set by the ActiveSetWeightingCoefficient parameter.

The soft handover reporting events are as follows:

1A. When one of the neighbouring cells becomes better than the reporting range – an offset from the best cell in the active set defined by addition of a window parameter. Event 1A is sent when the triggering conditions stay true for the time-to-trigger period defined by the addition time parameter. Event 1A requests the RNC to add the cell that triggered the event to the active set.
1B. When one of the active set cells becomes lower than the reporting range – this offset is defined by the drop window parameter and is typically 2 dB lower than the window addition. Event 1B is sent when the triggering conditions stay true for the time-to-trigger period defined by the drop time parameter. Event 1B requests the RNC to remove the cell triggering event from the active set.
1C. When there are already a maximum number of allowed cells in the active set and one of the neighbouring cells becomes better than the worst cell in the active set. There is an offset given for the worst cell in the active set using a replacement window parameter. The neighbouring cell must stay better than the worst cell in the active set + the given offset for a time-to-trigger period, defined by the replacement time. If these conditions are true the UE sends a measurement report requesting a replacement of the worst cell in the active set with a new cell that triggered event 1C.
1D. This is an event indicating that the best cell of an active set has changed. The best cell is used as the reference point for addition and the drop window. Event 1D is informative and allows the RNC to send new information, e.g. to update measurement control or the service area code.

Parameters to Optimise SHO

Addition/drop window size has a large impact on SHO areas. The addition window sets when the SHO is triggered; a small value for the addition window would trigger an SHO already close to the best cell in the active set. A drop window sets when the SHO branch is removed; a large value would keep the SHO connections far from the best cell in the active set. The relative difference between the addition and drop windows would set the size of the SHO area.

The cell individual offset (AdjsEcNoOffsett) can be used to extend or limit the SHO area for an individual neighbour. With a positive offset the SHO is taking place earlier and with a negative offset the SHO is delayed.

In the case where the cell is an unreliable reference point for SHO triggering with a disable effect on the reporting range (AdjsDERR) parameter, the cell can be set so that it does not have an impact on the reporting range.

Inter-RNC Soft Handover

In the case where a mobile crosses an RNC border the mobile needs to perform an inter-RNC handover. If an I_{ur} link is defined between the RNCs, a soft handover is possible, but in the case of a missing I_{ur} link the mobile makes an intrafrequency hard handover.

In the case of an inter-RNC SHO the mobile is controlled by a serving RNC (SRNC). The RNC that allocates resources but does not make decisions is called a drift RNC (DRNC). When the mobile crosses an RNC border the drift RNC becomes the serving RNC by a process called SRNC re-location. If SRNC re-location is not supported the original SRNC controls the DRNC resources, even after the UE has crossed the RNC border and no SRNC cells are in the UE active set. This process of not re-allocating RNC resources is called anchoring.

Intrafrequency Hard Handover

An intrafrequency hard handover is also possible a where an SHO is not possible. Usually intrafrequency handover is used as a 'last chance' in a case where an SHO is not possible. An intrafrequency hard handover is used, for example, when an inter-RNC SHO is not possible. This occurs if the I_{ur} interface is not implemented.

The decision to make an intrafrequency hard handover is made by the RNC. There is a threshold margin defined for a neighbouring cell. The neighbouring cell E_c/I_0 must be better than the serving cell E_c/I_0, with at least the amount of threshold margin to perform an intrafrequency hard handover. The margin is used to avoid the 'ping-pong' effect, where the UE selects back and forth between the two cells. Also the threshold margin has been defined such that the UE first tries to make an SHO.

The intrafrequency hard handover causes a temporary disconnection of the RT RAB. In the case of NRT RABs the disconnection is transparent to the user.

Interfrequency Handover and Intersystem Handover WCDMA $\rightarrow$ GSM

All the interfrequency and intersystem (or inter-RAT) handovers are hard handovers. The interfrequency and intersystem handover mechanisms follow a similar pattern. The decision to start interfrequency/RAT handover measurements is done by the network. Both of these hard handovers follow a similar network evaluated handover (NEHO) procedure.

The network evaluated handover procedure consists of three steps: measurement triggering criteria define when interfrequency/RAT measurements start; the RNC sends measurement parameters to the UE to command how measurements are done; the UE then performs measurements and reports according to measurement reporting criteria. Finally, the RNC commands handover when the handover criteria are fulfilled.

Measurement Triggering

Implementation of different measurement triggering criteria in the network is vendor specific. Possible measurement triggering criteria include:

- Best active set cell RSCP (UE measurement report)
- Best active set cell E_c/N_0 (UE measurement report)
- Uplink DCH quality (RNC measurement)
- Downlink DCH quality (UE measurement report)
- Cell load (RNC measurement)
- Distance
- UL TX power (UE measurement report)
- DL TX power (BTS measurement)

The UE can be ordered to do different measurements to support interfrequency/RAT measurement triggering:

- Event 1F. The best active set cell RSCP or E_c/N_0 falls below the set threshold.
- Event 5A. The DL Dedicated Physical Channel (DPCH) quality falls below the set threshold.

Parameter Planning

- Event 6A,6C. The UE transmission power is above the set threshold of the maximum transmitted power.

The threshold for each triggering event can be configured by the event triggering threshold and time-to-trigger parameters.

Measurement Control

In a measurement control the RNC sends measurement parameters to the UE to command how measurements are done. The UE then performs measurements and reports according to measurement reporting criteria. The measurement control for the interfrequency and intrasystem handover triggering mechanism is similar. There are different parameters for each type of handover, but the measurement triggering and measurement control functionality works in a similar manner. The interfrequency and intrasystem measurements require the compressed mode. The measurements on an adjacent carrier are done during the transmission gaps. The possible control parameters include the reporting interval, the number of measurements the mobile is allowed to make and penalty timers after an unsuccessful interfrequency handover (IFHO) or intrasystem handover (ISHO) attempt.

Handover Decision Criteria

The RNC commands handover when the handover criteria are fulfilled, but there can be different decision procedures depending on the triggering criteria. For example, if the triggering condition E_c/I_0 of the serving cell is poor the target cell E_c/I_0 is ranked against the serving cell, but if the target cell is better the RNC commands IFHO to take place. In the case where the source cell RSCP falls below the triggering condition the handover decision algorithm compares the source and target cell RSCPs. The RNC commands handover when the handover criteria is fulfilled.

Connected Mode

Power Control

Transmission power is one of limiting resources of the WCDMA system. Common channels (CCHs) are transmitted with fixed output power and the power control algorithm minimizes the transmission power for dedicated channels. The common channels are transmitted with fixed output power. They consume a fixed amount of transmission power resources. It can be seen in the example in Figure 2.43 that the used transmission power recourses for common channels can only take 6 W out of the 20 W total transmission power available in the WBTS.

Power Control Algorithms

Power control algorithms minimise the transmission power used for dedicated channels. The closed loop power control adjusts the transmission power.

Fast Closed Loop Power Control

The closed loop power control adjusts the transmission power 1500 times a second to keep the power at the minimum required level. The closed loop power control uses the signal-to-interference ratio (SIR) to adjust the transmission power. It measures the SIR independently for each connection and adjusts the transmission power an every timeslot basis. The measured SIR is compared with the SIR target. If the target is not reached the transmission power is increased; otherwise the transmission power is decreased.

In the downlink direction all the NodeBs participating with the SHO connection measure their SIR and determine their power control command. They independently adjust their transmission power according to the power control commands (TPC) received from the UE.

Total power used for common channels:				
Common channels		Power / CPICH	Power	on/off
CPICH	33 dBm		2.00 W	1
SCH1	30 dBm	-3 dB	1.00 W	1
SCH2	30 dBm	-3 dB	1.00 W	1
PCCPCH	28 dBm	-5 dB	0.63 W	1
SCCPCH	33 dBm	0 dB	2.00 W	1
PICH	25 dBm	-8 dB	0.32 W	1
AICH	25 dBm	-8 dB	0.32 W	1
		Total	6.62 W 38.21 dBm	during SCH
			5.25 W 37.20 dBm	during PCCPCH
			5.39 W 37.32 dBm	averaged

Figure 2.43 Total power used for CCHs

To determine the power control command in the uplink direction there is an SIR target set for each cell in the active set. The UE lowers its transmission power when all the SHO connections fail to reach the SIR target. This means that transmission power is increased only if all connections are below the set SIR target.

Outer Loop Power Control

The outer loop power control algorithm adjusts the signal-to-interference ratio (SIR) target. The SIR target is used by the closed loop power control as a reference value for transmission power adjustments. The outer loop power control measures the block error rate (BLER), indicating the transport channel quality. There are individual BLER targets set for each transport channel; if the BLER target is not reached, at least for one of the transport channels, the outer loop power control increases the SIR target for the dedicated physical channels. Adjustment of the SIR target would make the closed loop power control algorithm adjust the transmission power of the dedicated physical channels.

There are independent outer loop power control entities in both the uplink and downlink directions. The outer loop power control entity in the UE receives the BLER target, the initial SIR target and the minimum and maximum values for the SIR target in the radio bearer configuration message.

Packet Data Connection

Packet data is one example of non-real-time service. As currently the applications using packet data vary from email client, Internet surfing to music download the services use the packet data connection. Depending on the service requirements (QoS), the corresponding physical resources are allocated. In contrast to the real time services, the packet data as non-real-time services can momentarily lower their bit rate. Also in a case of inactivity the service can use common channels (Cell_FACH) to transmit small amounts of data or can, in the case of inactivity, use the idle mode, as in packet transfer states (Cell_PCH or URA_PCH) where no data are transferred.

Downgrading the bit rate for a packet data connection takes place when the network is having a high load, and resources used by non-real-time services are needed for real time users or higher priority

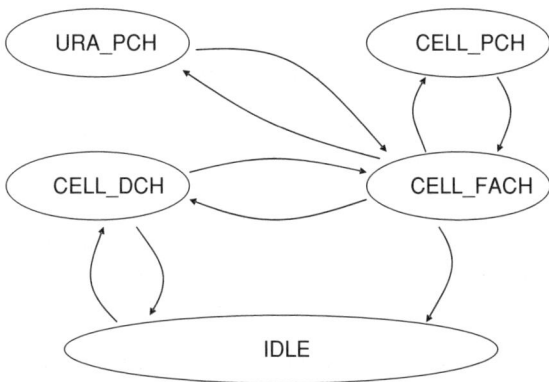

Figure 2.44 Packet data transfer states

non-real-time services. The target and overload thresholds define the operating space of the packet scheduler – the RRM functionality that controls allocated packet data bit rates. In the case of an overload situation packet data services are downgraded until the measured load reaches the load target. If the load is on a marginal load area no action is taken as the non-real-time services are not allowed to be upgraded and no downgrading takes place. On a feasible load area packet data services are upgraded if there is a capacity request indicating that one or more non-real-time services requires more capacity.

Figure 2.44 shows the different packet transfer states. In idle mode no service has been established and no radio bearer has been established. In Cell_DCH the dedicated physical channel carries the data and the throughput can vary according to allocated resources. In Cell_FACH the common channels (FACH on the downlink and RACH on the uplink) are used to transmit small amounts of data. The UE moves first to the Cell_DCH state when there is a date to be transferred; if there is a period of inactivity both in the uplink and downlink the dedicated channel is released and the UE moves to the Cell_FACH state. From Cell_FACH the UE moves back to Cell_DCH if the data buffer becomes full and a capacity request is sent. The UE moves from Cell_FACH to Cell_PCH or URA_PCH if a prolonged period of inactivity triggers an inactivity timer. The selection to move to URA_PCH is made if the mobile has been changing cells several times in the location update period; if the UE stays within a cell then Cell_PCH is selected. In Cell_PCH the UE camps on one cell area and indicates using a cell update message whether it has data to transfer or if it changes a cell. On URA_PCH the UE mows within a URA area and only sends an update if a URA area changes or it has data to transfer. In both Cell PCH and URA PCH the network indicates that it has something to transfer by sending paging to make UEs move to the Cell_FACH state.

Load and Admission Control

Load control functionality in every Node B measures the existing load. In the Nokia system the load measurements are total received UL interference on the uplink and total transmitted power in the downlink. The load control functionality in the WBTS transmits the load information about the UL and DL loads to the RNC. The frequency of how often the load information is sent by the BTS is controlled by the radio resource indication (RRI) period. The UL load is the total received power and the DL load is the total transmitted power.

Admission control (AC) functionality is designed to guarantee the quality for existing users at the expense of new users entering the cell. In the case of resource limitation AC prevents new CS users entering the cell. Admission control works independently, both in the uplink and downlink directions. In addition to the load radio interface, admission control is based on the available code, hardware channel and transmission resources. If any resource has a shortage the admission control takes preventive action.

The most important parameters related to load and admission control functionality are the load thresholds, which determine when the admission control actions take place. The wideband power based load approach defines the received uplink interference as the uplink load; the corresponding downlink load measurement is the total transmission power of the base station.

The load target in the uplink direction is defined as the offset from the noise floor. The uplink noise floor is the measured uplink interference when the system is considered to be unloaded. The noise rise is the amount of interference generated by the load. For example, if the uplink load target is set to be 4 dB the uplink interference can increase by 4 dB from the unloaded situation until the load control actions start taking place. In addition to the load target the overload threshold is also defined. The overload is defined as the load target + an overload offset.

The load target for the downlink direction is defined as the amount of transmission power of the base station. For example, if the total transmission power of the base station is 43 dB the load target could be defined as 41 dB; after the threshold has been exceeded the load control actions start taking place. A similar overload threshold as in the uplink direction is also defined in the downlink direction. This defines when overload actions start taking place.

The load thresholds divide the traffic into four classes in both the UL and DL directions. The area below the load target is called the feasible load, which can be further divided into two feasible load areas. The area above the load threshold but below the overload area is called the marginal load area. The load target + the overload offset define the overload area. The fifth traffic class is when the network is considered to be unloaded and the measurements can be used to tune the noise floor in the uplink direction.

Traffic Classes
Unloaded (network is unloaded)
Feasible load 1 (below the load target–overload threshold)
Feasible load 2 (below the load target)
Marginal load (above the load target)
Overload (above the load target + the overload threshold)

Admission Control
The admission control controls the new users entering the cell. In the case of resource limitation it blocks the new user entering the cell. This can be a user requesting a call to be established or a user trying to come in with a handover. The service of existing users is considered to be more crucial than letting a new user access a cell. If admission control is not in use a new user is allowed to drop existing users from the cell edge. Admission control functionality is needed to control resources in an organised manner. Admission control functionality has a central role in radio resource management: it controls the entry of a new user, provides the initial target values for PC and updates the load estimations for LC.

AC Process
- Quality requirements for the radio bearer
- Load (power) increase estimations
- Admission decision
- Load chance report to load control functionality
- Radio bearer establishment

Quality Requirements for the Radio Bearer
The UE requests a service with a particular quality-of-service (QoS) profile. The RNC forwards the request to the CN and the allowed QoS profiles for the user are verified. There is a QoS profile stored under the HLR. If the user is allowed to be allocated such a service, the RNC uses mapping tables to derive the target BLER for the service.

Parameter Planning

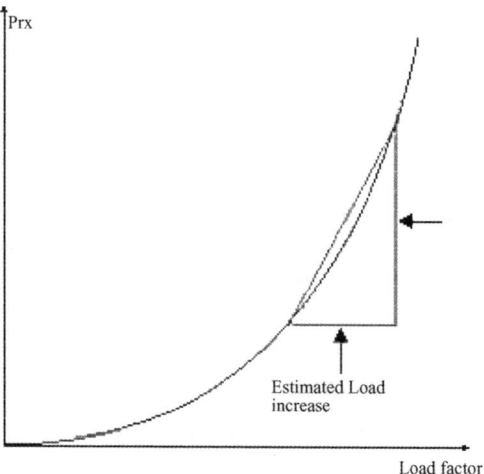

Figure 2.45 Load versus power curve

Load (Power) Increase Estimations
In the wideband power based load estimations the received uplink noise is considered as the uplink load and the total transmitted power is considered as the downlink load. The AC estimates the load increase, and based on the load increase and current load it admits or denies the establishment of a new radio bearer. This is done independently in the uplink and downlink.

Load Increase Uplink
The load increase for the uplink is estimated using a proprietary estimation algorithm. The change in the load factor is calculated by an AC based on the new bearer E_b/N_0 requirement. For an increase in load there is a corresponding noise rise, which that is caused by a noncontrollable load. The noise rise is calculated using the load versus power curve, as shown in Figure 2.45.

Load Increase Downlink
In the downlink directions the mobile measurements can be used to calculate the required transmission power of a new radio bearer. The required power is calculated using the following formula:

$$P_{tx} = \frac{\rho R}{W} \left(\frac{1}{\rho_c} P_{tx,CPICH} - \alpha P_{tx\_total} \right) \qquad (2.70)$$

where

$P = E_b/N_0$, required E_b/N_0 of the connection
$P_c = E_c/I_0$, Signal-to-interference ratio per chip of the CPICH measured by the UE
W = the chip rate
R = bit rate
P_{tx_total} is measured by the base station (and reported back to the RNC in an RRI)
$P_{tx,CPICH}$ = CPICH power (determined by PtxPrimaryCPICH)
α = orthogonality factor

Admission Decision
The admission decision is done independently for the uplink and downlink and the admission control algorithm has two conditions that must be true before a new bearer can be admitted:

1. The current load must not be on overload area above the load target + the load offset.
2. The noncontrollable load must not be on the marginal load area above the load target.

The noncontrollable load is the load that is not caused by controllable packet data users. The noncontrollable load is calculated from load control functionality. The noncontrollable load = the total measured load − the load of controllable packet data users.

If the radio bearer is admitted the admission control reports the load chance to load control functionality and the radio bearer is established. The admission control sets the initial SIR target and minimum and maximum SIR target values for the outer loop power control. It also sets the initial transmission power. The allowed Transport Format Combination Sets (TFCS) used by the packet scheduler is also determined by admission control.

2.9 Radio Network Optimisation
2.9.1 GSM Radio Network Optimisation Process

Once some hundreds of sites are on air, it becomes necessary to perform optimisation on the network in order to maximise benefits while minimising capital and operation costs for operators. This section, in fact, deals with all aspects of optimising a GSM network starting from standard operations and ending with specific trials, studies and fine-tuning. Before the network is commercially launched, the radio network optimisation process starts and then continues during the life of the network.

The main application in GSM technology is mobile telephony. Operators are worried about the quality perceived by the end user. Their competitiveness relies very closely on this aspect in parallel with the pricing policy.

GSM standards offer the possibility of monitoring the network by computing statistics. Depending on the type of network management system, either in the BSC or in the BTS, each cell reports thousands of statistics about all relevant behaviours (number of attempts, failures, successes, during call, handover, setup, etc.). These statistics are reported to the Network Management System (NMS) as counters. To facilitate interpretation of the behaviour, a set of key performance indicators (KPIs) is defined out of formulas using pure counters. Each operator chooses its own KPIs and sets, according to specific criteria, some objectives to be met in order to achieve a good end user perception of the service offered and also in order to benchmark one network with other operators.

Another aspect that is important in the optimisation phase deals with drive tests. In fact, while statistics give a general idea of the cell's behaviour at a certain period, field measurements give a one instant scenario of one area's behaviour during a call. Different tools can be used to perform drive tests. Each specific tool is able to report measurements such as MOS (mean opinion score), FER (frame erasure rate) and BER (bit error rate) in addition to standard reporting at the signal level, quality and site information (cell identity, BCCH, mobile allocation list, best neighbours, etc.).

Statistics and drive tests are the main methods used to monitor the network's performance. However, other specific methods can also be used.

Tracing catches one object's behaviour (TRX, cell, BTS or BSC) during a certain period and regarding a specific event (SDCCH allocation, conversation phase of a voice call, etc.) or a set of successful events (IMSI attach, paging, call setup, location update, etc.). Alarm monitoring, transmission network auditing and network switching subsystem (NSS) performance follow-up are also important in the sense that they give an idea of hardware problems or parameter errors, which definitely affect the end user perception of the network.

After deep analysis, actions are then taken to correct and improve performance. All the above-described methods help the optimisation engineers to identify the origin of the problem from the office while applying several analysis methods. Another aspect is, however, very important: field knowledge. Correct site re-engineering is the basis for a good performing network. Frequency planning review is also a key step in the process.

Optimisation is a running process that overlaps with roll-out activities. A radio network planning team may be requested to advise on other network aspects than pure performance monitoring. For example, with the introduction of new BSC elements, a BSC re-hosting plan requires an update. Another example is in order to save on CAPEX (capital expenditure), operators may resort to a re-structuring of TRX elements. TRX are then taken out from poor traffic areas and re-deployed in high traffic areas. The NWP (network planning) team is also concerned with the introduction and qualification of new features.

At the end, it is worth noticing that with complication of networks (GPRS, EDGE, UMTS, Dual Band, etc.) basic optimisation steps need an automation process in order to give free time for more valuable tasks for the engineers.

Statistics

Network planning optimisation consists of various operations, all leading to the improvement of KPIs. In this section, statistic aspects are tackled and only general notions are presented. More details are given later.

Each cell of the network should be operational intrinsically but also integrated in the network and interoperational with other cells. That is why two types of statistics are used to monitor: statistics on a cell level and statistics on an adjacency/area level.

The same statistics can be extracted at a cell level or at an area level with some specific adaptation for each case. For example, the drop call formula for a cell will contain a number of handovers in the denominator. For an area level, handovers are considered as null except in area borders (in average a handover goes from one cell of the area to another cell inside the same area), which is why the drop call formula contains only a number of calls in the denominator.

In fact, input data for starting optimisation are KPI values in a certain area. Depending an whether the area KPI is greater or less than the target, troubleshooting on a cell basis starts and statistics can be extracted weekly, daily or even on an hourly basis from the NMS. The call setup failure (CSF) and drop call rate (DCR) are the main KPIs relevant to operator losses. People who cannot access the network cannot call and then do not generate traffic and people who are dropped during a call are prevented from speaking more and generating more profit. These two rates can be computed according to various formulas according to what the operator wants to represent. In order to interpret properly and improve DCR and CSF values, other statistics should be monitored such as traffic, timing advance, TRX quality, SDCCH drop, handover distribution, handover failures, etc.

Statistics present the advantage of providing deep analysis to the cell level. On the other side, they present the disadvantage of not giving any geographical approach. This last issue is well mastered through drive testing.

Drive Testing

Performing a drive test is always the best way to localise geographically and analyse a problem. While statistics give an idea about the real behaviour faced by all end users regardless of their geographical location, drive testing or walk testing bring a simulation of end user perception of the network on the field from one call perspective. The only drawback of this method is complexity. In fact, drive testing can model the behaviour of up to five mobiles, but in order to analyse objectively a scenario, many attempts should be performed in the same conditions. For logistics reasons, this cannot be performed on the entire network, which is why drive tests and statistics are complementary methods. Both of them are necessary for optimisation engineers to assess the network performance.

Outdoor drive tests are also used to assess indoor coverage. Therefore, in a dense urban area, by assuming a certain building penetration loss (in general 20 dB is considered), the indoor signal is estimated to be 20 dB less than the outdoor level. If -80 dBm is then accepted as a good indoor level, the outdoor level should be at least -60 dBm.

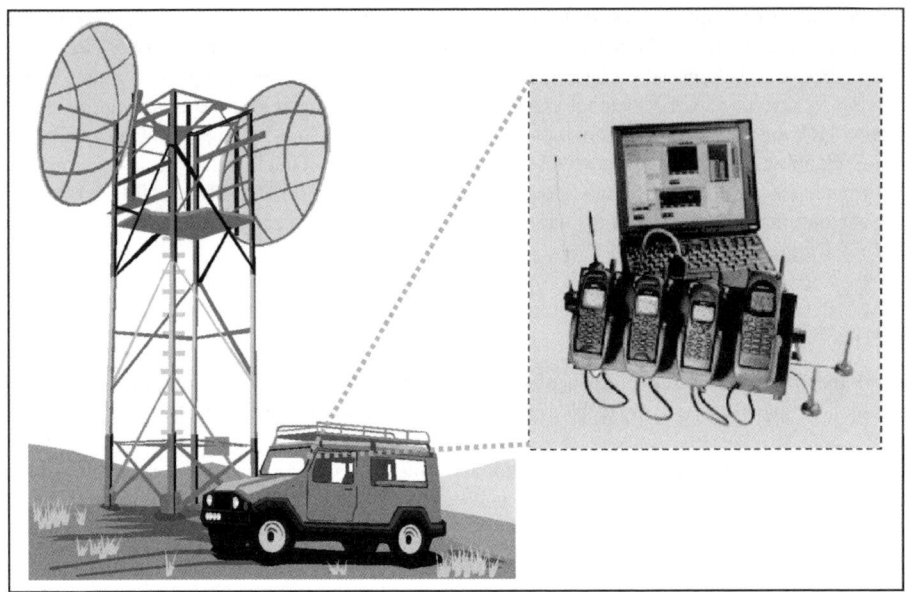

Figure 2.46 Drive testing system mounting

In the market, many tools exist where the idea is almost the same (see Figure 2.46). One or more handsets, configurable on different technologies like GSM 900/1800/1900, IS95, IS136 and AMPS, mounted on a car kit and connected via a cable to a laptop, report measurements for specific software. This application permits online visualisation on a map or a graph of different parameters (level, quality, serving cell identity, etc.).

A globol positioning system (GPS) is also needed to localise the position of quality degradation precisely. The overall system is mounted on a vehicle. The first installation of the equipment requires some operations like calibration of the GPS. In fact, manually calibrated GPSs contain an odometer and a gyroscope, which may require going through specific routes with various distances and various speeds. In some cases, an attenuator is used on the antenna chain in order to cancel antenna gain and have a 0 dBm chain.

The performance of the tool depends on the ability of the handset to report or not some specific data. Each mobile can be in one of three modes: idle, call generator or frequency scan. The same set can also be used for benchmarking several networks. As shown in Figure 2.46 (laptop screen), a lot of information can be displayed online. Some tools bring additional features such as channel lock, layer 3 message decoding and online sound warnings for events and GPS signal loss.

Log files of the measurements can then be recorded and reviewed later. They can also be used to plot some data on maps or graphs. Generally each tool has a specific software application to analyse the measurements further. Drive testing is necessary in many phases of network evolution. Before sites are installed and after BTS roll-out, frequency plan implementation, etc., every time that a change is performed on the network in terms of site engineering and/or parameters, drive tests should be performed.

To draw a drive test route, field knowledge is very important. The route should be drawn both by the measurements technician and radio planning engineer. The direction where a certain road will be driven is also important to identify, especially when interested in a certain handover sequence. KPIs used during drive testing are how given.

Table 2.19 BER to RXQUAL conversion table

RXQUAL	BER
0	BER < 0.2 %
1	0.2 % < BER < 0.4 %
2	0.4 % < BER < 0.8 %
3	0.8 % < BER < 1.6 %
4	1.6 % < BER < 3.2 %
5	3.2 % < BER < 6.4 %
6	6.4 % < BER < 12.8 %
7	12.8 % < BER

BER

The BER is an estimated number of bit errors in a number of bursts to which corresponds a value from 0 to 7 (best to worst) of the RXQUAL. After the channel decoder has decoded a 456 bits block, it is coded again using the convolutional polynom in the channel coder and the resulting 456 bits are compared with the 456 input bits. The number of bits that differs between these two 456 bits blocks corresponds to the number of bit errors in the block. The number of bit errors is accumulated in a BER sum for each SACCH multiframe and the result is classified from 0 to 7 according to the BER-RXQUAL conversion Table 2.19.

RXQUAL is still considered as a basic measurement. It simply reflects the average BER over a period of 0.5 s. However, listener speech quality evaluation is a complex mechanism that is influenced by several factors. Some of these factors that RXQUAL does not consider are:

- Time distribution of BER. For a given BER, if the rate fluctuates a lot, the perceived quality is less than if the BER is constant over the time.
- When entire frames are lost, speech quality is negatively impacted.
- Handovers generate some frame losses. It is not evident in RXQUAL measurements since, during handovers, BER measurements are skipped.
- Overall quality depends closely on the type of codec used.

In conclusion, RXQUAL does not capture many phenomena that affect the listener's perception of speech quality. That is why other metrics are defined.

FER

FER (frame erasure rate) range goes from 0 (being the best performance) to 100 %. This represents the percentage of blocks with an incorrect CRC (cyclic redundancy check). Since the BER is calculated before the decoding with no gain from frequency hopping, the FER is then used in this case. Being even more stable than the BER, the FER also depends on codec type. The smaller the speech codec bit rate, the more sensitive it becomes to frame erasures.

MOS

The MOS (mean opinion score) is a subjective KPI in the way that it scores human beings' perception of different calls. Scores go from 1 to 5 (bad, poor, fair, good and excellent voice quality). On the market some tools exist that have algorithms designed to measure speech quality objectively and deliver measures that are mapped into the MOS scale. Generally, audio quality depends not only on the radio link quality and codec capacity but also on the type of handset used, background noise, speech content and echo problems.

Table 2.20 Correlation between the BER and FER

BER and FER are very bad	Quality is confirmed to be bad
Either BER or FER are very bad	No conclusion can be made
BER and FER are good	Quality is good

Correlation
Practically speaking, the best KPI to monitor speech quality is the one that gives a combined indication on BER and FER. A certain correlation has to be considered between these two KPIs (see Table 2.20).

Tracing

The optimisation process is an ongoing operation, which becomes more and more complex as the network matures and becomes larger. The first steps of optimisation consist mainly of standard cleaning operations such as frequency correction, adjacency handling and site engineering review. At a certain phase, fine-tuning of parameters and a complex troubleshooting cell-by-cell start become necessary.

The principle behind tracing relies on analysing the flow of messages. This assumes certain a knowledge of signalling charts and call phases (explained later in the section). Several types of tracing exist; they can be summarisd into two categories:

- online tracing like interface tracing;
- offline tracing more commonly called observation.

A_{bis}/A Interface Trace
This type of tracing consists of plugging a special tool on to the wanted interface in order to intercept all layer 3 messages. It is a continuous tracing in the time. However, a review of the trace for analysis purposes is possible offline once completed. Some filtering is possible at the beginning in the setup parameter definition phase in order to reduce the size of the output log file. Figure 2.47 presents an example of the A_{bis} trace model.

The left-side text bloc refers to downlink messages and the right-side one refers to uplink messages. Possible identifiers of calls on an A_{bis} trace, for example, are called IMSI and ISDN numbers. When ciphering is active on the network, it is not possible to identify the calling party because of the use of the TMSI instead of the IMSI.

Offline Observation
Observations are handled at the BSC level in the same way as measurements. According to the setup definition, specific reports start to be updated only if the 'observed' event happens. For example, if the focus is on drop calls in one BTS, the observation report will give a history only on drop calls and only on this BTS. Observations help identify possible causes for the event. The main ones give reports on SDCCH/TCH seizures/releases, DCR and BSC outgoing handovers. The procedure idea of an observation is generation of reports immediately after an event is captured, such as a handover request or a channel seizure. In Figure 2.48 an SDCCH observation file is presented.

This example gives the statistical sharing between different SDCCH release causes in each call phase where the SDCCH is involved (example: basic assignment, internal handover, etc.).

Alarm Monitoring

The BSS holds a certain intelligence that helps operators to understand better and to dialog with the machine. Generally, it reports a set of alarms that help maintenance engineers identify, diagnose and

Radio Network Optimisation

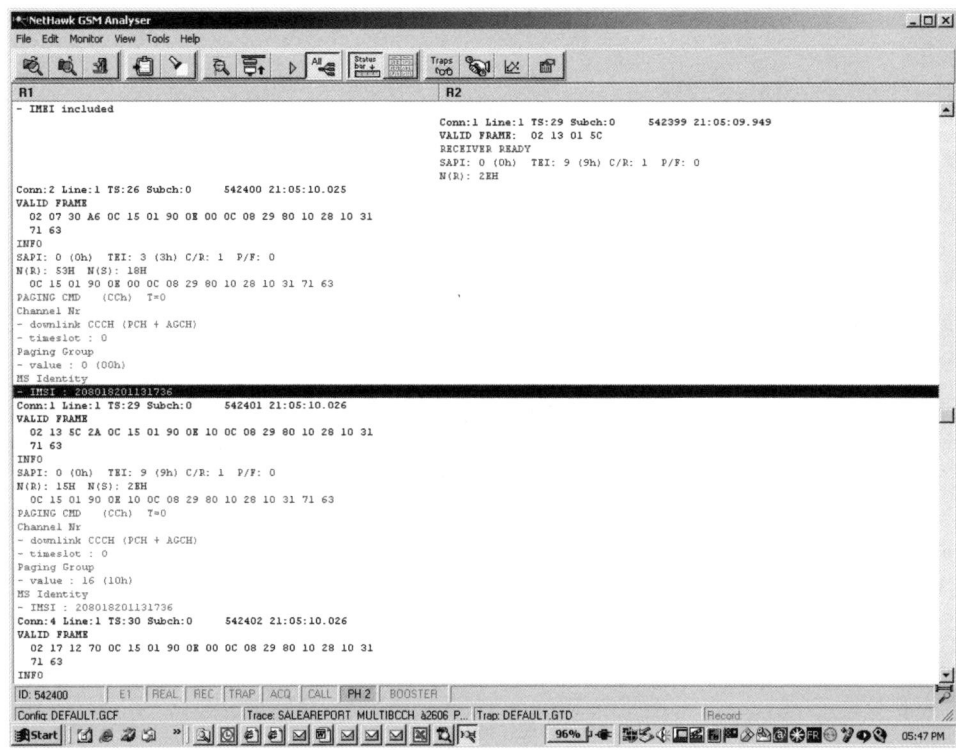

Figure 2.47 Example of an A_{bis} trace

analyse other types of problems on the network than radio related issues. An alarm is a warning, indicating that a certain counter has reached a predefined threshold. Not all the predefined thresholds are modifiable. Different categories of alarms exist. They can be classified into six fields:

- Switching equipment
- O&M (operation and maintenance) equipment
- Transmission equipment
- Power equipment
- External equipment
- Base station alarm

Each vendor has defined its own set of alarms. An example of an alarm description can contain the following parts:

- Alarm identifier. Alarms belonging to the same category will start with the same digits.
- Explanation. A brief description of the reason for the alarm and of its impact.
- Supplementary information fields. Interpretation of supplementary fields, if there are any.
- Recommended actions. There are instructions for all failure and diagnosis reports with urgency levels: one star (*) is a notification type of alarm, two stars (**) means a minor or major alarm according to the category) and three stars (***) is critical alarm.
- Cancelling instruction. Information on whether the user should cancel the alarm or whether the system does it after the fault situation is over.

SDCCH observations Statistics

BTS number	TRX number	Phase OUT	Cause OUT	Cause OUT count
BSC-xxx BCF-xxx BTS-xxx01	117	9	Basic assignment	306	5
		Basic assignment	328	1
		Basic assignment	352	63
		Basic assignment	614	1
		Basic assignment	622	1
		Basic assignment	909	3
		Ciphering	306	1
		Ciphering	317	86
		Ciphering	320	250
		Ciphering	334	3
		External HO, SDCCH or TCH source	43	1
		Internal HO inter, SDCCH or TCH source	40	1
		Internal HO inter, SDCCH or TCH source	42	34
		Internal HO inter, SDCCH or TCH source	43	18
		MM signalling	317	262
		MM signalling	320	2381
		MM signalling	334	22
		MM signalling	343	2
		MM signalling	604	1
		MM signalling	622	98
		Release	10	25170
		Release	40	32
		Release	42	3524
		Release	43	5191
		Release	47	538
		Release	320	1399
		Release	335	511
		SMS establishing SDCCH	317	2
		SMS establishing SDCCH	320	1
		SMS establishing SDCCH	334	2
		SMS establishing SDCCH	343	3
		paging/initial MS	321	12134

Figure 2.48 SDCCH observation example

Some alarms appear when another alarm has exceeded a certain threshold. Some other alarms have no signification if not associated with other alarms. Therefore why analysis requires a certain concentration to understand the origin of the problem properly.

An optimisation engineer uses, as described before, statistics and drive tests to identify and analyse a problem. However, he/she can resort to using alarms, but only as a confirmation of a suspected hardware anomaly. In fact, analysis is first done on radio settings (parameters, interference, site engineering, etc.). If nothing seems suspicious, then alarms can help to state hardware problems.

Thousands of alarms exist but only a few of them directly explain QoS problems. The most relevant ones to the Optimisation Team are those concerning a BTS hardware fault, a transmission link high error rate or intrinsic database incoherence. This latter one is especially related to adjacent cell configuration errors and is dealt with in the handover performance section later in the section.

The clearance of the origin of the alarm leads to its cancellation by replacement of the faulty hardware, where readjustment of the transmission link or correction of the database are the main actions that will be done in order to improve QoS from an alarm perspective.

Transmission Network Auditing

Although this subject is dealt with in Chapter 3, a brief reference is necessary at this point as transmission problems can affect KPIs in several ways due to three main limitation factors.

Capacity

Link capacity brings a limitation in terms of traffic channels that can be handled by a given site. For example, if a site with a 12 TRX configuration presents high TCH blocking rates it needs a TRX extension. Since one PCM (which refers to a 2 Mbps link) can handle only 12 TRXs, more transmission capacity (another PCM) is needed.

Quality/Interference

Quality is a relevant factor, especially for microwave media. Performance errors, availability and interference are the main aspects of quality. Bad frequency planning of hoppers, for example, negatively impacts on the quality of the network.

Synchronisation

Synchronisation is crucial for a good performing network. Transmission slips may not affect speech quality but will definitely affect the handover success rate (HSR). Most operators have a multivendor transmission network. In such a situation, a unique procedure to monitor the entire network is important. Two different configuration modes exist for a transmission link.

DBLF

The DBLF (double frame mode) is a configuration mode, which reports a few indicators on link quality. It is suitable only for short distances, as in the MSC-Transcoder links.

CRC4

When the CRC4 (4 bit cyclic redundancy check) mode is enabled, multiframes are transmitted and received. These frames are monitored for CRC4 errors and errors are reported in the CRC4 error counter. This mode therefore allows better monitoring of the transmission media by providing more indicators than the first one. It is recommended for long distances, as in the BTS-BSC and BSC-Transcoder links. When this mode is active, the BSS can report statistics on exchange terminals downtime, availability and synchronisation. Transmission experts then analyse the statistics, extract bad performing links and operate corrective actions.

NSS Performance

The NSS part plays a crucial role in interfacing mobile and fixed networks. The centralised architecture of GSM networks emphasises this point more (one MSC may handle several BSCs and some hundreds of sites). Therefore any MSC origin fault will have a sever impact on network quality. As mentioned in the transmission auditing section, radio optimisers are the ones concerned with end user QoS.

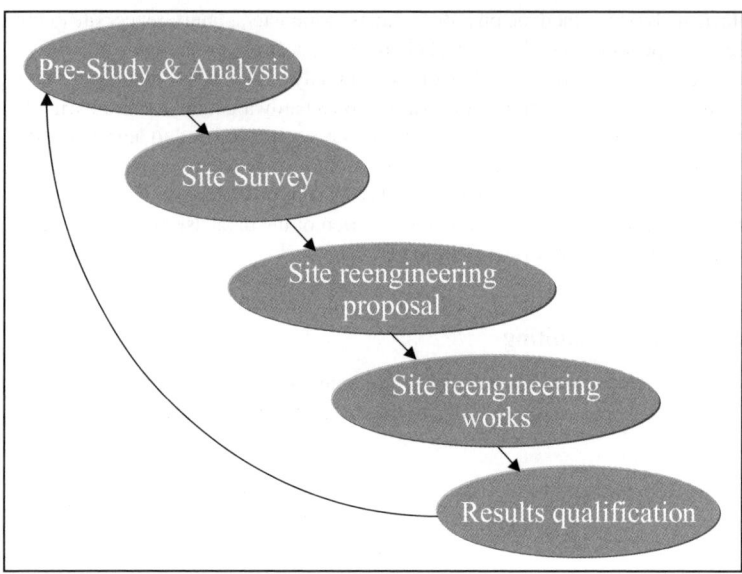

Figure 2.49 Site re-engineering process

Most operators proceed with multivendor networks. In such an environment, it is very important for operators to make sure that the system interoperability functions well. In fact, sets of timers and parameters from both sides (BSC and MSC) should be compatible and settled to the appropriate values at the beginning of the roll-out project.

Core optimisation processes are handled in detail Chapter 4. Radio engineers should work in close cooperation with NSS engineers to analyse problems together, especially when it is a matter of A interface tracing. Cooperation between radio and core is also mandatory during a site information update (after BTS re-hosting, LAC/CI modification, etc.) in order to ensure smooth impact on the quality of the network.

Site Re-engineering

At the beginning of a network deployment, the operator searches for rapid returns on investment. This is the reason why first sites are very high and do not have standard azimuths (0°, 120°; 240°). Sectors are directed towards hotspot areas. With network expansion and subscriber growth, it becomes necessary to review the engineering in order to use the frequency spectrum efficiently. Due also to lack of coverage, site physical re-engineering can be proposed. This is one way of optimising antenna coverage in order to improve site performance. This operation can follow the underdescribed process (Figure 2.49).

Prestudy and Analysis

The first step of the process consists of analysing the situation. Physical re-engineering is proposed for sites having quality problems. That is the reason why statistics and drive tests need first to be analysed. Definition of the drive testing path considers the site and its adjacent best serving areas. The aim of this optimisation solution is not to review all the network's engineering but only sites having problems. Statistical filtering (bad timing advance, high DCR, high interference outgoing handover percentage, traffic decrease, etc.) is then done to select the specific sites to visit. A traffic map of the area is also important to prepare in order to visualise the traffic distribution between sites. Before going on a site survey, the site history is considered (all operations that previously took place on the site).

Radio Network Optimisation

Site Survey

Visiting a site on air is a good opportunity to update the planning database and obtain newer panoramic pictures. A complete audit of the site is done (re-checking coordinates, azimuths, heights, tilts, antenna types, etc.). Urban changes affect the cellular network in the sense that a new building, for example, recently built in front of an existing antenna generates a signal propagation obstruction. Statistically speaking, this is translated into a traffic loss. The planners therefore check antenna horizontal and vertical spacing.

Site Re-engineering Proposal

The first reflex of the planning engineer when proposing a re-engineering solution is trisectorisation. In fact, with network expansion, in order to use the frequency spectrum efficiently, the network should be well designed to constitute better motifs (the Subsection on Frequency Planning Review deals in more detail with the motifs notion). Sectors of adjacent sites follow the same directions (0°, 120° and 240°, for example) to make it easy to use the manual and automatic frequency allocation. A re-engineering solution consists either of downtilting/uptilting, re-azimuthing or changing the antenna's height, location or type. Such operations have an impact on signal propagation. They make it larger or narrower and at the end improve coverage or reduce interference. Mastering propagation mechanisms in different types of areas is the key for correct site re-engineering.

Site Re-engineering Works

In this phase, the planners' recommendations are implemented on the field. Sometimes, the site needs to be reconfigured (in the case of re-azimuthing, for example). In this case radio parameters and adjacency network definitions may change.

Results Qualification

Qualifying results are always a crucial part of any optimisation process. This qualification consists of checking whether the optimisation actions helped achieve the objective or not. It is also the phase where the process can be launched again in order to propose better solutions.

Frequency Planning Review

Spectral efficiency is the key word in TDMA based networks. The frequency resource is rare and expensive, so it should be efficiently used. Frequency planning is an operation that is necessary each time it becomes difficult to find new frequencies for integrating new sites. In fact, while being optimised, the network still continues to expand with the addition of new sites. Therefore the frequency plan review, even though part of the planning process, also belongs to the optimisation phase. Frequency planning aptitude is a mandatory skill for good network optimisers. It is based on well-established procedures. However, field knowledge is also required, especially for manual tunings.

Process

Frequency plan implementation is a serious task that involves several departments. For instance, the NWP team is in charge of producing the radio parameters, the NMS team will implement the new database while the implementation team will be available on sensitive sites to intervene quickly where problems occur. Success in the implementation of a frequency plan on a multivendor area is a challenging issue. Good communication is very important in such a case. One efficient solution may be to use frequency band segmentation. Each vendor is allocated a separate frequency band that will not interfere with the others.

Frequency Planning in the Optimisation Phase

Establishing a good frequency plan consists of attributing frequencies to all TRXs using the optimum way. For this purpose, the regular design of sectors is crucial. In fact, the idea is to reallocate the same frequency to sectors shooting in the same direction and enough far from each other not to interfere. This means in practice to plan area by area rather than site by site (area is a set of 4, 7 or more sites).

Performing new frequency plans is a recurrent operation that takes place each time that adding new sites reaches a certain spectral limitation. At the launch phase, operators can handle the number of sites using the width of the frequency band. In this way dedicated channels can be attributed to each TRX. Nevertheless, at a certain stage, frequency hopping becomes necessary to introduce in order to spread interference. Two frequency hopping modes exist: synthesised frequency (SFH) and base band hopping (BBH).

Base Band Hopping
BBH consists of attributing fixed frequencies to all TRXs, and then the TCH hops from TRX to TRX within the same cell. This kind of hopping mode planning is quite difficult, since it is obligatory to allocate one dedicated frequency per TRX. Moreover, the maximum number of possible channels for a mobile to hop on is the number of hopping TRXs. This brings a certain limitation to the interference diversity. Hardware requirements for BBH are remote tune combiners at the BTS level.

Synthesised Frequency Hopping
SFH is more suitable for mature networks. It is easier to plan, since hopping TRXs use mobile allocation lists, so there is no need to plan them. However, it requires regular engineering for azimuths. It consists of planning the BCCH layer and then the hopping parameters (HSN, MAIO, etc.). SFH requires wideband combiners that have a higher insertion loss. A detailed method for planning an SFH based network is now presented.

BCCH Strategy
The BCCH layer is more sensitive to interference than the TCH as the broadcast channel transmits continuously in time. Therefore the planning has to be done carefully and with the highest rigour. For this, the BCCH and TCH bands are separated. Moreover, depending on the network topology, the BCCH band can be split between different site configurations. For example, in an area where a lot of hotspot micro cells are deployed, and since micro cell and macro cell propagation mechanisms are different, it is judicious to separate spectra for the two layers. The opposite occurs for a non-micro-dense area, as the same band can be reserved for both layers with a higher priority for macro and umbrella sites. The BCCH strategy also depends on whether synthesised frequency hopping is used or not. When SFH is used, a longer band can be allocated to the BCCH because the TCH layer uses MAList (mobile allocation list) random attribution (interference diversity).

There are two main possible ways for frequency planning: manual and automatic. This subsection deals with the manual method and the later subsection on automation tools approaches deals using the automated way. The manual method is especially justified when the network is well designed (small intersite distances, regular trisectorisation, homogeneous antennas and BTSs). Most European operators use the 7×21 model, which means distribution of 21 frequencies identically configured on areas of seven sites (see Figure 2.50). In this case, notice that channels 1 to 22 are used except 19 in order to avoid an adjacency interference with channel 18. For zones where it is difficult to apply this model without generating interference (relief, sea cost, high sites, sites with nonstandard azimuths, etc.), it is recommended that some additional joker channels should be raserved in order to tune sites manually. This hexagonal motif seems to give good results in theory and also in practice through several project experiences. Manual corrections should, however, be done according to field constraints/knowledge and planning tool-simulated interferences.

Theoretically speaking, if it is assumed in an urban area that intersite distances are around 400 m, the call's radius can be estimate a to be $\boldsymbol{R} = \frac{1}{2} \times \boldsymbol{400} \sim \boldsymbol{266}$ m. The number of model sites is $\boldsymbol{K} = \boldsymbol{7}$. Then the

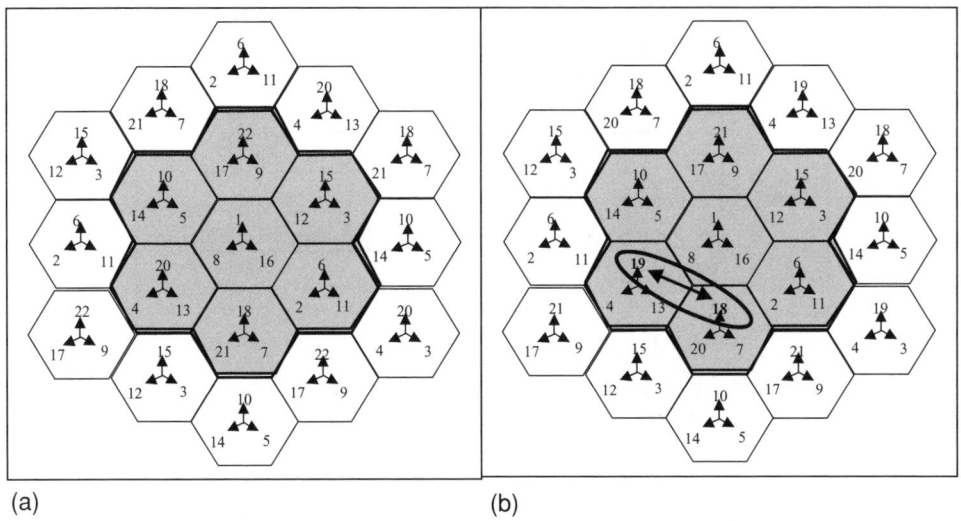

Figure 2.50 (a) SFH BCCH clean; (b) SFH BCCH with interference

closest re-use distance for a single channel is $D = R \times \sqrt{3K} \sim 266 \times \sqrt{21} \sim 1219$ m. *This distance is quite comfortable in a dense network.*

This applies in both urban and rural areas except that in rural areas the regular design is usually not respected. Site design follows revenue constraints for the operator; sites cover villages and roads regardless of any theoretical grid. Therefore, in rural areas the network planner may need to use joker frequencies.

The configuration illustrated in Figure 2.50 (b) shows the example of usage of the 1 to 21 band (19 instead of 20, 20 instead of 21 and 21 instead of 22). Unavoidable adjacent interference can be seen to be present between channels 19 and 18.

TCH Strategy
In the case of usage of SFH, the TCH layer is allocated sets of frequencies called MALists in a 1/1 (the same MAList for all sectors) or a 1/3 (one different MAList for each of the three sectors) configuration. How is MAList length calculated in each case? The frequency load calculation is the answer.

Frequency load = fractional load × *hardware load* (2.71)
where
Fractional load = number of hopping TRXs/MAList length
Hardware load = traffic (Erl) at busy hour/number of TCH timeslots

Generally it is recommended not to exceed a 30 % frequency load for the 1/3 model and 8 % for the 1/1 model. Therefore:

MAList Length_1/3 > (Traffic_BH × Nbr hopping TRXs)/(Number of TCH Timeslots × 0.3)
MAList Length_1/1 > (Traffic_BH × Nbr hopping TRXs)/(Number of TCH Timeslots × 0.08)

An example of an MAList organisation can be through interleaving in order to make the best use of the frequency band and maximise space between frequencies used at the same time (see Figure 2.51).

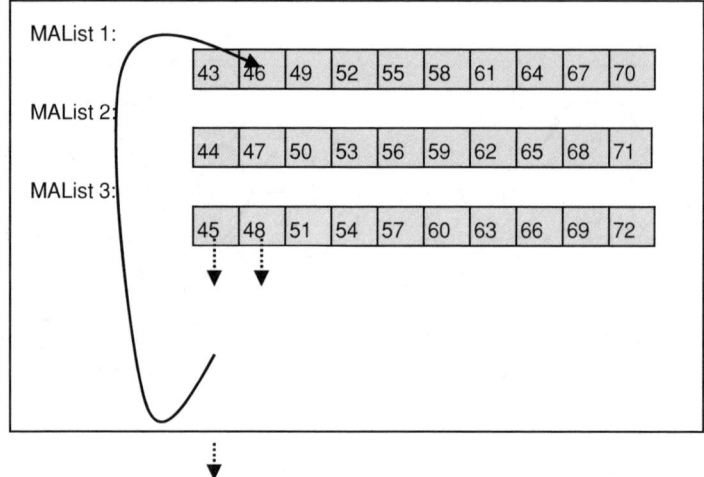

Figure 2.51 The 1/3 model MAList attribution example

HSN Planning

The hopping sequence number (HSN) is the parameter that differentiates the hopping algorithm between two cells having the same MAList. There exist 64 possible HSNs (0 is cyclic and 1 to 63 random). The principle of SFH (spreading of interference) is assured via the HSN. Two cells having the same MAList and different HSNs hop differently; they may at a single moment use the same frequency, but shortly after those two cells will camp on different frequencies. In the case of the 1/1 model, since all three sectors use the same MAList, three different HSNs should be attributed per site, whereas with the 1/3 model the same HSN can be used for all cells with the condition that sectors are synchronised. Some simulation results have shown that according to the hopping algorithm, behind each HSN value some pairs of HSNs produce a higher probability of interference than others. Table 2.21(a) and (b) give the probability values for the best and worst performing pairs of the HSN.

MAIO Planning

The mobile allocation identity offset (MAIO) is the parameter that allows two TRXs belonging to the same cell (with the same MAList and the same HSN) to traffic on different frequencies at each moment. In fact MAIO is the step between the positions of the first TRX and the second TRX of the initial frequency in the hopping list. For a MAList with N frequencies, MAIO can take values of 0 to $N - 1$. To understand better how orthogonality can be assured between two TRXs, an example will be considered.

Consider a cell with three TRXs (one BCCH and two hopping) for which the MAList is 43, 46, 49, 52 and the HSN is 5. The MAIO is fixed at 1. Then TRX3 will follow the same hopping sequence algorithm as TRX2 but with one step difference (see Figure 2.52). If the emission channel of TRX2 is in position 2 of the MAList (channel 46), then TRX3 will be in position 2+MAIO (channel 49). If the algorithm related to HSN 5 makes TRX2 hop, then the MAList is 46, 49, 46, 49, 52, 46, 49, 43, 52, etc. as TRX3 will hop to 49, 52, 49, 52, 43, 49, 52, 46, 43, etc. TRX 2 and TRX3 will never be co-channel.

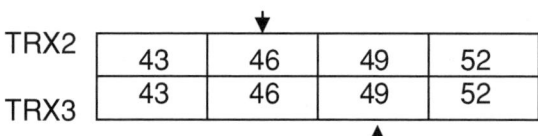

Figure 2.52 MAIO usage example

Table 2.21 (a) Worst pairs of HSN; (b) best pairs of HSN

(a)

HSN site 1	HSN site 2	Collision probability	HSN site 1	HSN site 2	Collision probability
1	17	0.136116	32	48	0.135541
2	18	0.136223	33	49	0.135539
3	19	0.136225	34	50	0.135568
4	20	0.136258	35	51	0.135571
5	21	0.136251	36	52	0.135540
6	22	0.136259	37	53	0.135545
7	23	0.136245	38	54	0.135511
8	24	0.136244	39	55	0.135509
9	25	0.136241	40	56	0.135547
10	26	0.136186	41	57	0.135541
11	27	0.136186	42	58	0.135523
12	28	0.136131	43	59	0.135522
13	29	0.136124	44	60	0.135494
14	30	0.136166	45	61	0.135497
15	31	0.136161	46	62	0.135538
16	36	0.123013	47	63	0.135532
17	37	0.123019	48	63	0.122307
18	38	0.122944	49	62	0.122312
19	39	0.122947	50	61	0.122239
20	32	0.123086	51	60	0.122227
21	33	0.123094	52	59	0.122253
22	34	0.123037	53	58	0.122257
23	35	0.123025	54	57	0.122253
24	44	0.122925	55	56	0.122258
25	45	0.122925	56	25	0.120764
26	46	0.122900	57	24	0.120752
27	47	0.122900	58	27	0.120789
28	40	0.122956	59	26	0.120737
29	41	0.122957	60	29	0.120822
30	42	0.122931	61	28	0.120792
31	43	0.122926	62	31	0.120740
			63	30	0.120776

(b)

HSN site 1	HSN site 2	Collision probability	HSN site 1	HSN site 2	Collision probability
1	35	0.068144	32	34	0.073267
2	32	0.068163	33	32	0.053682
3	2	0.053418	34	33	0.079270
4	38	0.068105	35	34	0.053662
5	4	0.053477	36	38	0.073247
6	36	0.068105	37	36	0.053719
7	6	0.053503	38	37	0.079217
8	42	0.068100	39	38	0.053698
9	8	0.053524	40	42	0.073259
10	40	0.068108	41	40	0.053742
11	10	0.053448	42	41	0.079245
12	46	0.068171	43	42	0.053700
13	12	0.053483	44	46	0.073297
14	44	0.068154	45	44	0.053682
15	14	0.053529	46	45	0.079239
16	50	0.067975	47	46	0.053728
17	16	0.053697	48	50	0.072844
18	48	0.068100	49	48	0.053800
19	18	0.053682	50	49	0.079304
20	54	0.067949	51	50	0.053798
21	20	0.053719	52	54	0.072842
22	52	0.068053	53	52	0.053823
23	22	0.053698	54	53	0.079281
24	58	0.068200	55	54	0.053775
25	24	0.053733	56	58	0.072804
26	56	0.068093	57	56	0.053828
27	26	0.053694	58	57	0.079296
28	62	0.068060	59	58	0.053801
29	28	0.053676	60	62	0.072846
30	60	0.068082	61	60	0.053784
31	30	0.053759	62	61	0.079285
			63	62	0.053815

The MAList is systematically ordered in an ascending way at the BTS level, which is why the MAIO allows the management of orthogonality of a cell and not a specific frequency ordering by TRX. Depending on the chosen frequencies in MALists of the three sectors, an offset may be needed to avoid adjacent interference between cells of the same site.

The example below is explicit for a $4 + 4 + 4$ *hopping TRX site. If the MAList configuration shown in Figure 2.53 is considered, it is assumed that assume HSN = 5, or HSN starts at the fifth position. A MAIO step of 2 at least is needed and the MAIO can be attributed as follows:*

- *Sector 1. MAIO = 0, MAIO step = 2*
- *Sector 2. MAIO = 1, MAIO step = 2*
- *Sector 3. MAIO = 2, MAIO step = 2*

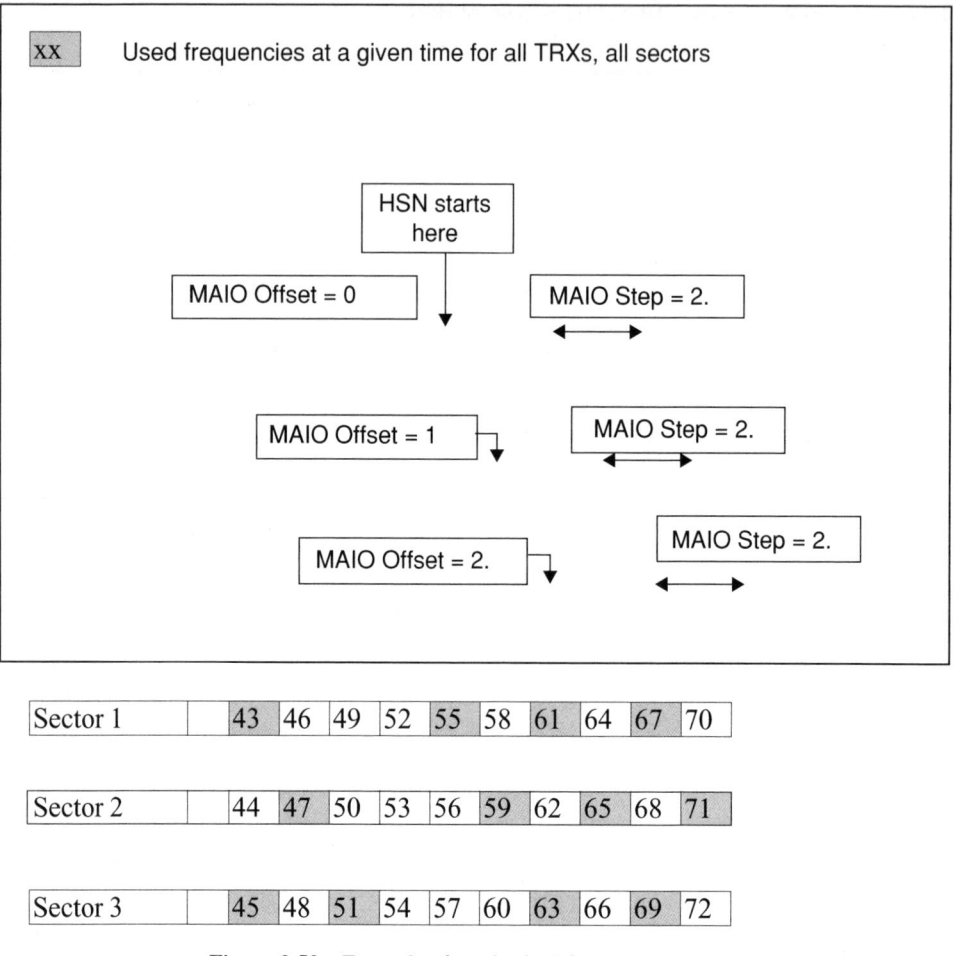

Figure 2.53 Example of synthesised frequency hopping

Other Procedures

Besides what is detailed in the previous sections, there are some operations that take place during the network life and that require advice from Network optimisation engineers.

BSC Re-hosting Plan Review

When adding new BSCs to the network, it is necessary to review the split plan (which site goes to which BSC with which Location Area Identity (LAI)). In defining a new BSC plan, it is fundamental to distribute sites properly depending on three criteria:

- Geographical distribution
- Intercell handovers
- BSC load

Remember that geographically speaking, it is not recommended to use highways or high traffic roads as a frontier between BSCs, especially if they have different Location Area Codes (LACs). This is to avoid a handover ping-pong for high speed car mobiles. For the same reason (to minimise ping-pong), BSC frontiers are drawn between cells to minimize possible handover attempts between them. In addition to these two conditions, at the end, the BSC plan should respect a certain balance in the BSC load, which is calculated according to several elements according to manufacturer limitations. In general, however, the BSC load will be calculated as a ratio between the actual number of considered elements (TRX, PCM, BCFs (base control function), BTS (base transceiver station), TCH (traffic channel), etc.) and the maximum capacity of the BSC.

Location area planning should also be considered. In fact, an LA constitutes a logical group of BTSs to which the same paging messages are sent instead of being sent to the entire network to inform one single mobile about an incoming call or SMS. Choice of the adequate LA size ensures high paging success rates. If it is too small, there will be high location update traffic and if it is too large, there will be some deleted paging messages due to the paging load.

It is worth noting that a site re-hosting from BSC to BSC can generate many problems in the network if the operation is not properly performed and followed. This type of operation always takes place at night in order not to disturb traffic. Starting from the following day, close monitoring should be done. Hereafter are some clues on what to check.

Drive Tests
A comparison is done between drive tests performed before the migration and just after. In the last one, attention is particularly given to handover performance. Adjacency definitions and handover success are verified.

Statistics
Another way to check the handover success rate from and to concerned sites is by statistics. In fact statistics launched for a long enough period (4 or 8 hours after the operation) will give a clear idea on 100 % handover failure.

Parameters
An adjacency definition is crucial for the handover mechanism. Any error in CI, LAC, frequency, BSIC-NCC or BSIC-BCC means that 100 % handovers fail. Therefore adjacency tables should be checked precisely.

Alarms
Finally, some alarms may appear after the operation (synchronisation, transmission link...). Their signification gives hints on possible corrective actions.

Network Elements Restructuring
It can be cost efficient for the operator to think of some physical re-structuring of the network in order to use the deployed capacity efficiently. A better balance between offered and used capacities can be provided through a TRX rearrangement taking account of traffic hot spots. In such an operation, the interaction between departments is high; clear procedures are very important.

List Definition
Potential cells are first identified (cells with high traffic and blocking require extension of TRXs and cells with low traffic need to dismantle TRXs). This operation requires the marketing team to provide the expected traffic increase rate in order to estimate the target traffic that will be carried out by each cell in the mid-term plans.

Site Survey
A site survey may be needed, especially for sites that are candidates for the expansion phase. Implementation teams check whether the addition of new equipment is possible (enough space, no extra load on the building foundations, energy requirements, site access, etc.). At the same time, transmission link

capacity, real TRX configuration and alarms are checked in order to constitute a clean database and plan logistics requirements.

Site Follow-up
Optimisation engineers deal with frequency preparation and KPI follow-up. It is their responsibility to check that the operation is transparent to network performance except for a positive impact on TCH blocking. Here, an important effect shoud be mentioned: some TRX extension/dismantling configurations generate a transmitting power increase or decrease depending on the BTS configuration (number of combining floors). This can explain some strange behaviour, such as a traffic increase after dismantling some TRXs.

New Features Introduction
What is interesting about optimisation is that it implicitly contains at the same time optimisation activities (for already running sites) and some planning activities (introduction of new sites, BSCs, features, etc.). New features are designed to boost network performance. Prior to the wide introduction of a new feature, operators generally want to test it beforehand so as not to generate any disturbance on the offered service. A trial or the more commonly used RSV (Regular Service Validation) procedure allows validation or the introduction of the new feature in the entire network. RSV mainly consists of the following steps:

- documentation on the hardware/software requirements of the feature;
- choice of a suitable area for testing;
- constitution of a reference, with drive tests and KPIs chosen according to what is relevant for the assessment of the feature;
- implementation;
- observation and follow-up;
- readjustments and assessment;
- setup of a generalisation planning.

The aim of pretesting a new feature is to control some unwanted behaviours better and find the most appropriate parameters set before introduction in the whole network.

Automation tools

In increasingly complex networks where new technologies are added to already existing GSM, return on investment is important to fulfil in a short period. To improve operational efficiency and free resources for value-added tasks, there is no way out of process automation. Usage of automatic programs varies from simple macros to very specific and complex tools. According to what is presented as important tasks in the previous optimisation process, many operations can be done by means of developing computer programs.

The first steps in automation enclose common tasks such as statistics visualisation, adjacency management, geographical illustration, troubleshooting follow-up and documents generation. These fields can be classified into 'administrative automation'. More complex tools are then used for other purposes, such as tasks that demand an analysis of huge amounts of input data such as frequency and adjacency planning, parameters optimisation, location area planning and cellular troubleshooting. Classification of these goes under 'technical automation'.

Administrative Automation
Every task that is repetitive in a network optimisation activity can be automated to a certain level. Administrative automation brings the advantage of minimising human error risks. An automation process is shown in Figure 2.54.

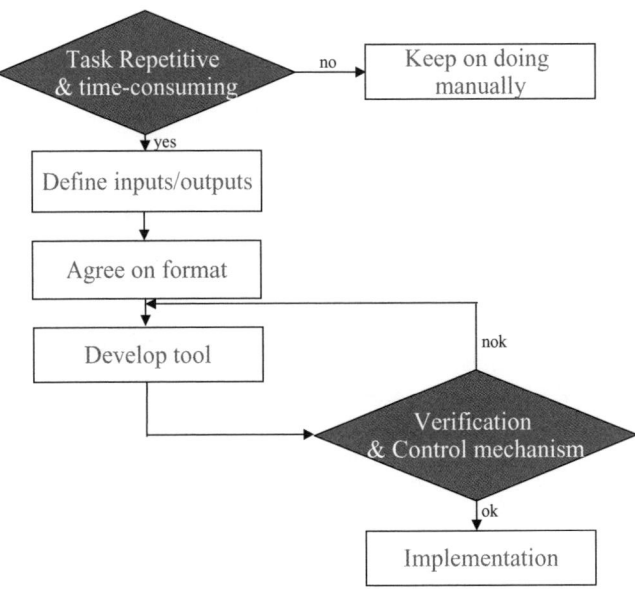

Figure 2.54 Administrative automation process

The process described here is based on a simple idea which says that, whatever the task needed to automate, the control mechanism should allow validation of the tool according to what is agreed before. Several fields are suitable for administrative automation.

Statistics Visualisation
Generally, pure counters are extracted from the NMS as text reports. There is then a need for conversion into a user-friendly interface. This means applying some formatting macros on the original files. Through various project experiences, it is recommended that Microsoft Access is used rather that Microsoft Excel for such applications building. In fact, Excel has a size limitation of 65 536 lines, which means that Excel cannot display, for example, the adjacency file of a 2000 macro sites network where each cell has an average of 12 neighbours ($2000 \times 3 \times 12 = 72\,000$). More details on statistics is given later in the section.

Adjacency Management
Dealing with the adjacency relationship is a regular part of an optimisation engineer's job. As the network grows (integration of new sites) and changes (implementation of new frequencies), adding new neighbours, deleting obsolete ones and updating existing ones become a daily concern and a repetitive task. In defining the Man–Machine Language (MML) commands of adjacency management, there is definitely a risk of errors due to human intervention when they are manually handled, whereas tools guarantee a certain level of reliability. In order to minimise error risk, automatic programs are the key. This method also helps to control the automation loop quickly. By automatically analysing the logouts, it is possible to identify whether the required changes have been implemented or whether some problems have been encountered.

Geographical Illustration
The field approach is very important in the optimisation process. Knowing the geography of the area helps a great deal in understanding interference, overshooting, weak coverage, etc. In this way, it is important

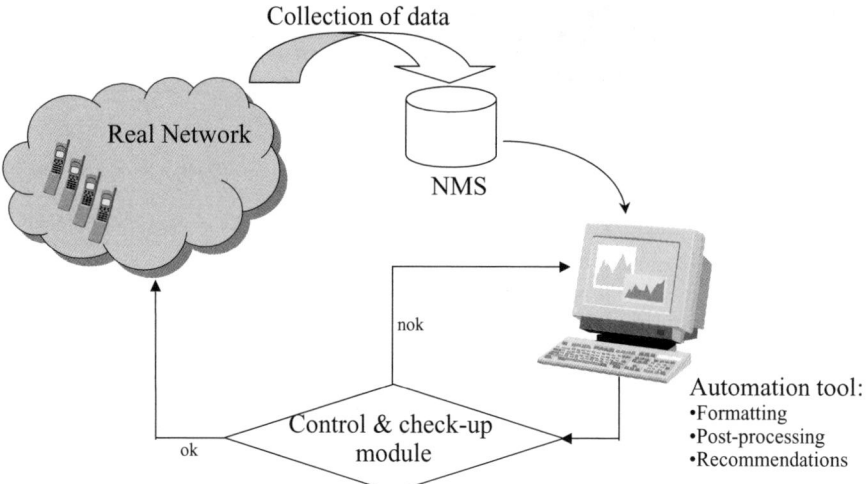

Figure 2.55 Technical automation process

to visualise planning parameters (BCCH, BCC, NCC, neighbours, etc.) along with statistics (traffic, blocking, DCR, CSF, etc.) in addition to geographical positioning and engineering parameters (antenna type, tilt, height, etc.). Some planning tools allow the import of statistics. Otherwise, such applications can always be built on MapInfo. MapInfo brings the advantage of also visualising high definition maps (details on roads, buildings, etc.).

Troubleshooting Follow-up
The optimisation process is a continuous procedure of analysing problems, making actions and assessing improvements. It is important to keep the trace history of problems and actions on a cell level. Such information fits within a general database. The filling task can be automated in order to improve operational efficiency. This may reduce typing effort by automatically filling in cell information, statistics and old/new parameters.

Documents Generation
Optimisation reports are regularly generated under a predefined presentation format. If manually filled in, these reports may have different formats from time to time. However, automation assures a unique output format.

Technical Automation
Real network subscribers generate a huge amount of statistics, which are then collected into the Network Management System. Specific tools allow postprocessing of these data and making recommendations accordingly. A control and check-up module is important to plan in order to verify recommended changes before implementation to the network (see Figure 2.55).

Frequency Planning
Before handling an automatic frequency planning module, it is helpful first to expose the manual frequency planning procedure. The manual method is based on interference prediction. Better accuracy of prediction depends on both the high resolution of digital maps and specific propagation models. In fact, the Okumura–Hata model, which is commonly used in planning tools, is still applicable for 1 km resolution maps. This combination of model/map resolution does not apply to small cells (10 m radius). As the network grows, more sites are added including of macros, micros and pico cells specifically designed for building

coverage improvement. In such a scenario, it is necessary to separate frequency bands between macros and to build micros with better master interferences. Adopting this technical choice becomes more and more difficult when the number of operators grows and frequency bands become shorter (sharing constraint). More efficiency in resource management is then required. The C/I (carrier/interferer) ratio gives the minimum acceptable level of a carrier compared to an interferer having the same or an adjacent frequency. Interference constraints between two cells are translated into high or low factors in the interference matrix. Field knowledge on interference factors between two cells can make frequency planning more accurate and of higher quality. The introduction of synthesised frequency hopping simplifies planning but requires rigorous sectorisation of cells.

Many planning tools now offer the option of automatic frequency planning (AFP). AFP does not need any subscriber distribution or radio wave path prediction. It uses real mobile measurements reported on subscribers' locations and an experienced interference level. It has, however, three limitations:

- quantisation and truncation;
- only six neighbours reported;
- BSIC decoding uncertainty.

AFP has shown in trials some 10 % improvement from the initial plan quality.

Adjacency Planning

Handover automatic handling is also based on mobile measurements. Deletion of an adjacency is based on NMS statistics. Some benchmark criteria need to be carefully chosen in order to help the system decide whether to delete an adjacency or not. For example, when a few attempts are performed between two cells with reference to the maximum handover attempts of each one, the tool can advise deletion of the relationship. This, indeed, can be done only after checking the geographical positioning of each cell according to the other in terms of distance. In fact, sometimes the cells are very proximate to each other but for some reason (low traffic, the presence of an obstacle, hardware anomaly, etc.) they perform few interhandover attempts. In this case the relationship should be maintained. In any case, after deletion is downloaded to the network, a control mechanism will assess the impact on quality.

Adding new adjacencies is a more complicated concern. It uses surrounding cells measurements that are then ranked from the strongest to the weakest. The number of measured samples of each cell is also to be considered in order to declare only useful new adjacencies. The add/delete mechanisms are launched on a regular basis and hand-in-hand in order to optimise the number of adjacencies. Each cell can support up to 32 neighbours, but to increase handover performance it is recommended not to exceed some 15 adjacencies per cell. Automation in adjacency management can also include a parameter incoherence check. A specific subsection presenting the types of checking that can be useful to do is included in the handover performance section.

Parameter Optimisation

At the deployment phase, operators tend to use standard parameter templates, as thousands of GSM parameter exist. However, due to differences between cells (propagation and interference environment), it is necessary to fine-tune some parameters on a cell-by-cell or adjacency-by-adjacency basis in order to get *optimum* performance. The essential idea behind optimisation is how to find the correct value for a specific parameter applied to a specific cell. Even this task, which requires optimisation expertise and is time consuming, can be automated. A complex mechanism based on four main principles allows such automation to be performed:

- definition of subsystem architecture;
- choice of adequate control techniques;
- consideration of control algorithm mapping to network elements;
- assessment of performance according to initial rules.

Cellular Troubleshooting
Troubleshooting consists of three main steps:

- detection of the faulty cells;
- analysis and diagnosis of potential origins for the problem;
- application of the estimated solution.

Automation can be applied to all three phases. Detecting faulty cells can be done using preestablished criteria and thresholds on the basis of comparison against some restrictive conditions. The main idea behind automatic diagnosis is the transfer of experts' knowledge to the system. To achieve this, troubleshooting models should be created and corrected. Modelling means equipping the machine with adequate reasoning algorithms to help it find out the possible origin of the problem and propose eventual solutions.

Location Area Planning
To inform a subscriber about an incoming SMS or call, the network uses paging to all BTSs of a given LAC. Location area planning consists of finding the appropriate size of LA in order to ensure that paging success is maximised and the number of location updates (LUs) is minimised. Paging success decreases due to deleted paging messages in the case of a very large LA. On the other hand, paging load increases due to the high number of LUs in the case of a too small LA. It is also worth noting that the capacity of an SDCCH is much lower than that of the PCH. Thus the aim of the automatic method is to establish the correct balance by defining the correct LA size.

Operators traditionally use one of two approaches for LA planning:

- Turning each BSC into an LA, in which case a huge amount of LU takes place due to inter-BSC handovers (note also that mobiles in the idle mode perform LU). This approach consumes much SDCCH resource.
- Turning each MSC area into an LA, in which case the paging load goes up, with the result that some mobiles are not informed about incoming calls or SMS. This approach saves the signalling resource, but still does not properly consider the actual capacity of the paging channel.

The main idea behind Automated Location Area Planning is to use Network Management System (NMS) data to fix an optimum LA size. Elements used can be:

- paging capacity of all BTSs in order to estimate the paging load that can be carried by the BTS;
- paging traffic of BTSs to use for estimating the paging traffic in the case of several BSCs having the same LA;
- handover statistics between BTSs in order to group BSCs having many handovers between them.

Based on these pieces of information it is possible to determine an optimum LA plan that fully utilises paging capacity while ensuring that paging success is maximised (deleted paging messages minimised) and LU traffic on the SDCCH is reduced.

At the end, optimisation can be assimilated with a medical check-up of the network. The optimiser first detects anomalies (symptoms) and then analyses the situation (diagnosis). After that, he/she proceeds by trying a first corrective action (treatment) and then sees the effects. If it does not work he/she tries to push more investigation by asking for a detailed analysis (scans). The optimisation process runs continuously as many operations take place in the network without interruption (adding new sites, BTS re-hosting, sites re-engineering, adding/removing TRXs, etc.). In most cases, operations are correlated between them because they are either interdependent or interexclusive.

Optimisation consists of addressing three main aspects: coverage, capacity and frequency planning. Actions can then be taken according to what is identified as a problem on a cell or an area in order to perform better. Frequency plan review, handover parameters change, site-engineering correction and specific parameter applications are only examples of common optimisation actions. Adding new sites, new BSC or reorganising TRXs are also possible improvement solutions. Through drive testing, the NMS and trace analysis information can be obtained, faults diagnosed, new recommendations tested, parameter change implemented and the network assessed.

It is important to note that performance is also dependent on handset capability. Depending on the manufacturer, different transmitting powers, different sensitivities and possible bugs many exist. All this definitely influences the subscriber's perceived quality.

To perform well in the optimisation phase, network optimisers should work in very close cooperation with other services. The operation and maintenance centre (OMC) team is the most concerned in the sense that all tunings proposed by optimisation personnel should be implemented via the OMC. The roll-out team is the one responsible for all site activity (site re-engineering civil works, site re-hosting from BSC to BSC, TRX redeployment, etc.). The care team ensures alarm monitoring, hardware fault management and maintenance interventions. The operator then comes at another level of interaction to take care of multivendor management, validation and work facilities (access, authorisations, etc.).

The success key in such an environment is the application of process where inputs, outputs, procedures and owners are identified. Optimisation engineers have to master each process well and always be ready to adapt the solution because they are often asked for advice on parallel operations taking place at the same time.

With the introduction of automation, optimisers are freed to address advanced issues. In more and more complex networks, the optimisation task takes different orientations from before. Effectively, previously the focus was on optimisers' knowledge of one specific technology and of manual optimisation practices. Today, competition in the market of telecommunications reaches a very fierce level where concentration on own technology is not sufficient any more. Today, what is asked from engineers is a multivendor competence. Multivendor competence means being able to address any optimisation or planning issue regardless of the equipment manufacturer.

Statistics and Key Performance Indicators

Quality of service in telecommunications means the level of usability and reliability of a network and its services. As mentioned before, statistics are the most efficient way to monitor the network's performance besides drive testing. Monitoring the network is a key element to achieve the best quality of service. QoS monitoring involves permanent observation, qualification and adjustment of various network parameters. The objective of this subsection is to present and detail all aspects related to statistics extraction, handling and exploitation.

Use of Statistics

The notion of statistics in mobile networks refers to a general set of metrics that helps the operator in three main directions:

- First, assess the performance of the network and benchmark it to other networks.
- Second, analyse faults and check improvements.
- Finally, dimension extensions for the network.

These metrics are directly generated through real subscriber traffic. Each event that occurs in the network (originating/terminating call, handover failure, location update, etc...) is reported to the NMS. With such a mechanism, the operator does not have to invest in any simulation tool to model the performance of the network towards a certain subscriber's profile.

In dealing with statistics, two elements should be distinguished:

- Pure counters (elementary performance indicators, or PIs), which are incremental values of events generally with no significant relevance if handled individually. They provide data on a specific aspect (number of calls, for example) but, practically speaking, it is hard to interpret their values.

- Key performance indicators (KPIs), which are computed formulas based on PIs, better translating the subscriber's experience.

Network Assessment

Most operators have chosen relevant KPIs to monitor. They have also established targets to be met in order to achieve the wanted level of end user quality of service. The idea is to check whether the relevant KPI meets the target terms. If not, then troubleshooting starts to identify the faulty element or the network bottleneck. Operators use KPI targets to benchmark their network with others and to benchmark different suppliers in their own network.

Fault Analysis and Improvement Check

Troubleshooting aims to identify and correct the faulty cell, which lowers the overall area performance. Here, two approaches are necessary: the approach of setting up thresholds and seeing whether the network performance meets the objectives or not and the approach of monitoring the performance variation (percentage of increase or decrease on a certain indicator). For example, a cell that dramatically lost traffic from one day to another should alarm the operator.

Network Dimensioning

Statistics also help to dimension the network. In fact, based on user experience, the operator can drive the strategy of network expansion. For instance, it is easy to identify hot spot areas (high traffic) and bad quality areas (high dropped call rate). Statistics indicate where to add capacity and from where to take out hardware. This assessment helps the operator to decide which capacity solution to deploy (dual band, micro sites, improving indoor distributed antenna systems, etc.).

In the following subsections, each aspect is presented in more detail.

Principle of Statistics Extraction

Process

In a macroscopic view, the collection mechanism of statistics is described in Figure 2.56. To proceed first with collecting statistics, measurement tables for the BSC should be configured and activated. Otherwise, no values are reported. Tables are organised by categories to allow the operator to reduce load on NMS processors and download only wanted measurements (traffic, resource availability, handover, power control, etc.). Mobile subscribers report the active wanted measurements to the BTS while in idle or connected modes. The BTS sends those measurements to the internal database of the BSC. According to the operator's settings, the BSC can either store the statistics or redirect them to the NMS. Via a specific tool installed at the NMS level, pure counters are computed into predefined formulas. The output KPIs are then grouped into generic reports and sent to the office. Then, some other specific tools, generally internally made, allow archives and visualising statistics to be stored in a friendlier interface than text reports.

Statistics are reported regularly during the day to allow the operator to monitor the network in a very reactive way.

Reports Generation

A KPI is the result of a formula that is applied to counters (called performance indicators). Using a specific tool, KPIs are extracted in the format of predefined reports. These reports are addressed to specific target groups who may use them for various purposes. This is why the content of the report is customised according to the needs. This mechanism is illustrated in Figure 2.57.

In the following subsections, key components of statistics processing are detailed.

General Notions

Here the basic notions used in statistics management are presented.

Radio Network Optimisation

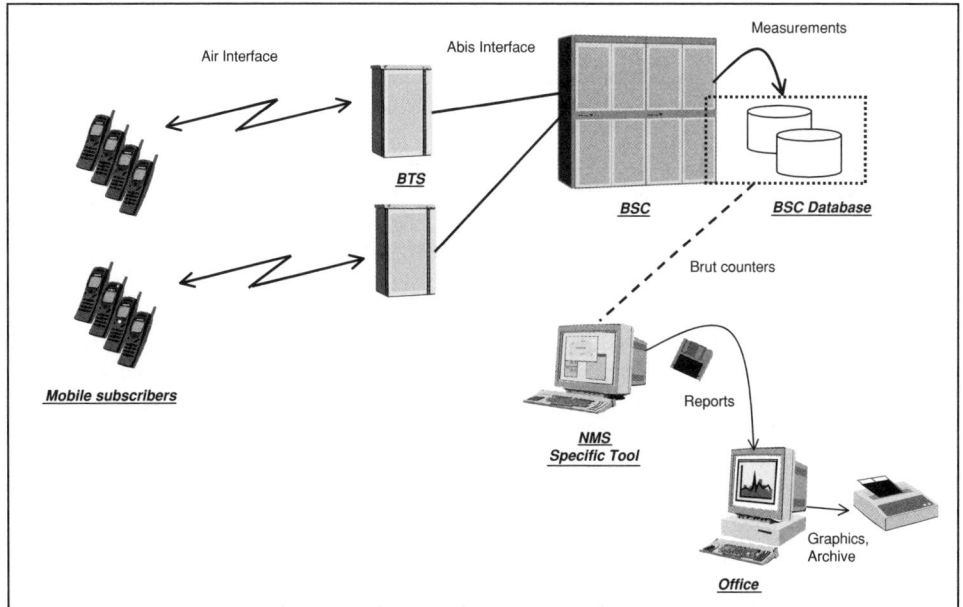

Figure 2.56 Architecture of KPI extraction

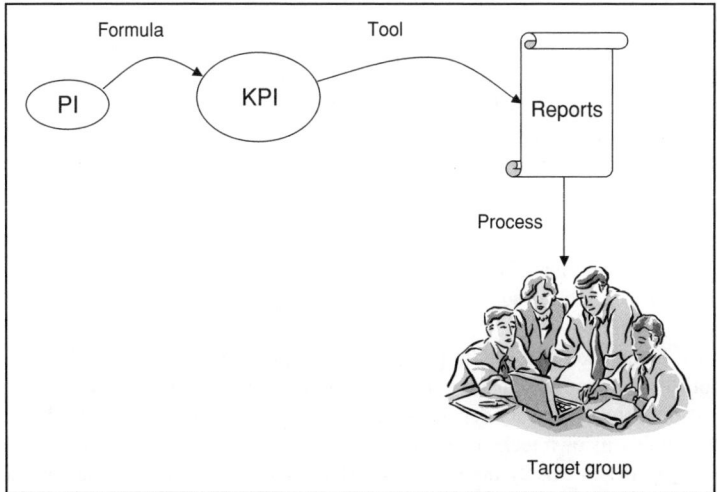

Figure 2.57 KPI extraction mechanism

Counters/Performance Indicators

Definition

A counter can be defined as an incremental value of a specific repetitive event. In GSM language, an event corresponds to a signalling message. Here, it is necessary to introduce the notion of signalling charts. A voice call is supported through millions of messages exchanged between the MS and the MSC. Therefore a signalling chart illustrates the flow of signalling messages between various network elements,

Table 2.22 Example of causes for an SDCCH drop

	Description
306	The A_{bis} interface receives an ASSIGNMENT_FAILURE message from the MS during a Basic Call
315	The A_{bis} interface receives a CONNECTION_FAILURE message
316	The A_{bis} interface receives a CONNECTION_FAILURE message and its cause is 'radio interface failure'
317	The A_{bis} interface receives a CONNECTION_FAILURE message and its cause is 'radio link failure'
320	The A_{bis} interface receives an ERROR_INDICATION message
323	The handover has failed and the MS returns to the source channel. The MS has sent an assignment failure or a handover failure message to the network during the handover
324	The A_{bis} interface receives an HO_DETECT message. The contents of the message have been corrupted or a timer expires while waiting for the message
334	The A_{bis} interface receives a RELEASE_INDICATION message
349	The A_{bis} interface receives a CH_MODE_MODIFY_ACK message and the mode has not been the same as the mode requested
350	The timer expires while waiting for the ASSIGNMENT_COMPLETE message
354	An ERROR_REPORT message is received from the BTS
365	The number of failed configuration changes due to configuration change reject or no answer from the MS
400	The call is released after the distance between the MS and the BTS has reached the maximum value allowed, i.e. the MS has exceeded the planned cell boundaries
503	Signalling with the internal processes has failed during handover signalling
507	The timer t3107, t3103 or t9113 has expired. The assignment complete or handover complete message has not been received. The MS has not established a channel to the target side
510	Failure in the target channel
511	Failure in the source channel

for instance MS, BTS, BSC and MSC. Many signalling charts exist, each one corresponding to a certain phase in call handling. The next subsection on call phase signalling charts should be referred to for signalling charts.

A counter will be updated at a certain point from exchanged messages. This is what is called 'the triggering point'. As millions of signalling messages exist, many counters may also exist. However, the number of available counters is very dependent on the equipment's supplier and his/her strategy. Each BSS vendor has in fact its own technology and infrastructure products. The counter's behaviour from one vendor does not match exactly the same triggering points of the other. In the absence of a GSM standardised definition, different BSS vendors have their own interpretation and hence different formula definitions result.

Due to the large number of counters and the basic level of their triggering, it is hard to make an objective analysis through them. They bring an idea to a specific event, the number of calls for example. In practice, however, they have no relevance as such in interpreting the quality of the network. The same counter can be updated after several messages in various call phases, and vice versa; one event can increase several counters.

Different individual signalling messages that update one counter can be called 'causes'. A cause describes in detail how the transactions have run. Table 2.22 presents an example of several causes that may affect the SDCCH drop counter.

Triggering

The supplier of each equipment has his/her own philosophy in conceiving counters. Therefore, when benchmarking is done, careful analysis is needed in order to highlight the difference of each vendor in reference to the GSM standard. As triggering points differ from one supplier to another, comparing different counters does not constitute an apple-to-apple comparison.

Key Performance Indicators

Key performance indicators (KPIs) can be defined as a set of results that measure performance in the busy hour or average period on the entire network. The KPI is the result of a formula that is applied to performance indicators (PIs). PIs can be extracted at an area, cell, TRX or adjacency level. Hundreds of KPIs exist. They use counters from one or several measurements and can be mapped directly to one counter or a formula of several counters. The period of observation refers to the duration of collected samples: hour, day, week, month, etc. The area indicates the location and the sites where the statistics are collected.

Formulas

Formula means a mathematical combination of counters that results in a meaningful indicator. Defining a formula using several PIs helps to identify a KPI. As explained before, the KPI gives more flexibility and clarity to the operator in interpreting the network's behaviour. For one KPI there can be several different formulas available, depending on three elements:

- the capabilities of the equipment to provide detailed level of statistics;
- the track request of the operator;
- the area where the formula is calculated (cell or group of cells).

Formulas, once chosen, should stay unchanged in order to observe the evolution of network performance in the time. In a multivendor environment, the operator sets a performance strategy and defines formulas for each KPI. Then as each equipment triggers its own counters, formula terms are mapped to the corresponding counter for each vendor.

Detailed examples of KPIs are presented later in the chapter. To approach this notion of using different formulas for the same KPI, an example will be considered.

Example: Drop Call Rate (DCR)

From a user's perspective, the dropped call refers to the case where the voice communication is interrupted by the network. In this scenario the formula is logically:

DCR % = 100 × Nbr drops / (Nbr calls + Nbr Inc-HO)

where

Nbr drops = number of call drops that occurred during the conversation phase
Nbr calls = number of calls initiated on the cell
Nbr Inc-HO = number of incoming handovers

In this formula, incoming handovers are only considered because the successful outgoing handovers refer to TCHs that are normally released. Dropped call triggering starts logically after the called party answers the call. However, DCR for some operators starts to be calculated from the point where a TCH is assigned (before the communication phase). Another option is not to consider incoming handover attempts in the denominator, assuming that all successful incoming handovers resulted in successful outgoing handovers. Figure 2.58 presents the DCR calculation results according to two different formulas.

Main KPIs

The main concern for an operator in terms of voice quality is accessibility to the network and call quality (audio, drop call, handover). Those two aspects are directly related to revenue generated by voice traffic.

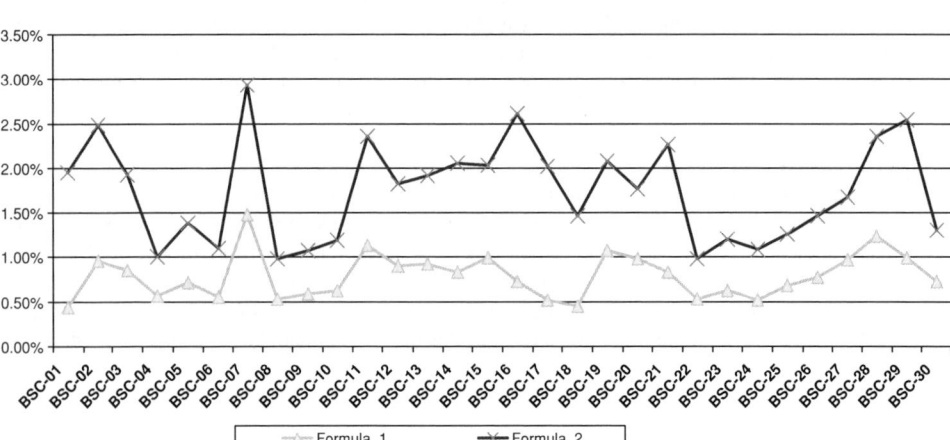

Figure 2.58 DCR example according to various formulas

In consequence, the main KPIs followed by operators are the call setup success rate (CSR) and the drop call rate (DCR). Occasionally, the handover success rate (HSR) may also be monitored.

Drop Call Rate
The DCR is logically defined as the ratio between the TCH drops occurring during the conversation phase to the number of successful seizures on the cell or area. Indeed, many definitions for the DCR exist according to the formula used to calculate it. Usage of a specific formula is operator/vendor dependent. Bad DCR performance can be due to capacity limitation, lack of coverage, interference, quality degradation, etc.

Call Success Rate or Call Setup Failure
The CSR measures the ability of a cell/network to provide traffic channel to call originating mobiles. Any failure in the assignment phase is charged to the CSR. Here also results may vary according to what the operator wants to track. This means that different formulas give different results. Some operators start considering CSF after the SDCCH has been assigned not to consider SDCCH blocking. Others want to model the user's perception and to consider all types of failures.

Handover Success Rate
The HSR is defined as the percentage of successful outgoing handover attempts. The total number of attempts, failures and blockings on incoming and outgoing handovers are available either to cell-to-cell or BSC-to-BSC levels. High rates of outgoing HO failures explain high values of the DCR. Moreover, in the troubleshooting process, engineers can resort to other KPIs to analyse anomalies in more detail. Hereafter, the most relevant KPIs used in the troubleshooting mechanism are presented.

TCH and SDCCH Availability
Resources availability is the first mandatory requirement for communications to proceed. If the system does not offer enough physical or logical capacity to handle the demand this impacts on the perception of unsatisfied users. If queuing or directed re-try, for example, are activated, it may help to handle more TCH requests with less access failure rates.

Traffic and Blocking
In a cell with low traffic, DCR and CSF rates may appear outside objectives but a result of the low number of attempts in the denominator. Such cells have the lowest priority in optimisation because their impact is neglected. Alternatively, a cell with high traffic and high blocking encounters has CSF problems. This is explained through congestion, which generates TCH requests to be rejected.

Timing Advance
Cells with high timing advance are those taking communications far from the site. Due to weak coverage and path unbalance, it is normal to have high DCR and CSF values on such cells. In this case a solution can be obtained by changing and/or down-tilting the antenna, reducing power or limiting access by specific parameters.

Path Balance
Generally, due to high power and gain from the BTS side, downlink power is greater than that reached by uplink. Such path unbalance impacts an uplink quality, which indicates a greater probability of a higher drop call and lack of setup success.

TRX Quality
TRX quality counters give an idea of the percentage of samples having quality 0 to 7 in the uplink and downlink directions. Quality scale 0 is the best. The more samples are in Q-0 the better is the performance of the cell. Low samples in Q-0 to Q-4 indicate an interference or hardware problem on the TRX.

SDCCH Drop
A high SDCCH drop rate explains the high values in CSF. It can be due to a bad radio link performance or any other hardware problem. High SDCCH traffic does not necessarily mean high TCH traffic. The SDCCH is used for other services that do not generate TCH traffic. KPIs for monitoring SDCCH performance in the SMS and location update also exist.

Handover Distribution
Handover distribution gives the percentage of outgoing handovers depending on the cause (UL/DL quality, UL/DL interference, UL/DL level, power budget, slow moving mobile station, etc.). For example, on a cell that does more handovers due to the uplink level, a problem of path unbalance (weak uplink signal) can be assumed. More details on handover performance are presented later.

Call Phase Signalling Charts

Physical Layer
The GSM communication protocols follow the open system interconnection (OSI) model with certain adaptations and additions due to the special characteristics of cellular networks. Therefore, some functions may involve more than one layer. This raises the necessity of shaping new protocol functions.

The physical layer is similar to layer 1 in the OSI model, except for the fact that it can be directly accessed by layer 3 for power control, handover control and roaming tasks where there is a need for estimating the signal level, C/I, error rate – to mention only a few. Figure 2.59 shows how the other layers can access layer 1.

Logical channels used on the air interface can be divided into two classes: common control channels that include paging, access grant and random access and dedicated channels containing associated control channels and traffic channels.

While a call is in progress, there is a need to carry signalling at the same time as user data. For that, an SACCH is allocated to each TCH. SACCH (Slow Associated Control Channel) is a bidirectional channel that transports two messages per second, mainly used for nonurgent procedures. Other more urgent procedures make use of a TCH, called an FACCH. FACCH (Fast Associated Control Channel)

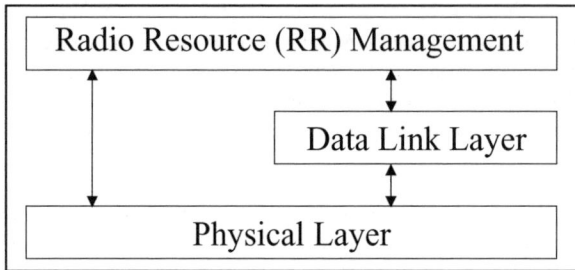

Figure 2.59 Layer 1 interfaces in the GSM

steals 20 ms of user data blocks to transmit signalling. The stealing flag, a kind of binary information transmitted on the TCH, allows the receiver to distinguish user data from signalling.

Outside a call, there is still a need for signalling (for location update or short messages, for example). Since there are only a few exchanged messages for this context, there is no need to waste a whole TCH resource; the SDCCH is used instead. The SDCCH uses a low rate equivalent to one-eight of a TCH. All these channels introduced here constitute the transport media on the air interface (physical layer in the GSM).

In the GSM, information flow mainly consists of traffic and signalling. Layer 1 is responsible for:

- constituting and multiplexing bursts into TDMA frames and transmitting frames over the available physical layer;
- searching for a BCCH using the MS;
- evaluating channel quality and signal level;
- error detection and correction;
- synchronisation with frame transmission;
- Encoding of the data stream.

For more details on the physical layer, the reader can refer to the GSM recommendations.

Linking Layer
Data link protocols depend on the interface. For an MS-BTS connection, a new protocol inspired from LAPD (Link Access Procedure on the D channel) but with some individual adaptation is used (LAPDm, where m stands for mobile). For the BTS-BSC interface, the LAPD protocol, which is oriented to the ISDN standards, is used. The BSC-MSC connection uses the Message Transfer Part (MTP) protocol, which is familiar from the SS7 standard (refer to Figure 2.60, where BTSM is the base transceiver station management and BSSAP is the BS Subsystem Application Protocol).

Most functions of the linking protocol fall in the LAPDm and LAPD, from which it is derived. The first function of the linking layer is to construct frames out of bits. On the LAPD and MTP, frames are delimited with start and end flags, whereas in the LAPDm, a frame length indicator allows frames to be distinguished. On several link interfaces, there is a limitation on the number of frames to be transmitted due to buffer size constraints. For this reason, segmenting and reassembling at the receiver end are other necessary functions of the linking layer.

In addition, the layer 2 main functions can be summarized as:

- frames construction;
- segmentation and assembly;
- error detection and correction;
- multiplexing of different flows at the same time (e.g. idle mode signalling and text short message);

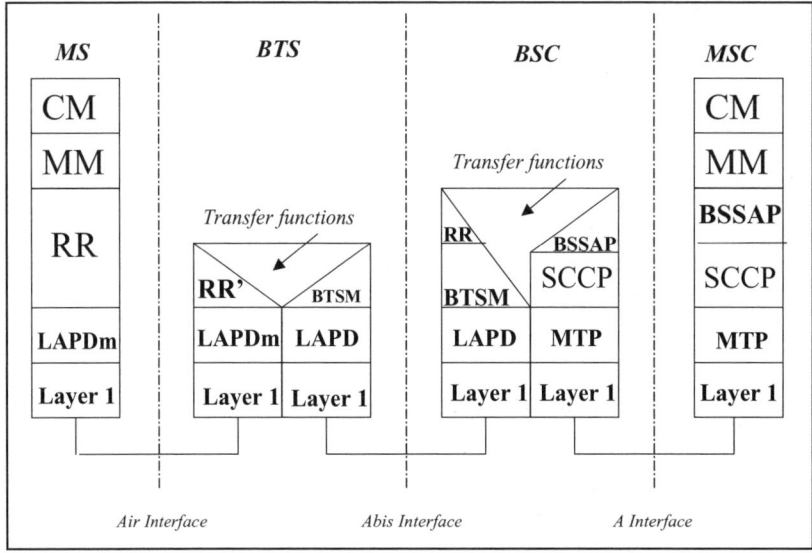

Figure 2.60 Architecture of the signalling protocols up to layer 3

- flow control, which is a kind of monitoring of throughput bottlenecks and reporting to the source sender to slow down the flow.

Data in the linking layer are transported on the air interface using the FACCH or SACCH (see the previous subsection on the physical layer). Signalling in the GSM between different network elements follows common channel signalling system 7. For more details on the link layer, the reader can refer to the GSM recommendations.

Network Layer
Networking allows end-to-end connection of two elements not necessarily physically interconnected. Its functions can be declined as:

- routing,
- multipath channelling,
- addressing.

The layer 3 protocol provides the functions for establishing, maintaining and terminating point-to-point (p-2-p) connections. A p-2-p connection is done to transmit data between two terminals via a channel. Each layer in the OSI model is responsible for some tasks (multiplexing/de-multiplexing, segmenting/de-segmenting, compressing/decompressing, etc.). Layer 3 is the most relevant for physical connection reliability. The radio interface carries bursts of the size of 114 bits; formatting is therefore very important. Layer 3 contains three sublayers.

Radio Resource (RR) Management
RR management is responsible for setup, maintenance and termination of a dedicated radio channel connection. It englobes procedures for managing shared transmission resources. An RR connection may be released and re-established on another physical layer several times during a single p-2-p connection between two terminals.

Mobility Management (MM)
This entity controls mobility of an MS and authorisation to access the network. MM manages the identification, authentication and allocation of new TMSI. It also mainly handles the procedures of IMSI attach, IMSI detach and location update.

Call Management (CM)
CM is useful for setup, maintenance and termination of circuit switched calls. It is in its turn divided into three entities (call control, short message service and supplementary services). Call control (CC) provides the transport layer with a p-2-p connection. CC is responsible for call establishment, call re-establishment and call clearing. For more details on the network layer, the reader can refer to the GSM recommendations. For optimisation purposes, this layer is the most commonly analysed as it detects network problems.

Protocol Trace Exploitation in Optimisation

Based on what has previously been explained in this section, here the practical applications are presented, and it is shown how daily optimisation uses signalling charts to improve network quality. The idea is first to present and explain the signalling charts of the main quality events of the GSM system (e.g. MTC, MOC, LU, etc.). An indications is then given of how to monitor and analyse network performance through protocol analysers. Some of the main events are studied that occur in a GSM network and that may encounter quality problems.

Connection Establishment
A circuit switched connection is established when either a terminated or originated call is launched. Before going in detail into the mobile originated call (MOC) scenario, a block diagram (Figure 2.61) is given that summarises the base scenarios. As illustrated in the figure, the following procedures occur in order:

- Paging takes place only for mobile terminated calls (MTC) on the PCH.
- A control channel is established between the MS and BSC for identification and service request (RACH from the MS side).
- An SCCP connection request is made from the NSS by the BSS (assignment of the SDCCH).
- Authentication is asked for by the NSS. IMEI checking may also take place if activated by the operator.
- Ciphering takes place on the air interface once authentication occurs.
- There is exchange of additional information between the MS and MSC. This is the last step of a successful LU.
- There is a TCH assignment on the A and air interface and waiting till an end-to-end connection is established (ringing or alerting at the called terminal).
- The called party answers; the connection is then established and charging starts.
- The MS and MSC release all resources after the conversation ends.

This figure can be split into two cases: the mobile originated call (MOC) and the mobile terminated call (MTC) (see Figure 2.62).

Location Update
When a mobile first switches on, it camps on the appropriate network and registers to a given VLR. Then the SIM card in the mobile station stores the LAC and starts to systematically compare it to the LAI broadcasted in the BCCH layer of the serving cell. When the MS detects a difference, it initiates a location update procedure. If implemented in the network, an MS can be asked to perform a location update initiated by the network at periodic times (typically every 4 hours). The location update uses signalling only; there is no traffic channel assigned. Hence, signalling messages exchanged over air, A_{bis}

Radio Network Optimisation

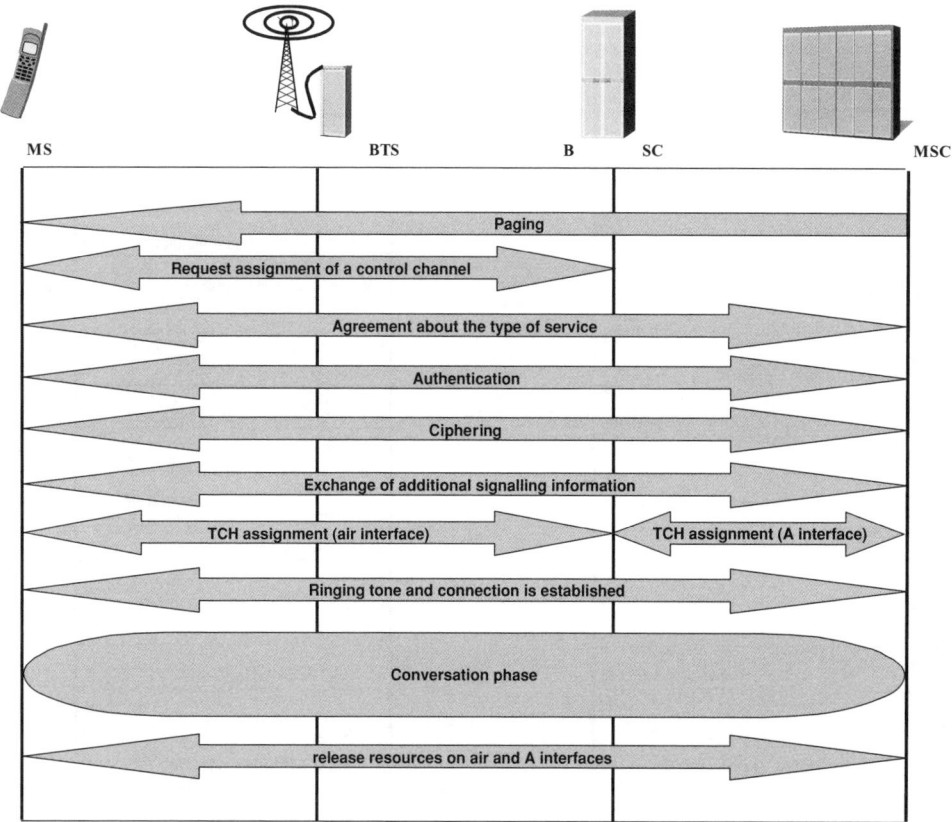

Figure 2.61 MTC and MOC establishment message flowchart

and A interfaces are quite similar to a call establishment except for the TCH assignment, which is present only in the call establishment.

Call Phases

Each of the steps in an MOC or MTC is classified within a certain call phase number. Signalling scenarios are divided into 15 phases. By understanding properly call phases along with failure causes reported in the signalling trace, the origin can be determined of an abnormal channel release, for example. A failure cause gives a description of the error that generated the failure. The 15 phases cover the following items:

- *Phase 1* covers the first part of signalling in an MOC or MTC over a dedicated channel, be it an SDCCH or a TCH.
- *Phase 2* covers the MM signalling between the MS and MSC in a transparent way to the BSC. It may occur on the SDCCH or TCH.
- *Phase 3* covers the assignment procedure step of the MS from the SDCCH to TCH by means of the basic call setup.
- *Phase 4* englobes clear command and release steps on an SDCCH or a TCH.
- *Phase 5* covers the FACCH call setup assignment signalling with channel mode modification (from Assignment_request to Alert); it is the same as phase 3 where the FACCH call setup is active. The dedicated channel can only be a TCH.

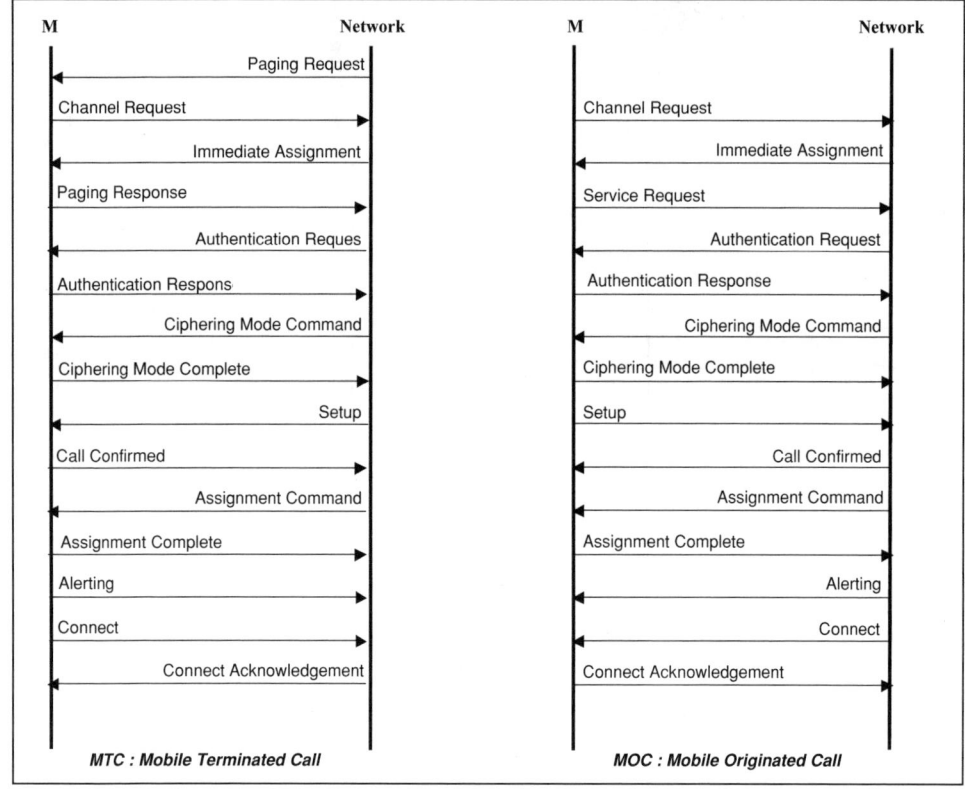

Figure 2.62 MOC and MTC connection establishment messages

SMS Establishment Phases
- *Phase 6* covers the SMS establishment signalling over a dedicated TCH and data request from the MSC.
- *Phase 7* covers the SMS establishment signalling over a dedicated SDCCH.
- *Phase 8* covers the ciphering signalling (from Ciphering_mode_command to Ciphering_mode_complete); it is a sort of 'bracket' inside phase 2. The dedicated channel can be an SDCCH or a TCH.

Handover Phases
- *Phase 9* covers the external handover signalling of the source side on an SDCCH and a TCH.
- *Phase 10* covers the internal intracell handover signalling of the source side on an SDCCH and a TCH.
- *Phase 11* covers the internal intercell handover signalling for the source side on an SDCCH or a TCH.
- *Phase 12* covers the external handover signalling for the target side on an SDCCH and a TCH.
- *Phase 13* covers the internal intracell handover signalling for the target side on an SDCCH and a TCH.
- *Phase 14* covers the internal intercell handover signalling for the target side on an SDCCH or a TCH.
- *Phase 15* covers the MM signalling between the MS and the MSC during conversation on a TCH.

The phases relevant to a call will now be highlighted along with an illustration of a mobile originated call scenario (Figure 2.63). The mobile terminated call uses the same signalling chart with some differences as compared to the mobile originated call:

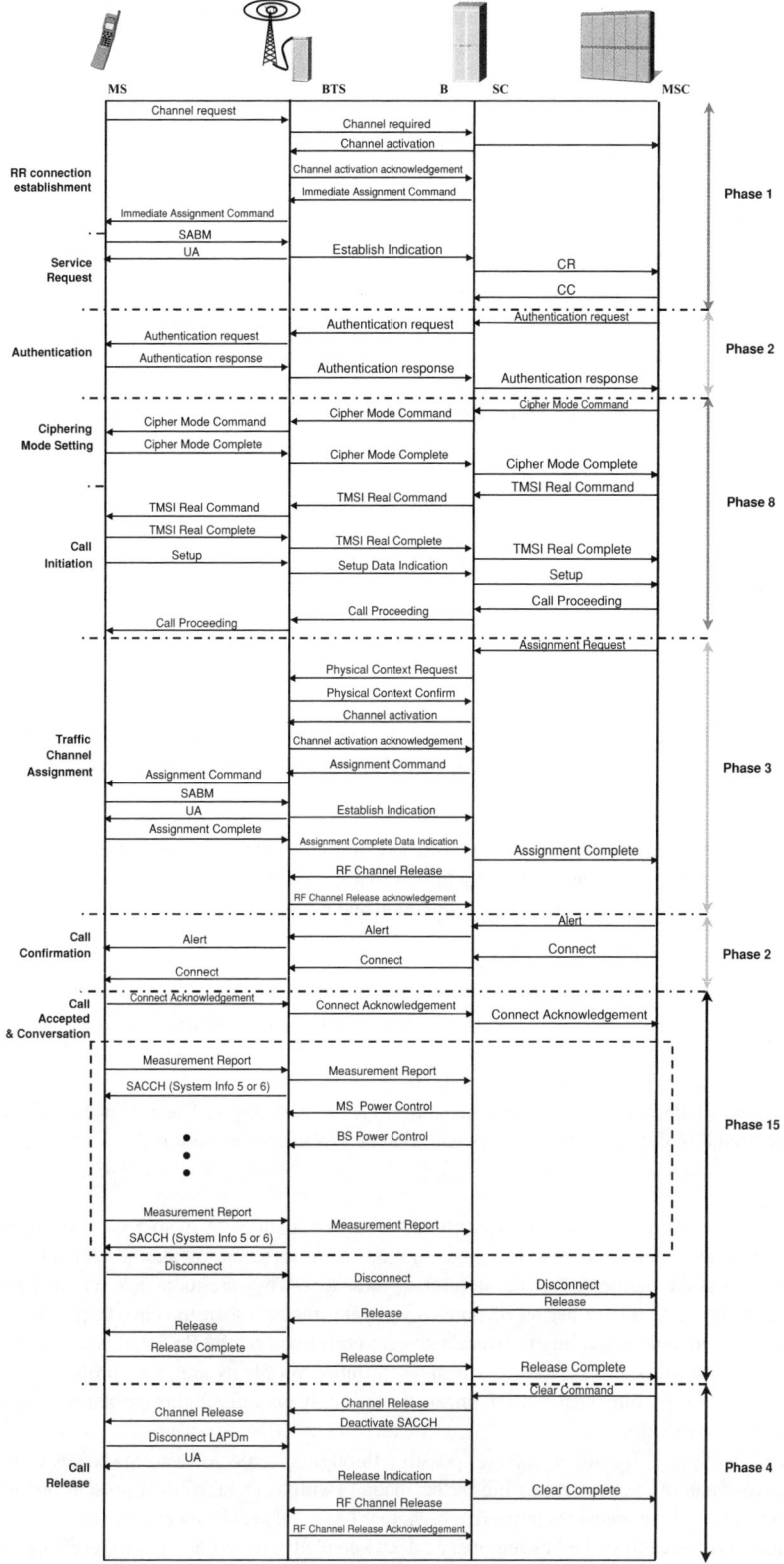

Figure 2.63 MOC scenario signalling illustrated with call phases

- First, the called party is paged; hence the channel_request is a response to the paging_request.
- The first message sent on the SDCCH is the paging_response and not the CM_service_request.
- The MSC-VLR initiates the setup message, not the MS.

Location update and handover signalling charts follow the same philosophy and can be found in the GSM literature.

Optimisation through Signalling Charts

Practically speaking, the air interface is impossible to trace via a tool. In the BSS subsystem, the most commonly traced and analysed interface is the A_{bis}. Many messages transit on the A_{bis}. Messages relevant to a unique BSC-BTS transaction should first be isolated. This means that transiting messages are discarded. Then, different MSs in uplink and different TRXs in downlink use different radio channels. How filtering should be performed on an A_{bis} trace has just been presented in order to diagnoses call per call. Now, a discussion follows on how the protocol test equipment can be used to improve network quality. There will also be a focus on most common GSM errors and their roots.

Quality of service in telecommunications, as a reminder, means the level of usability and reliability of a network and its services. QoS monitoring implies permanent observation, qualification and adjustment of various network parameters. As explained before, three complementary ways exist to measure and follow QoS.

QoS Measurement Techniques

Drive Tests

These present the advantage of reproducing a real subscriber's perception of the network. However, on another hand, they are very expensive and concentrate on a limited area within a small timeframe.

OMC

Statistics delivered by the OMC cover the whole network. However, they are abstract and do not really reflect what subscribers encounter. An advantage is that the OMC platform is free of charge since it is part of the GSM system and serves statistics collection and newly tuned parameter implementation at the same time.

Protocol Analysers

All events can be available for further analysis. Moreover, these are independent tools that do not need any specific vendor system knowledge. However, they have the disadvantage of being difficult and complicated to plug to the entire network all the time.

As illustrated, all three alternatives have advantages and disadvantages. That is why only a combination of the three methods allows the level of a network quality of service to be stated.

Tools

Protocol analysers are devices equipped with certain functionalities to enable them to intercept data signals on different interfaces, store the huge amount of information, display it later on a screen for analysis and assist the field engineer in interpreting messages. They are tools that may be connected to the BTS, BSC or MSC over a period of time to generate trace files. In the case of problems, manual analysis starts. Generally signalling measurements are performed on the BSS part where more network elements and parameters exist; hence, hardware and tuning problems are more probable in the NSS. When used in NSS signalling analysis, the protocol analyser mainly highlights problems in interworking of protocol implementation.

Automatic analysis of protocol traces is possible through specific software packages developed by the manufacturer. It offers postprocessing in the shape of graphs or tables that present one level up in interpreting and summing of events per cell or area. In the manual analysis scenario, the task is more time consuming and requires from the field engineer a deep knowledge of protocols and signalling charts. The

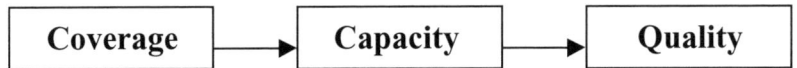

Figure 2.64 Network performance constraint evolution path

main task consists of making use of filters and search functionalities to detect errors and then interpret and analyse them.

There is a close interaction between signalling charts and counters. Step by step, while messages flow in both uplink and downlink ways, timers and counters are incremented and decremented. Therefore, the philosophy of upstep and downstep is dependent on the message. Timers are indicators generally expressed in seconds; they start counting down when the network sends a request and is waiting for the other side to answer. When they expire, a failure counter is incremented and the system itself takes the decision to either terminate the transaction, for example, or re-initiate a certain phase. On the other hand, most of the request, acknowledgement, nonacknowledgement, failure, rejected, seizure, assignment and complete messages result in an increase in several counters. Counters seen before are relevant items to estimate the quality of service before and after optimisation actions.

Network Coverage and Capacity Enhancement Methods

The performance of a mobile network is seen from several perspectives:

- Capacity. Each operator owns a limited frequency band which is shared by an increasing number of network users.
- Quality. Quality is measured by metrics such as the carrier-to-interferer ratio (C/I), bit error rate (BER), frame error rate (FER), dropped call rate (DCR) or call setup failure (CSF).
- Coverage. Service continuity is tremendously important in a cellular network.
- Cost. Technical requirement to achieve the best quality perception from the subscriber side should be balanced with costs.
- New services. Operators cannot differentiate and gain new customers if they do not offer enhanced data services on top of high data rate technologies (like GPRS, EDGE and WCDMA).
- Complexity and flexibility. Two very important criteria from an OPEX point of view (operation of the network) make it easy for maintenance teams to maintain the network.

All these aspects are correlated between them and are interdependent. Commonly, a network transits mainly through the path shown in Figure 2.64.

At the first roll-out stage, the most important thing is to provide coverage. Then, when the subscribers' base starts to increase and generate more traffic, capacity becomes the main issue for the operator. When a certain maturity of the network is reached in terms of subscriber behaviour and coverage/capacity levels, the operator should improve quality and end user perception. To achieve the best tradeoffs, especially between quality and costs, the operator tries to make use of several coverage and capacity enhancement methods.

Coverage Enhancement

Coverage improvement may result from different approaches. The first approach consists of improving the path balance. The second way can be through increasing radio performance. Boosting coverage also means expanding the indoor signal level.

Path Loss Improvement

Uplink and downlink transmissions should be balanced. This is why it is important to perform power budget calculations. The output path loss is very closely related to two parameters: transmitted power of the transceivers and sensitivity of the receivers. There is no need to increase output power if the sensitivity of the receiver from the other side is not high enough to detect this power.

Many mechanisms exist to enhance coverage in both uplink and downlink directions while maximising the path loss:

- increase output power using boosters;
- increase sensitivity using mast head amplifiers (MHAs);
- use higher gain antennas, preferably directional;
- reduce multipath fading using diversity at the receiver end;
- lower the received signal threshold;
- reduce cable loss using thicker and shorter feeders;
- bypassing combiners.

Boosting Power
Some features help to boost the power beyond the initial transmitter capacity.

Extended Cell
When there are environmental constraints preventing building on more sites, the extended cell feature can be used as a coverage enhancement solution. The serving area of a normal cell is 0–35 km while an extended cell serves the area of, for example, 33–70 km. This option is excellent in rural areas with low population and special coverage situations, such as archipelagos.

One main normal transceiver plus one extended transceiver are used to make the output power double. The implementation of the extended cell is based on one TRX serving the normal area and the other TRX (E-TRX) serving the extended area. This solution is cost effective in the sense that it saves new site deployment.

Smart Radio Concept
This feature is based on two parallel mechanisms: intelligent diversity in the downlink and four-way diversity in the uplink. The basis uses two physical TRXs to carry the same traffic in the DL and four antennas for reception in the UL.

Indoor Coverage
Indoor coverage improvement remains the most challenging planning achievement. In hotspot indoor locations, operators increase the coverage in order to intercept all possible traffic. Due to different attenuation factors, indoor and deep indoor coverage always suffer signal degradation. Therefore specific indoor solutions are planned.

A distributed antenna system is generally used to deploy a wide coverage by means of a single transmitter. The principle is to feed a large number of antennas through a network of feeders, splitters and couplers. Splitters divide an input signal into two or more equal signals while couplers distribute an input signal unevenly on different output branches.

Generally, the use of active elements such as boosters to pump the power is not needed. Therefore passive solutions are discussed.

Capacity Enhancement
Radio Channel Allocation Techniques
To increase the capacity of the system, the first step is to optimise the use of resources.

Directed Re-try
This is a useful feature used to distribute traffic load at the call setup. It is a kind of SDCCH to TCH handover where the communication is initiated on the SDCCH of a loaded cell and the call is then handed

out to the TCH of an adjacent cell that is less loaded. It is possible to tune and rule the algorithm according to defined parameters.

Load Sharing
This solution is useful during the call. In fact, calls belonging to a loaded cell and initiated by mobiles at the edge of the cell are handed out to less loaded neighbours.

Queuing
Enabling queuing is an efficient way to improve the satisfaction of the end user by letting the call attempt wait by some milliseconds till the system is freed.

All capacity enhancement features can be used simultaneously. This gives a possibility to combine the benefits of several solutions and multiply capacity gain.

Means to Limit Interference Influence
Multiplied interference in a system also impacts the capacity.

Discontinuous Transmission
The aim is to stop emitting in periods of silence, either from the mobile side or the base station side. It allows interference to be reduced on other communications. It can be performed either in the downlink or uplink.

Frequency Hopping
Frequency hopping is a way to use a given frequency spectrum more efficiently. It makes the interference spread on the whole network instead of being localised in the case of fixed frequency allocation. This solution is especially adequate in urban areas. The rural environment is noise limited; this means that a tight frequency re-use is not needed. The re-use distance is large, so there is no interference.

Uplink and Downlink Power Control
Power control in the uplink and downlink is the mechanism where the powers of the mobile station and the base station are monitored. Depending on the distance between the two stations, they are asked to increase or decrease their power. This also helps to reduce the overall interference in the network.

Smart Antenna
This is a switched beam system where MS hand-offs are between antenna beams rather than sectors.

Network Physical Extension
Instead of adding new sites, it is possible to increase the capacity of existing sites by adding other transceivers.

Dual Band Network
Using another frequency band is also an efficient way to separate traffic load. The majority of handsets can use GSM900 or DCS1800 bands. The traffic management between the two bands is done like a multilayer handling in GSM. Thresholds and decision parameters move the hand-offs from one band to another.

Half Rate
In a normal case, one communication requires one timeslot in the TDMA frame at 13 kbps speech coding. One timeslot can be used to carry two traffic channels at lower bit rates. This allocation is called a half rate. The idea is to remove user data in order to fit two parallel communications in the same physical channel. This has an impact on audio quality, which in the worst case is a matter of a half-rate mobile to half-rate mobile communication.

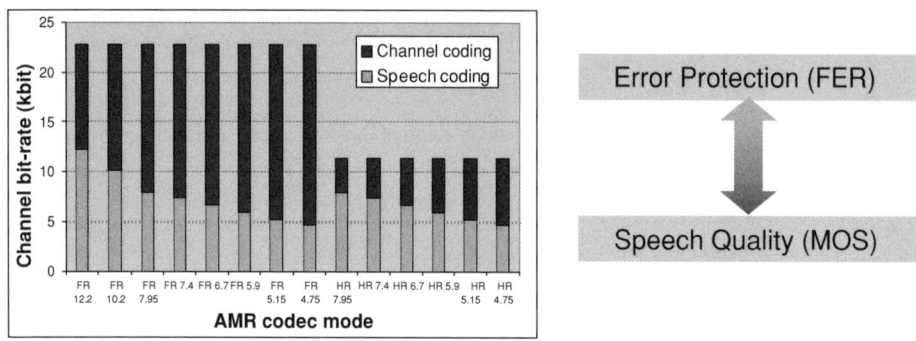

Figure 2.65 AMR tradeoff

Radio Performance Increase

Sectorisation and cell splitting are methods to enhance capacity for the macro cell case. Sectorisation means using three sector cells instead of omnidirectional antennas. This method helps to re-use the frequency pattern more efficiently by setting a smaller frequency re-use factor. Cell splitting, on the other hand, means adding a new site between two existing sites to share the traffic load with them. This solution is limited because at a certain site-to-site distance, it is not possible to add any more macro sites. Then, the other option is to add micros and in-building cells, especially at hotspot area levels.

Multilayering the network means structuring overlapping cells into upper and lower layer cells in order to allocate users to these layers according to their speed. Slow moving users camp on small cells and fast moving mobiles (i.e. vehicles) use the macro layer. This separation optimises the number of handovers and distributes traffic to reduce overload.

AMR for Both Coverage and Capacity Enhancements

The adaptive multirate (AMR) is a kind of tradeoff between source coding and channel coding in a TDMA frame. The AMR provides a set of codecs fitting in full rate (FR) and half rate (HR) physical channels. These codecs provide a variable tradeoff between error protection and speech quality (see Figure 2.65). The aim is to improve speech quality by adapting the most appropriate channel mode and codec based on radio conditions. It consists of two algorithms:

- link or codec mode adaptation (choice of codecs);
- channel mode adaptation (half rate or full rate).

The AMR takes into consideration the radio conditions (C/I) and the load conditions (EFL %). It uses 14 codecs, 8 in FR ranging from 4.75 to 12.2 kbps and 6 in HR ranging from 4.75 to 7.95 kbps.

In GSM FR/EFR, the channel bit rate is

$$22.8 \text{ kbps} = 13 \text{ kbps (speech coding)} + 9.8 \text{ kbps (channel coding)}$$

In the AMR case, the codec mode can be changed and more error correction bits can be used whenever the channel requires. The best codec choice is done through the link adaptation algorithm where, to each C/I value, the codec providing the best MOS is chosen.

The AMR has several benefits:

- Speech quality (MOS). Due to the use of new codecs that are more robust to interference than the EFR codec, the AMR can maintain good speech quality in poor C/I conditions. Audio quality is proved

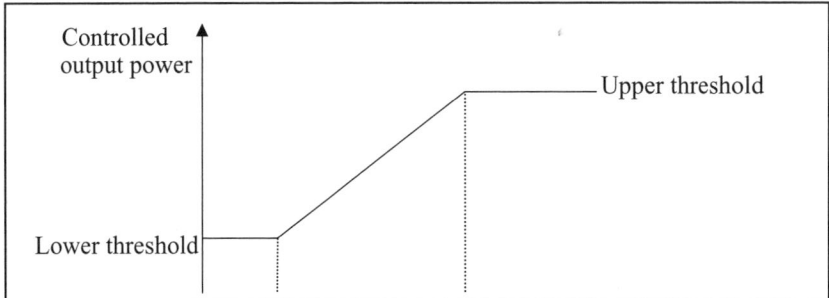

Figure 2.66 Variation of controlled output power

to be better than with the EFR in bad C/I conditions. In addition, AMR HR speech quality (MOS) is significantly better than GSM HR and is comparable with GSM FR in good link conditions.
- AMR coverage gain. Performance at the cell edge is improved thanks to the use of more robust codecs. The most robust codec (4.75 kbps) seems to provide, according to laboratory tests, a 5 dB gain compared to the EFR performance.
- Capacity gain. AMR robustness allows tightening frequency re-use patterns, increasing spectral efficiency. The AMR-FR can provide around 140 % capacity gain versus EFR in the hopping layer.

The AMR is at the end a speech codec, which adapts its operation optimally according to the prevailing channel conditions. The adaptation is based on the received channel quality estimation in both the mobile station and base transceiver station (BTS).

Conclusion
A multitude of methods exist to enhance capacity and coverage. They are therefore interdependent. The issue involves how to chose from several methods and when. Once implemented, these solutions increase the complexity of quality degradation analysis.

Handover Performance
The main idea of mobile cellular networks is enabling mobility. Mobility means allowing the subscriber to make or receive calls when moving within an own network or between different networks (e.g. different cities or countries, if a roaming agreement exists between the operators). To ensure this, two elements are important: firstly, cells constituting the network should overlap between them at an acceptable level and, secondly, there should be a software feature that handles mobility management. This section concentrates on explaining the most relevant issues on handover as the most important feature of cellular networks.

However, before that the main characteristics of power control are highlighted, as it is a feature that influences the handover algorithms and decisions.

Power Control
Power control is a feature of GSM that enables control of output power of either the MS or BTS in order to reduce interferences. In fact, the idea is that when the distance between the two stations is small, there is no need to transmit 'speak' at full power. An analogy to assimilate this idea is to state that when someone is close to you, you reduce your voice to the level that the other person can hear (there is no need to shout), and vice versa, when the person is far from you, you try to shout at your maximum power to reach that person. The same algorithm exists in power control. There are two level values that delimit the range where power can vary: upper and lower thresholds, as illustrated in Figure 2.66. Whenever the signal level goes beyond the upper value, the mechanism of PC brings back the power to the upper value by means of a power decrease step size. Whenever the signal is below the lower value the mechanism

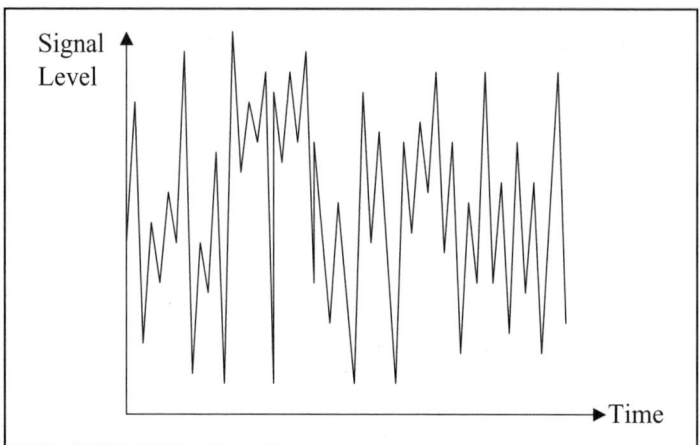

Figure 2.67 Variation of real signal level values

adjusts the power to reach the minimum lower value through progressive increment of a defined step size. Generally the upper and lower values are not the same in order to prevent the ping-pong phenomenon. As power control can work in both directions – uplink and downlink – there is a separate threshold range for each direction.

Uplink power control means that the BTS controls the emission of the MS. This is the one most commonly used in almost all GSM networks. Downlink power control is the one where the MS controls the power of the BTS, but is less commonly used because of its complexity. Especially when the DTX is activated, DL power control does not work properly. In fact, one BTS controls one MS but many MSs may control one BTS. Two separate parameters enable the activation or not of either of the two power control mechanisms.

Generally, signal strength real values fluctuate greatly over time (see Figure 2.67). In order for either DL or UL power control to work, the idea of averaging is introduced.

The mechanism of averaging is based on three parameters: signal level (L), window size (WS) and weighting (W). The samples of successive L measurements within WS are averaged with attribution of W to the last sample in order to produce one equivalent level averaged sample (LAS). Then the window is shifted in the time and the same calculation is repeated. When P_x (probability number) values of LAS out of N_x (total number) fulfil the criterion of being bigger or smaller than the upper or lower thresholds respectfully, power control is launched. The bigger the window size, the smoother is the equivalent output power. Figure 2.68 shows the difference between the two different periods.

In conclusion, power control has a direct impact on signal level. Since handover decisions take into account signal level, it can be said that there is a close interdependency between the power control and handover control mechanisms. Power control is, moreover, a feature that aims to reduce interference and thus influences many handover decisions. The idea is to reduce the power of the transmitter until the level is reached that is necessary and sufficient to the receiver to understand the signal. Practically speaking, if the MS is too close to the BTS, it asks it to reduce its power. In the power control mechanism, there is also a need for sampling and averaging the signal before any decision is taken. Power control can occur either in idle mode or connected mode. The PC algorithm has also a set of parameters, as mentioned earlier in the parameter tuning section.

Handover and Adjacency Difference

Each cell of the network should be operational intrinsically, but also integrated in the network and interoperational with other cells. Handover is the action of transferring one call from one source cell to

Radio Network Optimisation

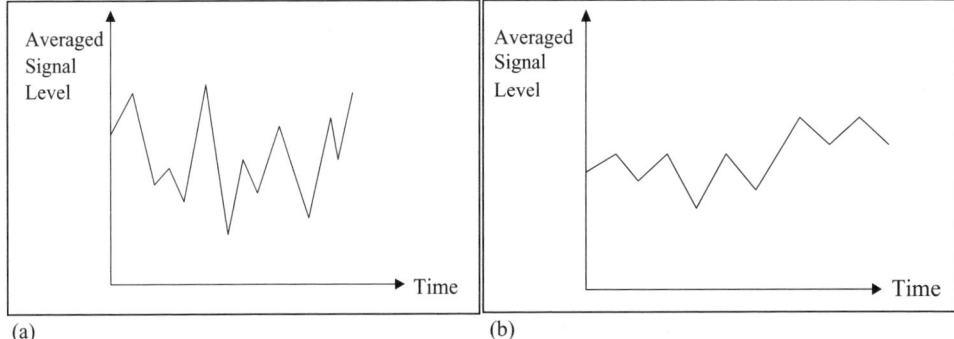

Figure 2.68 (a) Illustration of small window size; (b) illustration of big window size

a target one. Such an operation is complex and takes into account a lot of elements such as parameter setting, network conditions, architecture and resource availability. A handover is not possible between two cells unless they are defined as adjacents beforehand. An adjacency definition involves a couple of cells; the relationship should be defined from cell A to cell B and also from cell B to cell A. This requires the usage of some parameters that are explained in the following subsections. The notions of handover and adjacency are quite different in the sense that a handover indicates an action between cells while adjacency indicates the relationship between cells.

Handover Algorithm
A handover means changing the traffic channel during an ongoing call, either to another timeslot within the same cell or to another cell. The same operation in idle mode is called reselection rather than handover. The network performs radio resource management and triggers handover decisions according to measurements reported by the mobile station and according to predefined parameters and thresholds.

The handover process is initiated if the source cell fulfills some criteria and if a better adjacent cell exists to move to. Handovers can be made due to several reasons such as quality, level or interference. Handover causes follow the priority order given below:

- Interference (UL or DL)
- UL quality
- DL quality
- UL level
- DL level
- MS-BS distance
- Turnaround corner MS
- Rapid field drop
- Fast/slow moving MS
- Better cell (PBGT or umbrella)

Radio resource handovers are discussed when the cause is quality, level, interference, power budget or umbrella.

Quality
Quality handover is issued because the source cell experiences a decrease in quality.

Level
Level handover is done when the source cell signal level falls.

Interference
When a cell experiences quality degradation at a good signal level, an interference handover is performed.

PBGT
Power budget (PBGT) handover is also called the comfort handover. It means that it is a handover triggered in nonurgent conditions. This is why it has the lowest priority. It consists of moving from the current cell to a cell with less path loss. In general, and since almost all cells transmit at the same power, the PBGT threshold assessment finally becomes a received signal level comparison. Power budget handover is allowed between cells belonging to the same layer.

Umbrella
Umbrella handover indicates a passage from one layer to a lower one. Generally, very high macro sites have a specific parameter set that redirects traffic to 'normal' macro sites as soon as level conditions permit. The aim is to decrease the traffic load on umbrella sites.

Target Cell Evaluation
Target cell evaluation is based on the following criteria:

- Radio link properties
- Threshold comparison
- Priority levels of neighbouring cells
- Load of neighbouring cells
- R_x level comparison

When more than one cell fulfills the criteria, the one that has the strongest level and priority is chosen. In order to prevent the ping-pong phenomenon in handovers, the network does measurement averaging through the use of four notions: window size, weight, N_x and P_x (explained earlier in the power control subsection).

In umbrella handover, for example, adjacent cells are qualified according to radio link properties in addition to the power class of the mobile station, which determines the size of the cell. In fact two methods can be used to estimate the size of an adjacent cell, either the use of the adjacent cell layer parameter, which indicates lower, same and upper layers, or the maximum RF power an MS is allowed to use, which rules the power budget calculations. A list of candidate target cells is then set and ranked according to priority.

Adjacent cell layer management is also related to the MS speed detection. Firstly, whether an MS is moving slow or fast is decided and then the slow-moving MS is directed to lower layers. When a mobile is camping on a macro cell, for example, it reports downlink measurements of adjacent cells to the BSC. If the MS measures an adjacent cell with a 'good' level during a certain time, it is assumed that the MS is slow and an umbrella handover can be initiated towards the lower layer cell.

In layer management, careful attention drives the choice of averaging window parameters. A small window size accelerates handover decisions for a fast MS, while a large window size will prevent unnecessary oscillation for a slow MS.

Selection refers to the action of selecting a first cell when in idle mode and as soon as the mobile station is switched on. Reselection means the same action, but considers another cell when the mobile is already camping on a source cell either in idle or connected modes.

Handover from macro to micro sites induces the use of cell selection and reselection criteria. C1 can be assimilated to the received level minus Rxlevaccessmin. It is used in cell selection as a ranking criterion. It can also be used in reselection with the introduction of a hysterisis. C2 criterion is used in cell reselection and introduces the consideration of a time dimension. When time is less than a certain value, an offset penalises the level of a cell and once the predefined time elapses, C1 criterion is improved with a certain offset.

Finally, it is assumed that layer management is used in order to share traffic load in an optimised efficient way. Thus, when traffic is low on the macro layer, cell reselection offset is used to prevent unnecessary handovers to the lower layer.

Handover Features

Many handover features exist to allow faster handover decisions to be made and at the end contribute to call quality improvement. Turn around the corner MS and detection of rapid field drop features help to identify a mobile that quickly loses visibility with the micro cell antenna and commands an outgoing handover to the best adjacent cell. A decision is made according to the value of signal level variations between the two defined samples.

A traffic reason handover commanded either by the MSC or BSC allows outgoing handovers from high traffic cells towards less-loaded cells. The idea behind this feature is to reduce the power budget handover margin logically in the cell's edge. Therefore, mobiles in the border of high load cells are automatically handed-out to neighbours, enabling loads to be shared over the whole area. To be valid candidates, target cells should not be loaded. This criterion is also verified in the algorithm.

Handover due to an exceeded distance between the BTS and MS allows the repartition of traffic on all cells to be controlled, especially in rural area where wide-range cells are used. Also, a directed re-try feature explained before allows additional loads on already loaded cells to be controlled and reduced.

Parameters

As described above, handover and adjacency notions are to be separated in one's mind. Each aspect indeed involves a different set of parameters. This subsection explains the philosophy behind choosing parameters for both handover and adjacency mechanisms.

Handover Parameters

This type of parameter, also called handover control (HOC) parameters, is tuned on a cell basis. It means that cell by cell a different value can be set for a given parameter. Those parameters have the specificity to be cell dependent (e.g. a parameter that sets the level threshold in the downlink in order to perform an outgoing handover with a level cause). According to the handover algorithm, each parameter and its usability can be described explicitly.

Enablability

Most of the handover types described above are triggered only if they are enabled by setting the 'enable HO' parameter to 'yes'. Otherwise they do not occur.

Periodicity

Some parameters drive in terms of SACCH intervals the time difference between two successive similar handover events, such as handover attempts, handover failures, power budget and umbrella handovers.

Priority

In a lot of handover mechanisms, prioritisation is important in the selection of a target cell for handover. It is based on giving more priority to cells experiencing less traffic load. This issue is driven by three main parameters:

- a threshold in percentage starting from what a cell is considered as loaded;
- a load factor that gives in units the offset to apply to overloaded cells;
- a priority level that ranks primarily all of the adjacent cells.

The algorithm is simple. The final priority to rank adjacent cells is equal to the primary priority decreased by the load factor for cells loaded with more than the load threshold. At the end, cells with a high traffic load have the lowest priority and handover is then commanded to cells with no load.

Averaging

Averaging is important to prevent repetitive handover attempts between two cells. Four parameters drive the process:

- Window size gives the number of measurements that should be averaged.
- Weight indicates the importance given to the last measurement in the averaging formula.
- N_x is the total number of averaged samples that are to be considered.
- P_x is the required number among the N_x samples that should fulfil the criterion before a handover can be triggered.

Thresholds

Once the network performs averaging, it compares the N_x samples to a given threshold. Depending on the type of handover (quality, level or interference, for example), there is a special criterion that P_x samples should meet before the handover is initiated. Level and interference thresholds are in the 'RXLEV' terms whereas the quality threshold is given as 'RXQUAL' values.

Adjacency Parameters

An adjacency is defined on a cell-to-cell basis. These are in fact parameters that define the behaviour of a couple of cells linked by an adjacency relationship.

Adjacency Definition

An adjacent parameter set is useless unless the two cells are declared as adjacents. In defining an adjacency between two cells, each direction can have its own parameter set. Source and target cells are talked about specifically. Typically, the 'source-to-target' parameter set defines the handover criteria that will be used in commanding handovers from source to target and vice versa. Indeed, each cell should be able to recognise the other whether either the MSC or the BSC are handling the target cell. An adjacency is defined with a target cell location area code, cell identity, BCCH frequency, BSIC network colour code and BSIC broadcast colour code.

In idle mode, the MS listens to the defined adjacent cells BCCH list. It is, however, possible to improve and optimise the number of target BCCHs by using the double BA (BCCH allocation) list parameters. A number of preferred BCCHs can be set as the BA list.

Layers

Cellular networks can be split into different layers, each one targeting a special traffic profile. To guarantee mobility for end users, the system should ensure interworking between cells of the same layer and also between cells of different layers. This interoperability is assumed according to predefined parameters. A layer can be continuous (in the case of macro sites) or discontinuous (in the case of hotspot micro sites). The layer concept does not only include macro/micro splitting; it also englobes dual-band (900–1800 MHz) network interaction and multiradio access technologies (GSM-UMTS).

Cells belonging to the same layer have a layer relationship 'SAME'. In normal conditions, they basically perform power budget handovers between them. Cells that are among different layers have either a 'LOWER' or an 'UPPER' relationship. The idea behind layer management is to keep slow-moving subscribers on micro cells (hotspot traffic) and redirect fast mobile traffic to macro cells (Figure 2.69). Concretely speaking, a mobile station will camp on a macro cell. Handover is triggered in the lower direction only if the mobile station is moving slowly. If the MS experiences some radio problems, a radio reason handover will be done in the upper direction urgently.

Radio Network Optimisation

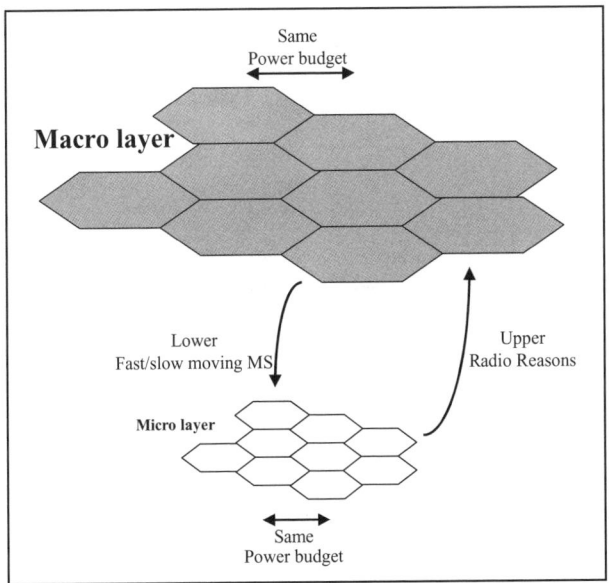

Figure 2.69 Example of layer handling

Layer Parameters

Layer
This defines the nature of the layer relationship from the source cell to the target cell. It can take one of four values (same, lower, upper, not used).

Power Budget Margin
This gives in dB how much the signal level of the target cell should exceed that of the source cell to perform a PBGT handover.

Fast Moving Threshold
This indicates the time (in SACCH frames) that a mobile should stay in a cell's coverage before deciding to hand it over to this lower layer cell.

Umbrella Level
This is the minimum acceptable level threshold that a mobile should measure to decide whether it is still in the cell's coverage area.

At the end, hundreds of parameters exist that handle handover and adjacency. Each one can be tuned separately according to the optimisation need. The parameter tuning section deals with some examples of handover control parameter tuning.

Adjacency Optimisation

Optimising an adjacency network means deleting the unnecessary relations and adding the missing ones so that the efficiency of the system increases. If each cell has to handle less adjacency the processing speed gets better. Each cell can have up to 32 adjacents, so as the network expands, the number of adjacencies increases exponentially. A rational management of adjacencies is the only way to optimise the network.

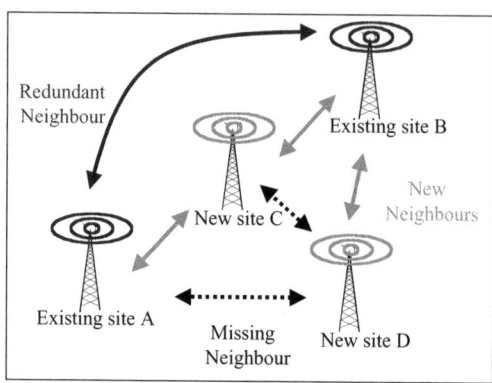

Figure 2.70 Neighbour topology in an expanding network

Figure 2.70 shows an explicit example of how adjacencies can become obsolete and at the same time shows whether there is a need to create new adjacencies continuously.

When a new site C is added and declared adjacent to existing sites A and B, the relation A–B becomes obsolete and can be deleted. A mobile camping on existing site A and moving towards new site D will perform a handover to a wrong site (C) because the neighbour relationship with the nearest cell (D) is missing. Being on a wrong server means causing interference, using high power due to distance and also means being on a frequency that, initially in the frequency plan, was not expected in that place.

Adjacent sites are continuously added and/or deleted. The add/delete task can be automated using of Man–Machine Language (MML) in order to minimise the risk of human error. MML is a text based command language with a standardised structure, designed to facilitate direct user control of a system through a user interface function.

An adjacency network is also used to monitor after BTS or BSC re-hosting. In fact, the network may require some reallocation of sites to new BSCs with a new LAC. Here, close coordination should be done with the MSC operation team in order to update the database (CI and LAC) at the BSC and MSC levels simultaneously. For instance, a bad adjacency declaration can cause a cell not to take any incoming handovers! This generally happens when LAC, CI, BCCH, BSIC_NCC or BSIC_BCC are false in the adjacent database. An alarm may indeed exist, which indicates a false declaration of target cell parameters in the source cell *adjacency* table. Such an anomaly has as a direct impact 100 % failure of outgoing handovers between those two cells. Therefore, at any change on a cell, attention should also be paid to this cell's incoming adjacency declarations.

Two elements to consider in optimising an adjacency network are the statistical aspect and the geographical aspect. As explained before, a subscriber's voice activity generates the necessary statistics used to measure the performance of the network. For handover and adjacency behaviour, it is possible to collect the number of handovers and the percentage of failure between each pair of cells in both directions over a period of time. Considering one source cell, by sorting the outgoing handover attempts in decreasing order, it is possible to calculate which relations are less useful for this cell. The operation is, however, not that simple. Before deciding to delete an adjacency, the opposite case should be considered to see whether the weight of this relation is not important from the target cell to the source cell. Consider an example to illustrate this. Table 2.23 shows that for cell 40472, the relation with cell 40298 has the lowest priority and could be deleted without affecting the performance. However, the adjacency table of the target cell 40298 is considered, illustrated in Table 2.24, it can be seen that the relation with cell 40472 is important for cell 40298 and so it is not advisable to delete this relation.

An adjacency relation is generally symmetric. This means that if cell A is declared towards B automatically cell B will have to be declared towards A. This is why if the relation is important in either

Radio Network Optimisation

Table 2.23 Adjacency table of the source cell 40472

From_LAC	From_CI	HO=>Att	HO=>Blck%	HO=>Fail%	HO<=Fail	HO<=Blck%	HO<=Att	TO_CI	TO_LAC
38	40472	15050	0	3	2	0	15804	40113	38
38	40472	5589	0	7	9	0	5485	40094	42
38	40472	3837	0	2	3	0	4518	40114	38
38	40472	3399	0	2	3	0	3016	40473	38
38	40472	2558	0	14	10	0	2317	40305	42
38	40472	2406	0	11	7	0	3104	40729	42
38	40472	2384	0	1	3	0	2070	49847	38
38	40472	2178	0	5	6	0	2845	40115	38
38	40472	1549	0	5	5	0	1858	40782	42
38	40472	1509	0	1	2	0	1500	49848	38
38	40472	1403	0	2	2	0	1260	40471	38
38	40472	1377	0	4	6	0	906	40781	42
38	40472	1058	0	3	3	0	862	40411	38
38	40472	1056	0	11	9	0	1025	40628	38
38	40472	996	0	7	3	0	941	40442	38
38	40472	691	0	86	8	0	105	40678	38
38	40472	632	0	15	8	0	704	40304	42
38	40472	589	0	8	10	0	706	40330	42
38	40472	495	0	4	5	0	405	40089	38
38	40472	410	0	5	9	0	307	54348	38
38	40472	403	0	2	6	0	442	40627	38
38	40472	380	0	9	5	0	522	49367	38
38	40472	317	0	9	18	0	371	40780	42
38	40472	56	0	46	8	0	48	40028	42
38	40472	35	0	11	3	0	834	40298	38

Table 2.24 Adjacency table of the target cell 40298

From_LAC	From_CI	HO=>Att	HO=>Blck%	HO=>Fail%	HO<=Fail	HO<=Blck%	HO<=Att	TO_CI	TO_LAC
38	40298	3440	0	3	4	0	3537	40299	38
38	40298	2560	0	2	2	0	2719	40627	38
38	40298	2068	1	2	2	0	2110	40629	38
38	40298	1300	0	4	2	0	1286	40297	38
38	40298	834	0	3	4	0	851	40472	38
38	40298	711	0	2	5	0	715	40648	38
38	40298	631	0	1	3	0	650	40445	38
38	40298	221	0	6	5	0	267	40628	38
38	40298	196	1	5	6	0	175	40411	38
38	40298	155	0	3	6	0	171	40485	38
38	40298	26	0	0	0	0	28	40622	38

direction the declaration is kept in both ways. However, for optimisation reasons, sometimes the optimisation engineer decides to keep some relations asymmetric. This is the case when for special reasons a certain handover sequence needs to be forced. The idea is to declare adjacency from unwanted cells in the sequence towards the wanted cells in one direction. The aim is that if the mobile camps on an unwanted cell can still handover towards the cell supposed to cover the MS location and if the MS camps on the right cell, there is no room to let it handover towards an unwanted cell.

The second criterion for adding/deleting unnecessary adjacencies is the geography and the location of sites. The case of adding new neighbours will first be considered. If a relation is missing between two close cells, the statistics will not detect this anomaly. In fact, statistics are picked on already existing neighbours and are blind towards missing declarations. This is why considering the use of a map is very important in the detection of missing neighbours. For the case of deletion of unnecessary relations, it has been exposed previously that the process of selection takes place through statistics. Therefore, after selection of unwanted neighbours and before the deletion, it is worth checking on a map to see whether there is enough overlapping with other neighbours to cover the whole geographical area.

Statistics

Handover performance is crucial for quality of service and mobility. Drive testing allows localising handover disfunctioning. The NMS gives statistics on handover performance to let operators identify bad performing cells and areas in the handover mechanism. The main KPI that qualifies handover on a network is the HSR. Hereafter, a clear definition of HSR will be given. Two other main notions are also important to understand: handover cause and failure cause. By identifying these two elements, troubleshooting becomes easier.

Handover Success Rate

The HSR indicates the percentage of successful handover attempts in accordance with the total number of attempts. Generally a value of 95–98 % is acceptable. Bellow this range, the optimisation engineer should look for the origin of the problem and try to find solutions. The HSR is measured on a cell or area basis considering all outgoing successful handover attempts. If the HSR is not good enough then there should be a detailed consideration of all adjacencies cell-to-cell in order to highlight the worse performing couples.

Handover Causes

In the handover troubleshooting process, causes are important to consider. In fact, for several reasons the network can initiate one handover. A high percentage ratio indicates the main cause for outgoing attempts from a certain cell. It also gives hints on possible optimisation needs. For example, a cell that performs with a high ratio of interference handovers may in reality be experiencing external interference. Therefore the frequency plan of the area needs to be reconsidered. Possible causes for outgoing handover can be power budget, quality, interference, level, slow-moving MS, congestion, turnaround corner, distance, etc.

Handover Failures

A handover failure refers to an unsuccessful handover attempt either while trying to move to the target cell or while returning back to the origin cell. Field drive tests highlight handover failure problems, especially when they simultaneously give information on other indicators such as the serving and neighbouring cell level, quality, base station identity code (BSIC) decoding and location area code. We talk about handover blocking when there is some factor preventing the call being transferred to the target cell.

Usual Checks and Corrections

As handover is a crucial operation that ensures mobility on cellular networks, more attention is paid to define coherence precisely. An adjacency is defined by the target cell location area code, cell identity,

BCCH frequency, BSIC network colour code and BSIC broadcast colour code. Any change in one of these identifiers requires an update in the adjacency definition from neighbour cells towards this target cell. Updating such information is intended to be transparent and automatic, but verification is better than assumption. One cell in the adjacent list may be declared with a false BCCH, BCC or NCC that differs from the actual values. In this case no handover can succeed towards this cell. The only solution remains to correct the parameters in the adjacent database.

In general, neighbouring relationships are symmetric. A handover may occur from cell A to cell B and from cell B to cell A. There may be an error if an adjacency is not symmetric unless explicitly wanted by the engineer for optimisation purposes.

Different check and control mechanisms help to localise and clear possible eventual incoherence in handover settings. Some specific warning alarms in the Operations and Maintenance System (OMS) may help engineers to identify adjacent cell configuration errors. Possible examples of this are:

- Same BCCH used in source and target cells. In fact, when measuring this frequency an MS cannot tell the difference between the cell it is camping on and the adjacent cell. In this case, one of the frequencies should be changed.
- Duplicate BCCH + BCC + NCC in adjacencies. If a cell has two adjacents carrying the same frequency and BSIC, the MS camping on this cell cannot differentiate between the two target cells and then cannot perform any handover towards them. There may be many handover failures because the MS may measure the first cell and the network may initiate handover towards the second one. One possible solution could be changing only the NCC or BCC.

In the end, correct neighbour topology saves a lot of optimisation time.

Parameter Management

The base station subsystem (BSS) along with the network switching subsystem (NSS) play a fundamental role in setting up calls to and from different users. Thousands of parameters exist that drive the GSM network functioning. From the BSS perspective it is possible to talk about parameters and from an NSS point of view to signify that more timers are related to the radio resource, mobility and call managements. Parameters and timers are of different types and concern many algorithms. This is why tuning is the trickiest part in the optimisation process.

As the number of parameters is huge and as performance targets are not fixed, a great number of combinations is possible. This complicates the tuning task. Operators usually start at the beginning of the site roll-out by implementing standard templates of parameters and then apply fine-tunings to specific sites/areas. It is important to understand first the parameter classification and the algorithms they are involved in and afterwards to deal with their handling.

BSS Parameters Classification

The network architecture shows that a BTS is the entry point for an MS to the GSM network and that the BTS is linked to the BSC. However, from the NMS perspective a BTS site is recognised in a certain hierarchy reflecting the hardware structure. Figure 2.71 illustrates the BSS hierarchy.

One BSC ID identifies each BSC on the network. The BCF ID is unique per BSC and identifies one site. One site may contain more than one sector; these sectors are recognised through the BTS ID. The BTS ID should also be unique per BSC. Finally, each sector transmits on several frequencies; one transmission/reception unit commonly called TRX carries each of the frequencies. The TRX ID is unique per BCF.

As explained before, parameters concern several algorithms and depending on these they may belong to one of the layers in the hardware hierarchy.

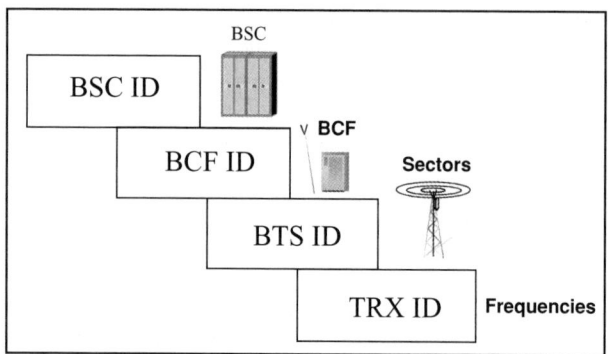

Figure 2.71 BSS hierarchy from the NMS perspective

Concerned Algorithms

Parameters can generally take one value among a range of possible values. Different values for one parameter generate different behaviours of the system towards a certain event. More explicitly, parameters are involved in determining the logic behind algorithms. The most important algorithms related to optimisation issues are power control, handover control and adjacency control.

Power Control

Power control parameters handle the thresholds for activating the algorithm, the steps for varying the power and the averaging mechanisms. It is interesting to note that power control parameters can be classified into four groups.

Enablability

Such a parameter indicates whether uplink and downlink power control mechanisms are allowed or not. Generally this type of parameter is Boolean; they take either 'yes' or 'no' values.

Interval

This group of parameters defines first the interval of the control range specifying the minimum and the maximum possible values for the power. Within this set is also included the size of steps for increasing and decreasing the power.

Averaging

Before being compared with a certain threshold, quality and level values are averaged in order to reduce the ping-pong phenomenon. In fact, quality and level samples may vary much of the time. This is why averaging is used to make the variation curve smoother. Two parameters handle the averaging process. One is the 'window size' which gives the number of consecutive level or quality samples that are averaged. The second is 'weighting', which is the factor used to increase the importance of the last sample. A weighting greater than 1 is multiplied by the last sample before the series of window size values are averaged.

Decision

Each averaged value is compared to either a level or a quality threshold. When a comparison of P_x with N_x fulfills the criteria of being greater than the upper limit or smaller than the lower limit, the power control mechanism is launched.

Radio Network Optimisation

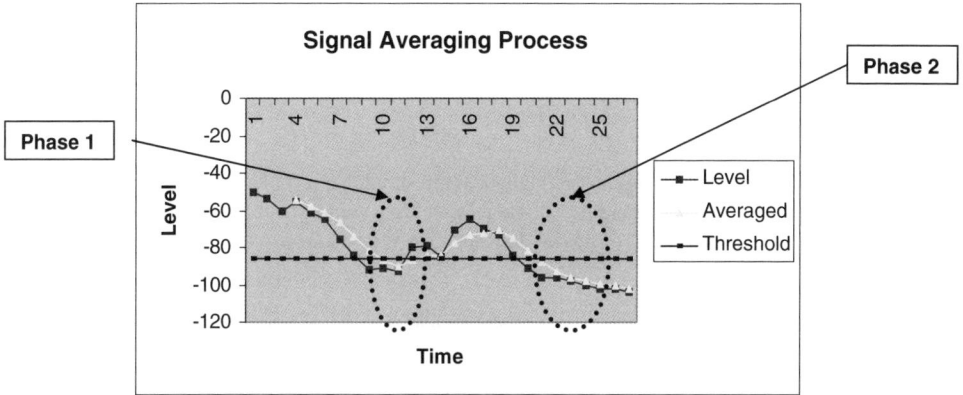

Figure 2.72 Signal averaging process and handover decision making

Handover Control

Handover control parameters handle the thresholds for activating the algorithm and the averaging mechanisms. Handover control parameters are also classified into four groups.

Enablability

As a power control algorithm, handover control can be either enabled or disabled. From the perspective of the parameter, an 'enable handover' indicator exists for each type. The following list mentions only a few:

- Intrahandover due to interference in the uplink and downlink
- Power budget handover
- MS distance process
- SDCCH handover
- Umbrella handover

Periodicity

This category of parameters rules the periods when various handover events occur. One parameter, expressed in seconds, defines the minimum interval between handover requests. Another fixes the same indicator between unsuccessful attempts. 'Handover period' parameters set the minimum allowed time between two handovers of the same type (e.g. the HO period for the PBGT).

Averaging

The notion of an averaging process in handover control is the same as for other mechanisms such as power control. The idea is to use smoother values for the comparison with the thresholds. Therefore a 'window size' of samples is averaged, while applying a 'weighting' to the last sample.

Decision

Regardless of the type of handover concerned, decision criteria remain almost the same. If P_x values out of N_x from the already averaged samples fulfil the condition, handover can take place. The condition consists of being either greater or smaller than a defined quality, level or interference threshold.

To illustrate the notion of decision making either in handover or power control algorithms, a level handover example is considered (Figure 2.72). In this example, the values set for various parameter are as follows:

- Averaging window size: 4
- Averaging weighting: 2
- HO level threshold: −86 dBm
- HO level/P_x: 4
- HO level/N_x: 6

In phase 1, only three averaged samples are below the threshold. No level handover is then performed. However, in phase 2, the signal decreases continuously and four samples out of six fulfil the criterion of being smaller than the threshold. At this point, the handover is performed.

Adjacency Control

As seen before, adjacency management has a special characteristic of handling pairs of cells and not only individual cells. This specification multiplies the complexity of the system because of the increased number of possible combinations. In fact, if an average of 12 neighbours per cell are considered, the number of items to consider in defining adjacent parameters is multiplied by 12.

An adjacent relation is defined in both ways from cell A to cell B and from cell B to cell A. Parameters used in both directions can be different. This is what brings flexibility to the system. Setting adequate values for adjacency parameters means defining the way one cell will hand over communications to the other. Adjacent parameters can be classified into four categories.

Layer

Layer handling refers to the classification of adjacent relationship types. In other words, the number of possible values for the layer parameter indicates the number of relationship types that can macroscopically be defined in a network, each type having its own set of parameters.

Margins

Regardless of the type of handover performed, a certain margin exists that should be kept between the source and target cells to prevent the ping-pong phenomenon. In a power budget handover, for example, if the margin is set to 6 dB, this means that cell A cannot hand out to cell B unless the power budget of B is greater than A by at least 6 dB. The same idea applies for level and quality.

Priority and Load

The priority notion is based on prioritising the target cells according to their load. The idea is to give the highest priority to the least loaded cells. This aims to balance traffic between cells. Three main parameters handle this algorithm: HO priority, HO load factor and load threshold.

Load threshold is a BTS parameter that sets the limit for considering a cell loaded. HO priority is an adjacent table parameter; it is a decimal that allocates a priority to each target cell. The higher the value the higher is the priority. The HO load factor is a step affected by the loaded cells in order to decrease their priority towards incoming handovers from other cells. The following example gives a better understanding of this mechanism. At first sight, cell C seems to be the one with the highest priority for cell X to make the handover (HO priority 3). However, after applying the load calculation, cell A gets the biggest priority (Table 2.25).

Threshold

Apart from specific thresholds to each type of handover, one threshold parameter exists that should always be overcome by the target cell level in order to make it eligible for taking an incoming handover. This parameter is commonly called rx_lev_access_min.

Parameter Handling

Two main operations concern parameter handling via the NMS: implementation and extraction. The first operation concerns how to change the value of a certain parameter. The second one deals more with how to recover the value of a certain parameter.

Table 2.25 Example of the load impact on handover priority

Source cell	Target cells	HO priority	HO load factor	Load threshold	Cell load	Final priority
Cell X	Cell A	2	1	70%	50%	2
	Cell B	2	1	70%	80%	1
	Cell C	3	2	70%	90%	1
	Cell D	1	1	70%	50%	1

Parameter Implementation

For implementation purposes, the definition of Man–Machine Language (MML) is introduced. This is a formatted way for man to give instructions to the machine. To change a parameter, the MML instruction should contain an identifier of the concerned element, the name of the parameter and the wanted value.

Another option is to visualise the wanted parameter graphically and to change its value. This method seems complex and heavy when many different parameters on different cells are concerned.

In conclusion, handling parameters can be done either through a graphical interface when a small number of operations is required or through MML commands and scripts when a large amount of data is concerned.

Parameter Extraction

Parameters can be either visualised at the NMS monitor or extracted from the database. The graphical interface allows the value of a certain parameter to be checked without giving the possibility of making this value available for further use. Hence, it is possible to extract the parameter database automatically through scripts and make it available independently of the availability of the NMS system. Extracting a parameter database is like making a snapshot of the network at a certain moment. It should be noted that parameters do not vary in time but between two moments, some of which may have been changed explicitly by the NMS team.

Once the database is extracted, some formatting can take place in order to make it easy for optimisation engineers to run customised queries for special analysis.

Parameter Tuning

The tuning parameter value is the most sensitive operation for optimisation engineers. It includes extracting parameters, analysing them, finding the appropriate new values and implementing them. As many parameters are interdependent, the tuning cycle never ends. After implementation of new values, monitoring takes place through observation of statistics and drive tests. According to the results, other analysis and tunings occur.

Network Care Activity

Care activity of the network is generally the biggest part of operation costs for an operator. In fact, the roll-out cost is fixed and known; once a site is on air, all operations that occur on this site fall into the category of care activities, even if it is matter of site re-engineering.

BTS Hardware Maintenance

Operators perform preventive maintenance on the BTS level to keep service quality conforming to certain norms. The idea is to clear hardware problems either by repairing or changing the faulty units. BTS maintenance means a site visit to perform various checks.

Hardware faults impact the quality of service. Optimisation engineers through statistics analysis can help to identify hardware problems. The most common problems encountered in the BTS level are classified either in hardware or inconsistency problems.

Faulty TRX or combiners require change or repair. Sometimes when the BCCH TRX output power becomes weak, a BCCH swap on a healthier TRX improves the quality. Before any unit replacement, it is worth checking the connections at all levels along the antenna line.

Another fault management aspect concerns inconsistency. Incoherence of parameters between the BTS database and what normally should be implemented (the software version, for example) requires correction. A branching table, which indicates the position where each timeslot is allocated, may differ from the BTS to the OMC.

First level maintenance consists of changing the faulty units on site and then setting up a failure report that describes the encountered problem on the module. The hardware service is responsible for repair/replacement of the units.

BSC and Transcoder Hardware Maintenance

Hardware Monitoring

BSC maintenance requires a site visit to perform some verifications and tests. Via the MML a connection is established between the BSC and an ordinary laptop at the BSC location. This laptop plays the role of a deported OMC. All operations are possible via this connection. MML tests on the main units of the BSC help to identify any dysfunctions.

The software version of the BSC, BTS and Transcoder is also checked and a backup mechanism is verified. Alarms investigation and analysis help to localise and define the problem. It is not always a matter of unit replacement. At the Transcoder level, the most important check is the exchange terminals labelling, which indicates to which BSC each card is connected. The environment of the BSC or the Transcoder is important to check: power supply, air conditioning, external alarms, etc.

BSC Load Calculations

Depending on the vendor, each BSC has special hardware and capacity limitations. This indicates a special handling and monitoring of the BSC load. The objective of such a follow-up is to prevent system overload and at the same time guarantee a good quality of service.

The BSC load calculation means identifying the percentage of use of each capacity item. Capacity items are:

- Number of TRX
- Number of BTS
- Number of traffic channels
- Number of exchange terminals

The idea of a good dimensioning is to estimate which of these items introduces the biggest usage. This percentage corresponds to the load of the BSC.

Antenna System Maintenance

The antenna is an active element that adds a certain gain to the received power from the BTS. In fact, the power at the BTS connector experiences cables and connectors loss along the antenna line. The connections are normally tied to a certain pressure. When the optimisation team suspects a power decrease, it is worth checking the connections. To detect such an anomaly a special tool measures the standing wave ratio (SWR). This method consists of replacing the antenna with a load and creating a standing wave along the antenna line. The SWR is the ratio of the maximum to the adjacent minimum magnitude of the signal level. When this ratio exceeds a value of 1.3, it means that feeder loss is high.

Site Re-engineering

One of the network maintenance operations is site re-engineering. This has been explained in previous sections.

NMS Maintenance

The NMS is an important entity as it is the supervision system that monitors the whole network. The NMS is used to change the settings of the network, collect statistics, detect alarms, etc. It runs over an operating system, so the NMS care activity consists of checking that this operating system works properly. The system administration task is required on either a daily, weekly or monthly basis.

Daily Tasks

Internal alarms are the first thing to check daily. They indicate what are the faulty processes or applications. Special database log files record all the errors that occurred during the previous day. Some alarms can even indicate the processes that are still running but will encounter some problems soon.

Incremental backup is performed on a daily basis, where only changes from the previous day are stored. The aim is to optimise the storage space and to activate the backup in the case of a system crash. The advantage of such backup is that it is performed online without cutting into the network. Another task is to check that this daily backup is well done.

In terms of parameter handling, a regular consistency check is done to compare the parameter values on the network (i.e. in the BSC and BTS local databases) and the OMC. So that the radio optimisation process runs correctly, it is also important to check the status of statistics transfer from the network towards the NMS databases and fix all relative problems.

Weekly Tasks

On a weekly basis, the team verifies the available disk size and cleans all temporary files and logs. It is the duty of the NMS care team to ensure that the system is ready to operate correctly all network operations.

Monthly Tasks

An online incremental backup is performed daily. In addition, what is called 'cold backup' exists and is performed offline. It consists of freezing the network and storing all the parameters and settings on external devices. This integral backup is time consuming and is done on a monthly basis.

Given the crucial importance of correct functioning of the NMS, an overall audit can be done systematically to check whether the daily, weekly and monthly tasks are performed well and according to the process.

Alarms Reporting in the NMS

Alarms monitoring constitutes the main source of information on the fault status of the site. It can be separated into different types, as explained previously. Not all alarms are critical or require clearing.

Impact of Transmission and NSS Faults

The end user experiences bad quality of service regardless of the problem location on the network. In this chapter, radio optimisation has been discussed but quality degradation may come from the NSS or from the transmission lines. Incompatible timer values between the NSS and BSS can impact on the quality. Synchronisation faults on the transmission network also degrade the quality. The next two chapters handle in detail all optimisation aspects of the transmission and core networks, respectively.

Multivendor Interoperability

The standardisation organism made some interfaces open in the GSM network. This means that the manufacturers of the two end equipments of such interfaces should follow the same protocol. Operators can then use different vendors to supply their infrastructure. A multivendor environment introduces verification of different equipments from different vendors that can interoperate with each other. For that, main vendors follow interoperability tests (IOTs) between their latest versions of products and software.

The IOT is a running program as vendors produce new hardware and software continuously. From the operator's perspective, this operation is crucial for evolution of the network towards the newest technologies.

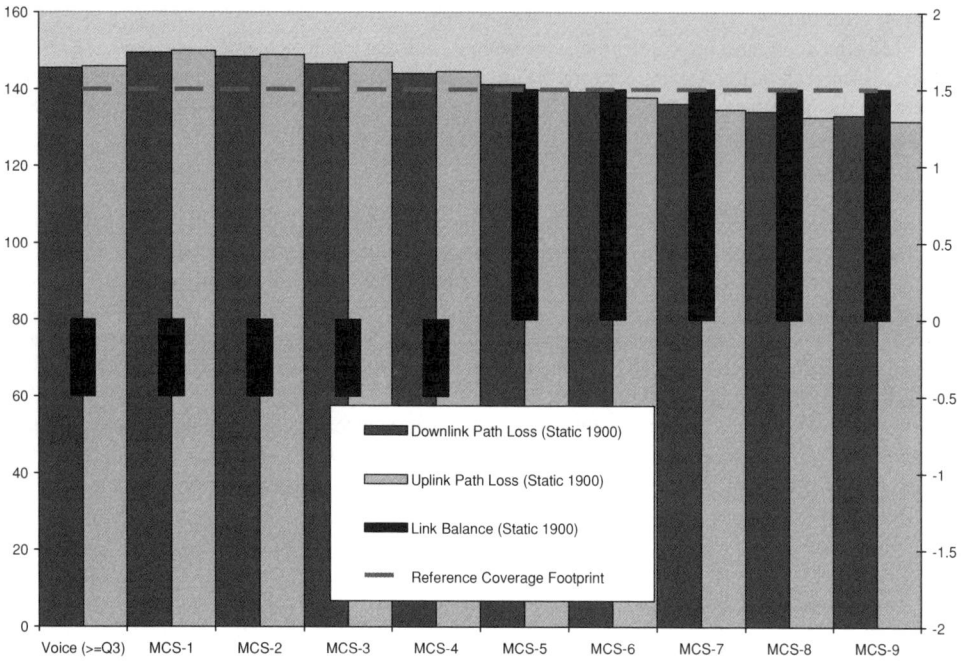

Figure 2.73 Voice versus EGPRS path loss

2.9.2 Optimisation in the EGPRS Network

The process of EGPRS optimisation can start by reviewing the link budget (see Figure 2.9, dimensioning in the EGPRS network). The graph in Figure 2.73 depicts the uplink and downlink path losses plus link balance based on a given reference coverage footprint.

While real world/practical path loss and link balance will vary dynamically based on the fading environment and equipment used (MS, BS), the following generalizations can be made with respect to all EGPRS link budget planning:

- MCS3 and voice link performance are comparable due to similar forward error correction (FEC) characteristics.
- As higher MCSs are used, the 'service area' is reduced due to a gradual decrease in receiver sensitivities.
- As higher MCSs are used, the link imbalance increases due to BS receiver performance always being better than mobile receiver performance.

Noting the above, it is very possible to achieve a 'pseudo-balanced' network assuming different coding schemes are independently link adapted on the uplink and downlink. In this scenario, link balance may be achieved for a given set of channel conditions if the subscriber's required quality of service is satisfied (assuming asymmetrical uplink and downlink bandwidth requirements for a given application).

As part of EGPRS link planning and optimization, the following key considerations need to be reviewed (the detailed reasons and significance of each of these on EGPRS performance is detailed in later sections):

- For 8-PSK, the average TX power is backed off. In the EGPRS link budget example included in this text, the site TRX backs off 2 dB while the MS backs off 4 dB.

- Receiver sensitivity gradually degrades for a higher MCS due to the decreasing forward error correction. The actual receiver sensitivity must be taken from the receiver Eb/N0 versus BLER 'waterfall' curve measurements for the given fading environment.
- Incremental redundancy recovers 1–2 dB on the highest coding schemes due to a hybrid automatic repeat request (ARQ) Type II, which essentially improves forward error coding though maximum ratio combining.

GMSK Versus 8-PSK Backoff

The introduction of 8-PSK modulation and coding schemes has introduced an inherent amplifier design constraint which must be made when planning and optimizing EGPRS radio performance. Specifically, the average power of 8-PSK driven timeslots must be decreased in order to meet the GSM specifications for adjacent channel power and spurious emissions. With this in mind, it is clearly easier to minimize this average power decrease when designing base station transmitter power amplifiers (PAs) compared with mobile transmitters. Therefore, mobile transmitters will typically require more backoff than their base station transmitter counterparts for maximum 8-PSK EGPRS transmission.

At this point, it is worth mentioning the distinction between backoff and average power decrease as referenced within this text. Backoff refers to the intentional decrease in PA output power while an average power decrease refers to the resulting average measured power of an 8-PSK signal in the time domain. Further explanation is discussed later in this chapter.

This section summarises the impact of 8-PSK timeslots requiring an average power decrease (APD) while operating on EGPRS capable transmitters. For reference in this section, it will be assumed that the requirement is 2 dB APD for the BS transmitter and 4 dB APD for the mobile transmitter during 8-PSK transmission.

There are three main considerations when one or more timeslots are operating at 8-PSK with 2 dB APD:

1. Impact to the current EGPRS user. The 'link budget' for the current user will be 2 dB lower when in the 8-PSK mode. This translates into a slightly lower grade of service (i.e. lower throughput) but not significant enough to have a major impact on the service area. Also, depending on the link balance, the mobile may use 2 dB higher transmitter power in the uplink, controlled by open loop power control. Finally, the 2 dB APD applies only when the TRX is set to maximum output power. If the entire TRX is set to the second highest output power, there is no difference between the average power of 8-PSK and GMSK signals.
2. Impact to the overall C/I. The overall impact on C/I to other users will be reduced since the 8-PSK timeslots will produce 2 dB lower average signal levels. This will 'improve' the performance of neighbouring voice and data users. However, at a maximum power of 8-PSK, the served user will have a 2 dB worse C/I so there is a higher probability of performance degradation experienced over the larger boundary zones, depending on the average C/I for the service area. Incremental redundancy and link adaptation minimize this degradation.
3. Impact to other subscribers (i.e. voice, BA list and HO evaluations). Other mobiles trying to evaluate BCCH will measure up to 2 dB lower signal strength, which is within the tolerance level for evaluating neighbours in the BA list. This has some small impact on cell selections and reselections. Adjusting the cell reselect hysteresis (CRH) high enough can be used to minimize unnecessary cell reselections.

The duty cycle of 8-PSK to GMSK on any given radio timeslot (RTSL) will be at most about 92% due to the 52 multiframe structure of a PDTCH (packet dedicated transmission channel) timeslot (which includes 2 × PTCCHs (packet transmission control channels) and 2 × idle bursts). Considering this and the fact that all PACCH (packet associated control channel), RLC signalling blocks, and dummy RLC blocks are in GMSK, the overall impact of APD across a given burst period will be minimised.

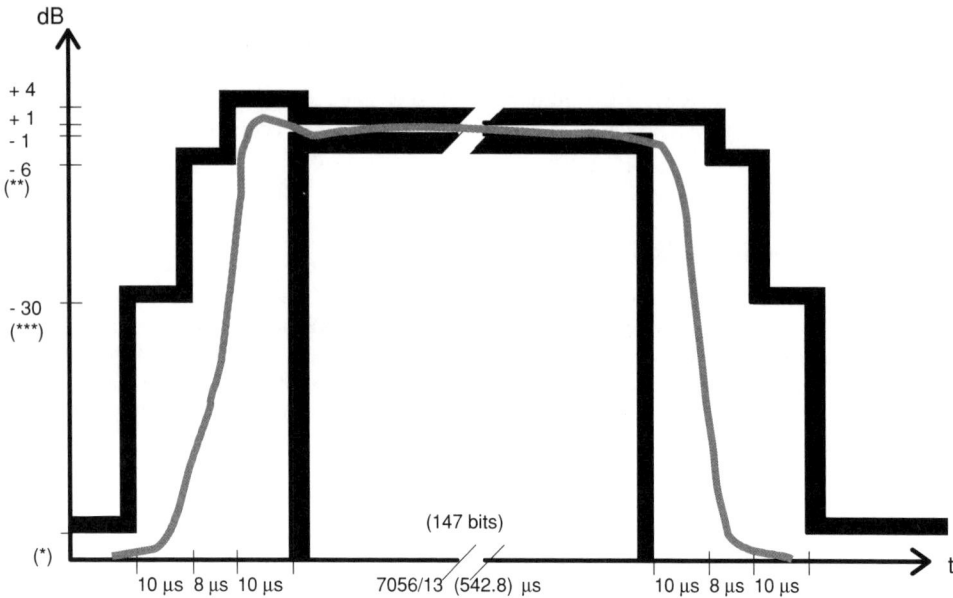

Figure 2.74 Time mask for normal GMSK burst

Burst Structure (GMSK Versus 8-PSK)

3GPP TS 05.05, Annex B identifies the following GSM/EGPRS burst structures for the transmitted power level versus time. Figure 2.74 shows the time mask for normal duration bursts (NB, FB (frequency connection burst), dB and SB (synchronisation burst)) at GMSK modulation. Figure 2.75 shows the time mask for normal duration bursts (NBs) at 8-PSK modulation. The grey 'envelope' shows a conceptual example of the appearance of a normal burst.

Average Power Levels over the Burst Period

Figure 2.76 shows an example of an existing GSM/GPRS BCCH TRX with eight normal bursts in GMSK (note that the average power is constant and the power down interval is exaggerated to illustrate timeslot separation). Figure 2.77 shows an example of a GSM/EGPRS BCCH TRX with a 3TSL EGPRS mobile active on the downlink with five normal bursts in GMSK (APD = 0 dB) and three normal bursts in 8-PSK (APD = 2 dB).

In this example, the average power decreases by 2 dB during the last three bursts due to 8-PSK. This has the following key impacts:

- slightly lower throughput near the cell edge or in a poor C/I environment;
- 2 dB lower signal level to neighbouring cells or GSM phones evaluating neighbours.

Figure 2.78 shows an example of GSM/EGPRS BCCH TRX with a 2 dB backoff and a 3TSL EGPRS mobile active on the downlink with five normal bursts in the GMSK (APD = 0 dB) and three normal bursts in 8-PSK (APD = 0 dB). Note that the average power remains constant since both GMSK and 8-PSK are operating in the linear range of the PA.

Radio Network Optimisation

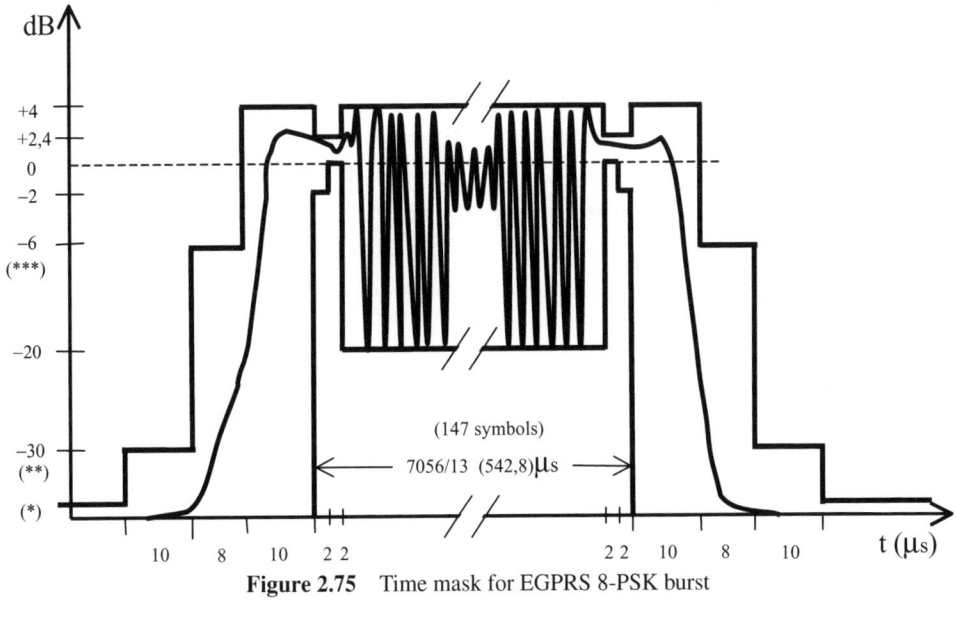

Figure 2.75 Time mask for EGPRS 8-PSK burst

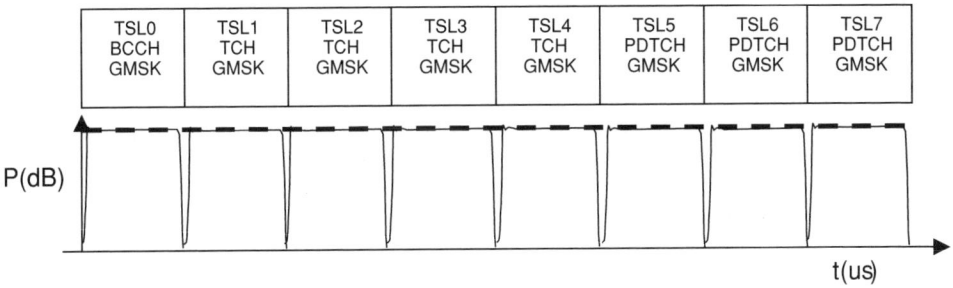

Figure 2.76 Average BCCH power with GMSK

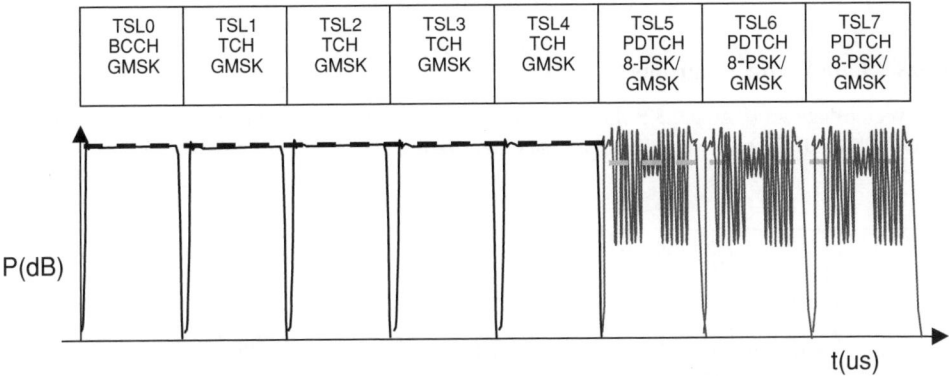

Figure 2.77 Average BCCH power with GMSK / 8-PSK mix

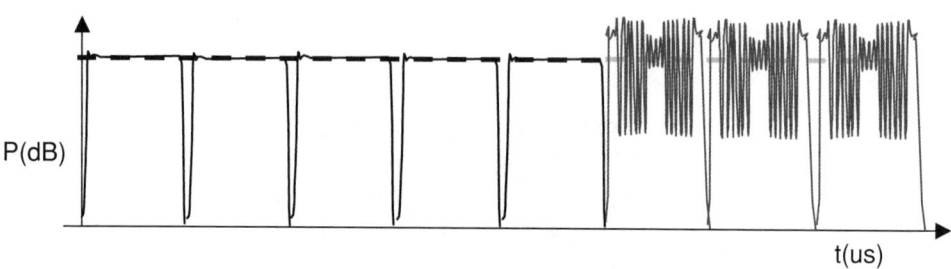

Figure 2.78 Average BCCH power with a 2 dB backoff

52 Multiframe Structure of the PDTCH

Figure 2.79 shows the 52 multiframe structure of a PDTCH. Note how the frame structure includes 2 × PTCCH and 2 × IDLE bursts that are in GMSK modulation. Therefore, the 'duty cycle' of TS 7 will always have at least four bursts in the GMSK every 240 ms. Considering this and that each EGPRS MS TS will also operate on PACCH blocks (estimating every second 52 multiframe for sustained data transfer), the average power decrease will be less than 2 dB, depending on the traffic type (i.e. the File Transfer Protocol (FTP) will occupy most RLC blocks with 8-PSK while other traffic will have 'breaks' of GMSK RLC dummy blocks or PACCH signalling blocks).

Average Power Decrease Versus Backoff in a Linear PA

Figure 2.80 illustrates the concept of a linear PA power amplitude response curve with 0 dB backoff and an average power decrease of 0 dB. Note that the 8-PSK bursts would operate beyond the 3 dB compression point (and therefore experience non-linear distortion) so an APD of 2 dB is require to ensure this does not happen. Depending on the linearity of the PA near the compression point, the APD at full power (0 dB backoff) would change. For example, mobiles are likely to have an APD of 4 dB or more near the 3 dB compression point due to the use of lower performance amplifiers (less linearity) compared to base station transmitters.

Figure 2.81 illustrates how operating at full power (left figure for a 0 dB backoff) requires a 2 dB APD for 8-PSK to operate efficiently (in the linear range) while operating at a 2 dB backoff (right figure) does not require any average power decrease for 8-PSK (APD = 0 dB).

Phase-State Vector Diagrams of APD

The vector diagram (Figure 2.82) illustrates how amplifier performance directly affects the APD and required backoff. In reviewing the vector state transition diagram, the peak-to-average 'overshoot' is

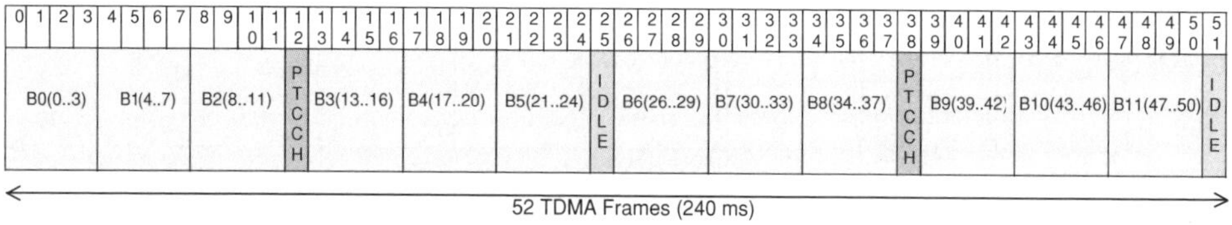

Figure 2.79 GSM 52 multiframes

Radio Network Optimisation

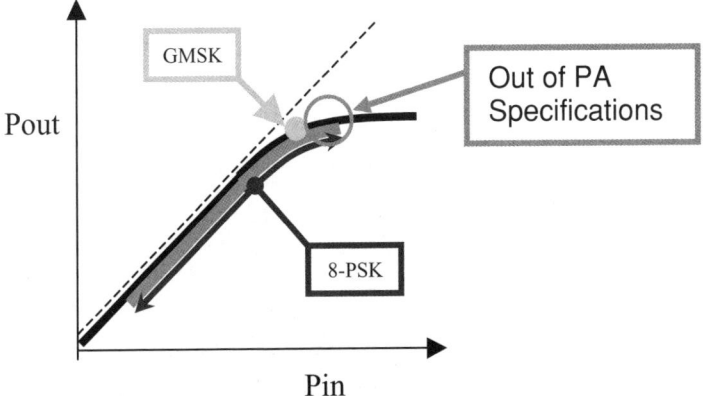

Figure 2.80 8-PSK operating beyond nonlinear compression

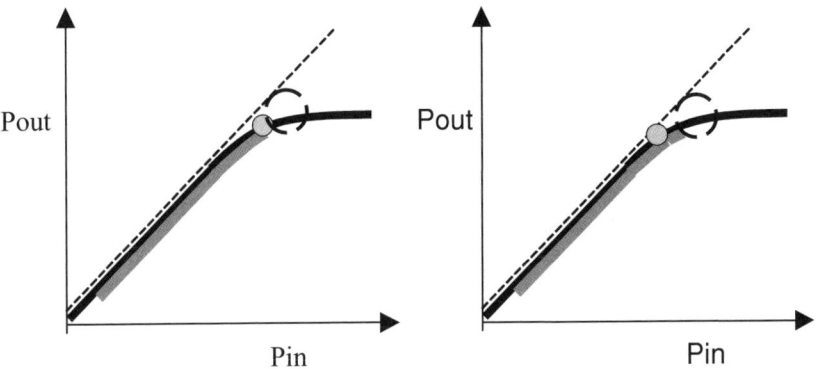

Figure 2.81 8-PSK operating with 2 dB average power decrease and backoff

required to ensure a smooth/continuous transition between phase states (as shown by the sample black trace).

Figure 2.83 illustrates the power 'envelope' in the phase-state domain. Note how the ± 1 dB power variation tolerance in the GMSK falls within the $+4$ dB tolerance for 8-PSK peak-to-average value. This illustrates that the tighter the peak-to-average 8-PSK power required for 'overshoot', the closer the average 8-PSK power approaches the average GMSK power. However, at the maximum power, 8-PSK will always require additional APD compared to GMSK due to the requirement for linear 8-PSK overshoot.

Cell Reselection

With the increased availability of higher multislot class EGPRS capable mobiles, it is important to understand the performance impact this has on the ability for the mobile to select efficiently, reliably and accurately the best serving cells in a network. At the time of writing, mobiles and networks are restricted to standard GSM idle mode measurements and much of the performance is dependent on advanced measurement algorithm techniques in each particular mobile. In the near future, Network Assisted Cell

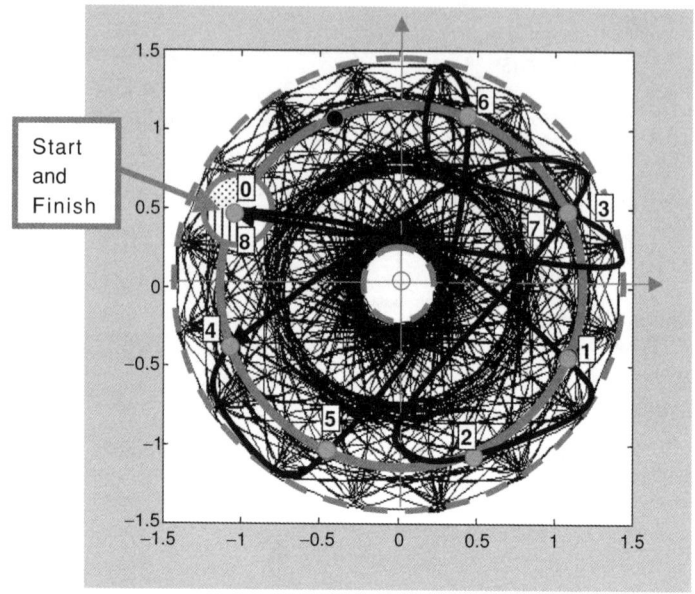

Figure 2.82 8-PSK vector state transition diagram

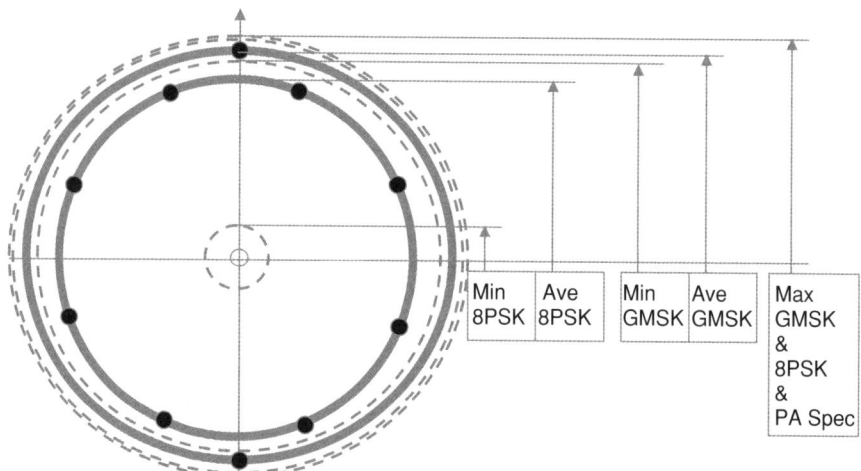

Figure 2.83 GMSK and 8-PSK power envelope

Change (NACC) and Network Controlled Cell Reselection (NCCR) capabilities will be added to the network and mobiles that support these features and enhanced measurements will become available. These enhanced features will significantly improve the cell reselection performance by reducing latency and interruptions in throughput during EGPRS cell changes.

Figure 2.84 illustrates the difference between existing GSM voice and EGPRS capable mobiles in how they evaluate neighbours for performing handover and cell reselections. During a circuit switched (CSW) call, the mobile is able to evaluate neighbours effectively every SACCH period (480 ms) since it can take full advantage of measuring during the idle time between reception and transmission on the

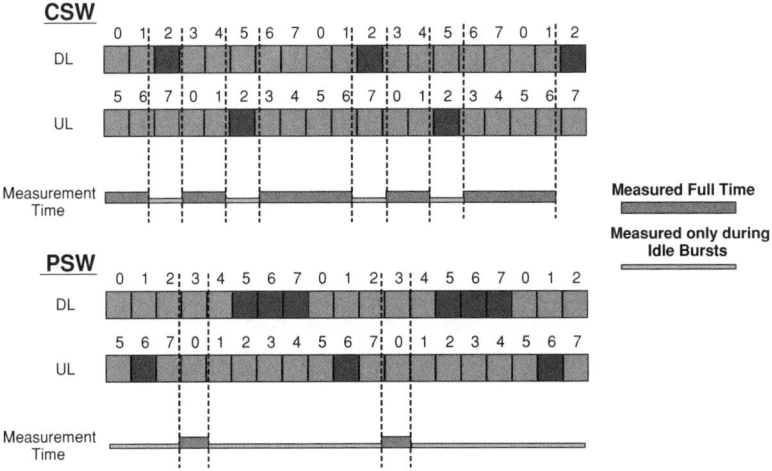

Figure 2.84 Cell reselection measurement cycle

assigned timeslot. It can also measure during the idle bursts in the TCH 26 multiframe of its assigned timeslots. The measured neighbours are reported to the network on the SACCH so the network can make an HO decision.

In packet switch (PSW) 'calls', the mobile evaluates neighbours in idle mode every 5 seconds and makes its own determination of when to make a cell change. During each measurement cycle, the mobile is significantly more limited in its available time to take neighbour measurements due to the higher duty cycle of its own downlink and uplink transmission times. In fact, the example above is a 3 + 1 (multislot class 4 or 6) device leaving only about one TSL burst period to spend full time evaluating candidates. Considering multislot class 10 (4+2) and 12 (4+4) devices, the primary reason they are limited to five simultaneous uplink and downlink timeslots is due to the single transceiver design limitation requiring at least one full timeslot dedicated to perform acceptable neighbour measurements during the 5 second idle mode cycle.

This illustrates the importance of an effective measurement algorithm within the phone to evaluate accurately and rapidly candidate cells in limited idle bursts and transmit/receive windows while balancing battery saving. The consequences of a slow or inaccurate measurement algorithm are clearly apparent and have been witnessed by the author. The symptoms can be as severe as application timeouts (loss of data RLC flow control or window stalling caused by a low signal level or poor C/I) or as mild as throughput degradation (latent cell reselection, forcing extensive link adaptation and then leading to congestion on the application layer).

The impact of the mobile measurement capabilities should not be underestimated for subscribers in mobile or static (deep fading) environments. The measurement speed and accuracy of mobiles will continue to play a significant role in future NACC and NCCR capable mobiles and will have an even bigger impact to the EGPRS throughput and latency performance for data subscribers.

The Anatomy of an EGPRS Cell Reselection

EGPRS performance is also affected by the inherent delays in idle mode cell reselection. Figure 2.85 shows the typical downlink RLC/LLC throughput response for an EGPRS cell reselection.

Note how the downlink throughput degrades after the cell reselections. The key to understanding why this is the case is that the EGPRS mobile is no longer 'listening' to the previous cell so it must re-tune

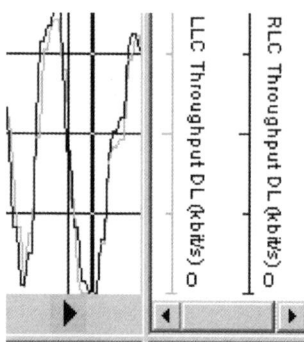

Figure 2.85 EGPRS cell reselection throughput response

to the new cell and then decode the system information messages (1 to 4 and 13). Following this, the EGPRS mobile will perform a RACH request to identify that it is on the new cell. This RACH is followed by a dummy packet control to alert the SGSN of the new base station virtual circuit identifier (BVCI) to route the LLC packets. Finally, when the packets are re-routed, the downlink sends a packet downlink assignment to re-establish full downlink data transfer on DL timeslots 5, 6 and 7.

Figure 2.86 highlights the Layer 3 messages that make up the EGPRS cell reselection procedure. In reviewing the procedure, it is noted that (in this cell reselection) the EGPRS mobile took about

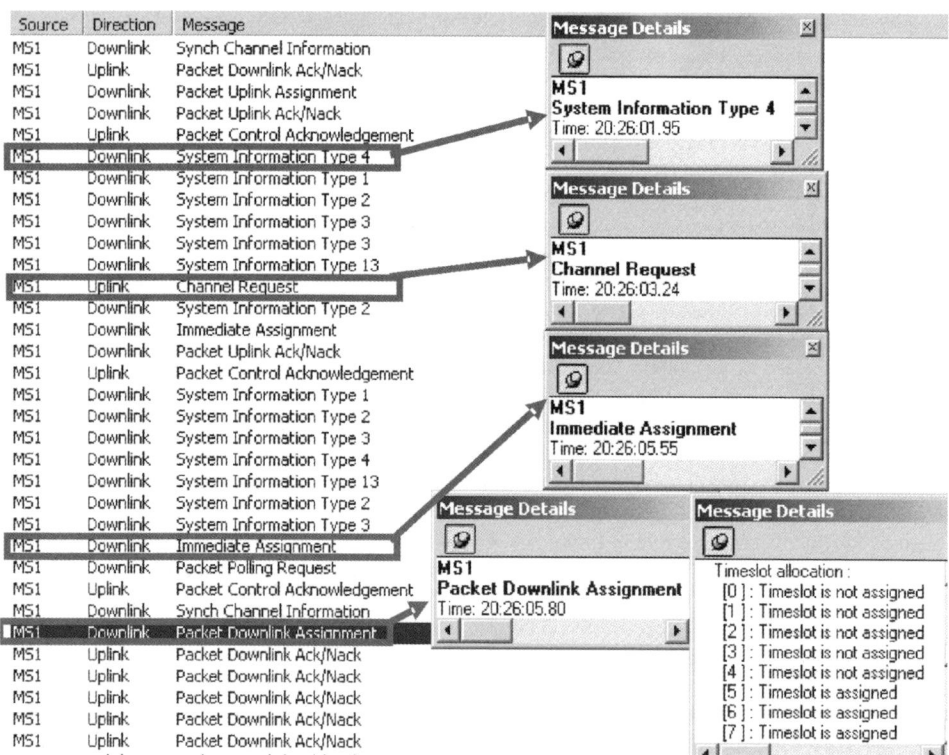

Figure 2.86 Layer 3 messaging sequence for EGPRS cell reselection

1.29 seconds to decode all required system information messages (1 to 4 and 13) and send a RACH (channel request). This time can change based on the frequency with which the BCCH/PBCCH (packet BCCH) broadcasts the system information messages along with the capability of the mobile to decode these messages (either based on MS activity or RF environment).

It should also be noted that it took about 2.31 seconds for the LLC packets to be re-routed from the old BVCI to the new BVCI. This time can change based on the level of LLC packet routing required (i.e. intraPCU, interPCU, interBSC, interSGSN), along with the loading/speed of each network element (PCU, BSC, SGSN) and 'inter element signalling efficiency'.

Finally, the packet downlink assignment message indicates that the PCU has segmented the LLC frames into RLC blocks and the EGPRS session is therefore re-established. Note that it takes some additional time (about 3 seconds) to ramp up and return to full downlink RLC throughput.

Packet BCCH

This text would be incomplete without addressing the packet BCCH (PBCCH) feature. Essentially, PBCCH utilises a timeslot on the existing BCCH TRX to provide additional dedicated packet system information and common control channels to separate control and signalling of voice and data calls. Packet BCCH can be used to reduce the latency during cell reselection since it broadcasts the packet system information messages at a higher frequency compared to standard BCCH. While this feature has some inherent benefits, it has not been deployed widely in US markets primarily due to the following four reasons:

- limited mobile support;
- additional radio timeslot is required and takes away from available TCH resources;
- additional operational complexity of maintaining separate BA and parameters for packet data;
- most mobiles have both voice and data capability and are likely to produce conflict when switching modes.

Having mentioned the above, PBCCH would be extremely useful for networks where the mobile devices are restricted to EGPRS data (e.g. data cards) and demanded very high performance with low latency requirements. PBCCH could be used in those areas where reducing cell reselection time would provide the greatest benefit while also supporting high data signalling load that would ordinarily congest a typical combined or noncombined BCCH channel.

EGPRS Throughput, Latency and Link Adaptation

EGPRS radio throughput and latency (and ultimately end user application throughput) are fundamentally influenced by the radio environment in which the mobile subscriber is immersed. EGPRS technology has the ability to take advantage of link adaptation (LA) to combat the impact of fading to improve the overall performance experienced by the user. As a reference, Figure 2.87 illustrates a typical fading environment and the mechanisms used to combat this fading. While the standard GSM forward error correction mechanisms (e.g. bit interleaving across TDMA bursts) are already applied in EGPRS, the focus here will be on the additional LA capability of EGPRS that significantly improves link performance and dynamically compensates for changing fading environments.

Given a typical fading environment (as illustrated in Figure 2.87), EGPRS throughput and latency will be affected based on the forward error correction (FEC) of its current MCS. As the fading changes, a different MCS may be more appropriate either to increase the throughput due to less bit errors or to increase the FEC to compensate for additional bit errors. The link adaptation of EGPRS allows a full range of nine modulation coding schemes to adjust to changing channel conditions. However, one of the most critical aspects in link adaptation is the change from GMSK to 8-PSK. As discussed previously,

Table 2.26 GPRS/EGPRS coding schmes, modulation, FEC and bit rate

	Coding scheme	Modulation	RLC blocks/ radio block	FEC code rate	User bits/20 ms	Bit rate [bps]
GPRS	CS-1	GMSK	1	0.45	160	8 000
	CS-2		1	0.65	240	12 000
	CS-3		1	0.75	288	14 400
	CS-4		1	n/a	400	20 000
EGPRS	MCS-1		1	0.53	176	8 800
	MCS-2		1	0.66	224	11 200
	MCS-3		1	0.85	296	14 800
	MCS-4		1	1.00	352	17 600
	MCS-5	8-PSK	1	0.38	448	22 400
	MCS-6		1	0.49	592	29 600
	MCS-7		2	0.76	448 + 448	44 800
	MCS-8		2	0.92	544 + 544	54 400
	MCS-9		2	1.00	592 + 592	59 200

there is a decrease in average power (assuming the initial GMSK link is at full power) with this change in modulation and this directly leads to increased TBF failures in threshold link level areas. Specifically, setting up the TBF in GMSK and then changing to 8-PSK could cause a link loss for each attempt at establishing a TBF.

Table 2.26 outlines the modulation, forward error correction and channel bit rate for all GPRS and EGPRS modulation coding schemes. Considering the objective of maximising the bit rate and minimising latency, the following key considerations must be made when optimizing link adaptation settings to compensate for channel performance (focusing on the MCS-4 and MCS-5 boundary):

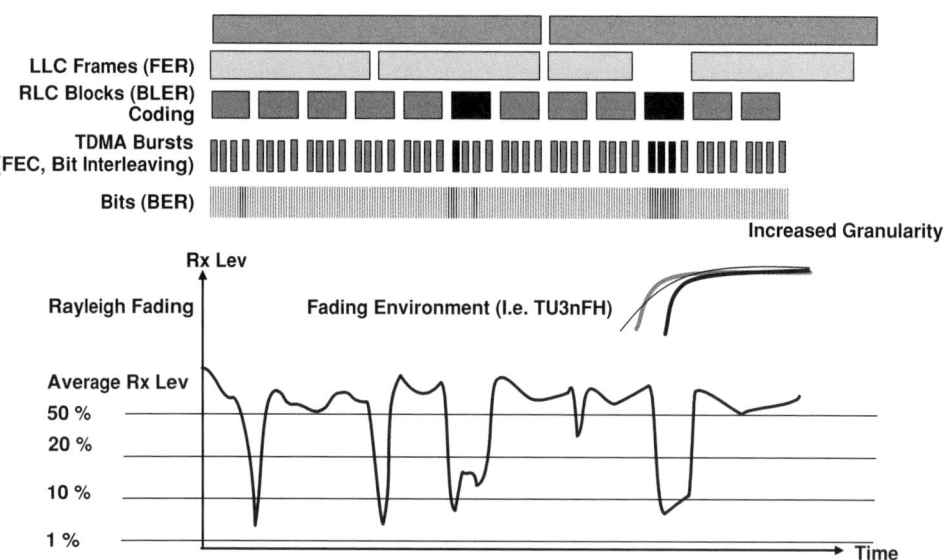

Figure 2.87 Impact of fading on BER, BLER and FER

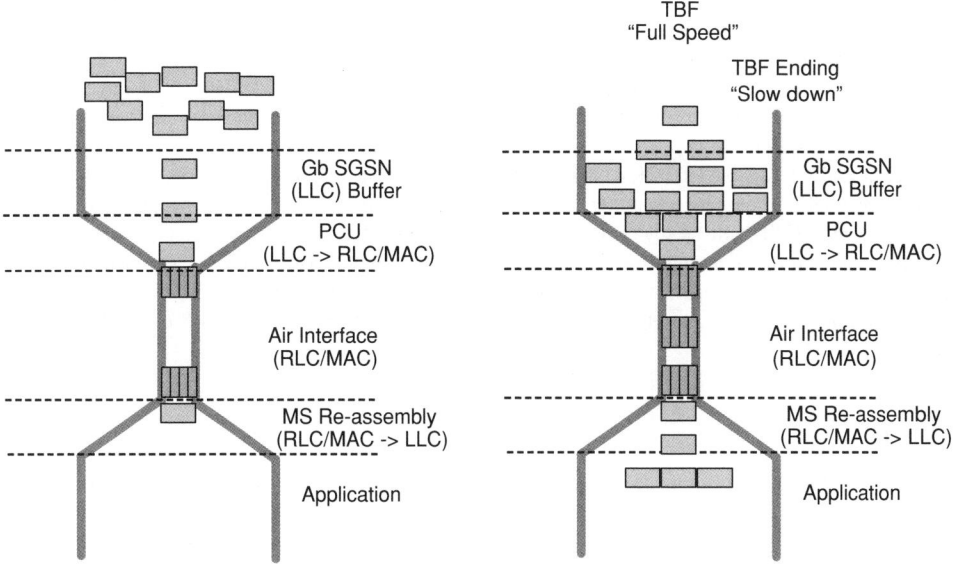

Figure 2.88 Data flow (segmentation and reassembly)

- Changing modulation from GMSK to 8-PSK increases the path loss so a reliable GMSK link may break down if too aggressive towards higher 8-PSK coding schemes.
- The FEC is more robust at MCS-5 than MCS-4 so some coding gain can be realised that compensates for the reduced E_b/N_0 in 8-PSK. Actual results show that the MCS-3 performance is better than MCS-5 for short TBFs.

With the increased dynamic range of bandwidth that EGPRS offers, it is very important to maintain smooth end-to-end flow of data so that the application layer does not experience buffer overflow or underflow. Figure 2.88 illustrates two examples of how application layer datagrams flow from the packet core to the mobile subscriber. The figure on the left shows how the slow/fragmented arrival of data packets results in fragmentation and short TBFs while the right figure shows a steady (even congested) flow of data that results in steady (maximum link adapted) TBFs. The significance of this on EGPRS is that suboptimum optimisation of transmission and radio link adaptation results in stalling and buffering issues. Either the PCU or mobile LLC segmentation/reassembly falls behind or ahead of the application layer and results in unsteady performance.

An important concept to consider in this discussion on EGPRS throughput is the actual desired profile of throughput and latency. The average and instantaneous throughput varies depending on the window size defined for calculating user throughput. With this in mind, it may be considered more important to ensure robust forward error correction (and a lower coding scheme) and minimum latency (reduced retransmission) rather than a higher coding scheme (less error correction) since retransmissions would lead to higher latency. This is a critical aspect when considering mobility management (GMM, or GPRS mobility management) and session management (SM) signalling performance of the packet core. For example, the routing area update (RAU) success rate along with attach and the PDP context success rate are adversely affected if too high an MCS is assigned in marginal channel conditions. In this case, it is more critical to reduce latency (and therefore signalling overhead) by maximising forward error correction than it is to attempt using the highest available coding scheme to gain throughput. The paradox of reducing latency by assigning higher throughput is resolved when considering the impact of retransmission time on latency. Radio network vendors use various algorithms to accomplish this balance. As a rule of thumb for very short data sessions, better performance is realized when lower EGPRS coding schemes are used (preferably in the GMSK range MCS-1 to MCS-4).

A key feature that significantly reduces EGPRS latency is one-phase access. Here, with the capability of the mobile and network to support an 11-bit PRACH access burst, the mobile can transmit the mobile capability (EGPRS and MS class) on the access burst while the network can directly assign an appropriate number of EGPRS capable PDTCHs. One-phase access improves the setup time by making mobiles stay at the lowest 'GPRS compatible' coding scheme (MCS-1) for initial signalling. The result is that EGPRS RLC throughput/TSL will be lower but more robust. While RLC throughput may be reduced (given a high proportion of signalling blocks), the subscriber performance is enhanced. The positive part of this feature is the reduced latency in ideal conditions compared to two-phase access, where the mobile must go through a second packet channel request round to be allocated an EGPRS capable MCS and timeslots.

MCS Distribution

The actual MCS distribution in a given network or service area will be biased, based on the following factors:

- the radio environment affecting RLC retransmissions and link adaptation;
- the link adaptation parameter settings affecting the LA algorithm and MCS selection thresholds;
- the ratio of short versus long TBFs;
- the ratio of actual user payload versus 'signalling' payload.

The stronger the signal level and the lower the interference, the higher will be the MCS utilised. This will ensure the best EGPRS performance (maximum throughput and minimum latency). To influence this factor, standard GSM optimisation techniques may be applied (e.g. additional sites, antenna downtilting, optimum GSM band selection 850/1900, 900/1800).

Carefully selecting link adaptation and MCS settings will influence the rate of change between the MCS and also the ability to respond to a changing RF environment. Optimum settings will depend on the dominant fading environment for a given coverage area. Being too conservative will result in a lower MCS and suboptimum throughput and latency. Being too aggressive will result in higher TBF failures, window stalling and a generally intermittent/unreliable service.

Short TBFs can be managed by either modifying TBF release delay parameters or adding performance enhancing proxy (PEP) servers connected to the appropriate service gateway or access point node (APN). A TBF release delay increases the TBF time and reduces latency by allowing normally short TBFs to be 'combined' so that new channel request and assignments are not required between the normally terminated TBFs. While this reduces latency, longer TBFs lead to an increased probability of TBF blocking and congestion, so care must be taken when optimising these parameters. Performance enhancing proxies act as an application buffer on the Gi/Internet backbone so they essentially encapsulate the desired (initially fragmented) data into a steady stream of data, causing radio allocation to be driven into longer steady TBFs. The gain in PEPs over TBF release delay parameters is significantly increased efficiency since the compiled data can be more effectively transferred at a higher MCS, leading to a shorter 'net' TBF time and therefore less TBF blocking and congestion. Depending on the network's mix of applications and services, a healthy balance of both functions may be employed to optimise EGPRS throughput and latency.

In practice, the last factor of user versus signalling payload heavily influences MCS distribution. This is because a significant proportion of the payload is driven by routing area updates (RAU) and GMM/SM signalling (attach and PDP context). It is therefore important to distinguish the user payload from the signalling payload so the performance engineer can tune the optimum EGPRS throughput and latency for the data subscribers. RAUs can be reduced and the attach success rate (ASR) increased by modifying the periodic routing area update (PRAU) timer and by careful RAC boundary planning using similar techniques to those applied in traditional GSM LAC planning (e.g. clearly defined boundaries using well-isolated geographical or highway transitions rather than ill-defined or ambiguous boundaries parallel to density populated/high mobility regions). GMM/SM (session management) signalling can also be optimised by modifying the SGSN and GGSN settings for releasing and retaining attached and PDP activated subscribers.

Figure 2.89 illustrates typical downlink and uplink MCS distributions on two representative BSCs respectively. Note that the BSC on the left contains more MCS-9 in both directions. Also, the distribution (on both the uplink and downlink) shows the first BSC having more MCS-3 and the second BSC having more MCS-5. The difference in distribution in this example is due to certain link adaptation parameters that bias the coding scheme on the first BSC to MCS-3 and the second BSC to MCS-5. The larger proportion of higher MCSs indicates a combination of more favourable channel conditions, less signalling and more opportunity of link adaptation to higher MCSs (due to a longer TBF transmission time).

TBF Performance

Temporary block flow (TBF) performance provides essential insight into the connection quality of data sessions and helps identify areas to focus on during EGPRS optimisation. The following TBF key performance indicators are useful for optimising EGPRS radio performance:

1. TBF setup success measures the rate at which TBFs are correctly established. The higher the TBF setup success rate the better the user experience is for reliably connecting to the network and minimising application latency.
2. TBF drops quantify the number of TBFs that have not completed. As a direct result of the cell reselection during the packet transfer mode, TBFs will timeout due to LLC flush (where the network continues to keep the original TBF open with no response from the mobile until it receives notification that the data session has been continued on another cell).
3. TBF blocking indicates when the packet data channel is unable to allocate radio resources due to maximum allocation of simultaneous TBFs. The limitation of simultaneous uplink and downlink TBFs is restricted to seven on the uplink (due to a 3-bit uplink state flag for identifying the unique UL TBF) and at least nine to a maximum of 31 on the downlink (due to either a vendor dependent restriction or a 5-bit temporary flow indicator for identifying the unique DL TBF).

The following sections provide additional details and examples of each of these three types of TBF performance indicators.

TBF Setup Success

Figure 2.90 identifies the downlink TBF setup/establishment success rate across several networks. Note that the range of the downlink TBF performance was originally from around 89–95 % and then suddenly improved for all networks near the end of the time period. This improvement was the result of modifying the link adaptation parameters to bias the initial MCS to 3, resulting in more robust forward error correction and taking advantage of GMSK modulation. What is actually occurring as a result of this change is that the TBF setups succeed more often when operating at full power GMSK compared to attempting immediately to establish a higher MCS and to fail due to decreasing average power for 8-PSK. By ensuring that a more robust MCS is initially allocated, the connection reliability is enhanced so that the data session can establish and continue smoothly.

Figure 2.91 shows the uplink and downlink TBF failure rates (inverse of TBF setup success rates) of several networks. In this example, the typical range for uplink TBF failures is between 6 % and 8 % while downlink TBF failures initially ranged from 3 % to 4 % and then began to stabilise below 2 %. Also note

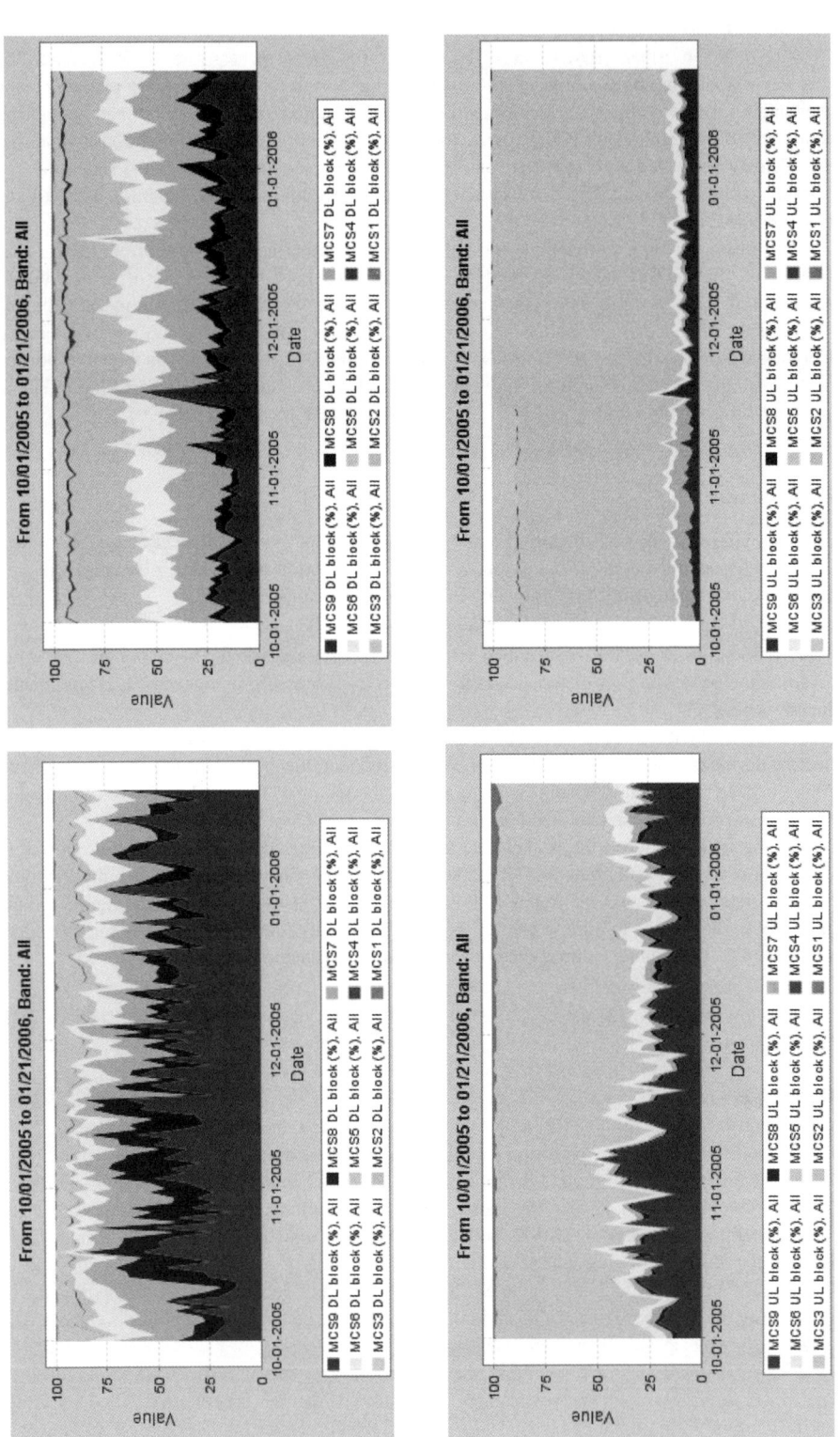

Figure 2.89 Measured uplink and downlink MCS distribution

Radio Network Optimisation

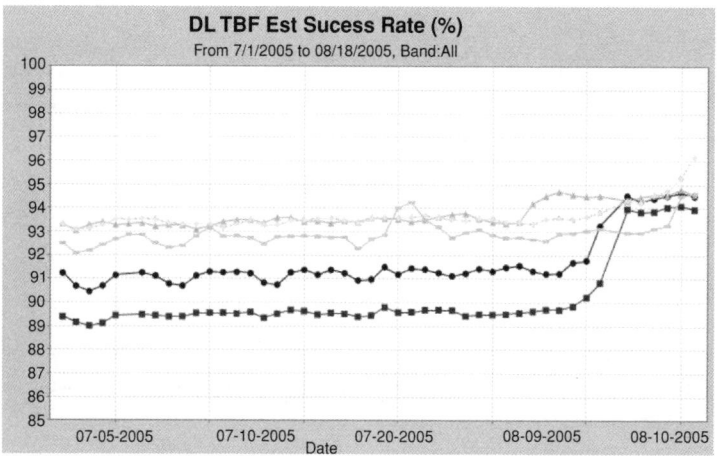

Figure 2.90 Sample downlink TBF establishment success rate

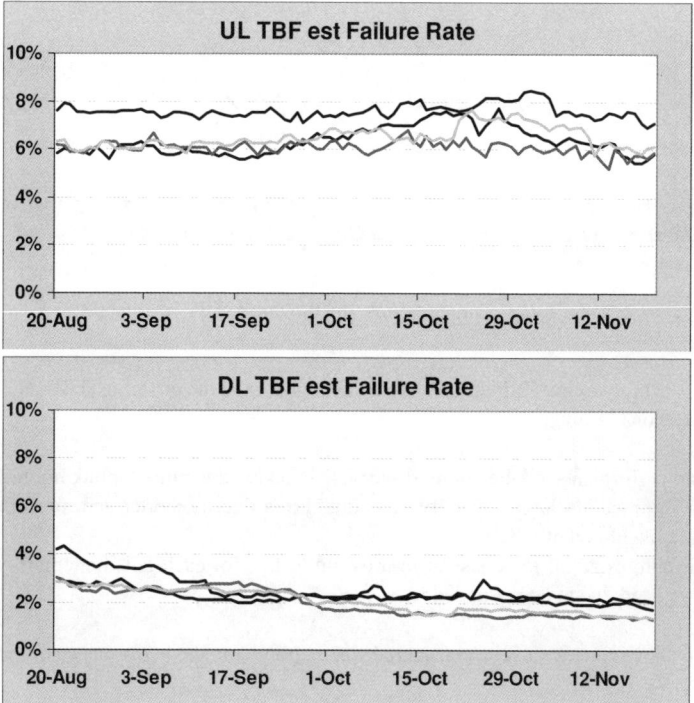

Figure 2.91 Sample uplink and downlink TBF failure rates

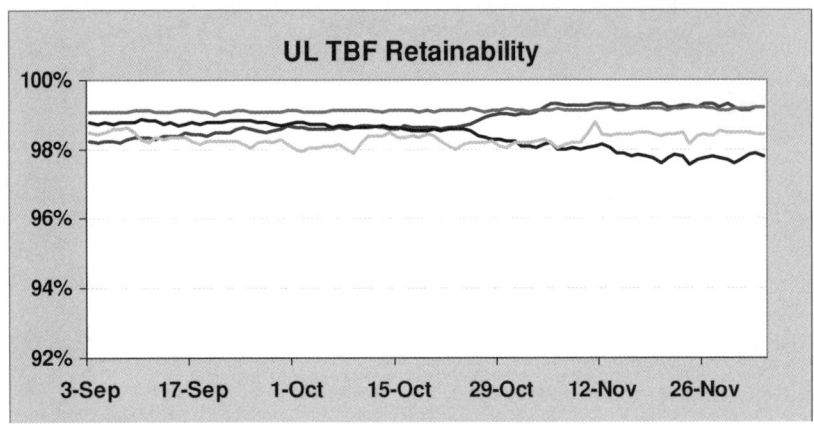

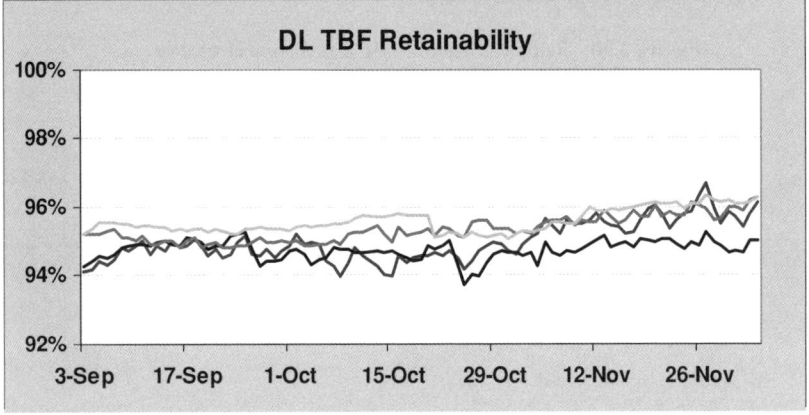

Figure 2.92 Sample uplink and downlink TBF retainability

in this example that the uplink TBF failures are much higher than the downlink TBF failures. There are three possible reasons for this:

1. Link loss and performance on the uplink is worse than on the downlink (uplink limited).
2. There are a significantly larger quantity of uplink TBFs due to shorter and more frequent TBFs, increasing the likelihood of failures.
3. Uplink TBF failures result in persistent re-tries (up to five for each failed attempt), so initial TBF failures are inflated due to compounded re-tries.

All three of the above causes must be considered when optimizing the performance of the TBF success rate.

TBF Drops

TBF drops are harmful to EGPRS network performance since they result in throughput degradation and increased application latency or even lead to application stalling. The following trends outline the uplink and downlink TBF completion rates (1-TBF drops) of four sample networks (Figure 2.92). This example shows uplink TBF retainability (completion rate) ranging from around 98 % to 99.5 % and downlink TBF retainability ranging from 94 % to just over 96 % (significantly lower than uplink). The uplink TBF completion rate is better than the downlink TBF completion rate in this example due to many short uplink TBFs successfully completing along with the inherent TBF drops (lost TBFs) that occur during cell reselection. Note the tradeoff between the TBF setup success and the TBF completion rate (drops): TBF failures can shift between the setup and transfer phase of the packet call. In this example, additional uplink TBF failures occur in the earlier setup phase.

Looking at the uplink TBF retainability trend again, there is a clear change in the middle of the time period where one network improves while the other degrades. The improvement was caused by decreasing TBF release delay parameters. This resulted in more rapidly releasing TBFs (typically 1 second) after the RLC flow timed out. The degradation on the other network was caused by an adjustment to the link adaptation algorithm that biased the MCS to 5 instead of 3. This change effectively increases the vulnerability of many short uplink TBFs since they became exposed to 8-PSK instead of GMSK and the forward error correction of MCS-5 could not compensate for the reduced signal level.

TBF Blocking

When packet data channels are overloaded, TBF blocking occurs and causes interruptions in the data service. The solution is to dimension the voice and data TRX capacity or groom the traffic such that data have enough room to allocate timeslots without blocking. On the one hand, the introduction of EGPRS does help alleviate this problem by increasing the transfer rate and reducing the probability of simultaneous access. However, on the other hand (and actual reality), EGPRS has made this 'problem' more prevalent (a good thing for growth!) due to the increased data demand on the network. Figure 2.93 outlines several examples of networks with high and nominal TBF blocking. In this example, both the uplink and downlink TBF blocking ranges from a residual < 0.5% to bursts of well over 2%. It is clear that the network with the high uplink and downlink TBF blocking requires special attention in re-balancing the traffic and is likely to involve expanding TRX capacity on select busy cells to permit additional allocation of packet traffic channels.

EGPRS Optimisation in the Field

This section briefly introduces some key concepts and knowledge that can be applied for field optimisation of EGPRS networks. One of the key aspects of field optimisation is to understand first the bottlenecks and performance in a sample test area in 'ideal' radio conditions (high signal strength and low C/I). The following considerations help to ensure reliable test results during field optimisation:

1. Ensure your mobile and test equipment are loaded with the proper software levels and support the applications desired to be tested.
2. Make sure your mobile measurement and application 'bus' can support the processing demand you are driving it so you are able to set up your repeated application driver plus take reliable measurements while the phone is still able to fulfil the EGPRS/GSM processing duties (e.g. reliable adjacent cell measurement).
3. Considering the above, always keep in mind that the actual device portfolio that your network supports will very likely have behaviours that differ (sometimes significantly) compared with your measurement/optimisation tool. This makes it important to try to make some device comparisons using 'engineering mode' flashed software on as many live deployed devices as possible.
4. While you are conducting your ideal radio condition test, you should benchmark the speed and latency of your reference APN to set an expected behaviour.

After the test equipment and ideal test have been benchmarked, mobile testing can proceed. During drive testing and analysis, it is recommended to focus on the following key layer 3 messaging information:

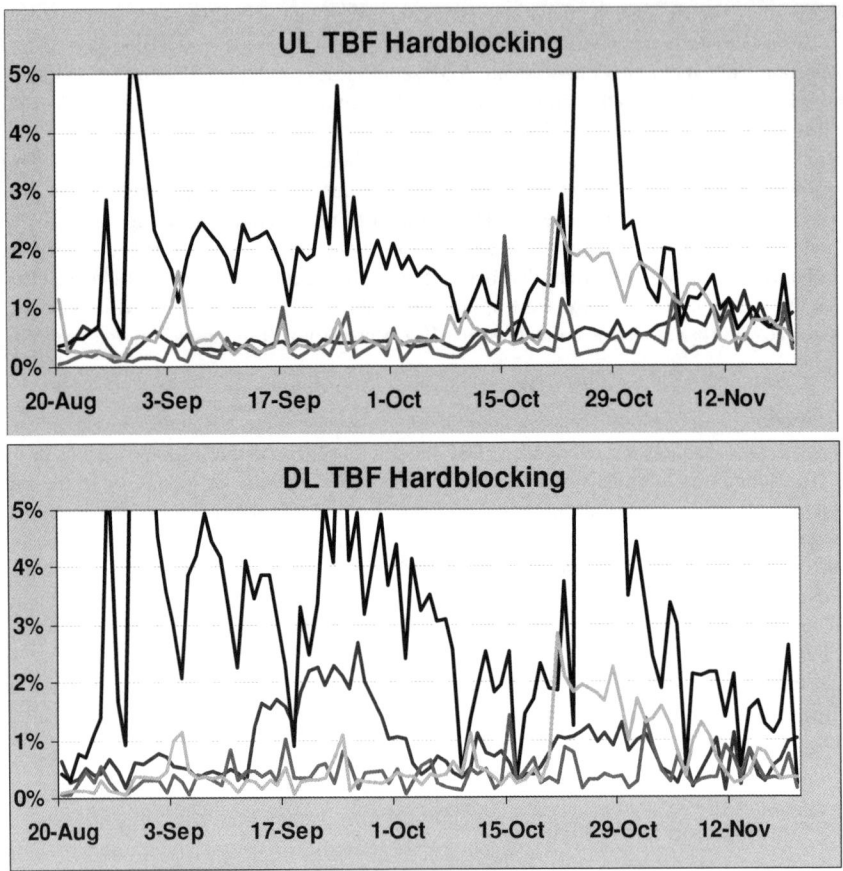

Figure 2.93 Sample uplink and downlink TBF blocking

During sustained packet data transfer:

1. Packet Downlink/Uplink Ack/Nack (PDAN, PUAN). These messages contain the essential information for flow control and measuring channel conditions. Two important contents include the Receive Block Bitmap (RBB) for identifying specific RLC block errors and final acknowledge indicator (FAI) to highlight the final RLC block in a TBF.

During cell reselections:

- System information messages 1 to 4 and 13. As indicated previously, the mobile must complete its scan of these five messages before it can lock on to the new channel and establish a TBF and then resume scanning for other adjacent cells in the BA list.

Upon completion of the drive test and collecting and filtering through the measurement data, the TBF performance can be mapped out and overlaid with the application performance. The graphs in Figure 2.94 show an example of TBF performance from a single FTP download. In this example, there were no block errors since the test was done in ideal conditions (the RBB does not indicate any missing blocks).

Radio Network Optimisation

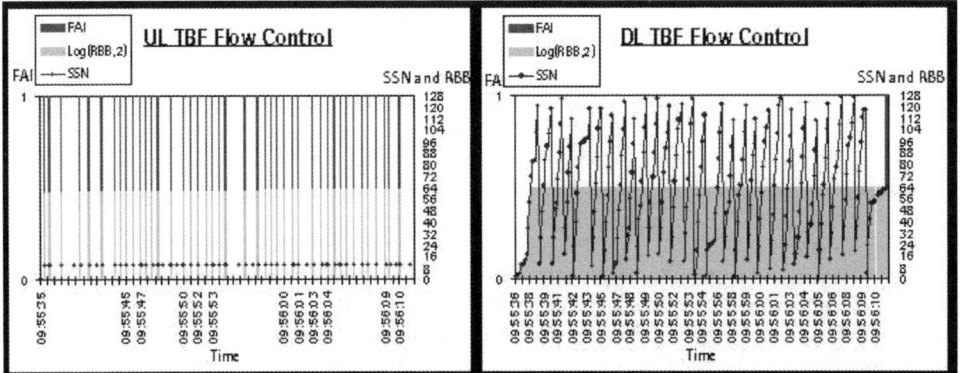

Figure 2.94 Measured uplink and downlink TBF flow controls

Nevertheless, the graph on the left clearly shows many short duration uplink TBFs while the right graph shows a single long TBF. Note how the starting sequence number (SSN) begins slowly, then rapidly repeats in a cyclical fashion until slowing down to the FAI.

The above technique provides a practical way of visualising the TBF behaviour of EGPRS data sessions from a drive test analysis. It helps significantly with the understanding of how different application types impact and drive the EGPRS radio performance. Routing area updates and other GMM/SM can be mapped during the drive test analysis and the performance engineer can effectively utilise and apply the knowledge gained in this text to maximise and optimise the radio network performance of an EGPRS network.

2.9.3 Optimisation in the WCDMA Network

Cluster Optimisation

The optimisation process starts from the stage where initial network planning has finished. The network has been built – at least the majority of sites have been integrated. No users are in the network and the network does not carry any traffic. Therefore the optimisation has a target to guarantee the quality of basic services when they are launched

The network is divided into geographical areas called clusters, each consisting of 10 to 20 sites. These geographical areas are each optimised to ensure QoS in that area. Geographical areas can be selected based of different methods, but the sites should be in geographical proximity and interference coming in to the cluster from neighbouring areas should be minimised. The quality targets are defined for each cluster in terms of, for example, overall RSCP and E_c/N_0, pilot pollution areas, drop call rate and call setup success rate.

Cluster Preparation

The cluster preparation phase consists of activities that can be done without drive test data. They should also take place to ensure that the drive test can take place and that it reveals planning related performance problems:

- Defining clusters
- Planning the drive route
- Radio parameter audit

- Site configuration check
- Neighbour list verification
- Fault management check
- Cell availability check

Prior to any drive tests a cluster should have the majority of its sites integrated and these sites should also be operational and available.

Defining Clusters

Clusters should be geographically contiguous areas comprising typically between 15 and 20 sites. The clusters are defined primarily based on geography: site locations, major roads, RNC borders, other geographical aspects, e.g. rivers. The secondary method for cluster definition is interference analysis, used to minimise the amount of external interference to a cluster from neighbouring clusters.

Planning the Drive Route

The drive route for a cluster includes all sites predicted for that cluster as well as sites not on air at the time of the drive but planned to be integrated before the network is launched. The drive route should cover a good percentage of main roads and motorways and also all areas of special interest, e.g. airport routes, corporate routes.

Before the drive test can be performed parameter and configuration checks are made to ensure correct settings for the sites within the cluster. Also the alarms indicating potential hardware or software problems are checked to ensure that no problems are due to true radio problems:

- Neighbour list generation
- Hardware configuration audit
- Radio parameter audit
- Fault management

Neighbour List Generation

The initial neighbour lists can be defined based on geographical proximity and antenna direction. Also as an initial step all intrasite cells should have the neighbour relation defined. The neighbour definition should include cells that are first-tier neighbours, as the cells should not provide coverage beyond the first-tier neighbours.

Hardware Configuration Audit

Hardware configuration audit checks on the correct configuration settings for MHA gain and cable loss can be done by using cell availability statistics and analysing the BTS commissioning file. A small number of sites can also be selected as a sample to perform site visits to find out if planned tilts and antenna directions have been implemented correctly. The potential candidate for site visits is the cluster on the most important service area. An indication of incorrect settings can be evaluated from PrxNoise statistics or real time values using online monitoring.

The uplink noise measurements are signalled to the UE in system information block 7 and used for an initial transmit power calculation of RACH preamble and DPCH initial transmit power calculation. Incorrect noise measurement has two effects on RACH procedure:

1. Too high PrxTotal. Too high initial transmit power for preamble and RACH message which causes uplink interference rise.
2. Too low PrxTotal. Too low initial transmit power which requires multiple preambles to be transmitted and in the worse case prevents the WBTS from detecting the RACH preamble due to too low transmit power.

Too low/high PrxTotal would cause too low/high initial power on the UL DPCH transmission and increase the UL power control loop settling time. Extreme values (±20 dB) can also prevent the UL synchronisation and thus cause connection setup failure. If correct cable loss values are not available, a default of 3 dB should enable normal network operation.

Base Station Hardware and Software Version Management

Base station hardware (HW) and software (SW) version management checks the HW and SW update status in the network, which is vital for optimum network wide performance. Both HW and SW versions can be audited with similar procedures, where the base station configuration files are transferred directly from the base stations by FTP and then the required information is gathered from these files.

Radio Parameter Audit

The radio parameter audit verifies that RAN parameters implemented on the network are set consistently in all objects and follow the default parameter set. The radio parameter audit is done to check the consistency of radio parameters by comparing the actual values and default recommended values defined for network objects. Typically in prelaunch networks the actual parameter values have the same value for all objects. The actual values should also follow the recommended parameter values. The radio parameter audit is performed to check the majority of parameters in RNC, WBTS, WCEL (WCDMA cell), ADJS (intrafrequency adjacency), ADJI (interfrequency adjacency), ADJG (intrasystem [WCDMA → GSM] adjacency), FMCS (measurement control set for intrafrequency adjacency), FMCI (measurement control set for interfrequency adjacency), FMCG (measurement control set for intrasystem [WCDMA → GSM] adjacency), HOPS (handover control parameter set for intrafrequency adjacency), HOPI (handover control parameter set for interfrequency adjacency) and HOPG (handover control parameter set for intrasystem [WCDMA → GSM] adjacency) object classes.

Parameter Consistency Check

Parameter consistency checks are done to verify consistency of the RAN parameters. Most RAN parameters are set to default values in prelaunch networks; later multiple templates can be defined for different types of cells. The default settings can be checked outside Nokia operations solutions system (OSS) by exporting the parameter information. Default parameters should be checked regularly in the prelaunch phase when new sites are integrated in order to verify that launched sites have correct parameters and avoid the negative impact of configuration errors. Checks should also be done prior to any optimisation activities or field test measurements.

Actual Versus Default Parameter Values

Most RAN parameters are set to default values in prelaunch networks. The actual values in the network should follow the default parameter values set. The default values are set according to operator strategy, network layout and available resources.

Scrambling Code Plan Audit

The main target of scrambling code planning is to ensure that the same scrambling code is not used in cells close to each other on the same carrier frequency. This is required to ensure that cells with the same scrambling code are not included in a combination of neighbours that an RNC sends to a UE from the neighbour lists that belong to cells that are in the active set. Intrafrequency cells in any neighbour list cannot be combined by the RNC with the neighbour list that is sent to a mobile in a soft handover area; trying to do so will cause a scrambling code conflict error.

The scrambling code plan audit will validate the minimum distance between the same carrier frequency cells with the same primary scrambling code. The scrambling code audit can also validate implementation of the operator nominal scrambling code plan if one exists.

Adjacency Audit

The purpose of the adjacency audit is to check the target cell parameter settings, number of adjacencies, adjacency distance and one-way adjacency definitions. A list of possible checks follows:

   ADJG (WCDMA to GSM adjacencies)
      ADJG distance > 20 km
      ADJG inconsistencies – incorrect LAC, NCC or BCC
      ADJG inconsistencies – incorrect MCC or MNC
      ADJG inconsistencies – invalid target 2G CID (cell identifier)
      ADJG one-way neighbours (3G ⇒ 2G)
      ADJG target DN (domain name) empty
   ADJS(/I) (WCDMA intra/interfrequency adjacencies)
      ADJS distance > 20 km
      ADJS inconsistencies – incorrect LAC
      ADJS inconsistencies – incorrect MCC or MNC
      ADJS inconsistencies – incorrect PriScrCode
      ADJS inconsistencies – invalid target CID
      ADJS one-way neighbours
      ADJS target DN empty
   ADJW (GSM to WCDMA adjacencies)
      ADJW distance > 20 km
      ADJW inconsistencies – incorrect LAC
      ADJW inconsistencies – incorrect MCC or MNC
      ADJW inconsistencies – incorrect PriScrCode
      ADJW inconsistencies – invalid target 3G CID
      ADJW one-way neighbours (2G ⇒ 3G)
      ADJW target DN empty
   CID (cell identity)
      Duplicate 2G CID
      Duplicate 3G CID
   Adjacency number checks
      WCEL with ADJG and very few ADJS (ADJG – ADJS > 15)
      WCEL with ADJG but no ADJS
      WCEL with ADJS and very few ADJG
      WCEL with ADJS but no ADJG
      WCEL with no ADJG
      WCEL with no ADJS

Fault Management

Fault management is a continuous process where faulty hardware units and software related faults are fixed. A special fault management effort should be given to the area that is about to be optimised prior to any optimisation activities. All critical alarms should be cleared and it should be verified that sites are available and in a working state.

RF Tuning

The radio environment is measured, benchmarked and analysed to verify and optimise the RF performance. The RF performance is also analysed to find out any parameter, service performance or potential site integration problem issues. In the initial RF tuning phase the cluster of cells is measured and the overall RSCP and E_c/N_0 values are benchmarked against predefined quality targets. The aim is to reach the performance target for RF performance. The performance target can consist of the coverage target (RSCP), the quality target (E_c/N_0), minimising pilot pollution areas and limiting soft handover area.

Radio Network Optimisation

As the first step the RF KPIs for the cluster are benchmarked. Benchmarking RF KPIs can be used to verify that the RF performance improves as a result of implementing the tuning recommendations.

Secondly, the overall RSCP is analysed. This can be done using an analysis tool that visualises the measurement samples. The poor performing areas are identified and the reason for poor coverage is analysed. For example, the coverage problem might be due to a missing site or a physical obstacle preventing the signal attenuation.

The overall E_c/I_0 is analysed to find out areas with potential quality problems. This is done in a similar manner to coverage analysis, using an analysis tool to find out problem locations. The E_c/I_0 problems might be due to dominance problems or coverage problems. In the case of coverage problems the RSCP also gives an indication of the problem. However, in the case of a dominance problem the RSCP is good on area problems. The cells providing coverage on an area should be analysed and some of them removed, in order to improve the quality of the remaining ones.

Recommendations to improve RF performance will be based primarily on physical changes to antenna tilts, azimuths, types and heights. In addition to RF performance in RF tuning phase the neighbour lists can also be verified. Also the potential scrambling code conflicts and site integration related problems e.g. swapped feeders should be identified. The first measurement results could also be used for initial check for service performance.

Neighbour List Analysis

Missing neighbours will be identified through comparison of the neighbour list generated based on scanner measurements and the actual ADJS, ADJI and ADJG objects uploaded from the RNC database for those sites. The RNC data upload should be performed on the same day as the drive test.

Voice Call Performance Optimisation

Call performance gives an indication about areas that have bad performance, e.g. areas with call setup problems and dropped calls clearly have something to investigate. The field measurements analysis is usually started after looking at (statistical) call performance data (counters) to evaluate how much the analysed area had problems. The next step is to identify where these problems occurred:

- Low E_c/I_0. The signal quality falls below −15 dBm and the call drops.
- Low RSCP. The Received signal code power falls below −115 dBm.
- System problems. E_c/I_0 and RSCP are at an adequate level but the call drops occur regardless of radio conditions.

When analysing 3G field measurements both scanner and call trace measurements should be available. By analysing scanner data and comparing the difference between the scanner and UE performance some common problem situations can be identified.

Dropped Call Analysis

Dropped calls give a reliable indication of areas where users are having problems. Each dropped call should be analysed to find out what are the reasons behind each drop (an example is shown in Figure 2.95). The first step is to analyse and quantify how many drops occurred during the drive test and where and on which scrambling code they occurred.

Symptoms: low E_c/I_0 in the UE but good E_c/I_0 in the scanner; call drops and the UE selects a new scrambling code not previously detected in the active set; E_c/I_0 improves.

Problem: a missing neighbour. There has been a strong interfering signal which the scanner could pick up, but because it had not been defined as a neighbour the UE was not able to pick it up and the call drops.

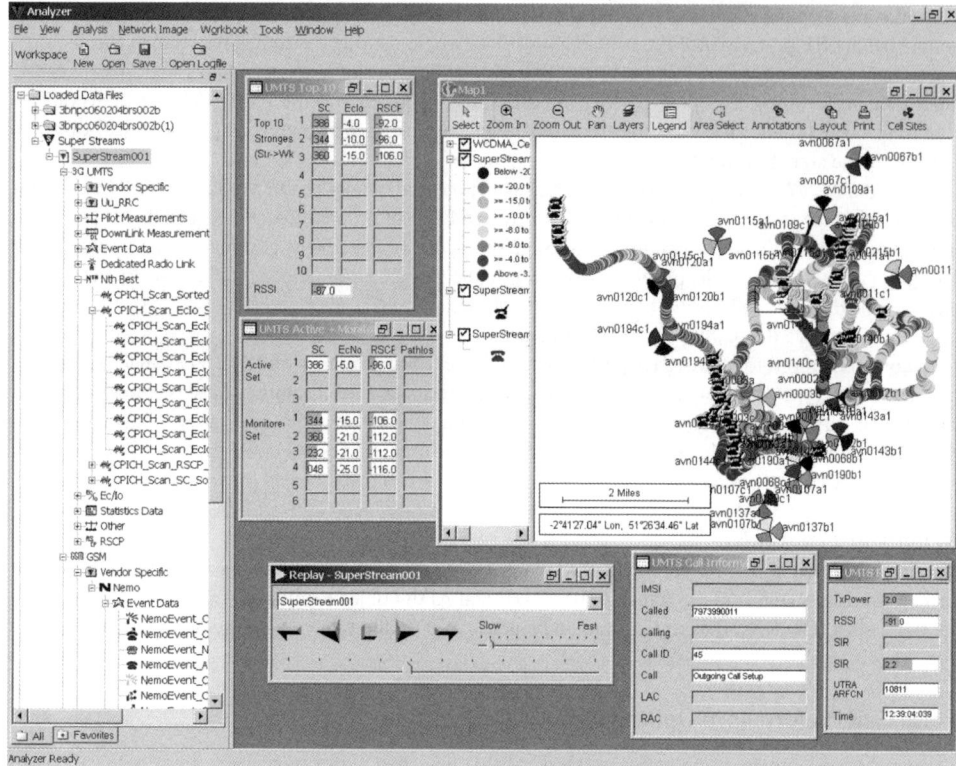

Figure 2.95 Drop call analysis

Symptoms: low E_c/I_0 in the UE and E_c/I_0 on the scanner; low RSCP in the UE and in the scanner; UE transmitting on the maximum TX power.
Problem: poor coverage area. The E_c/I_0 handover threshold should first trigger an interfrequency handover; if no 3G neighbour is detected the intersystem handover is triggered. If the UE is still not able to detect any neighbours the radio link to the serving cell is failing. The power control cannot improve signal quality due to maximum UE/NodeB transmission power already used. The call drops.

Symptoms: low E_c/I_0 in the UE and E_c/I_0 on the scanner; normal RSCP in the UE and in the scanner; several scrambling codes with similar E_c/I_0 are detected in the active and monitored sets; high RSSI.
Problem: pilot pollution area. Too many cells are covering the same area with equal RSCP. With many scrambling codes present the RSSI that measures the total received power has a high value. The RSCP should be within −15 dBm of the RSSI. If the overall interference level is high the requirement for the serving signal increases and thus even a normal RSCP does not provide the required E_c/I_0.

Analysing Poor Performing Cells

The poor coverage area and pilot pollution area problems described in the drop call analysis section also apply when analysing poor performing cells. The calls have not necessarily been dropping, but there could be performance problems on the cell level that could cause calls to drop, but this was not taking place at the time of measurement.

Symptoms: low E_c/I_0 in the UE and E_c/I_0 on the scanner; normal RSCP in the UE and in the scanner; small number of scrambling codes in the active and monitored sets; high RSSI.

Problem: external interference. Someone is transmitting on the WCDMA frequency band. The source of interference is to be investigated.

Symptoms: plotting the individual scrambling codes RSCP shows that the cell has a strong RSCP beyond the first-trie neighbours; plotting the E_c/I_0 shows that the site has a large area with a poor E_c/I_0 outside its first-tier neighbours, or it has dominance in the area that it is not intended to be the dominant server.
Problem: overshooting site. The antenna should be tilted to limit cell coverage within the first-tier neighbours on the intended dominance area. By tilting this type of site the performance of surrounding sites will improve.

Symptoms: plotting the individual scrambling codes RSCP shows that the cell has a small coverage area; plotting the E_c/I_0 shows that the site has a poor E_c/I_0; it is not a dominant server in the area in which it is supposed to have been a serving cell.
Problem: limited coverage. The antennas and hardware should be checked for problems. If the only terrain is limiting and there are only a few users in the dominance area of the cell, the cell could even be switched off to limit the interference.

System Issues
If the radio interface does not indicate any problems with RF conditions the problem is likely to be related to the UE, network or measurement system functionality. The L3 signalling messages before the dropped call should be analysed to find out if there were any messages sent that would indicate abnormal system behaviour.

In the case where an L3 message indicating problem is found the signalling flow is analysed to find out details of the system problem (shown in Figure 2.96). In addition to L3 messages, the power control commands, the bitch error ratio (BER), the block error rate (BLER), and L2 messages can be analysed to find out more details.

Measurement Equipment Failure
The measurement equipment is one source of potential problems. The dropped calls can be caused by failures in the measurement equipment or the mistakes made using the equipment. The drive test is prone to have errors, some of which can cause dropped calls.

Verification

The cluster optimisation analysis should be performed in two parts. Network verification is used to accept the network performance based on drive tests along 'reference routes' encompassing multiple clusters. KPIs are measured only in areas where 'acceptable' coverage has been achieved. Typically network verification measurements consist of larger areas than cluster measurements and are less detailed. They are also repeated to benchmark the performance over the time when the network develops and traffic patterns evolve.

Performance Monitoring

Statistics
High Level Performance Indicators
Analysing statistics is a top-down process. The high level (network or RNC level) performance indicators are monitored to find out if there is a problem in the network. The impact on high level performance indicators in greater depending on the number of network elements involved. Also, a severe problem on only a few network elements would have an impact on the RNC level performance.

The need for deeper analysis can also be triggered by constant variation of performance graphs or odd regularity of performance variation. The high level performance problems, anomalies and regular patterns of performance variation are triggers that indicate a need for further investigation.

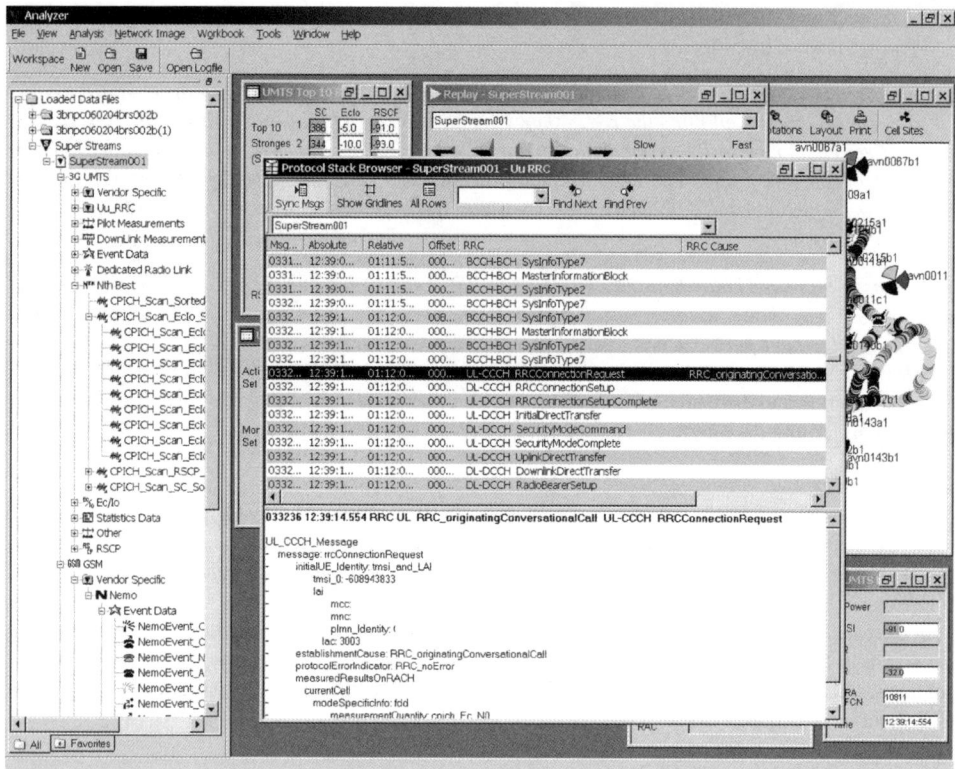

Figure 2.96 L3 signalling analysis

The high level performance indicators should be monitored regularly and on a constant basis (an example of cell availability monitoring is shown in Figure 2.97). Typical granularity of high level performance data is one day and the monitored period lasts for two months. Examples of high level performance indicators are:

- Cell availability
- Call setup success rate
- Drop call rate

Note that in 3G, instead of using a voice call there are different services, and therefore the KPIs could be broken down for different services, e.g. a call setup success rate for voice, a call setup success rate for video, a call setup success rate for PS background services, etc.

Low Level Performance Indicators
After identifying a problem in high level performance indicators in order to find out what is causing the problem, low level performance indicators need to be analysed. There are large numbers of performance indicators; thus the analysis of low level indicators is based on analysing large numbers of statistical counters. There are different approaches to start with. One approach is to calculate high level KPIs (drop call rate, call setup success rate) for individual cells and find out which cells should have a closer analysis. Another approach is to identify a problem type or a counter indicating a problem type, and only after that is the next step to identify which cells have this problem (shown in Figure 2.98).

Radio Network Optimisation

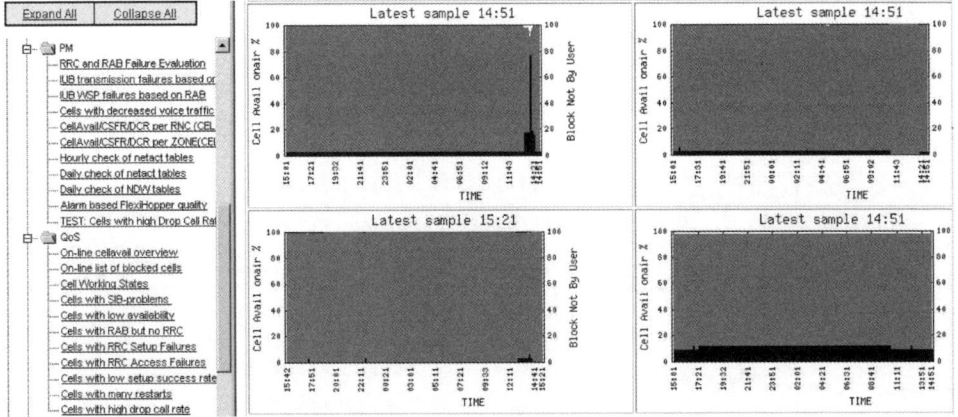

Figure 2.97 Example of cell availability monitoring

Analysing the geographical proximity of cells having problems might indicate something wrong with a whole cluster of cells, e.g. the HW/SW problem related to the transmission network, etc. When network elements having problems have been identified then the symptoms of the problem need to be identified. The analysis methods are different depending on whether the problem is already known or not. If the problem is already well-known there are probably ready made queries that can be used to identify the problem. If the problem is not known, queries are not available and the raw performance counters need to be analysed to classify the problem type.

Queries

Queries are used to speed up detection of already known performance problems. Already identified problem types do not require further investigation. They are to be detected and corrected as fast as possible (an example is shown in Figure 2.99). Due to the nature of this kind of 'find and fix' work, it is well suited to front line operation and maintenance team. They can run these queries regularly to find out the problems and then to correct them based on available solutions.

Analysing Performance Counters

There are large numbers of performance indicators. As the systems develop and new features emerge there are likely to be even more. The basic approach is to first look at counters that are used to calculate higher level KPIs. The drop call KPI consists of RAB active failures (see Figure 2.100) that are due to different reasons. The first step is to see what is the distribution of these reasons and whether there are different reasons incremented for the failures (RADIO, a radio interface failure; IU, an IU interface failure; RNC, an RNC internal failure). The cells that have the same reason incremented should be checked for similar problems. The same approach can then be used to analyse the RRC setup, RRC active and RAB setup phases.

Another example involves two counters that identify cells that have RAB active completions and no RRC connection attempts. This indicates that call setup is not possible in those cells but handovers are working.

More complex examples involve correlation analysis of many different variables. High numbers of RAB active completions (low pool capacity and RAB/RRC setup failures due to TRANS (transmission interface failure)/BTS or RAB/RRC setup failures due to AC) and high numbers of samples on traffic class 4 would indicate possible problems due to congestion.

Service Level PIs/KPIs on 08.11.2004 0-23 H

TIME	RNC_NAME	WBTS_NAME	WCEL_GID	WCEL_NAME	WCEL_ID	RAB_ACT_REL_CS_VOICE_SRNC	RAB_ACT_REL_CS_VOICE_P_EMP	RAB_ACT_FAIL_CS_VOICE_IU	RAB_ACT_FAIL_CS_VOICE_RADIO	RAB_ACT_FAIL_CS_VOICE_BTS	RAB_ACT_FAIL_CS_VOICE_IUR	RAB_ACT_FAIL_CS_VOICE_I_OHK	RAB_ACT_FAIL_CS_VOICE_RNC	RAB_ACT_COMP_CS_VOICE	RAB_ACT_COMP_RATE_CS_VOICE
08/11/20	RNUPF	871232	1722001	8712322	1000025	0	0	0	3	0	0	0	0	14	127,27
08/11/20	RNUPF	870842	1723001	8708421	1000030	12	0	0	1	0	0	0	0	71	88,75
08/11/20	RNUPF	870842	1724001	8708422	1000031	0	0	0	1	0	0	0	0	39	97,5
08/11/20	RNUPF	871002	1725001	8710021	1000033	0	0	0	1	0	0	0	0	10	90,91
08/11/20	RNUPF	871002	1726001	8710022	1000034	12	0	0	7	0	0	0	0	236	91,83
08/11/20	RNUPF	871002	1727001	8710023	1000035	4	0	0	1	0	0	0	0	58	98,31
08/11/20	RNUPF	87031A	1728001	87031A1	1000057	4	0	0	2	0	0	0	1	142	87,12
08/11/20	RNUPF	87031A	1729001	87031A2	1000058	0	0	0	0	0	0	0	0	44	102,33
08/11/20	RNUPF	87031A	1730001	87031A3	1000059	2	0	0	1	0	0	0	0	32	86,49
08/11/20	RNUPF	87032A	1731001	87032A1	1000060	6	0	0	1	0	0	0	0	215	102,38
08/11/20	RNUPF	87032A	1732001	87032A2	1000061	0	0	0	1	0	0	0	0	55	101,85
08/11/20	RNUPF	87032A	1733001	87032A3	1000062	20	0	0	11	0	0	0	0	179	82,49
08/11/20	RNVAS	88516A	1782001	88516A1	1023750	0	0	0	0	0	0	0	0	8	80
08/11/20	RNVAS	88516A	1783001	88516A2	1023751	0	0	0	3	0	0	0	0	109	99,09
08/11/20	RNVAS	88516A	1784001	88516A3	1023752	1	0	0	3	0	0	0	0	166	103,11
08/11/20	RNVAS	88524B	1785001	88524B1	1023753	1	0	0	1	0	0	0	0	109	103,81
08/11/20	RNVAS	88524B	1786001	88524B2	1023754	5	0	0	4	0	0	0	0	201	94,37
08/11/20	RNVAS	88524B	1787001	88524B3	1023755	9	0	0	3	0	0	0	0	306	97,45
08/11/20	RNVAS	88527A	1788001	88527A1	1023756	0	0	0	0	0	0	0	0	5	125
08/11/20	RNVAS	88527A	1789001	88527A2	1023757	0	0	0	0	0	0	0	0	5	71,43
08/11/20	RNVAS	88527A	1790001	88527A3	1023758	7	0	0	3	0	0	0	0	148	101,37
08/11/20	RNVAS	88502B	1791001	88502B1	1023759	24	0	0	10	0	0	0	0	728	96,81
08/11/20	RNVAS	88502B	1792001	88502B2	1023760	1	0	0	1	0	0	0	0	234	97,1
08/11/20	RNVAS	88502B	1793001	88502B3	1023761	4	0	0	2	0	0	0	0	306	101,32
08/11/20	RNVAS	88508A	1794001	88508A1	1023762	1	0	0	0	0	0	0	0	79	109,72
08/11/20	RNVAS	88508A	1795001	88508A2	1023763	10	0	0	13	0	0	0	0	513	93,78
08/11/20	RNVAS	88508A	1796001	88508A3	1023764	0	0	0	2	0	0	0	0	184	97,87
08/11/20	RNVAS	88529A	1797001	88529A1	1023777	0	0	0	1	0	0	0	0	17	121,43
08/11/20	RNVAS	88529A	1798001	88529A2	1023778	0	0	0	3	0	0	0	0	47	87,04
08/11/20	RNVAS	88529A	1799001	88529A3	1023779	1	0	0	1	0	0	0	0	6	60

WCEL_ID	RAB_STP_ATT_CS_CONV	RAB_STP_FAIL_CS_CONV_AC	RAB_STP_FAIL_CS_CONV_BTS	RAB_STP_FAIL_CS_CONV_TRANS	RAB_STP_FAIL_CS_CONV_RNC	RAB_STP_FAIL_CS_CONV_FROZBS	RAB_STP_COMP_CS_CONV	RAB_STP_SUCC_RATE_CS_CONV	RAB_ACC_FAIL_CS_CONV_MS	RAB_ACC_FAIL_CS_CONV_RNC	RAB_ACC_COMP_CS_CONV	RAB_ACC_COMP_RATE_CS_CONV	RAB_ACT_REL_CS_CONV_SRNC	RAB_ACT_REL_CS_CONV_P_EMP	RAB_ACT_FAIL_CS_CONV_IU
1000025	0	0	0	0	0	0	0		0	0	0		0	0	0
1000030	0	0	0	0	0	0	0		0	0	0		0	0	0
1000031	0	0	0	0	0	0	0		0	0	0		0	0	0
1000033	2	0	0	0	0	0	2	100	0	0	2	100	0	0	0
1000034	7	0	0	0	0	0	7	100	0	0	7	100	0	0	0
1000035	0	0	0	0	0	0	0		0	0	0		0	0	0
1000057	11	0	0	0	0	0	11	100	0	0	11	100	0	0	0
1000058	6	0	0	0	0	0	6	100	0	0	6	100	0	0	0
1000059	10	0	0	0	0	0	10	100	0	0	10	100	0	0	0
1000060	15	0	0	0	0	0	15	100	0	0	15	100	0	0	0
1000061	3	0	0	0	0	0	3	100	0	0	3	100	0	0	0
1000062	9	0	0	0	0	0	9	100	0	0	9	100	0	0	0
1023750	0	0	0	0	0	0	0		0	0	0		0	0	0
1023751	2	0	0	0	0	0	2	100	0	0	2	100	0	0	0
1023752	7	0	0	0	0	0	7	100	0	0	7	100	0	0	0
1023753	7	0	0	0	0	0	7	100	0	0	7	100	0	0	0
1023754	38	0	0	0	0	0	38	100	0	0	38	100	0	0	0
1023755	59	0	0	0	0	0	59	100	0	0	59	100	0	0	0
1023756	0	0	0	0	0	0	0		0	0	0		0	0	0
1023757	0	0	0	0	0	0	0		0	0	0		0	0	0
1023758	9	0	0	0	0	0	9	100	0	0	9	100	0	0	0
1023759	57	0	0	0	0	0	57	100	0	0	57	100	0	0	0
1023760	33	0	0	0	0	0	33	100	0	0	33	100	0	0	0
1023761	34	0	0	0	0	0	34	100	0	0	34	100	0	0	0
1023762	9	0	0	0	0	0	9	100	0	0	9	100	0	0	0
1023763	84	0	0	0	0	0	84	100	0	0	84	100	0	0	0
1023764	21	0	0	0	0	0	21	100	0	0	21	100	0	0	0
1023777	0	0	0	0	0	0	0		0	0	0		0	0	0
1023778	0	0	0	0	0	0	0		0	0	0		0	0	0
1023779	0	0	0	0	0	0	0		0	0	0		0	0	0

Figure 2.98 Raw counters

Radio Network Optimisation

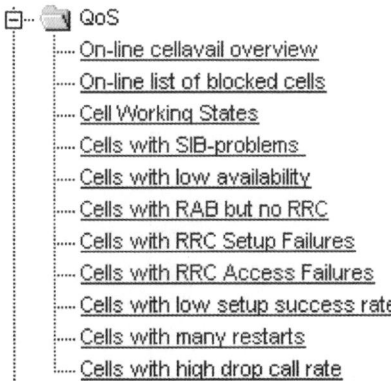

Figure 2.99 Example problem spotting query types

Using Statistics to Determine Cells to be Optimised

Selecting a Bad Performing RNC Area

To start an optimisation process the scope of the work needs to be defined. Even if the target is to improve overall network performance it is practical to start optimising a subset of network that is within geographical boundaries. An RNC area is a suitable geographical area to have a drive test team operating. Also the RNC level statistic is available for statistical analysis.

A bad performing RNC area can be selected based on high level performance criteria, e.g. the drop call rate for voice and video and the fluctuation of performance data over days. Different criteria might include a call setup failure rate or packet data performance.

It is important to have performance data from a longer period of time. In the example shown in Figure 2.101 the bad performing RNC area is selected using RNC performance trends. If the trend shows a large amount of fluctuation over a 2 week period or constantly bad performance, the RNC is selected for further investigation.

Selecting Cells to be Optimised

There are several ways to prioritise which cells are performing badly. Conventional practice is to select cells based on the main KPI that is to be optimised. If the target is to improve drop call rate cells are selected based on the drop call rate. However, this approach does not take into account the effect that cell has on the overall RNC performance. A more effective approach is to prioritise cells in an absolute number of drops. The cell causing the most drops has the strongest impact on RNC area performance (shown in Figure 2.102). If the target is to improve overall RNC area performance, the largest improvement can be achieved by selecting the cells that have the largest absolute number of drop calls. The drop call rate can be used as secondary criteria, ranking cells that are already performing well enough outside the scope of optimisation. A high number of drop calls shows the entry criteria for cells to start optimisation and a good enough drop call rate shows the exit criteria to determine when optimisation is no longer needed.

Longer time periods should be used to ensure reliability of performance statistics. For example, cells having the largest number of drops (video + voice) over the last 7 days could be selected to be analysed in detail. In networks where every cell carries traffic, the drop call rate (DCR) can be used to classify cells having the most performance problems. However, in a network where some of the cells are marginally loaded the absolute number of drops gives a more reliable classification.

The effect that each cell had on overall RNC area performance can be calculated. The number of cells needed to be investigated in detail can be decided by comparing the current drop call rate with the target drop call rate. For example, if the target drop call rate for the RNC was 2.5 % and the current drop call

Service Level PIs/KPIs on 08.11.2004 0-23 Hours

TIME	RNC_NAME	WBTS_NAME	WCEL_GID	WCEL_NAME	WCEL_ID	RAB_ACT_REL_CS_VOICE_P_EMP	RAB_ACT_FAIL_CS_VOICE_IU	RAB_ACT_FAIL_CS_VOICE_RADIO	RAB_ACT_FAIL_CS_VOICE_BTS	RAB_ACT_FAIL_CS_VOICE_IUR	RAB_ACT_FAIL_CS_VOICE_I_CHK	RAB_ACT_FAIL_CS_VOICE_RNC
08/11/200	RNUPPS	87067D	2594001	87067D3	1022128	0	0	0	0	0	0	17
08/11/200	RNOREI	88584C	5551001	88584C1	1032246	0	0	4	0	0	0	11
08/11/200	RNKRIS	74882A	9033001	74882A3	2005621	0	0	4	0	0	0	27
08/11/200	RNSKOY	73175A	1,3E+07	73175A2	1035621	0	0	0	0	0	0	11
08/11/200	RNJONI	77252C	1,6E+07	77252C1	2015436	0	0	0	0	0	1	13
08/11/200	RNNORI	87702C	2,7E+07	87702C2	1027755	0	0	0	0	0	0	13
08/11/200	RNKRIS	74803A	1,5E+08	74803A1	2007950	0	0	0	0	0	0	15
08/11/200	RNKRIS	74266C	1,5E+08	74266C1	2007965	0	0	11	0	0	0	22
08/11/200	RNNORI	88610A	1,6E+08	88610A3	1027960	0	0	4	0	0	0	11
08/11/200	RNJONI	79114A	4,6E+08	79114A1	2017855	0	0	0	0	0	0	11
08/11/200	RNOREI	86899A	4,7E+08	86899A1	1032721	0	0	3	0	0	0	19

Figure 2.100 RAB active failures

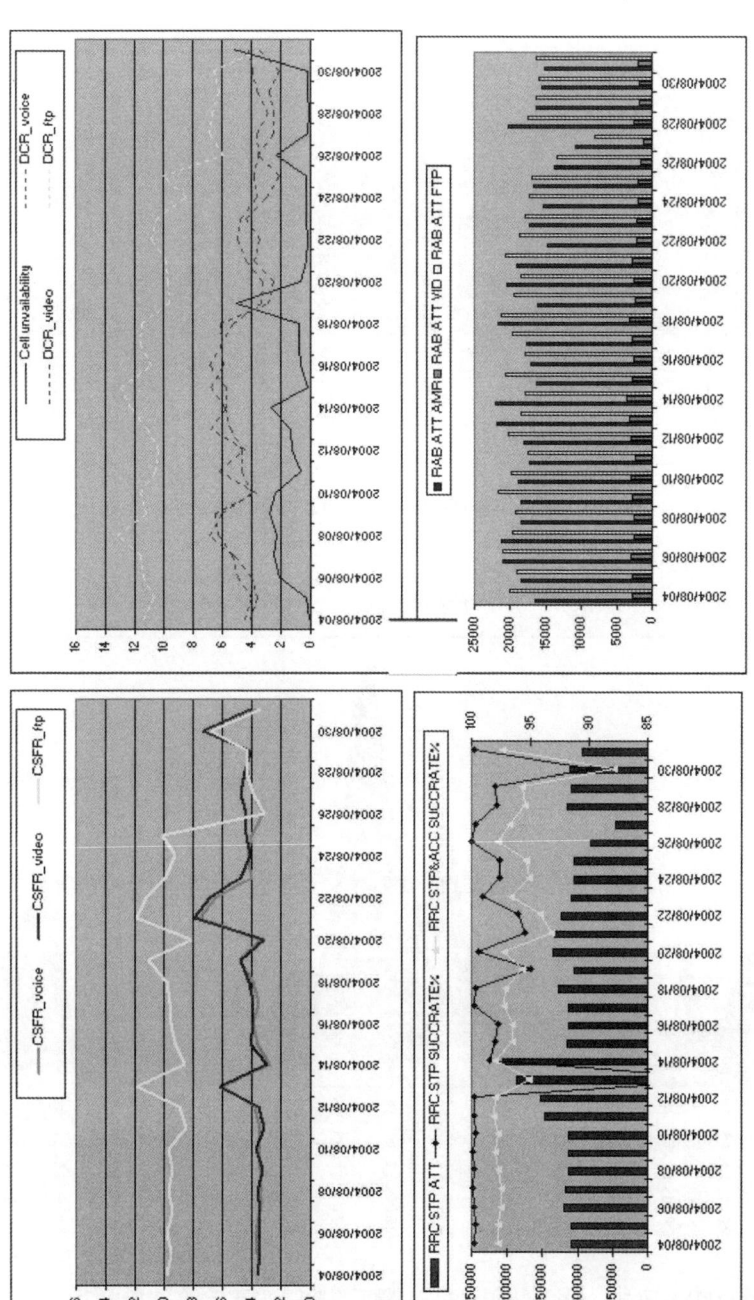

Figure 2.101 RNC level performance monitoring

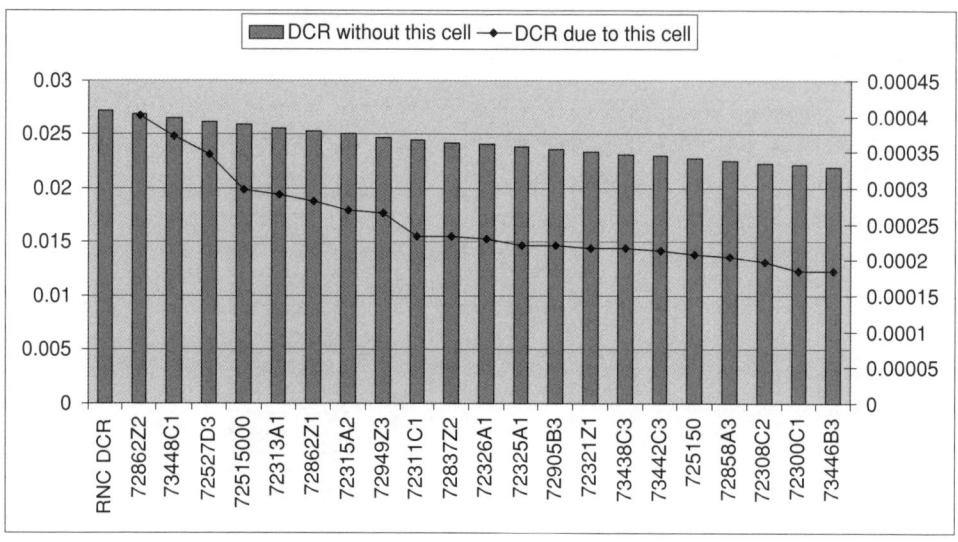

Figure 2.102 Selecting the worst performing cells

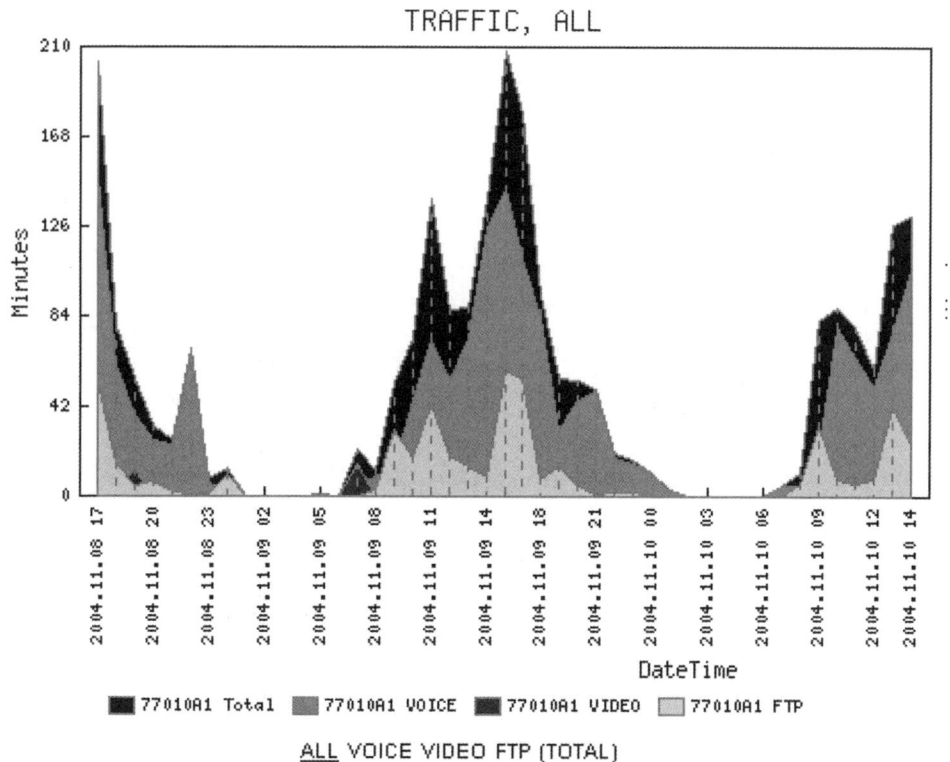

Figure 2.103 Traffic profile of the WCDMA cell (voice, video, PS)

rate was 2.7 %, then the eight worst performing cells need to be fixed to achieve the target. This includes the assumption that all cells can be fixed to a 0 % drop call rate, which means that in real life more cells need to be analysed.

The cells are optimised using the cluster optimisation process. In addition to the cluster optimisation process, the cell level statistics can be used to determine what kind of problem the bad performing cell is experiencing.

Traffic Profile Problems
The traffic profile of each problem cell is analysed as shown in Figure 2.103. The RRC and RAB failure reasons are analysed to determine whether the problem is on the radio interface on the transmission network or whether the SW/HW is the likely source of drops calls. If the problems are in the radio interface the drive test measurement and cluster optimisation process are used to locate the problem.

3

Transmission Network Planning and Optimisation

Ajay R Mishra and Jussi Viero

3.1 Access Transmission Network Planning Process

Access transmission networks are the ones that connect the radio networks and core networks. 'Traditionally' transmission network planning (TNP) was considered to be line-of-sight planning, i.e. if one site can 'see' another site and make a connection through microwave radios/optical cables/leased lines, etc. However, as the complexity of the networks increased, transmission network planning gained more significance. As the technology advances have gone from the GSM to the UMTS and beyond, the process becomes more complicated as the amount of interface with the radio planning teams increases considerably. In the GSM, the process of transmission network planning (shown in bold in Figure 3.1) can work almost stand-alone (with little impact on/from radio). However, this increases considerably in the EGPRS as the A_{bis} becomes a limiting factor for the air interface throughput. In the UMTS, E2E quality and delay makes sure that all three major domains of network planning, radio, transmission and core, are heavily dependent on each other.

The whole process does not really take place in the way shown in the figure. In any project most of the aspects start at a generic level right in the beginning. However, for the sake of understanding, the tasks can be subdivided into master planning and detail planning. Master planning includes dimensioning and other preplanning aspects. Detail plans consist of timeslot and 2 Mbit plans along with the synchronisation and network management plans. However, some tasks like microwave (MW) link planning are integrated in the transmission planning process so much that they can be considered to be a part of both master and detail planning.

3.1.1 Master Planning

Many tasks shown in Figure 3.2 are done in parallel during the initial phase of a network (NW) roll-out. In fact many tasks happen before actual roll-out takes place. The output of tasks such as dimensioning, equipments selection, etc., are utilised to order equipment for the network roll-out to start. This means that the the more accurate master planning is, the more accurate would be the equipment ordering, resulting in

---

Advanced Cellular Network Planning and Optimisation Edited by Ajay R Mishra
© 2007 John Wiley & Sons, Ltd

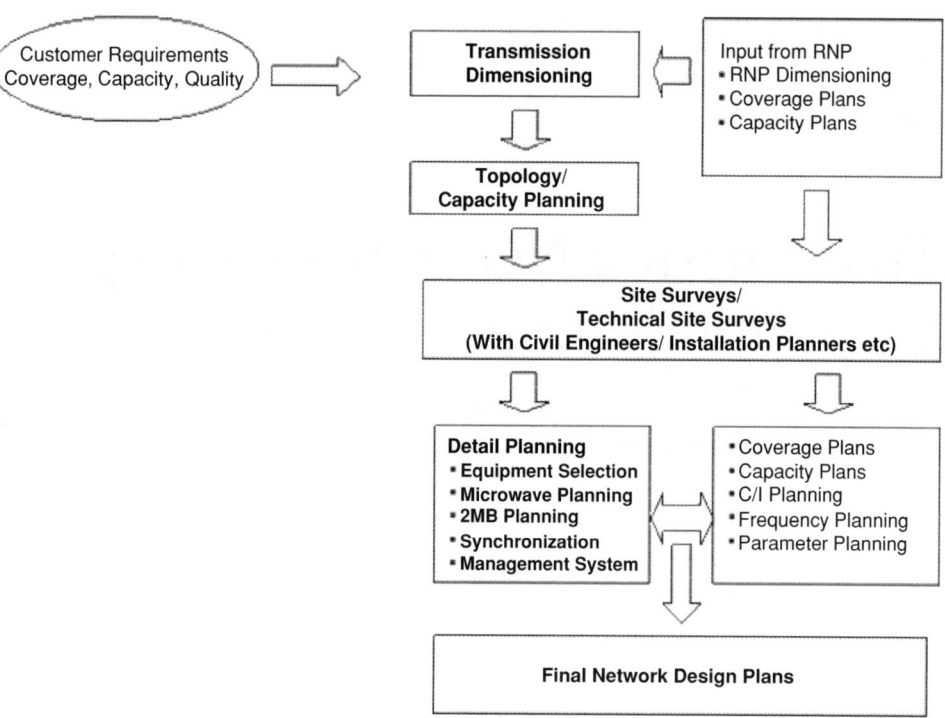

Figure 3.1 Transmission network planning process (along with radio network planning, or RNP)

higher cost efficiency of the roll-out process. Master planning generally includes tasks to the preplanning phase, i.e. dimensioning, planning for protection, deciding topologies, etc.

3.1.2 Detail Planning

A detail plan contains tasks that are related to the master plans but are in more concrete terms. These are the plans that are implemented with respect to the network. For example, a master plan can include a statement that 50 % of the hops are to be connected in loops, but in detail planning all such hops need to be identified. Also, detail planning includes how these hops will be connected, how many PCM lines need

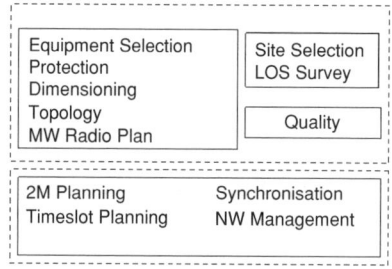

Figure 3.2 TNP tasks

to be connected to each site, timeslots that need to carry traffic for each site apart from synchronisation and management aspects related to the transmission equipment.

3.2 Fundamentals of Transmission

3.2.1 Modulations

The term modulation means to take the message-bearing signal and superimpose it on the carrier signal for transmission. The characteristics of the carrier signal such as the amplitude, width, etc., are varied in accordance with the information-bearing signal. The carrier signals are usually of higher frequency than the message-bearing signal. Pulse modulation is a process in which the characteristics of the individual pulse are modulated on to a carrier channel. The radio carrier can be modulated in terms of amplitude, frequency and phase, thereby giving three major modulation techniques: amplitude modulation (AM), frequency modulation (FM) and phase (or quadrature) modulation (PM). There are other modulations schemes such as amplitude shift keying (ASK), frequency shift keying (FSK) and phase shift keying (PSK), which are primarily variants of the analogue modulations. Depending on the discrete amplitude levels, frequencies and phase states, a variety of modulation methods can be derived from these basic methods. Both in analogue and digital modulations, the properties of the signal to be varied are similar, but the transmitted signal has a direct impact on the output of modulation.

3.2.2 Multiple Access Schemes

Multiplexing

The field of telecommunication has come a long way from just communicating between two individual users separately. Today on a single link many users seem to simultaneously communicate. This has been possible due to the efforts of what is now very commonly known as multiplexing.

With some 1000 million telephone connections in use around the world today and the number of Internet users continuing to grow rapidly, the network providers have been faced with the task of trying to deal effectively with increased telephone traffic. It was not feasible to give every user his/her own separate channel. Therefore, in response to the growing market needs, a number of methods and technologies have been developed within the last few decades to cater for the increasing marketing needs in as economic a way as possible. In the telecommunication field this resulted in the introduction of one of the very first multiplexing techniques, i.e. the frequency division multiplexing (FDM) technique.

In this technique, each individual telephone channel/signal is modulated with a different carrier frequency such that the carrier frequencies are sufficiently separated and so the bandwidths of the signals do not significantly overlap. These channels are separated by guard bands, which are unused portions of the channel spectrum, to prevent interference. These modulated signals are then shifted into different frequency ranges, enabling different telephone connections/signals to be sent over the same channel/telephone cable.

The FDM schemes used around the world are standardised to some extent. In the United States, AT&T has designated a hierarchy of FDM schemes to accommodate transmission systems of various capacities. Unfortunately this is not identical to the standards, which are internationally adopted by the International Telecommunications Union (ITU). The FDM standards adopted are listed in Table 3.1.

Multiplexing was not confined to the frequency domain only, and with the introduction of digital communication came the most successful multiplexing techniques, which today is used in most telecommunication networks – time division multiplexing (TDM). Attention will turn to TDM after discussing another multiplexing technique, which has been the outcome of the invention of the optical fibre – wavelength division multiplexing (WDM). Simply put, it multiplexes different light wave signals by a diffraction grating and sends them on a single optical fibre to a destination where the signals are split

Table 3.1 United States and international FDM carrier standards

Number of voice channels	Bandwidth	Spectrum	AT&T	ITU-T
12	48	60–108	Group	Group
60	240	312–552	Supergroup	Supergroup
300	1.232	812–2044		Mastergroup
600	2.52	564–3084	Mastergroup	
900	3.872	8.516–12.388		Supermastergroup

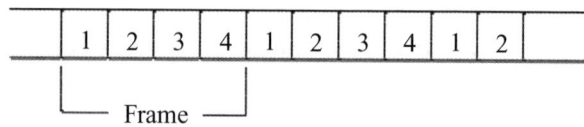

Figure 3.3 Bit interleaving

again. This type of multiplexing is independent of any protocol like Ethernet, etc. With the amount of bandwidth the optical fibre has to offer, this technique shows great potential for the years to come.

Time Division Multiplexing

Time division multiplexing (TDM) is by far the most commonly used and effective means of subdividing the capacity of the digital transmission service/channel among a number of sources/paths. Put in a different way, it can also be described as combining more than one signal such that each piece of a particular signal gets the channel only for a small amount of time, called its timeslot. This switching of different signals is done so quickly that the service users feel that the entire channel is entirely dedicated to them only.

In TDM, capacity/channel allocation may be done either bitwise or wordwise. In bitwise allocation, each source is assigned a timeslot corresponding to a single bit, as Figure 3.3 shows for a four-signal multiplexing. In wordwise allocation, a time slot corresponds to some larger number of bits (normally 4 or 8 bits), referred to as a word. The TDM frame structure for the wordwise interleaving of a four-signal system is shown in Figure 3.4.

With the advent of semiconductor circuits and the exponential growth of users of the telephone network, a new type of digital transmission method, pulse code modulation (PCM), was developed in the 1960s. With PCM, an analogue channel could be transmitted using the intermediate steps of sampling, quantisation and encoding. Thus an analogue telephone voice signal is first sampled at a rate more than the Nyquist criterion rate, i.e. at 3.1 kHz, then is quantised to set certain predefined levels for transmission and finally these predefined quantised levels are encoded, usually by the HDB3 (high density bipolar 3) encoding technique. The signal is then transmitted at the rate of 64 kbps. This is the basic rate of a single telephone voice channel more popularly/technically known as digital signal 0 (DS0).

This basic voice channel is a very low bandwidth channel and thus a whole channel cannot be dedicated to it as it would mean a total waste of the bandwidth. There is a need of multiplexing these voice signals,

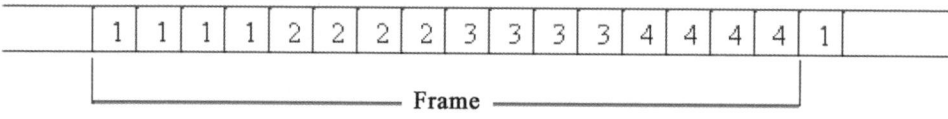

Figure 3.4 Four-bit word interleaving

which could make proper utilisation of the available bandwidth. The level/order of multiplexing them and the method of doing so categorises the TDM technique into three major types, which are technologies in themselves. These types of multiplexing in the time domain are:

- Plesiochronous digital hierarchy (PDH)
- Synchronous digital hierarchy (SDH)
- Asynchronous transfer mode (ATM)

3.3 Digital Hierarchies – PDH and SDH

3.3.1 Plesiochronous Digital Hierarchy (PDH)

The plesiochronous digital hierarchy (PDH) is a technology used in telecommunications networks to transport large quantities of data over digital transport equipment such as fibre optic and microwave radio systems. The multiplexing of basic voice signals is just not enough for the high channels that are used these days. With the advent of optical fibres the capacity/bandwidth of the channel has considerably increased. It is thus required to multiplex the signals at different levels and not only once. Thus the type of application required for a service decides the level/order of multiplexing. For example, services such as the integrated services digital network (ISDN) require more bandwidth for running applications like voice, video and data and thus the bandwidth requirement is greater than required by an ordinary telephone/voice service. The more the order of this digital hierarchy, the greater is the bandwidth of the channel.

The term 'plesiochronous' is derived from the Greek word *plesio*, meaning near, and *chronous*, meaning time. This indicates that PDH networks run in a state where different parts of the network are almost, but not quite perfectly, synchronised. Simply put, the data streams, also called the tributaries, have the same nominal frequency but are not synchronised with each other; i.e. the rising and falling edges of the pulses in each tributary do not coincide.

The European and American versions of PDH systems differ slightly in their data rates, but the basic principles of multiplexing are the same. In the North American hierarchy, four primary systems are combined/multiplexed to form an output having 96 channels. This is the second order of multiplexing. Seven 96-channel systems can be multiplexed to give an output of 672 channels (third order of multiplexing). Six 672-channel systems are multiplexed to give an output of 4032 channels (fourth order). Higher orders of multiplexing are also available but they employ another technique called the synchronous digital hierarchy (SDH).

The first level of multiplexing can be looked at in more detail along with the construction of its frame. Famously known as T1 in North America and Japan, the digital signal 1 (DS1) consists of 24 basic voice channels (64 kbps) multiplexed together to form a single frame. A frame consists of $24 \times 8 = 192$ bits, plus one extra bit for framing, making it a total of 193 bits in all in every 125 µs. The 193rd bit is used for frame synchronisation. Thus the data rate of the T1 is calculated as 193 bits /125 µs = 1.544 Mbps. This rate is also known as the primary rate.

When a T1 system is only for data just 23 frames carry the data/real information and the last frame is used for synchronisation and a signalling pattern. The frame structure of the T1 (DS1) system is shown in Figure 3.5.

Details of the higher order PDH systems in North America are shown in Table 3.2.

Europe and the rest of the world (excluding North America and Japan) use the PDH hierarchy in which 30 basic voice channels (64 kbps) are multiplexed together to form a primary system (first order of multiplexing). Four primary systems are combined/multiplexed to form an output having 120 channels (second order of multiplexing). Similarly, four 120-channel systems can be multiplexed to give an output of 480 channels (third order of multiplexing). Four 480-channel systems are multiplexed to give an output of 1920 channels (fourth order of multiplexing) and four 1920-channel systems are multiplexed to give

Table 3.2 PDH for North American systems

Level/order	Number of channels	Bit rate [Mbps]
DS1	24	1.544
DS2	96	6.312
DS3	672	44.736
DS4	4032	274.176

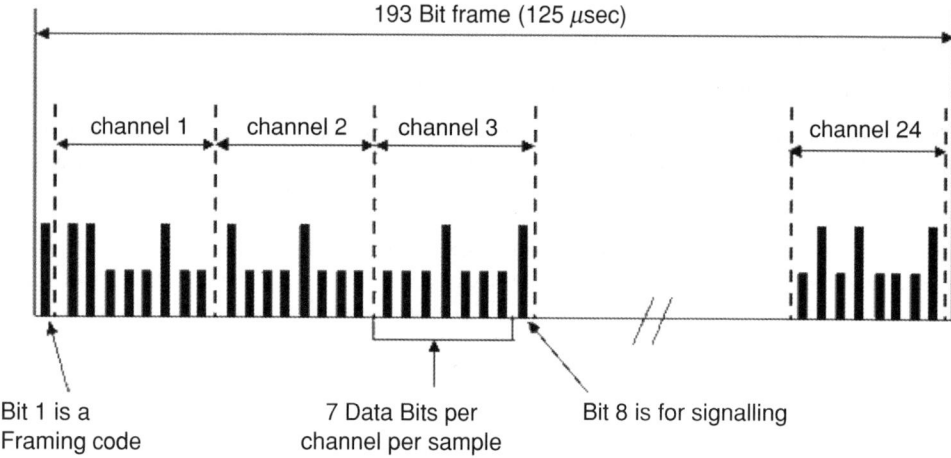

Figure 3.5 The T1 carrier frame structure

an output of 7680 channels (fifth order of multiplexing). All levels beyond the fifth level are now levels of SDH systems and are taken care of by their multiplexers. The bit rates for each PDH order are shown in Table 3.3.

Thus a summary of the PDH rates can now be depicted diagrammatically (Figure 3.6) to show a clear picture of the various hierarchies in different regions. Major disadvantages faced by the PDH are:

1. Pulse stuffing or positive justification. As the name suggests, plesiochronous systems are not perfectly synchronous. Hence the input data stream rates are very close but not identical so the pulses do not arrive in a synchronised manner. In order to multiplex different PDH signals at higher orders, bit stuffing is required to match the rates of the data streams. Pulse stuffing involves intentionally making the output bit rate of a channel higher than the input bit rate so that the data rate of that particular stream matches with the other tributaries and is thus de-multiplexed/received without any errors.

Table 3.3 PDH for European systems

Level/order	Number of channels	Bit rate [Mbps]
E1	30	2.048
E2	120	8.448
E3	480	34.368
E4	1920	139.264
E5	7680	565.992

Digital Hierarchies – PDH and SDH

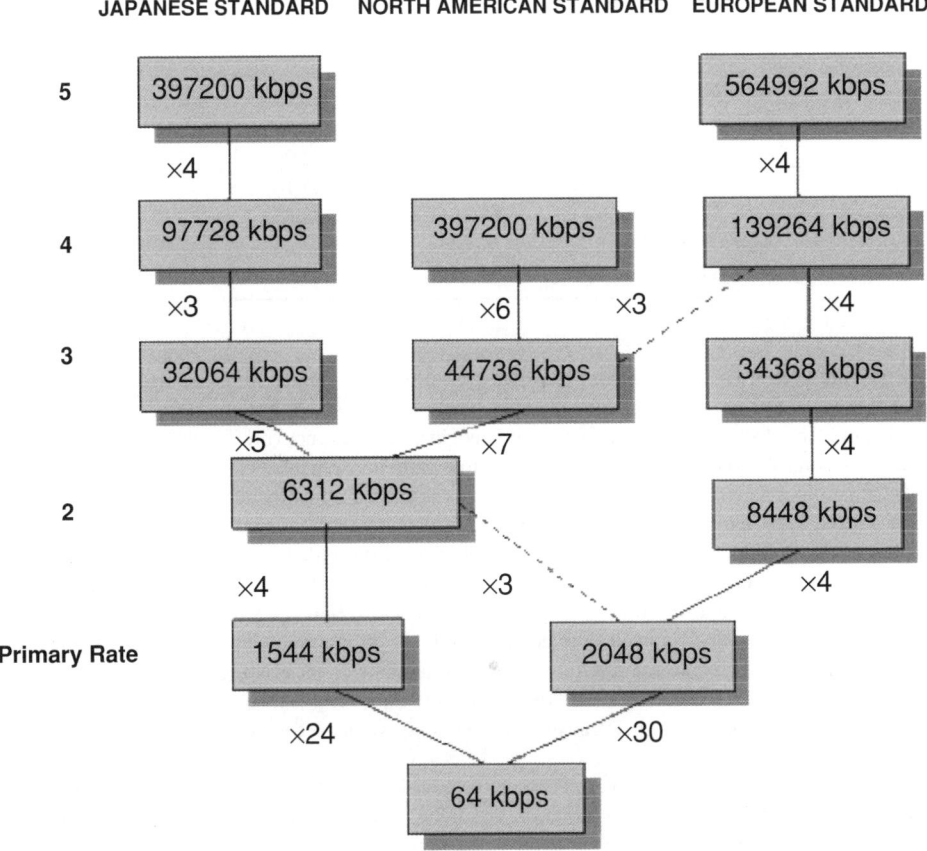

Figure 3.6 Summary of plesiochronous transfer rates

2. Add/drop facility not available. Whenever any signal is needed to be extracted from a higher order multiplexed signal, it cannot be just dropped from the signal. Instead the signal will be de-multiplexed till the needed signal is separated. The same applies when adding a signal to an already multiplexed signal. Thus no adding and dropping of signals can be done, which is a very essential part of a multiplexed system, as will be seen in the next section.

3.3.2 Synchronous Digital Hierarchy (SDH)

The data rates in the plesiochronous digital hierarchy went up to as high as 140 Mbps approximately. That might seem to be satisfactory but with the advent of the optical fibre channel capacity went up to tens of gigabits and the PDH for multiplexing was not able to satisfy this huge increase. The demand for greater channel capacity for 'bandwidth-hungry' applications and services, however, meant that more stages of multiplexing were needed throughout the world. This was one of the major reasons for looking out for another technology, which could not only provide more multiplexing stages and thus more channel bandwidth/capacity but also a synchronous service that PDH failed to give.

Another major disadvantage was that standards already existed for electrical line interfaces at PDH rates, but there was no standard for optical line equipment at any PDH rate, which meant that it was specific

Table 3.4 SONET and SDH transmission speeds

SONET signal	SDH signal	Bit rate [Mbps]
STS-1		51.84
STS-3	STM-1	155.52
STS-12	STM-4	622.08
STS-24		1244.16
STS-48	STM-16	2488.32
STS-192	STM-64	9953.28
STS-768	STM-256	39814.32

to each manufacturer. This implied that fibre optic transmission equipment from one manufacturer might not be able to interact/interface with other manufacturers' equipment. As a result, service providers were often required to select a single vendor for deployment in areas of the network and were forced into using the network control and monitoring equipments of that particular vendor.

The situation of reconfiguration of PDH networks was particularly difficult in North America, where a plesiochronous system (T-carrier) was in place. Understanding the scenario, Bellcore (the research affiliates of the Bell operating companies in the United States) decided to utilise the technological advances and associated reductions in cost since plesiochronous systems had already been introduced. This made them propose a new transmission hierarchy in 1985. Bellcore's major goal was to create a synchronous system with an optical interface compatible with multiple vendors, but the standardisation also included a flexible frame structure capable of handling either existing or new signals and also numerous facilities built into the signal overhead for operations, administration, maintenance and provisioning (OAM&P) purposes. The new transmission hierarchy was named the synchronous optical network (SONET).

The CCITT, now the International Telecommunication Union (ITU), also came into the matter and established an international standard based on the SONET specifications, known as the synchronous digital hierarchy (SDH), in 1988. The SONET is the subset of SDH as SONET is only used in North America and Japan, as compared to SDH, which is used in the rest of the world. The frame formats and thus the data rates of both systems are not the same but are compatible due to their synchronous nature. The hierarchies of both these systems along with their data rates are shown in the Table 3.4.

What is 'Synchronous Digital Hierarchy'?

The synchronous digital hierarchy refers to the group/layers of transmission rates or standards that can transport digital signals/data of different capacities through high bandwidth mediums like optical fibres or radio waves. The hierarchy starts from the basic SDH rate of 155.52 Mbps (which is greater than the highest transmission rate of the PDH system) and goes up to as high as 40 Gbps, which is more than sufficient for 'bandwidth-hungry' applications like video conferencing. The further up in the hierarchy the more the level/order of multiplexing keeps on increasing, with the higher order signals being made by multiplexing the lower order signals in the time domain.

Another aim for creating a new transmission hierarchy was the plesiochronous behaviour of the PDH. It is known that the data streams/tributaries in PDH systems were out of phase with each other, which made the processing of information and multiplexing more tedious. Bit stuffing/justification provided a solution along with the problem of extra bandwidth wastage and processing time. This demanded a synchronous system, giving the name 'synchronous digital hierarchy'. Thus, due to the synchronous nature of the SDH, the average frequency of all slave clocks in the system is the same and can be referred back to the highly stable master clock (e.g. the PRC/SSU/SEC).

Digital Hierarchies – PDH and SDH

Advantages of the SDH

The basic advantage of the synchronous nature has already been discussed in detail. There are other reasons why the SDH has overshadowed its plesiochronous counterpart:

- High transmission rates. Transmission rates of up to 10 Gbps can be achieved in modern SDH systems, making it the most suitable technology for backbones – the superhighways in today's telecommunications networks.
- Simplified add/drop function. Compared to the older PDH system, low bit rate channels can be easily extracted from and inserted into the high speed bit streams in the SDH. It is now no longer necessary to apply the complex and costly procedure of de-multiplexing and then re-multiplexing the plesiochronous structure.
- Reliability. Modern SDH networks include various automatic backup circuits and repair mechanisms, which are designed to cope with system faults. As a result, failure of a link does not lead to failure of the entire network.
- Future platform for new services. The SDH is the ideal platform for a wide range of services including ISDN, mobile radio and data communications (LAN (local area network), WAN (wide area network), etc.). It is also able to handle more recent services such as video-on-demand and digital video broadcasting via the ATM.
- Interconnection. The SDH makes it much easier to set up gateways between different network providers and to SONET systems. The SDH interfaces are globally standardised, making it possible to combine network elements (NE) from different manufacturers into a single network, thus reducing equipment costs.

The Layered SDH Model

All the network technologies and architectures are based on the layered model so that they can abstract the working of the entire structure in various parts. Hence the SDH system is modelled in a layered manner to separate the functionality of its various components, making understanding of the system easier.

SDH networks are subdivided into various layers that are directly related to the network topology. The lowest layer is the physical layer, which represents the transmission medium. This is usually a fibre optic path or a satellite or radio wave link. Then follows the regenerator section, which is the basic segment of the SDH network. It is the smallest entity that can be managed by the system. Each repeater monitors defects such as the loss of signal, loss of frame, etc. Part of the regenerator section overhead (RSOH) is fully calculated by passing the signal through a regenerator. The next section is the multiplex section, which covers the part of SDH link between multiplexers. The multiplex section overhead (MSOH) is used for this section to detect any errored blocks and defects and generates a special alarm in the forward and backward directions. The basic architecture of the SDH path is shown in Figure 3.7 (where MUX is the multiplexer and IP the Internet Protocol).

The two virtual container (VC) layers represent a part of the mapping process. Mapping of the signals is done to pack the various incoming tributaries like ATM and PDH into the SDH transport modules. VC-4 mapping is used for packing high capacity signals like the ATM signals, whereas VC-12 mapping is for low capacity signals like the PDH signals. This layered model can be represented as shown in Figure 3.8.

SDH Frame Format

The SDH signal is transmitted in the form of frames, with each frame of each signal containing a specific amount of information. The higher the hierarchy of the signal, the higher the speed of the signal and the higher will be its information carrying capacity. The SDH frame in general consists of two basic areas, the transport overhead and the synchronous payload envelope (SPE). The transport overhead comprise

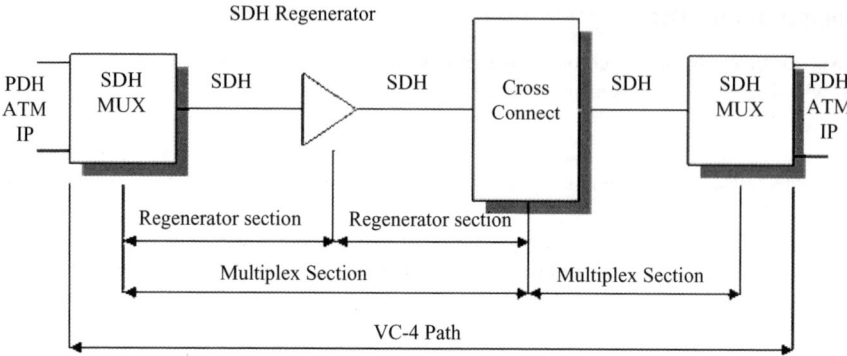

Figure 3.7 SDH architecture basics

the starting bits used for housekeeping of the real information, which is also called the payload. This is stored in the majority of the frame called the payload envelope. The SPE can further be classified into two parts: the path overhead and the payload.

For monitoring and controlling the SDH network, additional information is transmitted together with the traffic data (payload). This additional information, called overhead, is divided into two main groups, i.e. the section overhead and the path overhead. The section overhead is a minicontainer holding different types of information required for transmission. The section overhead offers free capacity, which can be used for additional information. The section overhead always starts at the beginning of the transport frame.

The transport overhead is basically divided into three parts:

1. Regenerator section overhead (RSOH)
2. Administrative unit (AU) pointer
3. Multiplexing section overhead (MSOH)

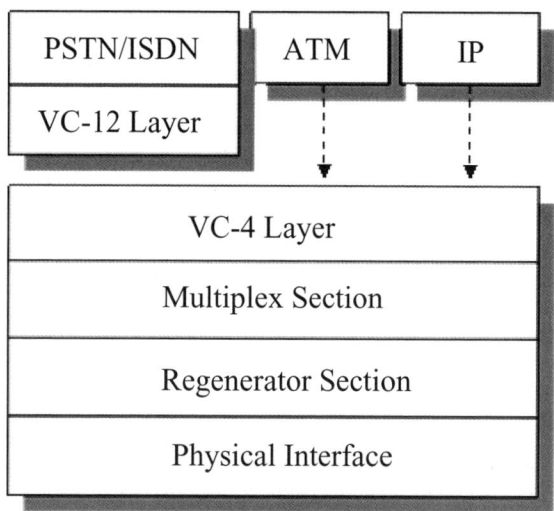

Figure 3.8 SDH layer model

Digital Hierarchies – PDH and SDH

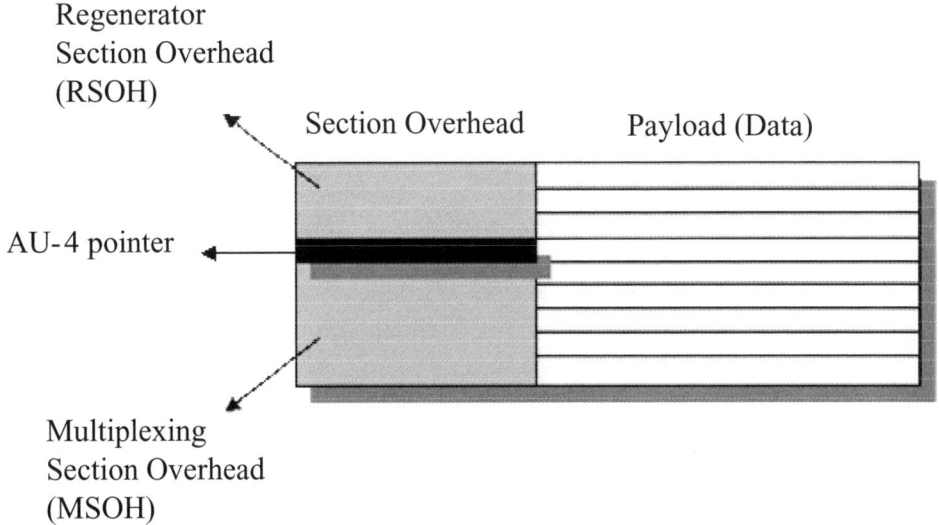

Figure 3.9 SDH frame structure

While the RSOH is terminated (i.e. disassembled, evaluated and newly generated) at each regenerator point, the MSOH passes the regenerator without being modified and is only terminated at the multiplexers (where the payload is assembled or disassembled).

The section overhead also includes a pointer defining the position of the containers in the payload area. The pointer value, also referred to as offset, indicates the offset of the container with respect to a reference point of the frame. The pointer, however, is not part of the section overhead.

Together with a container (C), the path overhead (POH) forms the virtual container (VC). The POH capacity depends on the path level. While the higher order POH is composed of 9 bytes (one row), only 4 bytes are available for the lower order POH. The higher order POH is located in the first column (9 bytes) of VC-3 or VC-4. It is formed on generation of the VC-3 (VC-4) and remains unchanged until the VC is disassembled in order to be able to monitor the complete path.

The SPE is completed with a POH, which is used to monitor and control the correct addressing as well as to identify the payload contents, whereas the payload is the revenue producing traffic being transported and routed over the SDH network. The SDH frame may contain more than one low speed digital signal such as 2/34/140 Mbps or some part of the high speed signal. A general SDH frame format can be represented as in Figure 3.10, where N is the level of hierarchy of the SDH signal. The number of columns remains nine in every level. The building block of the SDH hierarchy is the STM-1 (synchronous transport module, level 1). This is the lowest level SDH signal where $N = 1$; thus the total number of rows is 9 and the total number of columns is 270, out of which 9 are used for transport overhead and the remaining 261 columns include the synchronous payload envelope. At the macro level SDH data are transported frame by frame, with the basic frame being the STM-1.

Considering bitwise data transport, the MSB (most significant bit) of each byte is transported to the frame first, followed by the others. Then 1 byte is stored in every cell of the 9×270 matrix, which are transported to the frame till one frame fills up. All bytes of row 1 are first transmitted starting from column 1 and move off to the first byte in row 2 and so on till the end of the frame. This frame is then transported in the STM-1 container whose frame period is 125 μs (Figure 3.11). Figure 3.12 depicts the order of transporting nine rows one by one.

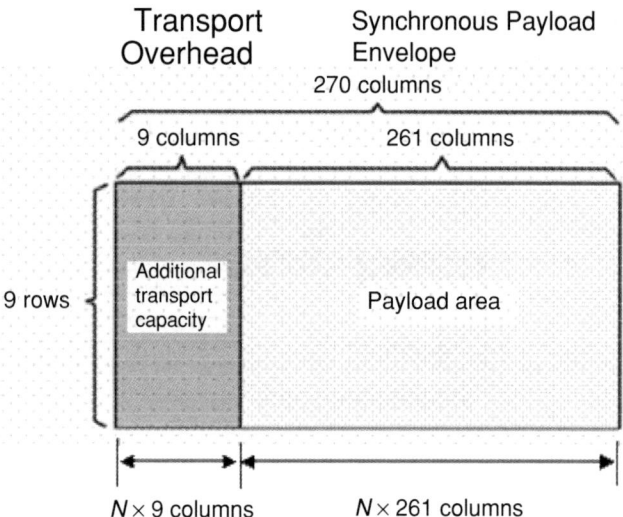

Figure 3.10 Transport frame of the STM-1 signal

The frame capacity is thus the number of bits contained within the single frame:

$$\text{Frame capacity} = 270 \text{ bytes/row} \times 9 \text{ rows/frame} \times 8 \text{bits/byte}$$
$$= 19\,440 \text{ bits/frame}$$

The bit rate of the STM-1 signal is thus calculated using the formula

$$\text{Bit rate} = \text{frame rate} \times \text{frame capacity}$$

As the frame period is 125 μs, the frame rate is 8000 frames/second. Thus the bit rate is

$$8000 \text{ frames/second} \times 19\,440 \text{ bits/frame} = 155.52 \text{ Mbps}$$

This is the basic rate of the SDH. Other rates in the hierarchy are STM-4/16/64/256 whose rates are defined in Table 3.5.

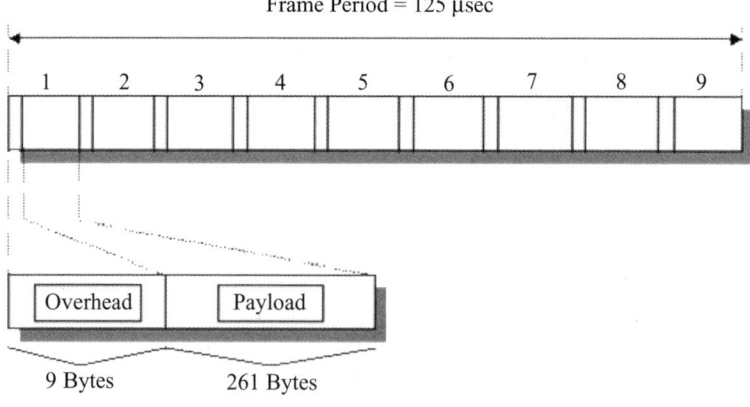

Figure 3.11 Transport frames in one frame period

Digital Hierarchies – PDH and SDH

Table 3.5 The SDH signal

SDH signal	Bit rate [Mbps]	Number of total columns	Number of payload columns
STM-1	155.52	9	261
STM-4	622.08	36	1044
STM-16	2488.32	144	4176
STM-64	9953.28	576	16 704
STM-256	39814.32	2304	66 816

Analogy of SDH Transmission

Containers

The transmission of SDH signals can be compared with the transmission of containers (Figure 3.13) on a conveyor belt. The payload is transported in containers of certain sizes. Since the payloads have different volumes, containers with different capacities have been defined. If the payload is too small, it is filled up with stuffing information. For transporting the information, the container needs a label (Figure 3.14).

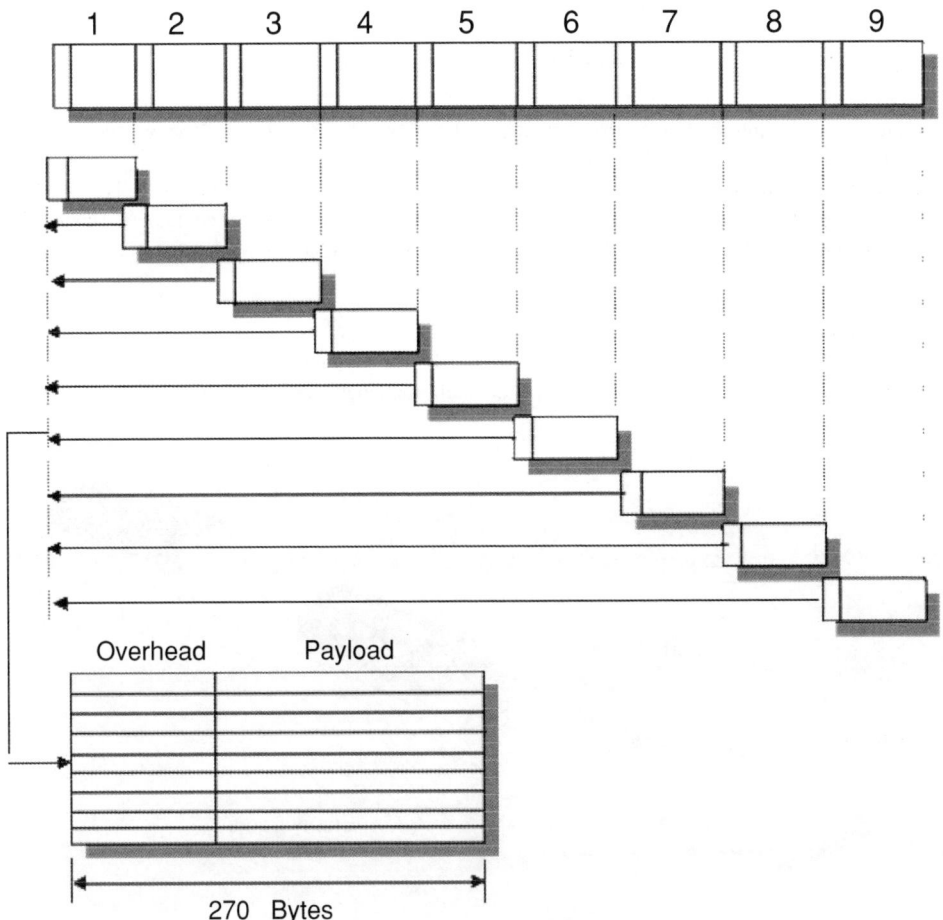

Figure 3.12 Transporting the STM-1 frame

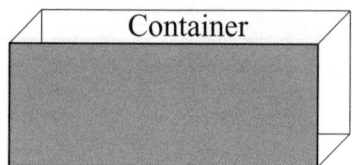

Figure 3.13 A container

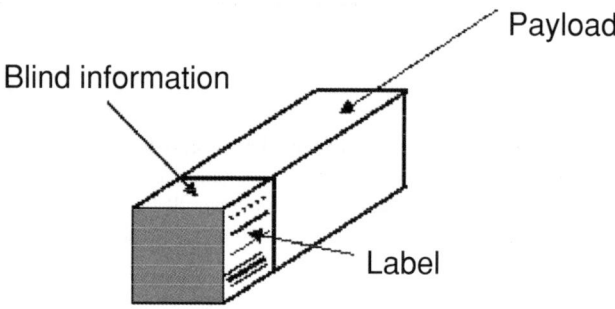

Figure 3.14 A container with a label

The latter includes information on the container contents, monitoring data, etc. The receiver evaluates this information.

The complete containers are then put on a kind of conveyor belt, as shown in Figure 3.15. This conveyor belt is divided into several frames of identical size. They are used to transport the containers. The position of the containers in the frame is arbitrary; i.e. a container does not have to start at the beginning of the frame.

Groups of Containers
Sometimes the information to be transmitted is small as compared to the container to be used for transmission. Thus before transportation, several small containers can be combined to form a group.

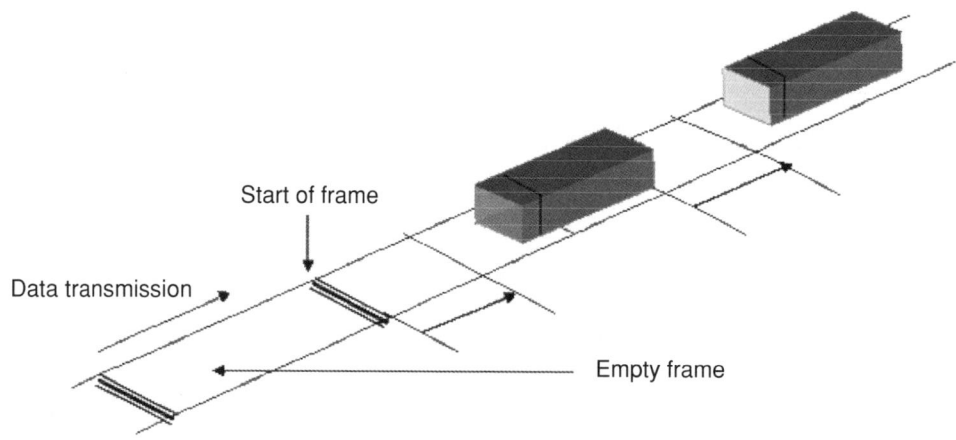

Figure 3.15 Transport frame

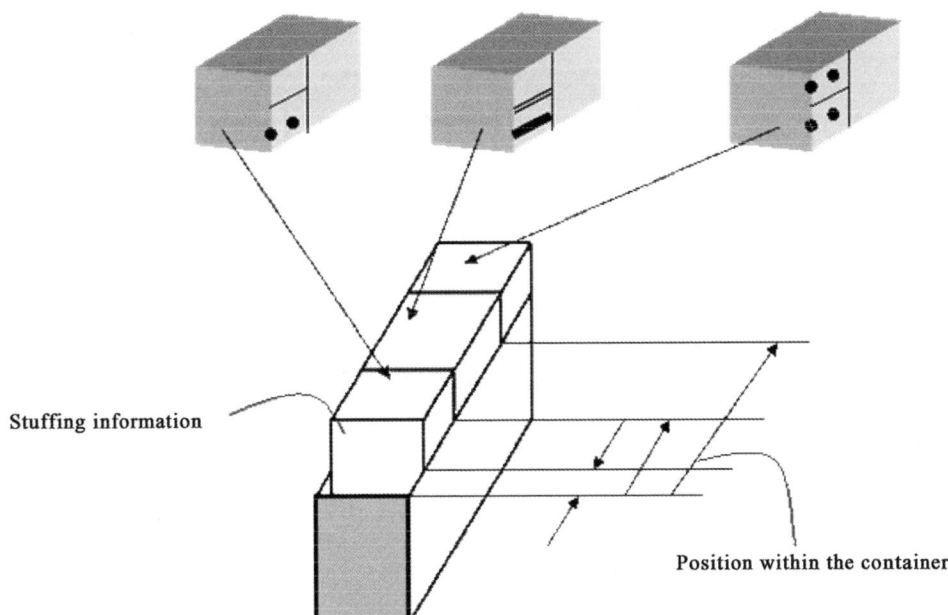

Figure 3.16 Conbining container in a container group

This group is then packed into a larger container, as shown in Figure 3.16. Each of these containers includes a label, which is evaluated by the receiver. Whenever necessary, stuffing information is added.

The individual containers are assigned a certain position within the group. The position number determines the start of the respective container. The type of payload in the containers is unimportant for transportation as it is just the raw information that is transmitted. The stuffing information, which is included to fill up the unutilised space, can be regarded as part of the payload.

Concatenation

The above description was based on the assumption that the payload is smaller than the container available. If the payload to be transported is larger than the container available for it, several containers can be concatenated. They then form a continuous container chain, as shown in Figure 3.17. In this case, the payload is distributed on this container chain. An example will now be given to illustrate this. The source signal is 599.04 Mbps (broadband ISDN). Since the largest container defined can transport a signal only up to 140 Mbps, four such containers have to be concatenated. The position of the container chain on the conveyor belt is defined for the first container. The first one determines the position of all other containers 2, 3 and 4.

This SDH system is analogous to a real world application. A company's production unit producing items and packing them to deliver to the market will be considered. The company is analogous to the SDH system, which is managing the hierarchy of steps taken to get the final result. The final result can be thought of as the multiplexed signal to be sent over the channel.

All the data streams to be sent over the channel are first collected. These data streams may be of different capacities, e.g. an ATM signal, a PDH stream or an ISDN BRI (basic rate interface) channel. In order to send these signals over the same channel they need to be multiplexed using the SDH system. In the SDH system the signals are put in their respective containers according to their capacities just like items of different sizes are put into boxes that are meant for them only. Next, on the outside of the boxes is written the name, etc., of the respective item it contains, which makes it easier for the receiver

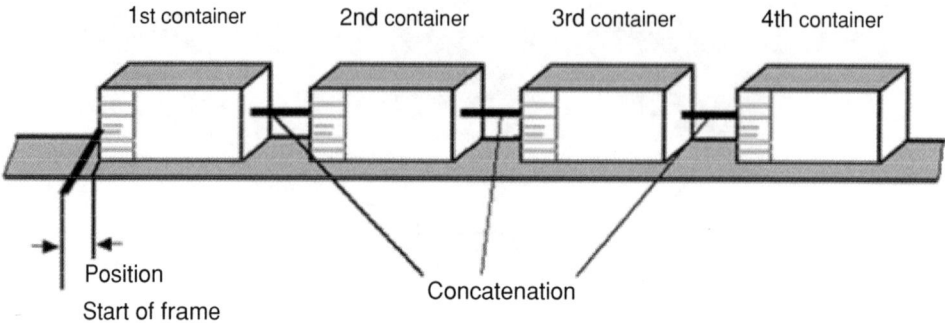

Figure 3.17 Concatenated containers

to decipher the item inside the box. Similarly, the signal containers are labelled using certain overheads for easy unpacking.

If the box is big enough nothing further needs to be added as the box carrier is just meant for a particular capacity and can carry a box of only a particular capacity. However, if there is still some space left inside the carrier, more boxes can be clubbed together in a bigger box, which is again labelled and put on the carrier. In this way the capacity of the carrier is used to its fullest and is how the SDH proceeds. If, say, two PDH signals along with an ATM signal need to be sent then the two PDH signals would not occupy the full channel. Thus to utilise the channel properly the PDH and ATM signals are put in their respective containers and are then combined at the next level of hierarchy. This bigger container with all the overheads included can then be shipped to the destination where it is unpacked.

SDH Multiplex Elements

An analogy between the SDH system and the real world clears the picture of higher order multiplexing, but the way these containers are made, labelled and moved up in the hierarchy is still hazy. In the following subsections there is a discussion on different types and capacities of containers and their hierarchy, making a complete STM-n signal. The discussion starts with a simple container and then moves on to the next hierarchies.

Containers
Containers are fixed size blocks that contain the incoming signals. As the signals are smaller than the capacity of the container they need to be stuffed with extra bits. This filling of the source signal with additional bits is called mapping.

Containers are denoted as C-nx (where $n = 1$ to 4 and $x = 1$ or 2), where n is the PDH level and x indicates the bit rate (i.e. $x = 1 \Rightarrow 1.544$ Mbps and $x = 2 \Rightarrow 2.048$ Mbps). Different container sizes (e.g. C-11, C-12, C-2, C-3, C-4) are available for the different source signal bit rates. C-2 is used for a 6 Mbps signal, C-3 for a 34 Mbps signal and C-4 for 140 Mbps signal (see Figure 3.18).

Virtual Containers
The path overhead (POH) is used to establish a path in the network by monitoring the addressing. When this component is added to the container, it becomes a virtual container; i.e. when POH is added to C-n, it is referred to as VC-n (where $n = 1$ to 4), as shown in Figure 3.19.

Administrative Unit
The administrative unit (AU) is a chunk of bandwidth used for communicating in a telecommunication network. Addition of the administrative pointer in the virtual container makes the VC the administrative

Digital Hierarchies – PDH and SDH

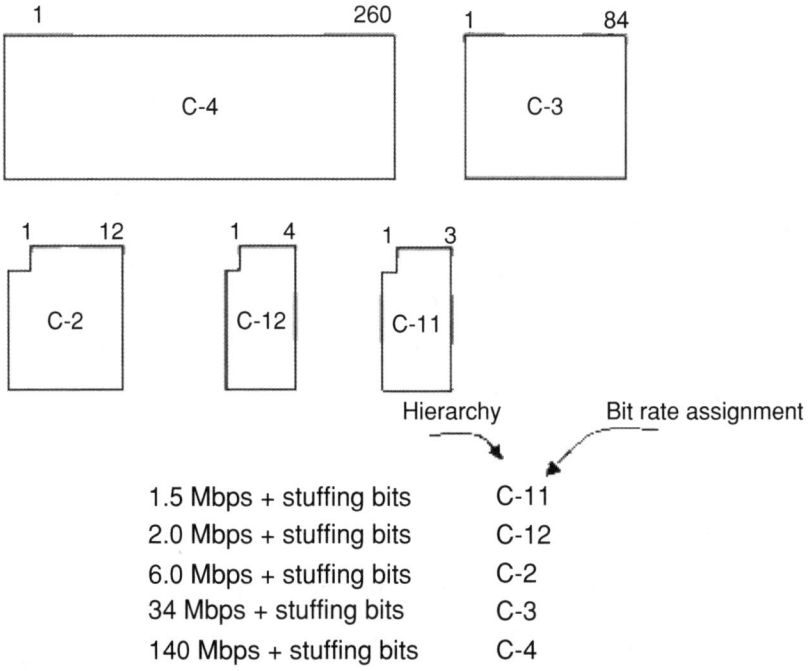

Figure 3.18 Containers with different capacities

unit; e.g. addition of the AU pointer to the VC-3/VC-4 makes it the AU-3/AU-4. The payload of an STM-1 signal consists of one AU-4 or three AU-3s, as shown in Figure 3.20.

Tributary Units

The next block is the tributary unit (TV), denoted as TU-nx (where $n = 1$ to 3 and $x = 1$ or 2). The virtual containers VC-11, VC-12 and VC-2 are completed to make a tributary unit by adding the pointer. The pointer specifies the phase of the VC, which is said to be mapped with respect to the TUs. In TU-11, TU-12 and TU-2, there is only space for one pointer byte (shown in Figure 3.21).

Administrative Unit Group

On multiplexing the AU-n into STM-n, an administrative unit group (AUG) is formed by three AU-3s or one AU-4. The three AU-3s are interleaved byte by byte. The AUG represents an information structure composed of nine rows each consisting of 261 columns plus 9 bytes in row 4 for the AU pointers (shown in Figure 3.22).

SDH Multiplexing

The basic elements given in the previous section are used in multiplexing signals of various capacities and are sent through the same common channel simultaneously. What actually happens while packing a signal into an STM-n signal will become clearer once the stages of hierarchy are gone through using a particular signal capacity.

Using these objects, the multiplexing of smaller signals is done stepwise. Consider a DS-4 signal (140 Mbps); the STM-1 can be generated in a sequence of steps. Firstly, a 140 Mbps signal is added with additional bits, making it fit in the C-4 exactly. The container is then added with the POH, converting

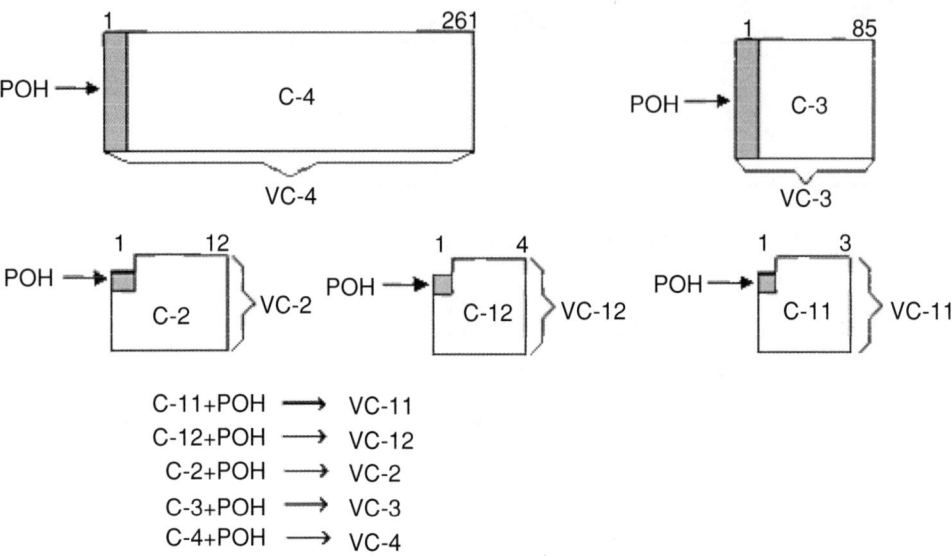

Figure 3.19 Virtual containers with different capacities

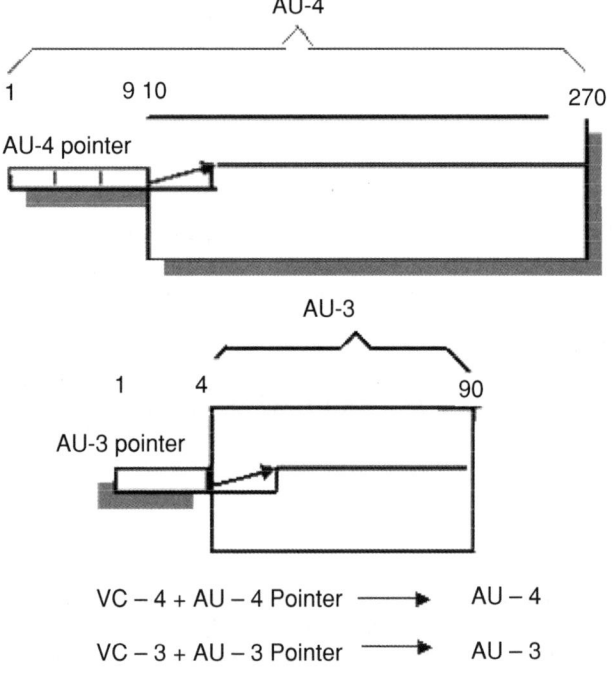

Figure 3.20 Administrative unit

Digital Hierarchies – PDH and SDH

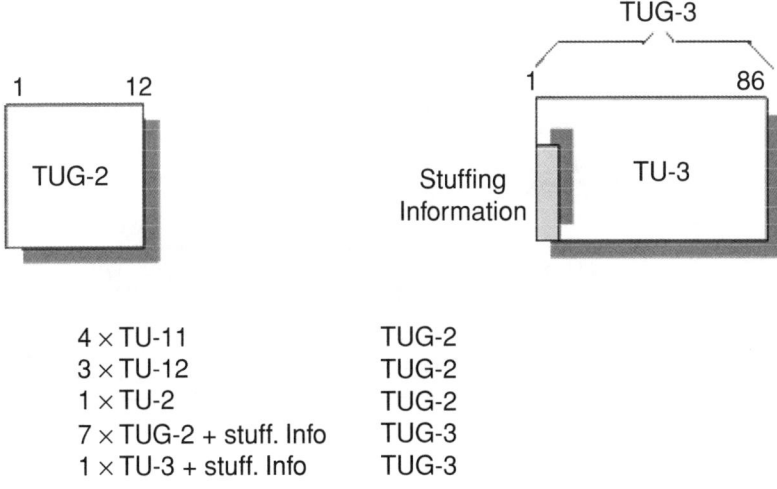

Figure 3.21 Tributary unit group

it into a VC-4. The next stage is to convert it into the AU by addition of the administrative unit pointer (AU-4). The addition of section overhead finally makes it a complete STM-1, to be transmitted over the network.

A very low speed signal can also be converted into an STM-1, but the intermediary multiplexing stages increase in order to maintain the proper hierarchy. Using this multiplexing hierarchy, any signal with less capacity or more can be transported very comfortably using the SDH standard. No compatibility between the signal and the hardware, i.e. the network equipment, is required. Moreover, the signal does not need to be compatible in any sense as it is being carried in a container.

The container just keeps the signal and adds, if necessary, any redundant bits. In a similar context this can be thought of as someone filling up his/her knapsack until it fills completely. In this way many signals of less capacity can be added to the container, but there is a set standard up to which the number of signals can be filled. The payload, which carries all the information, fills up the majority of the space but the housekeeping bits, known as the overhead, take up some amount. Generating the STM-1 when the source incoming signal is 2.048 Mbps can be pictorially represented as in Figure 3.23.

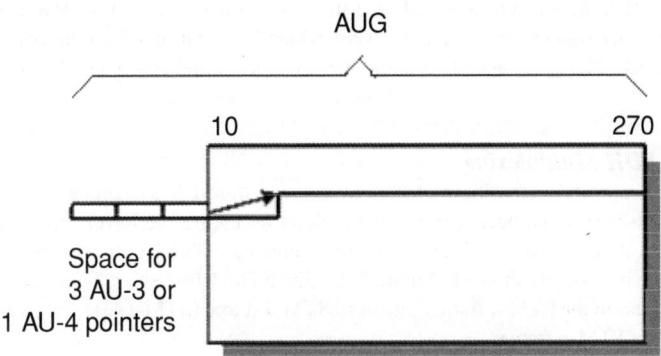

Figure 3.22 Administrative unit group

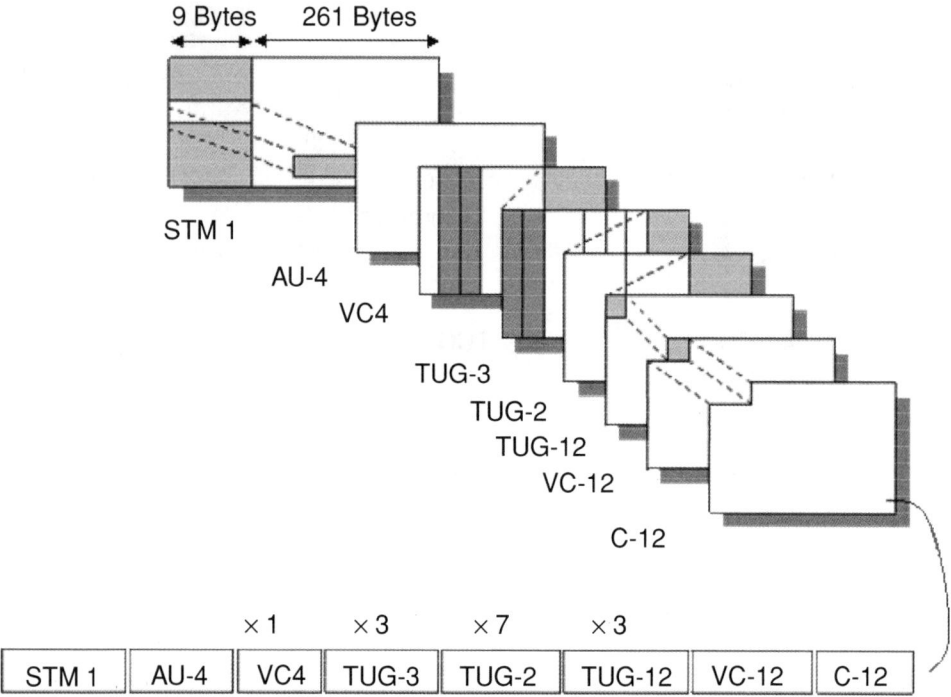

Figure 3.23 Generation of an STM-1 signal from a 2.048 Mbps signal

Multiplex Paths in the SDH

There is not a particular path that any signal has to follow to be transported as an STM-n signal. Instead signals of different speeds take up different paths on their way to the final signal. The source signals received are assembled in the corresponding containers, provided with the POH and pointer and converted into an STM-1 signal via the different multiplex steps as shown in Figure 3.24. Source signals with bit rates higher than 139.264 Mbps are multiplexed into the STM-1 frame in one step, while those with lower bit rates are multiplexed in two steps. This multiplex scheme complies with ITU-T G.707 and includes optional multiplex paths. The VC-3 can, for example, be multiplexed via TU-3 into VC-4 or the AU-3 path can be selected. A distinction is made between the lower order and higher order paths. For SDH signals there are two levels, which are used to set up the phase relation using pointers, TU-11, TU-12, TU-2 and TU-3 being the lower level and AU-3 and AU-4 being the higher level.

Higher Order SDH Multiplexing

With the channel capacity increasing multifold every year, the SDH system can multiplex signals up to 40 Gbps (STM-256) if the need arises. This is done by taking the lower order STM signals and multiplexing them in the same way. Interleaving the individual STM-1 generates the STM-n multiplex signal frames byte by byte, as shown in Figure 3.25. The STM-1 frames are numbered in the sequence in which they appear in the STM-n frame. The third STM-1 frame (STM-1#3) starts, for example, in the third column of the STM-n frame.

The de-multiplexing procedure (disassembling the multiplex signal into the individual STM-1 frames, TUG-3, TUG-2) is performed in the same way, but in the opposite order.

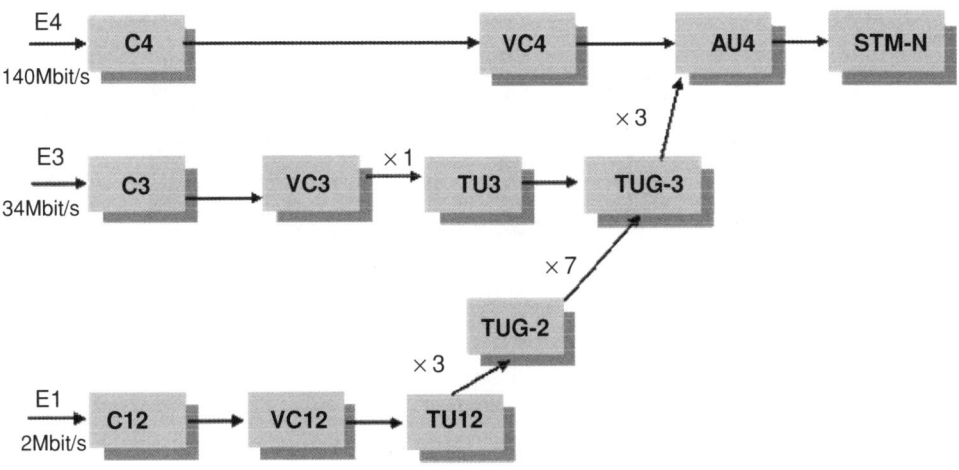

Figure 3.24 Synchronous multiplex structure

3.3.3 Asynchronous Transfer Mode (ATM)

ATM is a high speed network protocol for fast cell relay/transmission where different services like voice, video and data are transmitted in a special fixed size cell format. It is a dedicated connection-oriented switching and multiplexing technology which forms fixed size cells of 53 octets and transmits them over the physical medium using a digital signal technology (PCM, etc.) or multiplexing hierarchies like PDH and SDH. Speeds on ATM networks can reach up to 10 Gbps and such networks are sometimes also known as 'fast packet' networks.

ATM is also used as another multiplexing technique like the PDH and SDH, but the major difference lies in the way their layers are implemented. The ATM systems mostly work in the data link layer as compared to the PDH and SDH, which are implemented in the physical layer. More will become known about the ATM layers when the broadband (B)-ISDN ATM protocol reference model is discussed.

ISDN and B-ISDN

With the exponential growing demands for different services at good enough bandwidths, an integrated service network was badly needed. Though SDH multiplexing hierarchies turned the dream of channel capacity in a few Gbps into a reality, it was still uncertain whether any service could practically use this much bandwidth and transmit almost every service known, using the same channel. This idea of an integrated service network led to the creation of an end-to-end digital network, which had the capability of providing various services like voice, video and data on the same channel and was called the integrated services digital network (ISDN).

Original recommendations of ISDN were in ITI-U (CCITT) Recommendation I.120 (1984), which described some initial guidelines for implementing the ISDN. The ISDN is comprised of digital telephony and data transport services offered by regional telephone carriers. It involves the digitisation of the telephone network, which permits voice, data, text, graphics, music, video and other types of information to be transmitted over existing telephone wires. The emergence of the ISDN represents an effort to standardise subscriber services, user-to-network interfaces and network-to-network services. ISDN applications include high speed image applications, high speed file transfer and videoconferencing. Voice service is also an application for the ISDN, which has two types of services according to the number of channels

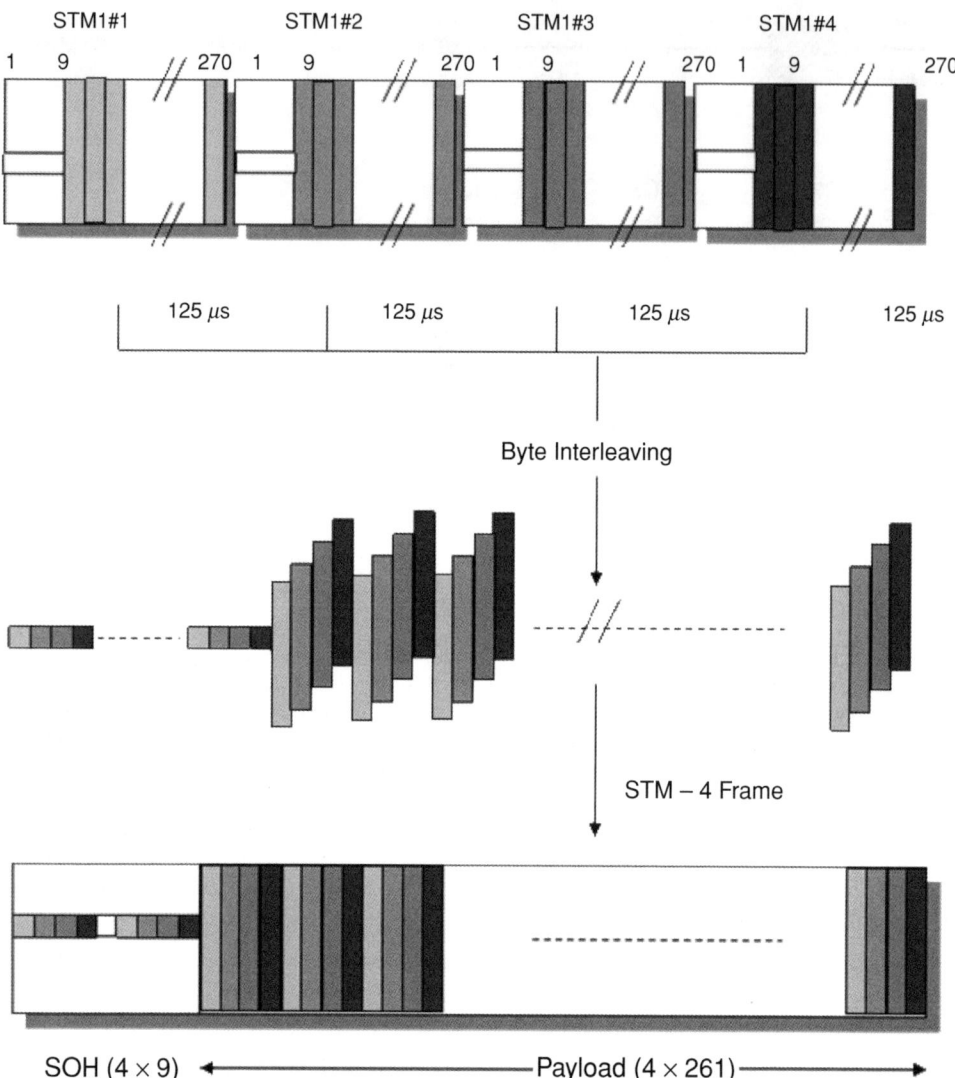

Figure 3.25 SDH multiplexing procedure

and speeds it had to offer. The first one is the *basic* service having a speed of 192 kbps while the other one has rates of 1.544/2.048 Mbps and is known as the *primary* service.

As the ISDN services were after not good enough, a new concept for integrated services came up as an enhancement over the ISDN; it worked at rates greater than the primary rates (1.544/2.048 Mbps) and covered the broadband aspects of the present ISDN. This service was thus called the broadband ISDN (B-ISDN) and forced the previous service of ISDN to be named the narrowband ISDN (N-ISDN). With greater rates than the N-ISDN, B-ISDN has become more popular among residential and business customers.

The primary triggers for moving towards the B-ISDN concept includes an increasing demand for high bit rate services, especially image and video services, and most important of all, the evolution of new technology, which has stood up perfectly to support these high bandwidth services with minimal transmission errors. These important developments are:

1. The fibre optic transmission system offers transmissions at high data rates with a very low probability of transmission error for subscriber lines and the network trunks. With widespread use the cost has also gone down, making it the perfect wired medium to transmit high capacity data.
2. The VLSI and ULSI developments in the integrated circuit technology has also made it possible for switching and transmitting circuits to do their respective jobs at a faster rate comparable to the fibre optic speeds.
3. Apart from the voice and data services on the same channel, B-ISDN has brought with it the much-awaited video service. Thus video conferencing and online video services have been deployed. This has been possible due to high quality video devices with good refreshing rates.

As part of its I series of recommendations on ISDN, ITU-T (formerly CCITT) in 1988 issued the first two recommendations relating to B-ISDN:

(i) I.113: Vocabulary of Terms for Broadband Aspects of ISDN
(ii) I.121: Broadband Aspects of ISDN

These recommendations provided a preliminary description of the future standardisation of B-ISDN and formed the basis of the development work, but this proved to be time consuming. The delays were due to standards being made which would help in B-ISDN transmission and would play a very significant role in this area. This new technology was the asynchronous transfer mode (ATM). Due to the importance of this new arrival in the field of telecommunications a separate committee called the ATM Forum was formed and was given the job of accelerating the development of ATM standards.

Broadband ISDN Services

The services provided by the B-ISDN are classified by the ITU-T into two categories, the interactive services and the distribution services. Interactive services are those where there is a two-way exchange of information, other than control signal information, between the two end users or between a user and a service provider. Examples of this type of service include conversational services, messaging services and retrieval services. Distribution services are those in which the information transfer is primarily one way, i.e. from the service provider to the B-ISDN subscriber. These include broadcast services for which the user has no control over the presentation.

Some examples of the services enjoyed by a B-ISDN user are telephone, tele-education, video conferencing, communication via mailboxes (e-mails and video mails), high resolution image and document retrieval, electronic newspaper, TV program distribution and digital video library. The large and diverse services of B-ISDN produced a huge load on the circuit switching hardware and for this very reason another switching technique was employed. This fast switching ATM technique thus became the basic switching technique for B-ISDN.

ATM Concepts

The transfer of information across the user–network interface for B-ISDN uses the asynchronous transfer mode (ATM). This technology is embedded into a protocol reference model that defines the B-ISDN user–network interface. As discussed earlier, the ATM is a form of packet transmission across the user–network similar to the X.25. The basic difference however lies in the signalling between the two. The X.25 includes control signalling on the same channel as data transfer, whereas the ATM makes use of common channel signalling. The ATM packets are of fixed size, referred to as cells, as compared to

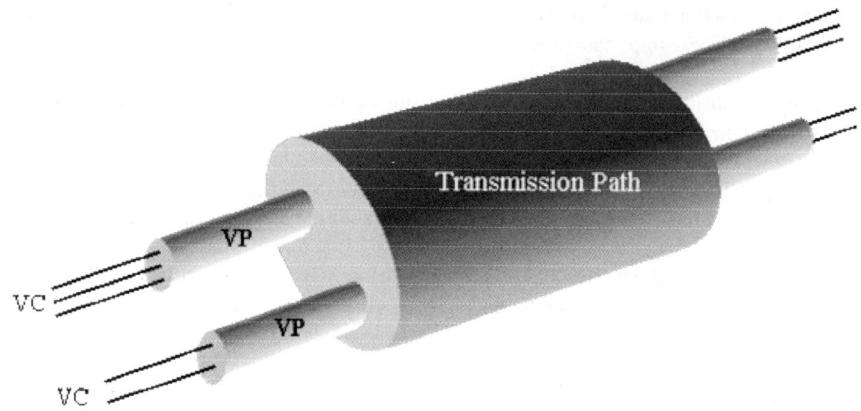

Figure 3.26 VCs concatenate to create VPs

variable length packets in the X.25. The main reason for the fixed size cells was the reduction in the queuing delay. If the packet size had been variable then a bigger video packet might have kept the smaller voice or data packets waiting in the queue for a longer time than usual.

The term 'asynchronous' shows that the multiplexing carried out is not synchronous, as was the case in B-ISDN. There were two reasons for this. Firstly, there could be a variety of applications with different data rates that needed to be switched. Also many data applications like voice and video are bursty in nature and can be handled more efficiently with some sort of packet switching technique. Secondly, the use of the synchronous approach for high speed transmission would complicate the switching system as compared to the N-ISDN, where just the 64-kbps data stream had to be switched.

The ATM is a virtual channel technology, i.e. connection is made before the communication commences, just like in connection-oriented circuit switching. Thus, although the fixed size cells are transmitted, they are carried using the virtual channels. The ATM therefore enjoys the advantages of both the circuit as well as packet switching. Packet switching ensures fast switching of user information and circuit switching makes sure of the cell integrity (ordering of cells at the destination) and the quality of service (QoS).

Virtual Channels and Virtual Paths

A very important feature provided by the ATM is the combination of both circuit and packet switching. The question arises as to how both types can exist for the same service. It is known that packet switching is done using fixed size packets called cells. The other type of switching, i.e. circuit switching, is accomplished by what is known as virtual channel connections (VCCs) and virtual path connections (VPCs), as shown in Figure 3.26.

The concept of these virtual paths (VPs) and channels is taken from virtual circuits/logical connections in the X.25/frame relay. This logical or virtual connection is established before the communication commences using a signalling protocol so that the route from the source to the destination is chosen well before time and all the packets, in this case the cells, are routed through the same path in the same order. Therefore the ATM enjoys the flexibility provided to it by the virtual channel concept. At one end there is a fixed path established for the cells to be transferred and thus no need for the intermediary nodes to calculate the route that the cell will take in order to reach the next node. This feature, as will be seen later, would provide good QoS and would also not disturb the arrangement of cells at the destination. At the same time this does not mean that there is a dedicated path present for a particular source–destination combination. The path is established only when a request arises. Each link or the node is shared by a set of virtual channels. A route will take a particular channel identified by its number called the virtual

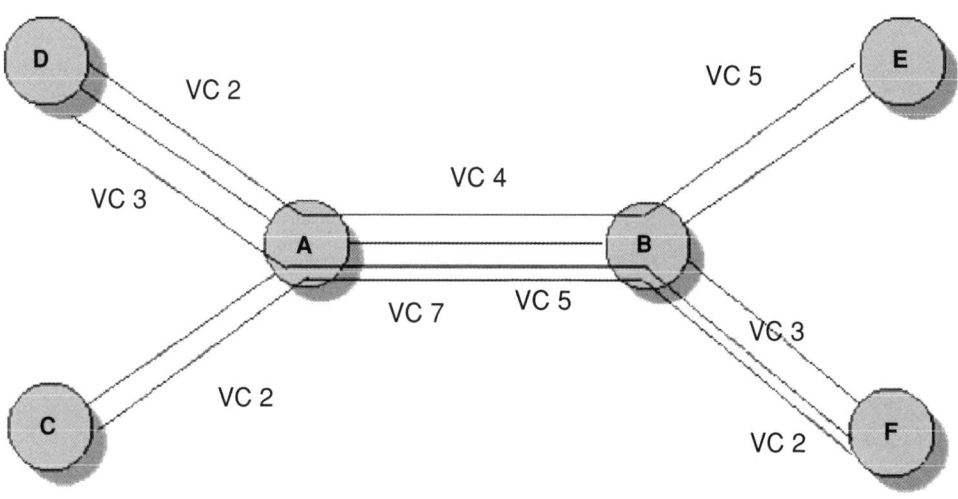

Figure 3.27 Virtual channels at node A

channel identifier (VCI), as shown in Figure 3.27. Also each link will maintain a table of routes (as shown in Table 3.6) filled by the signalling protocol, which the incoming channel will take.

The second sublayer in the ATM layer is the virtual path level. A VPC is a bundle of VCCs that have the same end points. This implies that all of the cells flowing over all of the VCCs in a single VPC are switched along the same route. This concept emerged as a result of a heavy control cost of the network as compared to its overall cost. With this technique the paths sharing the common paths through the network may be grouped together into a single unit called a virtual path and the network management actions can then be applied to a small number of groups of virtual paths rather than a huge number of virtual channels.

Some basic advantages of the virtual paths are:

- Greater reliability and increased efficient performance of the network due to a lower number of entities to manage in totality.
- Reduced connection setup time and hence less processing involved as much of the work is done when the virtual path is being set up. Once the virtual paths have been set up, the new virtual channel connections can be established by simple signalling. The processing is required at the end nodes only and not at every node. This not only reduces the processing delay but also the total connection setup time.
- A better organised and simple network architecture is a result of the combination of many virtual channels into single virtual paths.

Table 3.6 Table at node A

Link in	VC in	Link out	VC out
DA	2	AB	4
DA	3	AB	5
CA	2	AB	7

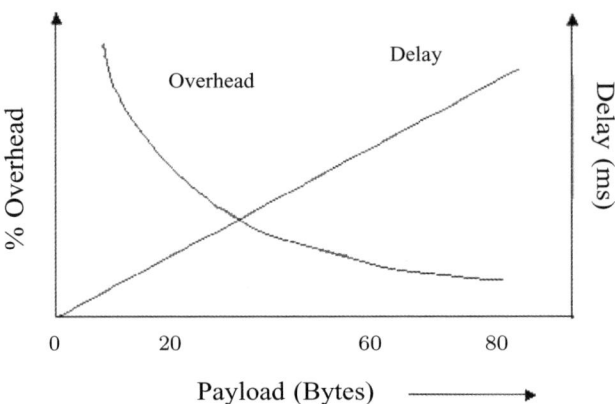

Figure 3.28 A graph indicating the importance of small sized cells

Now that there is familiarity with the virtual connections, it is possible to take a look at how a call becomes established in an ATM system using the virtual paths and channels. Two important things should be kept in mind:

- The virtual path connection setup includes calculating routes, allocating the total capacity that the virtual channels might use, etc.
- For the virtual channel connection setup, control mechanisms are needed that check a few things before a channel is set up. These functions involve checking that there is a virtual path connection to the required destination node with sufficient available capacity to support the virtual channel, with the appropriate quality of service, etc.

Consider a request for VCC establishment. This originates a call, which further checks whether a VPC exists. If a VPC does not exist then a new VPC is established and the capacity along with the quality of service is checked. If the capacity or the QoS is not guaranteed then the request of the VCC is blocked and more capacity is requested. If the capacity is not granted then the VCC request is rejected. However, if the QoS is guaranteed in the first stage or capacity is granted in the second stage, then a new VCC is established for communication. Several other aspects of the virtual channels and paths will be covered in the ATM layer of the B-ISDN ATM protocol reference model.

ATM Cell Format

ATM transfers information in fixed size units called cells. Each cell consists of 53 octets, or bytes. The first 5 bytes contain cell header information, and the remaining 48 contain the 'payload' (user information). Small fixed length cells are well suited to transferring voice and video traffic because such traffic is intolerant of delays that result from having to wait for a large data packet to download. This eliminates the need to wait for long for packets of some services. Thus services like video and voice queuing delays should virtually be absent, taking advantage of the small fixed size cell technology. Another important factor in favour of small size cells is that they can be switched more efficiently as compared to large packets, thus supplementing the very high data rates of the ATM. The importance of the small size cells can be depicted using the graph shown in Figure 3.28.

Of all sizes why was only a 53-byte cell used? Initially two cell sizes of 32 and 64 bytes were put forward. The 32-byte cell had less queuing delays whereas the 64-byte cell had better transmission efficiencies. A compromise, between Europe who wanted a 32-byte cell as the standard and the US and Japan who were in favour of a 64-byte cell, resulted in a 48-byte cell for the ATM, which was extended to a 53-byte cell after the header information was attached to the cell. A small sized header meant that

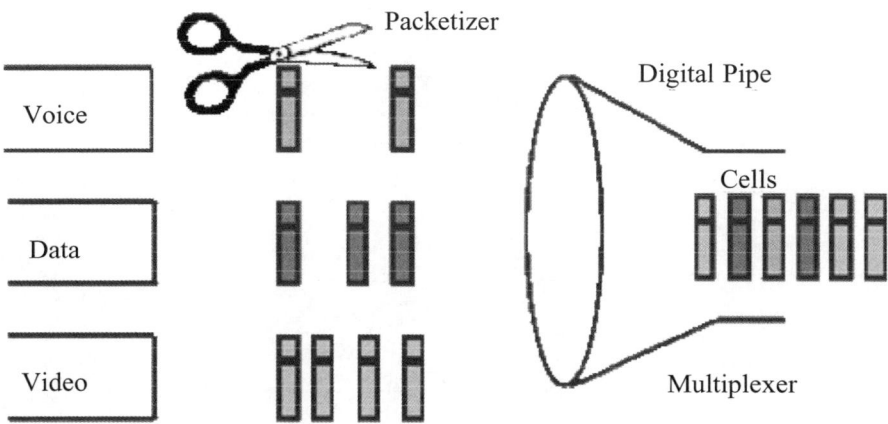

Figure 3.29 The multiplexing process of the cells

there would be a few provisions and limited functions for multiplexing information and error control. Due to virtual connections the information about cell number and the source–destination address was not needed in the header.

The information of different types of services is broken into fixed size packets called cells, using a packetiser, and is then multiplexed and sent through the same digital channel (shown in Figure 3.29).

B-ISDN ATM Protocol Reference Model

The ATM architecture uses a logical model to describe the functionality it supports. This logical model is sometimes also referred to as the B-ISDN ATM protocol reference model. Two layers of this architecture relate to the ATM functions. One is the ATM layer common to all services and provides packet/cell transfer capabilities. The second is the ATM adaption layer (AAL), which is service dependent. This layer converts higher layer information into ATM cells to be transported over the B-ISDN, and collects information from incoming ATM cells for delivery to higher layers.

As seen from the OSI reference model shown in Figure 3.30, ATM functionality corresponds to its physical layer and part of the data link layer. The ATM reference model is composed of the following planes, which span all layers:

- *Control*. This plane is responsible for generating and managing signalling requests, i.e. call controls and connection control functions.
- *User*. This plane is responsible for managing the transfer of data along with error and flow control.
- *Management*. This plane contains two components:
  (a) Layer management manages layer-specific functions, such as the detection of failures and protocol problems.
  (b) Plane management manages and coordinates functions related to the complete system and also provides coordination between all the planes.

The ATM reference model is composed of the following ATM layers:

- *Physical layer*. Analogous to the physical layer of the OSI reference model, the ATM physical layer manages the medium dependent transmission. This layer consists of two sublayers: the physical medium sublayer and the transmission convergence sublayer.

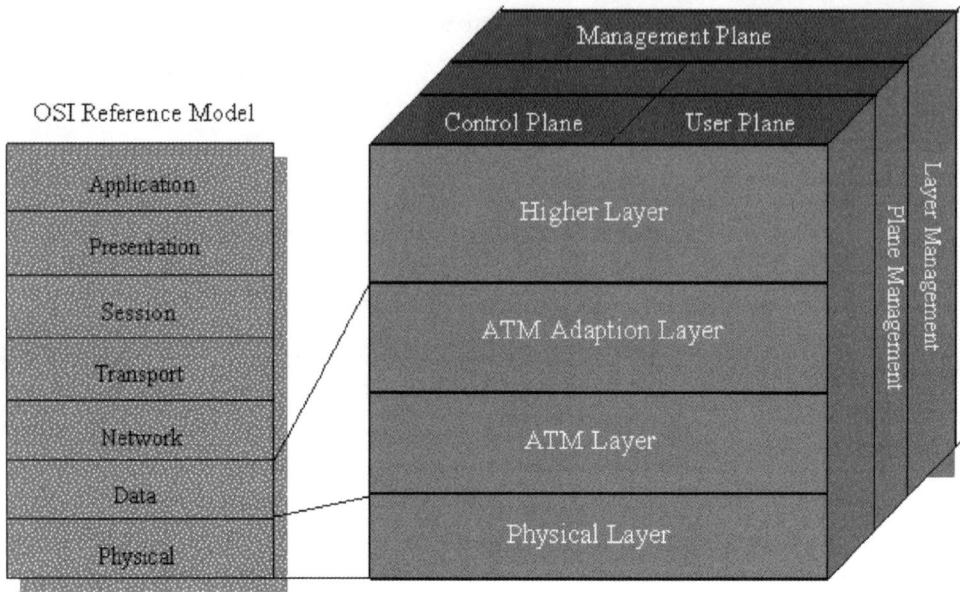

Figure 3.30 B-ISDN ATM protocol reference model

- *ATM layer.* Combined with the ATM adaptation layer, the ATM layer is roughly analogous to the data link layer of the OSI reference model. The ATM layer is responsible for establishing connections and passing cells through the ATM network. To do this, it uses information in the header of each ATM cell.
- *ATM adaptation layer (AAL).* Combined with the ATM layer, the AAL is roughly analogous to the data link layer of the OSI model. The AAL is responsible for isolating higher layer protocols from the details of the ATM processes. Finally, the higher layers residing above the AAL accept user data, arrange it into packets and hand it to the AAL. Figure 3.30 illustrates the ATM protocol reference model.

The Physical Layer

The physical layer of any architecture model takes care of the transfer of information from the higher layers above it to the physical medium the system uses. The ATM physical layer has four basic functions: bits are converted into cells, the transmission and receipt of bits on the physical medium are controlled, ATM cell boundaries are tracked and cells are packaged into the appropriate types of frames for the physical medium.

The ATM physical layer is divided into two parts:

1. Physical medium dependent (PMD) sublayer
2. Transmission convergence (TC) sublayer

The PMD sublayer provides two key functions. Firstly, it synchronises transmission and reception by sending and receiving a continuous flow of bits with associated timing information. Secondly, it specifies the physical media for the physical medium used, including connector types and cable. Examples of physical medium standards for the ATM include synchronous optical network/synchronous digital hierarchy

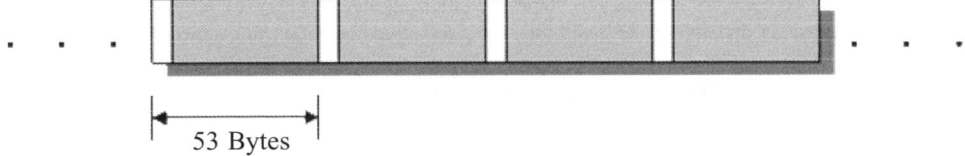

Figure 3.31 Cell based ATM cell transmission

(SONET/SDH), DS-3/E3, 155 Mbps over a multimode fibre (MMF) using the 8B/10B encoding scheme and 155 Mbps 8B/10B over shielded twisted-pair (STP) cabling.

The TC sublayer has five functions:

1. Cell delineation. The cell delineation function maintains ATM cell boundaries, allowing devices to locate cells within a stream of bits. This also helps to de-scramble the cells at the destination, which were scrambled at the source for transmission purposes.
2. Transmission frame adaptation. The transfer of information at the ATM layer is always in terms of the ATM cells. Transmission frame adaptation packages these ATM cells into frames acceptable to the particular physical layer implementation. The cells can also be simply transmitted and received without framing them.
3. Transmission frame generation and maintenance. This function generates and maintains the frame structure appropriate for a given data rate for the physical layer.
4. Header error control (HEC) sequence generation and verification. Each cell header is protected by a header error control (HEC) code. HEC sequence generation and verification generates and checks the HEC code to ensure valid data.
5. Cell rate decoupling. Cell rate decoupling maintains synchronisation and inserts or suppresses idle (unassigned) ATM cells to adapt the rate of valid ATM cells to the payload capacity of the transmission system.

After the functions of different sublayers of the physical layer are known, it is possible to dealt with its most important job: transmission of ATM cells. The ITU-T Recommendation I.432 specifies that the ATM cells may be transmitted at one of the several data rates: 25.6, 51.84, 155.52 or 622.08 Mbps. According to this Recommendation, two approaches used for transmitting can be followed: a cell based physical layer or an SDH based physical layer.

Cell Based Physical Layer

In this approach, no external framing is imposed and the 53-byte cells are transmitted as a continuous stream of data, as shown in Figure 3.31. However, no framing would lead to asynchronous receiving of these cells as a continuous stream and thus there should be some means of synchronising them. Synchronisation is achieved on the basis of the HEC field in the cell header. Cell delineation is performed at different times at different stages to confirm the correct occurrence of the HEC field. Once a cell undergoes these transitions it is assumed to be correct. The advantage of using a cell based transmission system is that the simple interface of framing and transfer are based on a common structure, which is the 53-byte cell.

SDH Based Physical Layer

This method of transmission employs time division multiplexing while packing the low capacity SDH containers into high capacity ones. This type of transmission also imposes a structure on the ATM cell stream; i.e. the cells are packed in the basic signal STM-1 frame format whose capacity is 155.52 Mbps. The payload can be offset from the beginning of the frame, indicated by the pointer in the section overhead of the frame. The synchronous payload envelope (SPE) consists of a 9-byte path overhead and the remaining 2340 bytes are used to store the information; in this case they are the ATM cells. As the payload is not exactly divisible by 53 a cell may cross a payload boundary to occupy space in the next container.

The advantages of the SDH based approach are discussed below:

- It can be used to carry either the ATM based payloads or the STM signals using the same deployed transmission systems. This means that if a high capacity fibre based transmission is initially deployed for the transmission of STM payloads, then the transmission can easily be migrated to transfer ATM cells over the same media to support the B-ISDN.
- Some specific connection can be circuit switched using an SDH channel. For example, a connection carrying constant bit rate video traffic needs a good quality of service at constant bit rates. Thus that signal can be mapped into its own exclusive payload envelope of the STM-1 signal, which can be circuit switched.
- Using the SDH interfaces, many ATM streams can be combined to provide higher bit rates than 155.52 Mbps. For example, four separate ATM streams with a bit rate of 155.52 Mbps can be combined to build a 622 Mbps signal. This could prove more useful than a single 622 Mbps ATM stream.

The ATM Layer

The ATM layer is known to be responsible for establishing connections and passing cells through the ATM network. To do this, it uses information in the header of each ATM cell. From the above lines it is transparent that understanding this layer would need the knowledge of the two basic entities of this layer, which are the virtual connections and the ATM cell format. Virtual connections have already been discussed in detail, so here they will briefly be revisited before moving on to the ATM cell structure in detail.

The basic terminology of the virtual channel and path connections as given in the ITU-T Recommendation I.311 can be looked at to obtain a formal definition of these terms:

- *Virtual channel (VC)*. This is a generic term used to describe the unidirectional transport of ATM cells associated by a common unique identifier value.
- *Virtual channel link*. This is a means of unidirectional transport of ATM cells between a point where a VCI value is assigned and the point where that value is translated or terminated.
- *Virtual channel identifier (VCI)*. This is a unique numerical tag that identifies a particular VC link for a given VPC.
- *Virtual channel connection (VCC)*. This is the concatenation of VC links that extends between two points where ATM service users access the ATM layer. VCCs are provided for the purpose of user–user, user–network or network–network information transfer. Cell sequence integrity is preserved for cells belonging to the same VCC.
- *Virtual path (VP)*. This is a generic term used to describe the unidirectional transport of ATM cells belonging to the virtual channels that are associated by a common unique identifier value.
- *Virtual path link*. This is a group of VC links identified by a common value of VPI, between a point where a VPI value is assigned and the point where that value is translated or terminated.
- *Virtual path identifier (VPI)*. This is a unique numerical tag that identifies a particular VP link.
- *Virtual path connection (VPC)*. This is the concatenation of VP links that extends between two points where the VCI values are assigned and the point where that value is translated or terminated. VCCs are provided for the purpose of user–user, user–network or network–network information transfer.

The definitions above give a clear picture of the range/extent of each link or path. Thus with these terminologies in mind it is safe to move to the switching of VCCs and VPCs. VP switches terminate VP links. As shown in Figure 3.32, a VP switch translates incoming VPIs to the corresponding outgoing VPIs according to the destination of the VPC, whereas VCI values remain unchanged. VC switches terminate

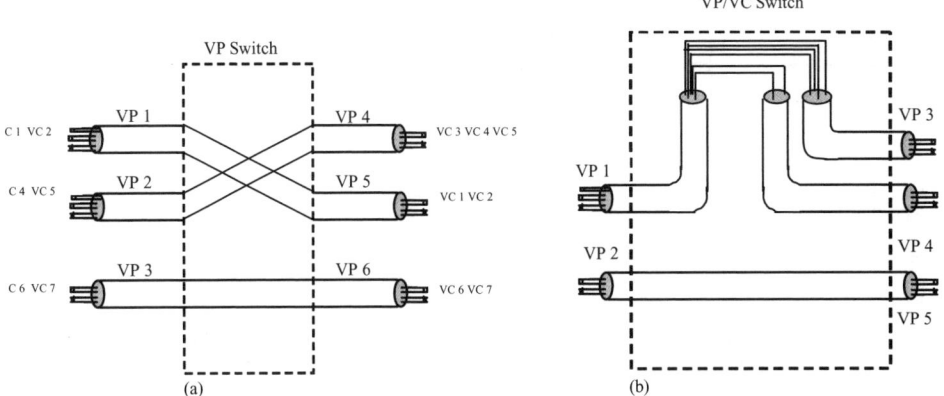

Figure 3.32 (a) A VP switch; (b) a VP/VC switch

VC links and necessarily VP links. A VC switch must therefore switch virtual paths and virtual channels and thus both the VPI and VCI translation is performed.

The various characteristics of the virtual path/virtual channel connections as given in the Recommendation I.150 are:

- *Quality of service.* A user of a virtual channel is provided with a quality of service specified by parameters such as the cell loss ratio (ratio of cells lost to cells transmitted) and cell delay variation.
- *Switched and semi-permanent virtual channel connections.* Both switched connections, which require call control signalling, and dedicated connections can be provided.
- *Cell sequence integrity.* The sequence of transmitted cells within a virtual channel is preserved.
- *Traffic parameter negotiation and usage monitoring.* Traffic parameters can be negotiated between a user and the network for each virtual channel. The network to ensure that the negotiated parameters are not violated monitors the input of cells to the virtual channel.

The ATM Cell Structure

The asynchronous transfer mode makes use of small fixed size cells of 53 bytes, consisting of a 5-byte header field and a 48-byte information payload field. An ATM cell header can be one of two formats: user–network interface (UNI) or the network–network interface (NNI). The UNI header is used for communication between ATM end points and ATM switches in private ATM networks. The NNI header is used for communication between ATM switches.

Figure 3.33 depicts the basic ATM cell format, the ATM UNI cell header format and the ATM NNI cell header format. Unlike the UNI, the NNI header does not include the generic flow control (GFC) field. Additionally, the NNI header has a VPI field that occupies the first 12 bits, allowing for larger trunks between public ATM switches.

ATM Cell Header Fields

In addition to GFC and VPI header fields, several others are used in ATM cell header fields. The following descriptions summarise the ATM cell header fields shown in Figure 3.33:

- *Generic flow control (GFC).* This provides local functions, such as identifying multiple stations that share a single ATM interface. This field is typically not used and is set to its default value.

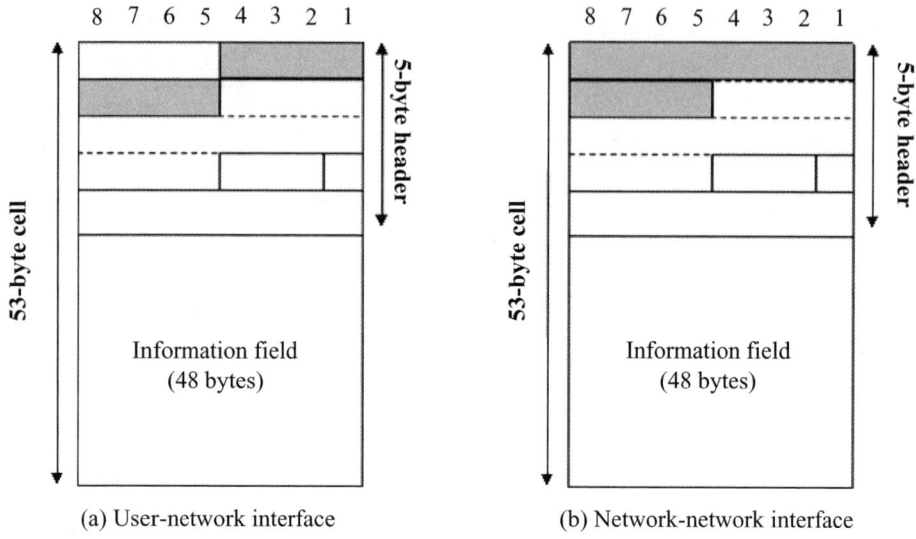

Figure 3.33 ATM cell format

- *Virtual path identifier (VPI)*. In conjunction with the VCI, this identifies the next destination of a cell as it passes through a series of ATM switches on the way to its destination.
- *Virtual channel identifier (VCI)*. In conjunction with the VPI, this identifies the next destination of a cell as it passes through a series of ATM switches on the way to its destination.
- *Payload type (PT)*. This field indicates the type of information in the information field. A value of 0 in the first bit indicates user information. In this case, the second bit indicates whether congestion has been experienced. The third bit, known as the service data unit (SDU) type bit, is a one-bit field that can be used to discriminate two types of ATM SDUs associated with a connection. The SDU is the 48-byte payload of the cell. A value of 1 in the first bit of the payload indicates that this cell carries network management or maintenance information. This indication allows the insertion of network management cells on to a user's VCC without impacting the user's data, thus providing the in-band control information.
- *Congestion loss priority (CLP)*. This indicates whether the cell should be discarded if it encounters extreme congestion as it moves through the network. If the CLP bit equals 1, the cell should be discarded in preference to cells with the CLP bit equal to zero.
- *Header error control (HEC)*. This calculates the checksum only on the header itself.

The ATM Adaptation Layer (AAL)

With so many services requiring greater speeds it would be suitable if ATM provided them with a common platform to be transferred. However, in spite of using different versions of ATM layers for different services, another layer was introduced above the ATM layer, which provided mapping of different types of applications to the ATM layer. This layer is thus service dependent. The basic functionality can be interpreted by having a look at the diagram below (Figure 3.34).

Thus the use of the ATM makes it mandatory to require an adaption layer to support other information based protocols not based on the ATM. The various services provided by an AAL are:

- handling of transmission errors;
- segmentation and reassembly, to enable larger blocks of data to be carried in the information field of ATM cells;

Digital Hierarchies – PDH and SDH

Table 3.7 AAL protocol classification as given by the ITU-T

	Class A	Class B	Class C	Class D
Timing relation between the source and destination	Required	Required	Not required	Not required
Bit rate	Constant	Variable	Variable	Variable
Connection mode	Connection oriented	Connection oriented	Connection oriented	Connectionless
AAL protocol	Type 1	Type 2	Type 3/4, type 5	Type 3/4

- handling of lost cell conditions;
- flow control and timing control.

Different types and quality of services would require different AAL protocols to support the service. To minimise this wide variety of protocols, ITU-T has defined four classes of service that cover a broad range of applications. This classification is based on three categories, which are the timing relation between the source and the destination, the bit rate and finally the connection mode. Based on these properties different AAL protocols have been devised, which come under the four classes defined. Table 3.7 shows this service classification as given by the ITU-T.

AAL protocols mentioned above work in two logical sublayers of the major AAL (shown in Figure 3.35). These are the convergence sublayer (CS) and the segmentation and reassembly (SAR) sublayer. The convergence sublayer provides the functions needed to support specific applications using the AAL. Each application attaches to the AAL at a service access point, which is the address of the application. This sublayer is service dependent.

The segmentation and reassembly sublayer is responsible for packing the information received from the CS into cells for transmission and unpacking the information at the other end. The SAR sublayer

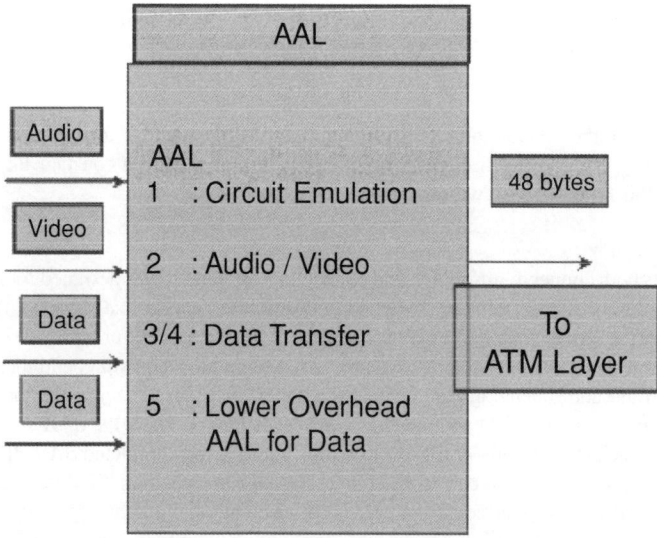

Figure 3.34 The ATM adaptation layer

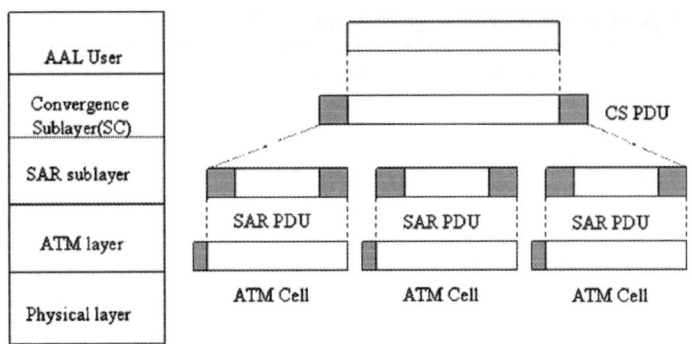

Figure 3.35 Working of the ATM model hierarchy

must pack any SAR headers and trailers plus the CS information into the 48-byte blocks and send them to the ATM layer. A closer look at the packing will make the process and the AAL working clearer.

When a higher layer application block of user data is encapsulated in a single protocol data unit (PDU) consisting of the higher layer data and a header and trailer containing protocol information at the CS layer, this CS PDU is then passed down to the SAR sublayer and segmented into a number of blocks. Each of these blocks is encapsulated in a single 48-byte SAR PDU, which may include a header and a trailer in addition to the block of data passed down from the CS. Finally, each SAR PDU forms the payload of a single ATM cell.

AAL Type 1
The AAL type 1, a connection oriented service, is suitable for handling circuit emulation services, such as voice and video conferencing. AAL1 requires timing synchronisation between the source and destination. For this reason, AAL1 depends on a medium, such as SONET, that supports clocking. The AAL1 process prepares a cell for transmission in three steps. Firstly, synchronous samples (e.g. 1 byte of data at a sampling rate of 125 µs) are inserted into the payload field. Secondly, sequence number (SN) and sequence number protection (SNP) fields are added to provide information that the receiving AAL1 uses to verify that it has received cells in the correct order. Thirdly, the remainder of the payload field is filled with enough single bytes to equal 48 bytes.

AAL Type 2
The type 2 deals with the variable bit rate information and is intended for analogue applications, such as video and audio, that require timing information but do not require a constant bit rate. An initial specification for this type has been withdrawn.

AAL Type 3/4
AAL3/4 supports both connection-oriented and connectionless data. It was designed for network service providers and is closely aligned with the switched multimegabit data service (SMDS). AAL3/4 is used to transmit SMDS packets over an ATM network. AAL3/4 prepares a cell for transmission in four steps. Firstly, the convergence sublayer (CS) creates a protocol data unit (PDU) by prepending a beginning/end tag header to the frame and appending a length field as a trailer. Secondly, the segmentation and reassembly (SAR) sublayer fragments the PDU and attaches a header before it. Then, the SAR sublayer appends a CRC-10 trailer to each PDU fragment for error control. Finally, the completed SAR PDU becomes the payload field of an ATM cell to which the ATM layer appends the standard ATM header at the start.

An AAL 3/4 SAR PDU header consists of type, sequence number and multiplexing identifier fields. Type fields identify whether a cell is the beginning, continuation or end of a message. Sequence number fields identify the order in which cells should be reassembled. The multiplexing identifier field determines

AAL Type 5
AAL5 is the primary AAL for data and supports both connection-oriented and connectionless data. It is used to transfer most non-SMDS data, such as classical IP, over ATM and LAN emulation (LANE). AAL5 also is known as the simple and efficient adaptation layer (SEAL) because the SAR sublayer simply accepts the CS-PDU and segments it into 48-octet SAR PDUs without adding any additional fields. AAL5 prepares a cell for transmission in three steps. Firstly, the CS appends a variable length pad and an 8-byte trailer to a frame. The pad ensures that the resulting PDU falls on the 48-byte boundary of an ATM cell. The trailer includes the length of the frame and a 32-bit cyclic redundancy check (CRC) computed across the entire PDU. This allows the AAL5 receiving process to detect bit errors, lost cells or cells that are out of sequence. Secondly, the SAR sublayer segments the CS-PDU into 48-byte blocks. A header and trailer are not added (as is in AAL3/4), so messages cannot be interleaved. Finally, the ATM layer places each block into the payload field of an ATM cell. For all cells except the last, a bit in the payload type (PT) field is set to zero to indicate that the cell is not the last cell in a series that represents a single frame. For the last cell, the bit in the PT field is set to one.

Performance of the ATM (QoS Parameters)
The following are classes defined by the ATM forum:

- *Constant bit rate (CBR)*. Under this class, the user traffic is continuous and steady. Supported by AAL1, this class has tight delay and delay variation bounds. Voice and video conferencing are typical examples of CBR traffic.
- *Real-time variable bit rate (rt-VBR)*. Supported by AAL3/4 and AAL5, the nature of traffic is bursty and insensitive to smaller delay variations.
- *Non-real-time variable bit rate (nrt-VBR)*. This is similar to rt-VBR, except that the traffic is non-real time. This class is used for frame relay interworking. Both rt-VBR and nrt-VBR have a variable bandwidth.
- *Unspecified bit rate (UBR)*. This is the best effort transmission. The user may request a maximum traffic rate, but the network provides the best rate possible. As in the previous two cases, AAL3/4 and AAL5 support this type of traffic.
- *Available bit rate (ABR)*. Under this, there is no guarantee of bandwidth. Also there is no timing relationship between the source and destination.
- *Sustainable cell rate (SCR)*. This is the upper bound on the average rate of the conforming cells of an ATM connection, over time-scales that are long relative to those for which the PCR is defined.
- *Maximum burst size (MBS)*. This is the maximum number of cells that can be transmitted on the peak cell rate.
- *Minimum cell rate (MCR)*. This is the rate negotiated between the end systems and the network(s), such that the actual cell rate sent by the end system on the ABR connection need never be less than the MCR.

The performance of the ATM layers can be measured by monitoring certain parameters. These parameters can be divided into two groups: negotiable and non-negotiable. Parameters that are negotiated are:

- peak-to-peak cell delay variation (CDV);
- maximum cell transfer delay (maximum CTD);
- mean cell transfer delay (mean CTD);
- cell Loss Ratio (CLR).

Cell Delay Variation

The CDV parameter describes variability in the pattern of cell arrival (entry or exit) events at an MP (measurement point) with reference to the negotiated peak cell rate $1/T$ (see Recommendation I.371). It includes variability present at the cell source (customer equipment) and the cumulative effects of variability introduced (or removed) in all connection portions between the cell source and the specified MP. It can be related to cell conformance at the MP and to network queues. It can also be related to the buffering procedures that might be used in AAL1 to compensate for cell delay variations.

Cell Transfer Delay

Cell transfer delay (CTD) is the time, $t_2 - t_1$, between the occurrence of two corresponding cell transfer events, CRE1 (cell reference event 1) at time t_1 and CRE2 at time t_2, where $t_2 > t_1$ and $t_2 - t_1 \leq T_{max}$. The value of T_{max} is for further study, but should be larger than the largest practically conceivable cell transfer delay.

Mean Cell Transfer Delay

The mean cell transfer delay is the arithmetic average of a specified number of cell transfer delays.

Cell Loss Ratio

The cell loss ratio (CLR) is the ratio of total lost cells to total transmitted cells in the population of interest. Lost cells and transmitted cells in severely errored cell blocks are excluded from the calculation of the cell loss ratio.

QoS parameters that are not negotiated are:

- cell error ratio (CER);
- severely errored cell block ratio (SECBR);
- cell misinsertion rate (CMR).

Cell Error Ratio

The cell error ratio (CER) is the ratio of total errored cells to the total of successfully transferred cells, plus tagged cells, plus errored cells in a population of interest. Successfully transferred cells, tagged cells and errored cells contained in severely errored cell blocks are excluded from the calculation of the cell error ratio.

Severely Errored Cell Block Ratio

The severely errored cell block ratio (SECBR) is the ratio of total severely errored cell blocks to total cell blocks in the population of interest.

Cell Misinsertion Rate

The cell misinsertion rate (CMR) is the total number of misinserted cells observed during a specified time interval divided by the time interval duration 2 (equivalently, the number of misinserted cells per connection second). Misinserted cells and time intervals associated with severely errored cell blocks are excluded from the calculation of the cell misinsertion rate.

Some other parameters of interest may be the availability ratio and the mean time between outage.

Availability Ratio

The availability ratio (AR) applies to ATM semi-permanent connection portions. The service AR is defined as the proportion of time that the connection portion is in the available state over an observation period. The service AR is calculated by dividing the total service available time during the observation period by the duration of the observation period.

Digital Hierarchies – PDH and SDH

Table 3.8 Service class related parameters

Service class	CBR	rt-VBR	nrt-VBR	ABR	UBR
Traffic parameters					
PCR	Specified	Specified	Specified	Specified	Specified
CDVT	Specified	Specified	Specified	Specified	Specified
SCR	n/a	Specified	Specified	n/a	n/a
MBS	n/a	Specified	Specified	n/a	n/a
MCR	n/a	n/a	n/a	Specified	n/a
QoS parameters					
CLR	Specified	Specified	Specified	Specified	Unspecified
CTD	Maximum CTD	Maximum CTD	Mean CTD	Unspecified	Unspecified
CDV	Specified	Specified	Unspecified	Unspecified	Unspecified

Mean Time between Outage

The service mean time between outage (MTBO) is defined as the average duration of a time interval during which the portion is available from the service perspective. Consecutive intervals of available time during which the user attempts to transmit cells are concatenated. The network MTBO is defined as the average duration of a continuous time interval during which the portion is available from the network perspective.

The service class related parameters are given in Table 3.8.

ATM Addressing

Like IP, the ATM has a scheme by which the ATM end points, such as end nodes, can be given a unique address. The address can be public and private, of which the former are defined by the ATM Forum while the latter are defined by ITU-T. Public addresses are based on the telephone networks (E.164) while the private addresses are based on the International Organisation for Standardisation (ISO) network service access point (NSAP) format. E.164 basically consists of country code (CC), national destination code (NDC) and the subscriber number (SN):

$$E.164 = CC + NDC + SN$$

The country code (CC) identifies the country where the subscriber is while the national destination code identifies how the country is divided into ISDN networks. The subscriber number identifies the subscriber (end user). Though the international number may be of variable length, the maximum international number length is 15 digits.

A private address, also called an ATM end system address (AESA), is always of 20-octet length and is composed of an initial domain part (IDP) and a domain specific part (DSP). The IDP consists of the authority and format identifier (AFI) and the general domain idex (IDI). The DSP is subdivided into the high order DSP (HO-DSP), the end system identifier (ESI) and the selector (SEL):

$$AESA = IDP + DSP = (AFI + IDI) + (HO - DSP + ESI + SEL)$$

The AFI describes the type of the address (DCC (Data Country Code), ICD (International Code Designator) or E.164 ATM format). The IDI tells the network addressing domain where values of the DSP are allocated; the network addressing authority is responsible for allocating values of the DSP from that domain. The HO-DSP describes the hierarchy of the addressing authority and also conveys topological significance. The ESI is 6 octets long and identifies an end system. Although SEL (1 octet long) is not used for ATM routing, it may be used by end systems.

3.4 Microwave Link Planning

Microwave links generally operate between frequencies of 2 and 58 GHz. Initially analogue links were used, but now far superior digital microwave links are used. Digital microwave links have many advantages over the analogue microwave links, some of which are:

- high tolerance against interference;
- high tolerance against deep fading;
- high signal carrying capacity ranging from 2 to 155 Mbps;
- high frequency range (2–58 GHz);
- easy, rapid (and hence economical) installations.

As the frequency increases, the length of the link decreases. Due to the high frequency range (2–58 GHz), the microwave links can be classified into three main categories:

(a) Long haul
(b) Medium haul
(c) Short haul

Long Haul

The frequency of operation of these links is usually 2–10 GHz. In the best of climatic conditions and frequency of operation, the distance covered by the links could range from 80 km to 45 km. These links are affected by multipath fading (explained later).

Frequency band 2 GHz

- Maximum path length 80 km
- Multipath fading
- Antenna diameters up to 370 cm for an antenna gain of 36 dB
- Both vertical and horizontal polarisations used

Frequency band 7 GHz

- Maximum path length about 50 km
- Multipath fading
- Antenna diameters up to 370 cm for an antenna gain of 46.8 dB
- Both vertical and horizontal polarisations used

Frequency band 10 GHz

- Maximum path length about 45 km
- Multipath fading
- Antenna diameters 60–120 cm for a gain range of 34–40 dB
- Both vertical and horizontal polarisations used

Microwave Link Planning

Medium Haul

The frequency of operation of these links is usually from 11 GHz to 20 GHz. Depending upon the climatic conditions and frequency of operation, the hop length can vary between 40 km and 20 km. These links are also affected by multipath fading and rain fading.

Frequency band 13 GHz

- Maximum path length about 40 km
- Multipath fading
- Antenna diameters 60–120 cm for a gain range of 36.4–42.4 dB
- Both vertical and horizontal polarisations used

Frequency band 15 GHz

- Maximum path length about 35 km
- Multipath fading
- Antenna diameters 60–120 cm for a gain range of 38–44 dB
- Both vertical and horizontal polarisations used

Frequency band 18 GHz

- Maximum path length 20 km
- Rain and multipath fading
- Antenna diameters 60–180 cm for a gain range of 39–49 dB
- Both vertical and horizontal polarisations used
- Atmospheric attenuation 0.1 dB/km.
- Attenuation due to rain about 1 dB/km at a rain rate of 20 mm/h

Short Haul

These links operate in high frequency ranges (23–58 GHz) and thereby cover shorter distances. At lower frequency ranges in this band, links are affected by both multipath and rain fading. At higher frequencies, when the hop length is only a few kilometres, the multipath phenomenon does not have a significant effect. However, the impact of rain is quite severe.

Frequency band 23 GHz

- Maximum path length about 18 km
- Rain and multipath fading
- Antenna diameters 30–120 cm gain for a gain range of 35.5–47.3 dB
- Both vertical and horizontal polarisations used
- Atmospheric attenuation 0.1 dB/km
- Attenuation due to rain about 3 dB/km at a rain rate of 20 mm/h

Frequency bands 26 and 27 GHz

- Maximum path length is about 15 km.
- Rain fading
- Antenna diameters 30–60 cm
- Both horizontal and vertical polarisations used
- Atmospheric attenuation 0.1 dB/km
- Attenuation due to rain about 3 dB/km at a rain rate of 20 mm/h

Frequency band 38 GHz

- Maximum path length about 10 km
- Rain fading
- Antenna diameter 30 cm for a gain of 39.66 dB
- Only vertical polarisation used
- Atmospheric attenuation 0.12 dB/km
- Attenuation due to rain about 5 dB/km at a rain rate of 20 mm/h

Frequency band 55 GHz

- Maximum path length only few kilometres
- Rain fading
- Antenna diameter 15 cm
- Only vertical polarisation used
- Atmospheric attenuation 5 dB/km
- Attenuation due to rain about 7 dB/km at a rain rate of 20 mm/h

Frequency band 58 GHz

- Maximum path length only one or two kilometres
- Rain fading
- Antenna diameter 15 cm
- Only vertical polarisation used
- Atmospheric attenuation 12 dB/km
- Attenuation due to rain about 7 dB/km at a rain rate of 20 mm/h

Before an into explanation is given of various factors that affect the performance of the microwave links, an attempt will be made to explain the fundamentals of functioning of microwave links.

3.4.1 Microwave Link

Figure 3.36 shows a simple microwave link. Major components that constitute a microwave link are:

(a) Indoor unit
(b) Outdoor unit
(c) Antenna
(d) Waveguide
(e) Microwave tower

Indoor Unit (IU)

Also called IDU, it usually contains the radio modem; i.e. it acts as a termination point for the digital signals from the end user equipments and subsequently converts them into a signal that forms the basis of a radio signal sent across the microwave link, using modulation schemes. Also, it demodulates the carrier to the digital signal in the receive direction. Apart from the modulation/demodulation of the signal, other IU functions also include: forward error correction (FEC), multiplexing of user data, control unit (monitoring and controlling the radio unit through the NMS) and acting as a communication channel

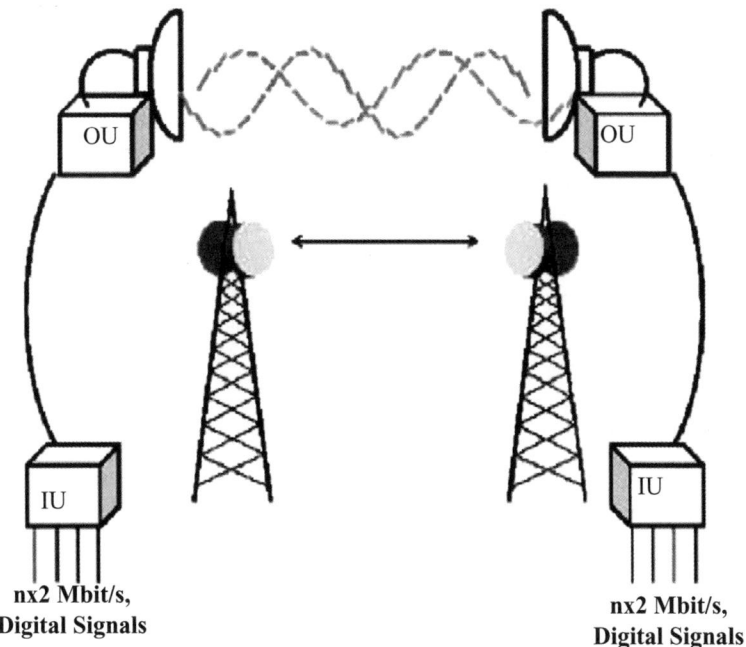

Figure 3.36 Microwave link

between the NMS and OU. Power to the microwave radio is fed through the IU. Indoor units are usually located in a protected environment, i.e. within cabinets inside a building or similar structure. They are not exposed to the environmental conditions like the outdoor units.

Outdoor Unit (OU)

Also called ODU, it converts the modulated low frequency digital signal into a high frequency radio signal. The OU contains radio frequency transmitters and receivers. Due this feature, it is also known as a radio transceiver. When the signal is received from the antenna, it usually passes the low noise amplifier (LNA), which strengthens the weak signals. This is followed by automatic gain control (AGE), which ensures equality of the signal strength when entering the radio receiver. The difference between the LNA and AGC is that while the former is used for received signal clarity and quality maintenance of the received signal, the latter compensates the signal power attenuated during transmission due to unpredictable climatic conditions. This signal is then demodulated into an intermediate frequency (IF) or a base band (BB) signal before it moves to the IU. The OU gets electrical power and the low frequency modulated signal from the IU through coaxial cable connections.

Antenna

An antenna is the unit of the microwave link that interacts with free space; hence its understanding becomes critical for the microwave link planning engineers. An antenna is defined as the structure that transfers the electromagnetic energy from the free space into transmission lines and vice versa. There are many types of antennas: horn, parabolic, flat or planar, lens, yagi, array, etc. The application of the antenna depends upon its electrical and mechanical characteristics. Some of the important characteristics of an antenna are now given.

Antenna Directivity

The directivity $D(\theta, \phi)$ is defined as the ratio of the radiation intensity $P(\theta, \phi)$, which is the power radiated per unit solid angle in the direction (θ, ϕ), to the average radiation intensity (the average power radiated from an antenna per unit solid angle):

$$D(\theta, \phi) = \frac{P(\theta, \phi)}{P_t/4\pi} = \frac{P(\theta, \phi)}{P_{av}} \qquad (3.1)$$

where

P_t = total radiated power
θ = elevation angle
ϕ = azimuth angle
P_{av} = average radiation intensity

Antenna Gain

The power gain $G(\theta, \phi)$ of an antenna in the direction (θ, ϕ) is defined by

$$G(\theta, \phi) = \frac{P(\theta, \phi)}{P_i/4\pi} \qquad (3.2)$$

where

$P(\theta, \phi)$ = radiation intensity
P_i = total power supplied to the antenna
$P_i/4\pi$ = radiation intensity produced in all directions by a hypothetical loss-less isotropic radiator

This gain may also be defined as the ratio of radiation intensity produced by the antenna to that produced by a loss-less isotropic radiator with the same input power. The expression $G(\theta, \phi)$ is called the gain function to distinguish it from its maximum value G_m, which is called the gain of the antenna, as shown by the following equations:

$$D(\theta, \phi) = \frac{P(\theta, \phi)}{P_t/4\pi} \qquad (3.3)$$

$$G(\theta, \phi) = \frac{P(\theta, \phi)}{P_i/4\pi} \qquad (3.4)$$

$$P_t = \eta P_i \qquad (3.5)$$

$$G(\theta, \phi) = \eta D(\theta, \phi) \qquad (3.6)$$

Note that:

- Good directivity allows dense packing.
- A large diameter gives better directivity.

Although high gain antennas are better for radio system performance, a higher gain would mean better directional use leading to a finer path of the radio signal. This has an implication during installation, as the antenna needs to be more accurately aligned.

Front-to-Back Ratio

The front-to-back ratio (FBR) is defined as the ratio of the power radiated in the desired direction to the power radiated in the opposite direction:

$$\text{FBR} = \frac{\text{power radiated in desired direction}}{\text{power radiated in opposite direction}} \tag{3.7}$$

The FBR becomes quite important when doing an interference analysis. The higher the front-to-back ratio, the better it is as the interference from the side lobes will be less. The FBR changes if the frequency of operation of the antenna system shifts. Its value decreases if spacing between the elements of the antenna increases. The FBR depends on the tuning conditions of the electrical length of the parasitic elements. Typical values of the FBR are 35–50 dB.

The higher the gain of the antenna, the higher is the FBR. However, an increase in the value of the FBR can be achieved by diverting the gain of the opposite direction (i.e. backward) to the forward or desired direction. This is done by adjusting or tuning the length of parasitic elements. Hence a higher value of FBR is achieved at the cost of sacrificing gain from the opposite direction.

In practice, for receiving purposes more adjustments are made to acheive a maximum front-to-back ratio than to achieve maximum gain.

Polarisation

Polarisation is defined as the direction in space along which the electric vector points. The polarisation of an antenna is described by the polarisation of its radiated field. Basically polarisation is of two types, vertical and horizontal, though some antennas are also called dual polarised antennas, i.e. the design of the antenna permits it to transmit in both polarisations.

Cross Polarisation Discrimination (XPD)

Suppose a transmitting antenna is oriented vertically and a receiving antenna has a similar orientation. Then in the absence of fading the transmitted signal will be completely received at the receiving antenna. However, due to the presence of multipath fading effects and atmospheric fading there is an apparent change in the phase of the signal. The greater the phase angle, the lower is the power received in the vertical component. Therefore,

$$\text{XPD} = \frac{\text{Power in desired orientation}}{\text{Power in undesired orientation}} \tag{3.8}$$

Parabolic Antenna Gain

The most widely used antenna used to design microwave links is the parabolic antenna. The gain of the parabolic antenna is given by

$$G = 17.5297 + 20\log(D_\text{m}) + 20\log(F_\text{GHz}) \tag{3.9}$$

where

D_m = antenna diameter (m)
F_GHz = frequency (GHz)

Beam Width

The beam width of an antenna is the angular separation between the two half power points on the power density radiation pattern. It is also the separation between the two 3-dB down points on the field strength radiation pattern of the antenna. The term is used more frequently with narrow-beam antennas than with others and refers to the main lobe.

Standing Wave Ratio (SWR)

The standing wave ratio (SWR) can be defined as the ratio of the maximum current (or voltage) to the minimum current (or voltage) along a transmission line. The SWR is a measure of the mismatch between the load and the line. In the case of perfect matching, the SWR is unity.

Antenna Mounting

Antenna mounting is required for easy alignment of the antenna. It is provided by the antenna manufacturer and is used to secure the antenna firmly on the tower (or pole, etc.). Two aspects that are needed during the alignment and mounting phase are the *azimuth* and *bearing*, the former being the angle of the antenna relative to the ground and the latter the compass orientation.

Weight of the Antenna

This figure is handy when many antennas are being mounted on one pole/tower. Engineers should know the strength of the pole/tower before mounting the antennas. If it is a green-field site, the type of pole/tower can be suggested according to the size and/or number of antennas that need to be mounted on it. If it is an area of high wind speed, then the structure supporting the antennas should be stronger, so that the antennas do not lose their alignment or fall down.

Waveguide

Minimising losses is one of the key aspects of microwave link design. Cables and waveguides also contribute to the losses. Hence cables and waveguides that give minimum resistance to the signal are used. Usually for frequencies below 2 GHz, coaxial cables are used and for frequencies above 2 GHz, waveguides are used. Coaxial cables are used because they give low losses and are economical. Generally, coaxial cables that are used are foam dielectric cables that are unpressurised, having diameters of 1/2, 7/8 and 5/8 inch. The smaller the diameter, the higher is the attenuation. If the feeder losses are critical, then air dielectric coaxial cables should be used as they have lower attenuation than the foam dielectric cables. Waveguides can be rectangular, circular and elliptical.

Rectangular Waveguide

These are most commonly used with oxygen-free, high conductivity copper. Some examples of these waveguides are:

    3–5 GHz: WR-187 (loss: 0.07 dB/m)
    6–7 GHz: WR-137 (loss: 0.114 dB/m)
    10–11 GHz: WR-90 (loss: 0.217 dB/m)

Usually, manufacturers specify the losses per metre or per inches. The losses are calculated for the whole length of the cable. Thus, microwave engineers should try to design the systems in such a way that cable lengths are as small as possible to reduce losses.

Circular Waveguides

These have the lowest loss of all the waveguides and can support two orthogonal polarisations in one guide. They can carry more than one frequency band in the same guide; e.g. WC-281, which is used with the horn reflector antenna, can provide two polarisations at 4 GHz and 6 GHz. However, these circular waveguides have the following disadvantages:

- Practically they can only be used for small runs.
- They can have problems related to waveguide modes.

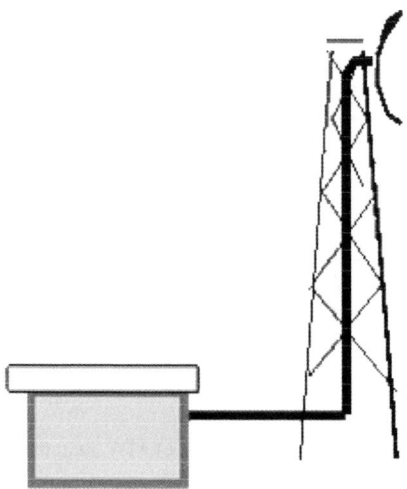

Figure 3.37 Feeder connected to the antenna

Elliptical Waveguides

Semi-flexible elliptical waveguides are available in sizes comparable to most of the standard rectangular waveguides. The most distinctive feature is that it can be installed as a single continuous run with no intermediate flanges. It provides good voltage standing wave ratio (VSWR) performances, but relatively small deformations can introduce enough impedance mismatches to produce severe distortions. A couple of examples can be:

11 GHz: EW-107 (loss: 3.70 dB/100 inch)
12–13 GHz: EW-37 (loss: 4.50 dB/100 inch)

Due to their characteristics, elliptical waveguides are usually used in microwave links. However, they are more expensive than rectangular waveguides.

Calculation of Feeder Loss

Consider a microwave cable connected from the IU to the OU as shown in Figure 3.37. The feeder loss in this case can be calculated as

$$\text{Feeder loss} = a(L_{\text{hor}} + L_{\text{ver}}) + L_{\text{con}} \tag{3.10}$$

where

$a = $ loss (dB/m)
$L_{\text{hor}} = $ horizontal length of the feeder
$L_{\text{ver}} = $ vertical length of the feeder
$L_{\text{con}} = $ connector losses

The size and attenuation of the waveguides are determined by the frequency band: the higher the frequency, the smaller the waveguide and the higher its attenuation for each unit length. The number of bends, twists and flexible sections should be kept to a minimum with great care taken during the installation.

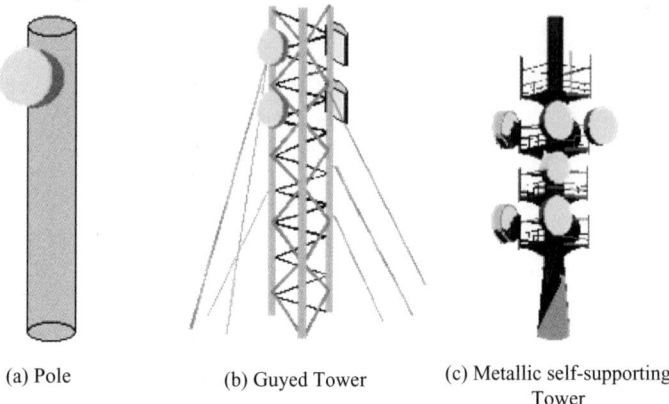

(a) Pole   (b) Guyed Tower   (c) Metallic self-supporting Tower

Figure 3.38 Different types of MW antenna towers

3.4.2 Microwave Tower

There are a few different types of towers that are used for mounting microwave (MW) links (shown in Figure 3.38). For smaller antennas that are mounted on top of a building in a city access network, strong poles are used. These usually have a height of 5 m. Simple guyed metallic structures and self-supporting towers are used when a high number of antennas are being mounted (thereby increasing the total weight of the mounted antennas). Engineers should ensure that the number of antennas should not exceed the load bearing capacity of the tower.

3.4.3 Microwave Link Design

The main purpose of the link design is to make sure that the microwave network gives high performance in all types of atmospheric conditions. A link design contains four major steps:

- Path calculations
- Antenna height calculations
- Frequency planning and interference calculations
- Performance calculations

However, as the atmosphere is highly unpredictable, all possible loss inducing factors should be taken care of. This makes an understanding of the atmospheric behaviour necessary for microwave engineers.

Refractivity and the *k*-Factor

Earth's atmosphere plays a vital role in the propagation of microwaves. Due to the properties of the atmosphere, communication between two microwave links is possible beyond the optical horizon. In a standard atmosphere, pressure, temperature and water vapour contents change (decrease) with (increasing) height. This means that the atmosphere contains several minute layers, each having its own pressure, temperature and water vapour content (which of course keeps on changing). This means that a microwave signal passes through many different layers or media. Each of these layers or media has different refractivity, defined by the refractivity index.

Refractivity η of a particular media can be defined as the ratio of the velocity of a signal in free space to the velocity of a signal in that medium:

$$\eta = \frac{\text{velocity of a signal in free space (vacuum)}}{\text{velocity of a signal in the medium}} \quad (3.11)$$

The velocity of the signal will be different in different media. Refractivity of the earth's atmosphere has been found to be approximately 1.000 3000 n units. As this number is difficult to use for mathematical calculations, another unit N for radio refractivity is used in propagation theory.

As pressure, temperature and humidity all change with height, the refractive index also changes with height, as refractivity varies with pressure, temperature and humidity. This also means that the radio refractivity N also changes with height:

$$N \propto \eta \quad (3.12)$$
$$\eta = 1.000\,300 \text{ n units} \quad (3.13)$$

Hence, N is defined as

$$N = (\eta - 1) \times 10^6 \text{ N units} \quad (3.14)$$
$$N = 77.6\frac{P}{T} + 3.73 \times 10^5 \frac{e}{T^2} \quad (3.15)$$
$$N = \frac{77.6}{T}\left(P + \frac{4810\,e}{T}\right) \quad (3.16)$$

where

P = pressure (hPa)
T = temperature (K)
e = water vapour pressure (hPa)

Under average conditions, $P = 1017$ hPa, $e = 10$ hPa (50 % relative humidity) and $T = 291.3$ K(18 °C). This equation has essentially two components, dry and wet, with the former representing N_{dry} and the latter N_{wet}. N_{dry} has a fairly constant value of about 256 N units, but it is the N_{wet} term that is responsible for the variability of the equation.

When a microwave signal travels in the atmosphere, these electromagnetic waves travel faster in the medium of the lower dielectric constant and the upper part of the wavefront tends to travel with greater velocity than the lower part, causing a downward deflection of the beam. In a homogeneous atmosphere vertical change in the refractivity is gradual; hence the microwave also bends gradually away from the rarer medium towards denser media, thus making the beam bend to follow the earth's curvature, as shown in Figure 3.39.

However, the radius of the microwave signal (kR) from the transmitting antenna to the receiving antenna makes an arc whose radius is different from the radius of the earth (R). In a real scenario, both the radio signal and the earth surface have a curvature, which makes the path profile calculations difficult. The possible solution of this problem can be modification of either of the two curvatures in such a way that one of them becomes a straight line. As the destination of the radio signal is an unknown factor, the only solution is to modify the earth's curvature so that the radio signal becomes a straight line. Thus, after changes, the new model of the above stated actual scenario would look like the one shown in Figure 3.40.

The amounts by which the curvatures of the radio signal and the earth surface are changed in order to make the radio signal a straight line is called the 'k-factor'. It is defined as the ratio of the effective earth radius to the actual radius of the earth:

$$k = \frac{kR}{R} \quad (3.17)$$

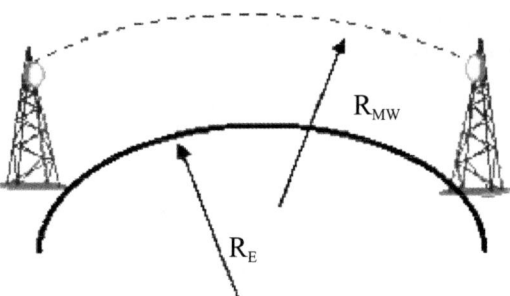

Figure 3.39 Microwave path

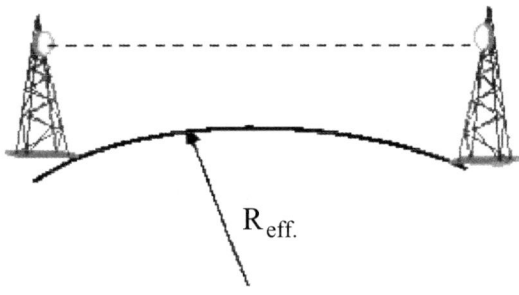

Figure 3.40 Microwave path

where R is the radius of the earth. The microwave signal varies with climatic conditions, with the variation of any one or all factors like pressure, temperature and humidity. These factors vary with height. This will result in a variation of the radius of the radio signal and hence the k-factor. Thus, the k-factor varies with the changes in height. Thus, with every small variation of height (dh), there will be small variations of refractivity ($d\eta$), which will result in a variation in the radius of the microwave signal. This is represented mathematically as

$$\frac{1}{R_{MW}} = \frac{dn}{dh} \qquad (3.18)$$

where R_{MW} is the radius of the microwave signal and dn/dh is the vertical gradient of the refractive index. This expression is valid when the angle of launch of the microwave signal is zero degrees (to horizontal) and refractivity is unity.

Hence the k-factor is dependent upon the refractive index gradient and is mathematically represented as

$$k = \frac{1}{1 + R\,(dn/dh)} \qquad (3.19)$$

$$\frac{dn}{dh} = \frac{dN}{dh} \times 10^{-6} \qquad (3.20)$$

In standard atmospheric conditions, the refractive index gradient is $-40\,\text{N/km}$. The radius of earth is approximately $6400\,\text{km}$; putting these values in the above equation

$$k = \frac{1}{1 + 6400(-40 \times 10^{-6})} = \frac{10^6}{10^6 - 256\,000} \approx 1.33 \qquad (3.21)$$

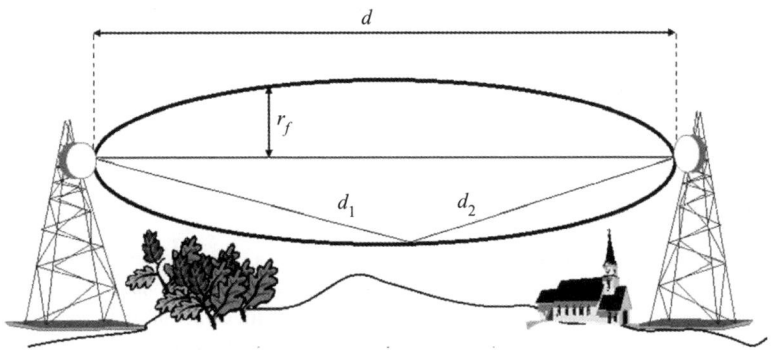

Figure 3.41 First fresnel zone

This means that (for standard atmospheric conditions), to find the effective radius of the earth, the radius of the earth should be multiplied by 1.33. This introduces a bulge on the surface of the earth, the curvature of which can be calculated from

$$h_B = 0.078 \frac{d_1 d_2}{k} \quad (3.22)$$

Fresnel Zone

Once the variation of the signal in the atmosphere is known, the next step is to ensure that the maximum strength of the signal reaches the receiving antenna. This is possible if the designer makes sure that there is no obstruction between the transmitting and receiving antennas. This is done by keeping the first fresnel zone (FFZ) clear of any obstruction.

As shown in Figure 3.41, the first fresnel zone can be defined as the volume contained in a three-dimensional ellipsoid between the transmitting and receiving antennas, in such a way that any ray drawn from the transmitting to the receiving antenna touching any point on the ellipsoid will be a half-wavelength longer than the direct ray between the two antennas. If the two lines originate from a point on the ellipsoid towards the centre of the transmitting and receiving antennas, then the combined length of the two lines will be one wavelength longer than the direct ray between the two lines. Thus, for subsequent fresnel zones, the combined length of the indirect signal will increase by a half-wavelength from the previous one.

Mathematically, this can be represented as

$$d_1 + d_2 = d + n\frac{\lambda}{2} \quad (3.23)$$

Thus, the radius of the fresnel zone is given by

$$R_{nF} = 12.75 \sqrt{\frac{n d_1 d_2}{f d}} \quad (3.24)$$

where

n = number of fresnel zone (1,2,3, ...)
f = frequency (GHz)
d, d_1, d_2 = distances (m)

This shows that as the frequency increases, the radius of the fresnel zone decreases.

Ideally, the fresnel zone should be clear of all obstacles. A 100 % clearance criterion for the fresnel zone is followed in temperate climates. However, for lower frequencies of about 2 GHz and long paths, i.e. over 100 km, the criteria becomes stricter, i.e. $k = 1$ or $k = 0.66$. The impact of the variations of k-values will be given later in this chapter.

3.4.4 LOS Check

As seen in the section above, a clear line-of-sight (LOS) is necessary for the microwave links to be approved and installed. The primary consideration is site selection. The radio planning engineers go for sites with lower heights, e.g. 12 m, which is contrary to the requirements of microwave link planning engineers for high rise sites. Low sites help in interference reduction while high sites increase the probability of getting connected to other sites due to a better LOS. When doing LOS, the microwave planning engineers need to have good coordination with radio planning, installation planning and civil engineering works because the same site may be used as a radio site, should have proper space for installation and should be strong enough to hold the equipments. To minimise site visits, LOS activity should be planned with each site visit. However, if the site is a transmission site only, then LOS activity could be planned separately. Surveyors should have all equipments that are necessary for doing the LOS, such as the global positioning system (GPS), binoculars, compass, single lens reflex (SLR) camera, mirror, searchlight or flashing beacon, large rubber balloons and gas, cellular phone (or very high frequency (VHF) radios), measuring tape (at least 50 m), maps (road and topographical), inclinometer for measuring heights of objects and telescopic mast (preferable 15 m). Software tools with digital maps are a good starting point. The results, however, are not very accurate because of the inaccuracies related to the digital maps, but they can be useful for rejecting cases where LOS does not exist, e.g. a hill's presence between two sites, etc. For sites that are within a few hundred metres of potential sites, a site visit would be sufficient for confirming or not confirming the line-of-sight. Complexity arises when sites are a few kilometres from each other. In this case, the usual techniques do not work, especially if the site is a greenfield site. In such cases, balloons are used for determining the LOS. When the digital maps are not available, paper maps are used. When using paper maps, the two sites are marked on the map using the coordinates obtained by the site visit using the GPS. A straight line is then drawn on the map and points at each kilometre are marked on this line. The surveyor then drives along the marked line (as much as possible), goes to each of these points to verify the height above the average mean sea level (AMSL) and cross checks with the map. A check is also mede to see whether there are any obstructions that might not be marked on the map. Then these actual AMSL readings are put in a software tool (or mathematical formulas are simply used, which are given elsewhere in the chapter) to find what minimum antenna heights will be needed to achieve the desired clearance. The creation of the profile is explained in the next subsection. The engineers/surveyors should, however, go through tower climbing training before they climb high towers. The outcome of the LOS survey is an LOS report that clearly marks whether there was an LOS or no LOS. Apart from this, the report consists of 360° photographs, feeder lengths (optional), exact location and minimum height of the microwave antenna(s), hop bearing and distance, future possible obstructions (photographed or listed), type of tower, mast or pole to be used, etc.

Path Profile

The path profile is a graphical representation of the path between the transmitting and receiving antenna sites. The path profile in conjunction with the fresnel zone calculations helps to define the antenna heights. Also, antenna heights can be adjusted based on the path profiles in order to avoid obstructions, reflective surfaces, etc.

Microwave Link Planning

Table 3.9 Measurements for a path profile

Distance [km]	0	2	4	6	8	10	12	14	16	18	20	22	24	26	28	30
Terrain height [m]	3	6	9	7	20	20	30	31	29	20	25	26	27	29	30	35

There are mainly three methods used to draw a path profile:

1. Fully linear method
2. 4/3 earth method
3. Curvature method

Of these three, the fully linear method is the one recommended so it will be discussed in detail here.

Fully Linear Method

This method is the recommended method because it is possible to represent many variations of the k-factor on one chart, apart from the fact that simple linear graph paper can be used to plot the path and straight lines are used instead of curved lines, making plotting of the path profile easy.

The path profile is drawn on a linear graph paper, where a straight line is drawn from the transmitter to the receiver, giving tangential clearance of equivalent obstacle heights. Bending of the radio beam is represented by an adjustment to each obstacle height by an equivalent earth bulge (calculated from Equation (3.22)), as shown in Figure 3.42.

Example. Draw a path profile for a 30 km terrain. The terrain profile at the standard atmosphere condition ($k = 4/3$) is drawn for a hop between sites A and B in such a way that radio wave propagation is shown as a straight line. The height of the terrain after every 2 km is given in Table 3.9.

Based on this information, the earth bulge should be calculated for each of the distance points mentioned. On top of that clearance is added, which will take into account the trees, buildings, etc. Depending upon the region's obstacle heights, the clearance factor can be changed. Once this is done, a straight line is drawn and the antenna heights are found. In this case, an antenna at a height of 14 m at both the sites would be sufficient (the maximum height to be cleared is 23.1 m) (see Table 3.10).

3.4.5 Link Budget Calculation

The link budget calculation is one of the most important aspects of microwave engineering. Through link budget calculations, the amount of power received at the receiving antenna is calculated, which forms the basis of calculation of fade margin and availability. Before link calculations are discussed it is important to understand the term 'free space loss'.

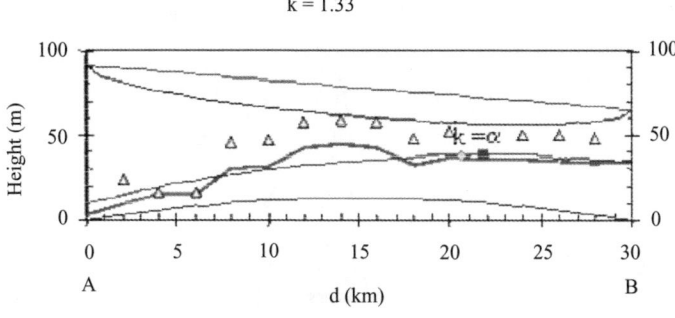

Figure 3.42 Path profile

Table 3.10 Results of drawing a path profile

D [km]	0	2	4	6	8	10	12	14	16	18	20	22	24	26	28	30
Height [m]	3	6	9	7	20	20	30	31	29	20	25	26	27	29	30	35
d_1 [km]	0	2	4	6	8	10	12	14	16	18	20	22	24	26	28	30
d_2 [m]	30	28	26	24	22	20	18	16	14	12	10	8	6	4	2	0
k-factor	1.33	1.33	1.33	1.33	1.33	1.33	1.33	1.33	1.33	1.33	1.33	1.33	1.33	1.33	1.33	1.33
Earth bulge [m]	0.0	3.3	6.1	8.4	10.3	11.7	12.7	13.1	13.1	12.7	11.7	10.3	8.4	6.1	3.3	0.0
Cleaarance [m]	10	10	10	10	10	10	10	10	10	10	10	10	10	10	10	10
Total height	10.0	13.3	16.1	18.4	20.3	21.7	22.7	23.1	23.1	22.7	21.7	20.3	18.4	16.1	13.3	10.0

Free Space Loss

Consider an isotropic antenna radiating in free space (as shown in Figure 3.43). The point source is located in the centre of the sphere. As the antenna is isotropic, it can be safely assumed that the power is radiated uniformly in all directions, forming an imaginary sphere.

The surface area of this sphere can be given as

$$S = 4\pi R^2 \qquad (3.25)$$

At any given point at a distance R from the antenna, the power density S can be given as

$$S = P \times \frac{G}{S} \qquad (3.26)$$

where

$P =$ power transmitted by the antenna
$G =$ gain of antenna

The received power P_r at a distance R is given as

$$P_r = PG_tG_r(\lambda\sqrt{4\pi R})^2 \qquad (3.27)$$

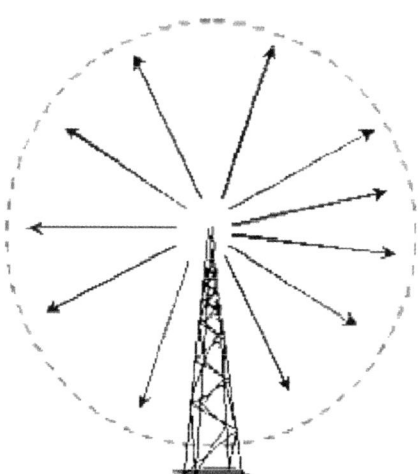

Figure 3.43 Istropic antenna

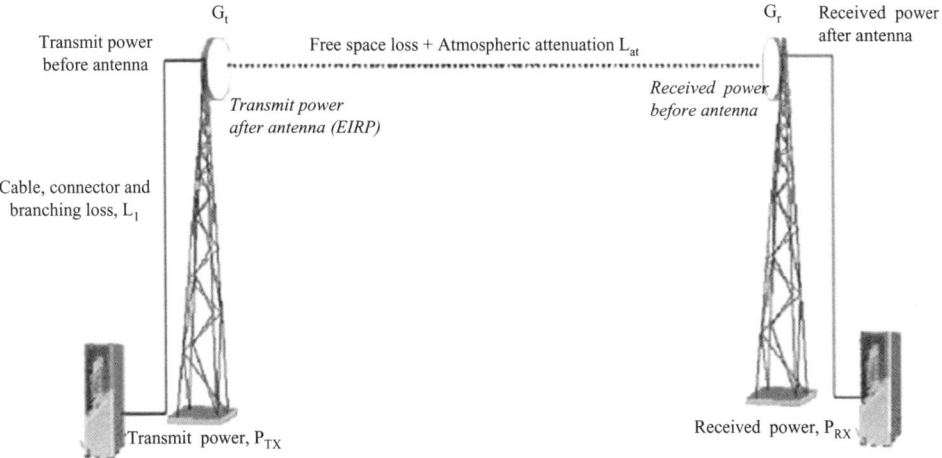

Figure 3.44 Microwave link

where G_t and G_r are the gain of the transmitting and receiving antennas respectively. When converted to dB, the above equation will be

$$P_r(\text{dB}) = P(\text{dB}) + G_t(\text{dB}) + G_r(\text{dB}) + 20\log\left(\frac{\lambda}{\sqrt{4\pi}}\right) - 20\log(d) \tag{3.28}$$

The last two terms in Equation (3.28) are called the path loss in free space or the free space loss (FSL). The first two terms, i.e. P and G_t, combined are called the effective isotropic radiated power (EIRP). Thus,

$$\text{FSL}(\text{dB}) = \text{EIRP} + G_r(\text{dB}) - P_r(\text{dB}) \tag{3.29}$$

Thus, the free space loss can be given as

$$L_{dB} = 92.5 + 20\log(f) + 20\log(d) \tag{3.30}$$

where f is the frequency in GHz and d is the distance in km.

Now consider the microwave link shown in Figure 3.44. The hop loss is the difference between the gains and losses on a microwave link. The gains constitute the antenna gains on either side while the losses constitute the sum of free space loss and losses due to extra attenuation and atmosphere (water vapour and oxygen):

$$L_h = L_{dB} + L_{ex} + L_{at} - (G_t + G_r) \tag{3.31}$$

where

G_t, G_r = antenna gains at transmitting and receiving ends respectively
L_{dB} = free space loss
L_{ex} = extra attenuation loss (e.g. radome loss)
L_{at} = loss due to water vapour and oxygen

When the signal travels from the transmitting to the receiving antenna, the signal power reaching the receiving antenna P_{rx} is calculated as

$$P_{rx} = P_{tx} - L_h \tag{3.32}$$

Once a microwave link is designed and installed, it is there 'forever'. Hence, during the design stage itself, the microwave engineer should make sure that even under worse conditions, the signal received at the receiving antenna is above the threshold. Thus, the concept of fade margin (FM) comes into the picture, which is defined as the difference between the received signal and the receiver threshold:

$$\text{FM} = P_{\text{rx}} - P_{\text{rxth}} \tag{3.33}$$

The fade margin is necessary as it gives the necessary guard band to prevent link failure against unpredictable atmospheric conditions.

The lower the frequency of operation, the longer is the hop length and the, higher the probability of atmospheric unpredictability, resulting in a higher fade margin during the planning phase. At lower frequencies of 7 GHz, the FM is kept at about 35–40 dB.

Outage Calculation (P_w)

Once the received power and fade margin have been calculated, the next step is to calculate the outage. To calculate the outage (based on ITU-R 530), the following inputs are required:

- Fade margin (FM)
- Frequency of operation f (GHz)
- Hop length d (km)
- Geoclimatic factor K
- pL factor
- Magnitude of path inclination εp

From the calculations above, the FM can be calculated. The frequency (in GHz) and hop lengths are known inputs. However, the pL factor needs to be found from the ITU-R Recommendations and the geoclimatic factor calculated.

pL factor

This is the percentage of time that the average refractivity gradient in the lowest 100 m of the atmosphere is less than 100 N units/km.

Geoclimatic Factor

This is dependent upon the location of site, i.e. latitude and longitude, pL factor and antenna heights. The following are the empirical relations.

For overland links (TX and RX antenna heights < 700 m AMSL)

$$K = 10^{-(6.5 - C_{\text{Lat}} - C_{\text{Lon}})} P_L^{1.5} \tag{3.34}$$

For overland links (TX and RX antenna heights > 700 m AMSL)

$$K = 10^{-(7.1 - C_{\text{Lat}} - C_{\text{Lon}})} P_L^{1.5} \tag{3.35}$$

For links over medium sized water bodies, coastal areas, lakes, etc.

$$K = 10^{-(5.9 - C_{\text{Lat}} - C_{\text{Lon}})} P_L^{1.5} \tag{3.36}$$

For links over large bodies of water or coastal areas besides such bodies of water

$$K = 10^{-(5.5 - C_{\text{Lat}} - C_{\text{Lon}})} P_L^{1.5} \tag{3.37}$$

where the coefficients of latitude ξ are given by

$$C_{\text{Lat}} = 0, \qquad 53°S \geq \xi \leq 53°N$$
$$C_{\text{Lat}} = (-53 + \xi)/10, \qquad 53°\text{Nor}°S < \xi < 60°\text{Nor}°S$$
$$C_{\text{Lat}} = 0.7, \qquad \xi \geq 60°\text{Nor}°S$$

and the longitude coefficients C_{Lon} are given by

$C_{\text{Lon}} = 0.3,$ Longitudes of Europe and Africa
$C_{\text{Lon}} = -0.3,$ Longitudes of North and South America
$C_{\text{Lon}} = 0,$ All other longitudes

The month that has the highest value of pL is selected from the four seasonally representative months of February, May, August and November for which maps are shown. However, for latitudes greater than 60° N and S, only the maps of May and August must be used.

Magnitude of Path Inclination

If the antenna heights h_e and h_r metres above sea level or some other reference heights are known, the magnitude of path inclination in milliradians (mrad) is

$$\varepsilon_p = \frac{|h_1 - h_2|}{d} \qquad (3.38)^*$$

where the path length d is in kilometres. The percentage of time p_w that fade depth FM (dB) is exceeded in the average worst month is

$$p_w = K d^{3.6} f^{0.89} \left(1 + |\varepsilon_p|\right)^{-1.4} \times 10^{-\text{FM}/10} \qquad (3.39)^*$$

where f is the frequency in GHz.

Example. Error performance calculations for a radio relay system based in ITU-R 530 for a 15 km hop operating at a frequency of 7.4 GHz. The parameter values are given in table 3.11.

3.4.6 Repeaters

The path between the transmitter and receiver is a straight line clearing all obstacles. As already seen, a microwave signal requires more than just an optical line-of-sight clearance to allow for both fading and for certain interference characteristics of the beam itself at microwave frequencies. To achieve desired clearance, the antenna systems on both ends can be increased in height, which of course results in increased costs. Another way to overcome the obstructions is by the use of passive repeaters or radio reflectors. As microwaves follow most of the rules of conventional optics, a system of 'mirrors' similar to an optical mirror system can be constructed to reflect the beam over or around an obstruction. Passive repeaters are used when a tower cannot provide clearance over an obstruction. For example, if two sites are separated by a hill or very tall building, the microwave beam may have to be redirected at one or more intermediate points to get it around or over the hill/building. Although active repeater stations could be used at these points to amplify and retransmit the signal, passive repeaters may be used to merely change the path direction without amplification. As passive repeaters add no signal amplification and require no power and very little maintenance, they can be located in places where access is very difficult. A passive repeater is a highly efficient device used to re-radiate intercepted microwave energy without the introduction of additional electronic power. Passive reflectors may be used in microwave systems in both the near field and far field.

Double passive repeaters are used when the deflection angle is less than 50° or where no site exists from which one reflector could see both terminal stations; this is usually the case when transmitting over a mountain ridge. This problem can be solved by using two parabolic antennas placed back to back.

Table 3.11 Parameter values for error performance calculations

Parameters		Site A	Site B
Radio frequency	7.4 GHz vertical		
Hop length	15.0 km		
Latitude	6° N		
Longitude	80° E		
Percentage pL	50.0 %		
Geoclimatic factor	1.41E-003		
Rain rate (0.01 %)	120.0 mm/h		
Station heights (reference levels)		200.0 m	200.0 m
Antenna heights (above station reference levels)		50.0 m	50.0 m
Feeder lengths		60.0 m	60.0 m
Feeder loss/100 m		4.9 dB	4.9 dB
Feeder type		EW 64	EW 64
Antenna diameters		1.2 m	1.2 m
TX output power P_{tx}	27.0 dBm		
Free space loss (FSL)	133.4 dB		
Additional terrain loss L_{ad}	0.0 dB		
Antenna branching loss L_{br}	5.0 dB (SD, HSB)[a]		
Feeder losses L_{c1}	3.4 dB		
Connectors loss L_{c2}	3.4 dB		
Antenna gains G_1	36.8 dBi		
G_2	36.8 dBi		
Hop loss in nonfaded state L_{ho}	71.7 dB		
Received unfaded power P_{rx}	−44.7 dBm		
Receiver threshold power P_{rxth} at BER 10-3	−76.5 dBm		
Flat fading margin (FM)	31.8 dB		
Calculated flat outage time p_{fm}	0.0051 %		
Total outage time p (nondiversity)	0.0051 %		
Annual unavailability due to rain	0.0001 %		
Flat outage with diversity p_{fmd}	0.0001 %		

[a] SD, space diversity; HSB, hot standby.

Microwave power is received from one antenna through a short piece of transmission line and re-radiated. Since these parabolic antennas may be pointed individually to the transmitting and receiving antennas, the projection factor is unity.

The gain of the repeater with back-to-back antennas is

$$G_R = G_{A1} + G_{A2} - A_c \qquad (3.40)$$

where

G_{A1}, G_{A2} = gains of the two antennas
A_c = coupling loss

The expression for the total free space loss with two parabolic reflectors is given by

$$L_0 = 184.88 + 20\log(d_1 d_2) + 40\log f_{GHz} - G_{r1} - G_{r2} + L_{cr} \quad (3.41)$$

The total free space loss in a two hops transmission with a single passive repeater is

$$L_0 = 141.98 + 20\log(d_1 d_2) - 20\log A_R \quad (3.42)$$

where

d_1 = distance between the transmitting antenna and the passive repeater
d_2 = distance between the receiving antenna and the passive repeater
A_R = projected area

3.5 Microwave Propagation

The link budget example shown above is the first step towards a final microwave plan. In any given microwave link, there will be variations in the received signal level about its median value, even when the transmitted power is constant. These variations are sometimes so high that the link may go down, which means that the received signal is below the threshold value (a condition also known as 'blackout' as no signal is received by the receiving antenna). This is called as fading. Microwave engineers should be able to predict this fading as close by as possible and hence design the link accordingly.

Fading can be broadly divided into two types: slow fading and fast fading.

3.5.1 Slow Fading

As the name suggests, the signal may fade in hours. The signal variations that result in fading are mainly due to changes in atmospheric conditions. Due to the changes in temperature, pressure and humidity, the radio refractivity changes, which varies the k-factor. This results in a variation of the path taken by the microwave signal and then variations in the angle of arrival, resulting in less (or no) power being received by the receiving antenna. This phenomenon is also called k-fading. Although the standard value of k is taken to be 4/3 or 1.33 for the path calculations, microwave planners should be aware of the atmospheric variation for the region when they design microwave links (shown in Figure 3.45).

There are two types of refractive conditions: subrefractive and superrefractive. When the variability becomes greater than −40 N/km, the k-value decreases and hence the signal becomes a straight line. This would result in the microwave signal getting closer to the surface of earth, increasing the chances of surface or vegetation obstructing the signal and decreasing the received power at the receiver. If the variability increases further, the k-factor becomes less than unity, thus making the signal travel further away from the surface of the earth.

When the variability of radio refractivity reduces much more than −40 N/km, the microwave signal makes an arc that is much smaller in radius (shown in Figure 3.46). This condition increases the clearance between the obstacles on the earth's surface and the microwave signal and is called the superrefractive condition. If the variability of refractivity increases to −157 N/km, the k-factor becomes infinity and the radius of the microwave signal becomes equal to that of the earth's surface. Graphically, this would mean both the microwave signal and the earth's surface behaving as straight lines. Under these conditions, the signal may not reach the surface of earth as the phenomenon of ducting takes place.

In both the subrefractive and superrefractive conditions, the angle of transmission and angle of reception may change, depending upon the severity of change in atmospheric conditions. This leads to poor signal strength at the receiving antenna.

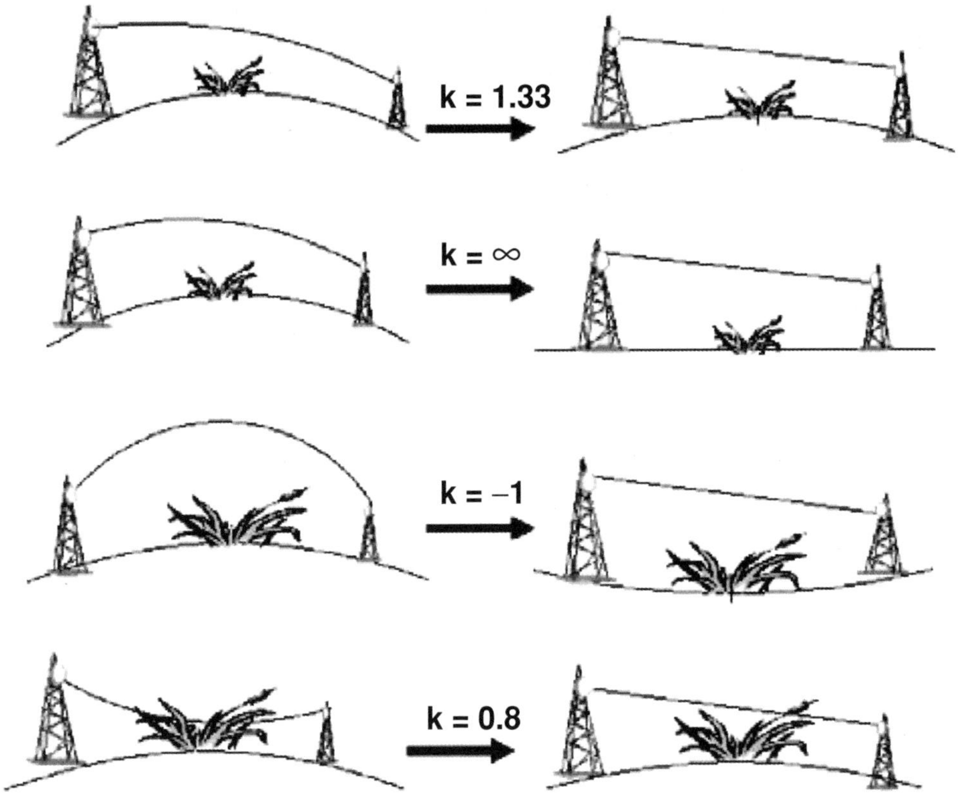

Figure 3.45 k-fading

Figure 3.46 Superrefraction

Diffraction Fading

Due to the variation in the atmospheric conditions that lead to variations in the k-factor (deviations from the standard atmospheric conditions), especially in the subrefractive domain, the signal bends in a way where the earth's surface starts to obstructing the direct path between the transmitter and receiver. The losses that happen due to this phenomenon depend upon the type of obstruction. There are basically two kinds of diffraction (shown in Figure 3.47):

- Smooth sphere diffraction
- Knife-edge diffraction

Microwave Propagation

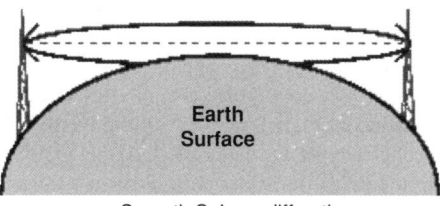 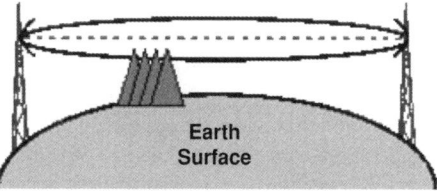

Smooth Sphere diffraction   Knife-edge diffraction

Figure 3.47 Type of diffraction

Smooth sphere diffraction takes place when the earth's surface (e.g. water) acts as a diffraction source itself. The losses are quite high in such cases, e.g. 15–20 dB. However, knife-edge diffraction takes place when the signal grazes over pointed objects such as trees, etc., making a loss of approximately 6 dB. The diffraction loss depends upon the path and type of vegetation and will vary between the maximum of the smooth sphere diffraction and minimum of the knife-edge diffraction.

There are various methods used to calculate diffraction.

Terrain Averaging Model

If the obstacle is neither sharp nor rounded, this method is used as it gives quite a good estimation of the diffraction loss. The loss can be calculated as

$$A_d = -20\frac{h}{F_1} + 10 \tag{3.43}$$

where

h = difference between the path trajectory and the most significant obstacle
F_1 = radius of the first fresnel zone

Knife-Edge Model

This is used when the object is sharp and is obstructing the first fresnel zone. These models are used when there is more than one obstacle inside the first fresnel zone. The diffraction loss is given as

$$L = 20\log(l) \tag{3.44}$$

where

$$l = 1 \text{ for } v < -0.8$$
$$l = 0.452 - \sqrt{(v - 0.1)^2 + 1} - (v - 0.1) \quad \text{for} -0.8 \leq v \tag{3.45}$$

and the fresnel–Kirchoff diffraction parameter can be calculated as

$$v = h\sqrt{\frac{2(d_1 + d_2)}{\lambda d_1 d_2}} \tag{3.46}$$

where λ is the wavelength, d_1 and d_2 are the distance to the sites from either side of the obstacle and h is the height of the obstacle.

There are some models that are used to calculate the knife-edge diffraction loss. For obstacles that are separated widely the Epstein–Peterson model is used; for obstacles too closely placed the Japanese Atlas model is used; for simple calculations the Bullington model is used; and for many obstacles the Deygout model is used to calculate the knife-edge diffraction loss.

Another factor that may cause slow fading, especially in a longer path above frequencies of 10 GHz, is rain. Attenuation due to rain is lower at lower frequencies (8–10 GHz) and increases when the frequency of operation increases. However, other factors such as the rain rate, link length and polarisation play an equally important role in the attenuation due to rain. A rainfall-like atmosphere is quite unpredictable, varying with time and location. Although for planning calculations rainfall rates mentioned by the ITU standards are used (Figures 3.48 to 3.50), it is always beneficial to use the data collected from previous years. Researches have proved that the accuracy of the prediction increases with larger amounts of collected data. Error reduction can vary from 0.001 % to 10 % when collected data varies from 10 years to 1 year. The attenuation due to rainfall is not only dependent upon the rain rate but also upon factors such as the drop size distribution, terminal velocity, canting angle and properties of water. The drop size distribution is further dependent upon the drop shape, precipitation type and atmospheric temperatures. Precipitation is dependent upon the type of clouds. The canting angle is the angle that the raindrops make with the vertical. The terminal fall velocity of the raindrops is dependent upon the air pressure, humidity and temperature. Dielectric properties depend upon temperature and frequency. For $l > 1$ mm, the dielectric properties are due to the polar nature but for $l < 1$ mm the dielectric properties depend upon the resonant absorption of the molecules.

Based on long-term statistics of the rain rate, the attenuation due to rain can be calculated using the following steps:

1. Find the rain rate exceeded for 0.01 % of the time (see Table 3.12) or the graphs given in ITU-R PN837 (Figures 3.48 to 3.50).
2. Calculate specific attenuation γ_R (dB/km) based on frequency, polarization and the rain rate. The specific attenuation can be calculated using the following equation:

$$\gamma_R = kR^\alpha \qquad (3.47)^*$$

where k and α are frequency dependent constants whose values are given in Table 3.13.

3. Calculate the effective path length d_{eff}:

$$d_{\text{eff}} = dr = d\left(\frac{1}{1+(d/d_0)}\right) \qquad (3.48)^*$$

where $d_0 = 35\ e^{0.015 R_{0.01}}$. This is valid for $R_{0.01} \leq 100$ mm/h. For a rain rate greater than 100 mm/h, the value of 100 mm/h should be used instead of $R_{0.01}$.

4. Calculate the attenuation due to rain (A_{rain}). The product of γ_R and d_{eff} give the attenuation due to rain. The path attenuation exceeded for 0.01 % of the time is given by

$$A_{\text{rain}} = \gamma_R d_{\text{eff}} \qquad (3.49)^*$$

As seen in the equations above, attenuation due to rain is dependent upon two frequency dependent constants k and α. Both of these constants are defined for spherical droplets.

3.5.2 Fast Fading

In this type of fading, the time for which the signal fades ranges from a fraction of a second to a few minutes. The main cause of such fading is the multipath phenomenon. In a microwave link, ideally a signal takes one single path from the transmitting to the receiving antenna. However, it has been observed that a signal takes a few different paths from the transmitting to the receiving antenna (as shown in Figure 3.51). Thus the signal received by the receiving antenna consists of a direct signal and an indirect

Microwave Propagation

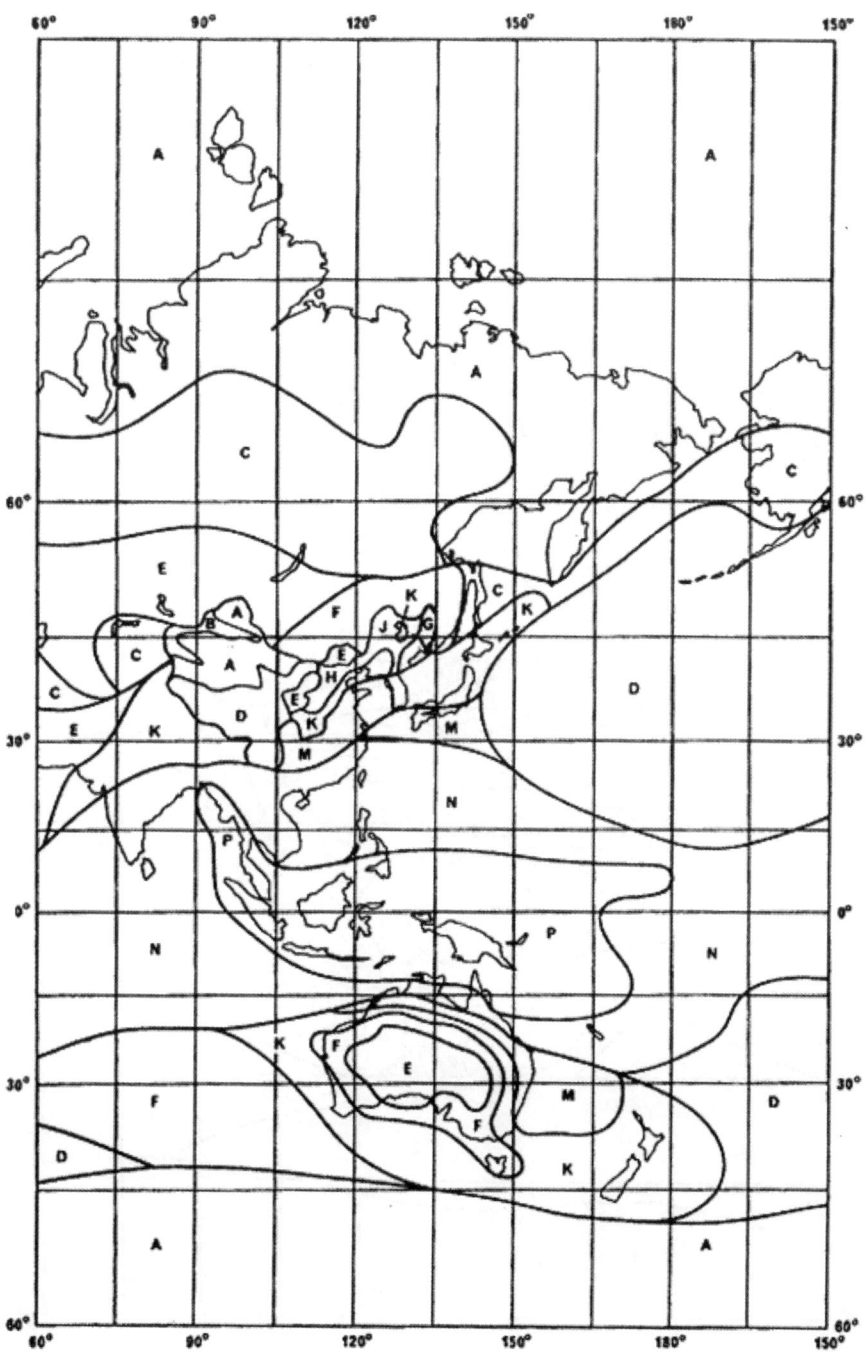

Figure 3.48 Rain charts for Asia Pacific (APAC) (Source: ITU-R PN 837-1. Reproduced with the kind permission of ITU.)

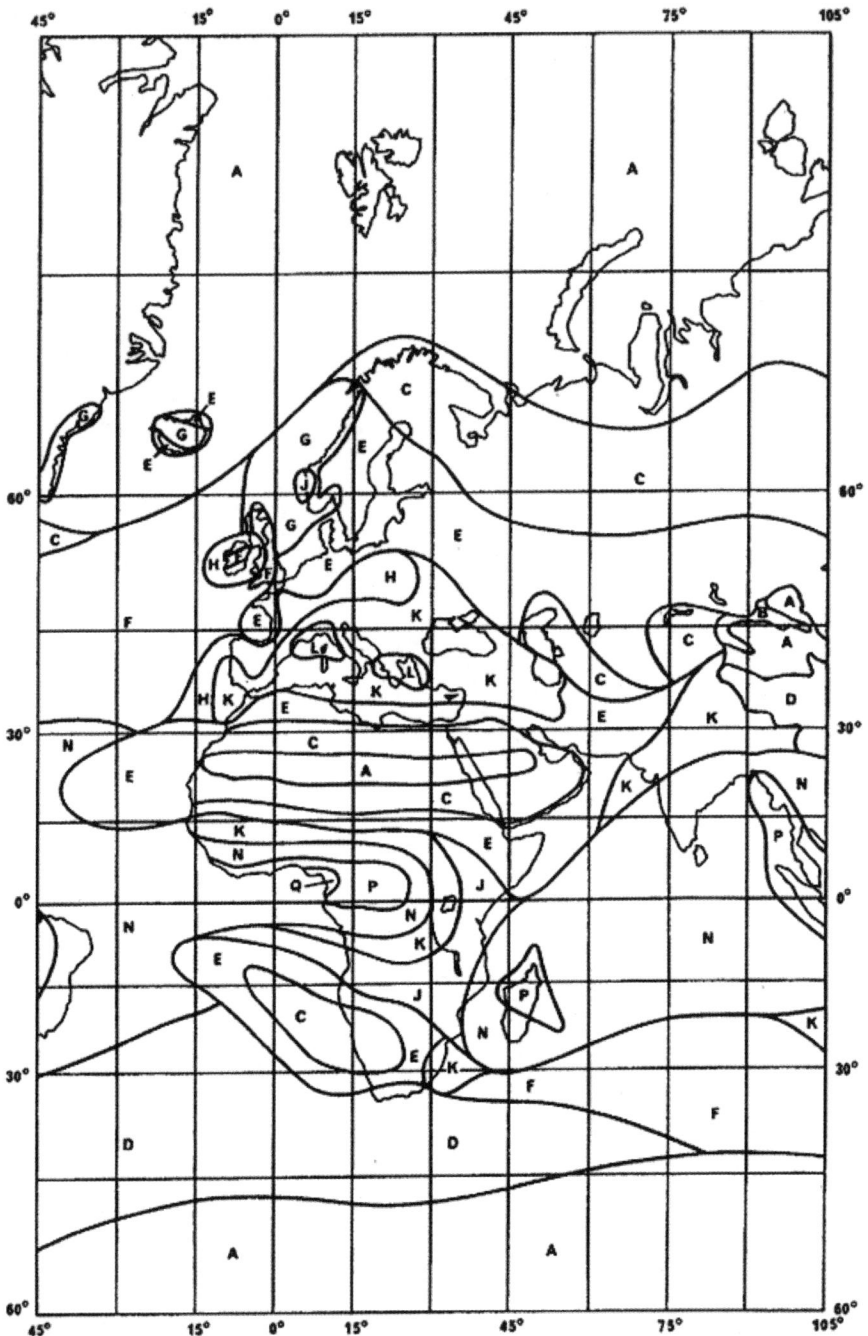

Figure 3.49 Rain charts for Europe and Africa (Source: ITU-R PN 837-1. Reproduced with the kind permission of ITU.)

Microwave Propagation

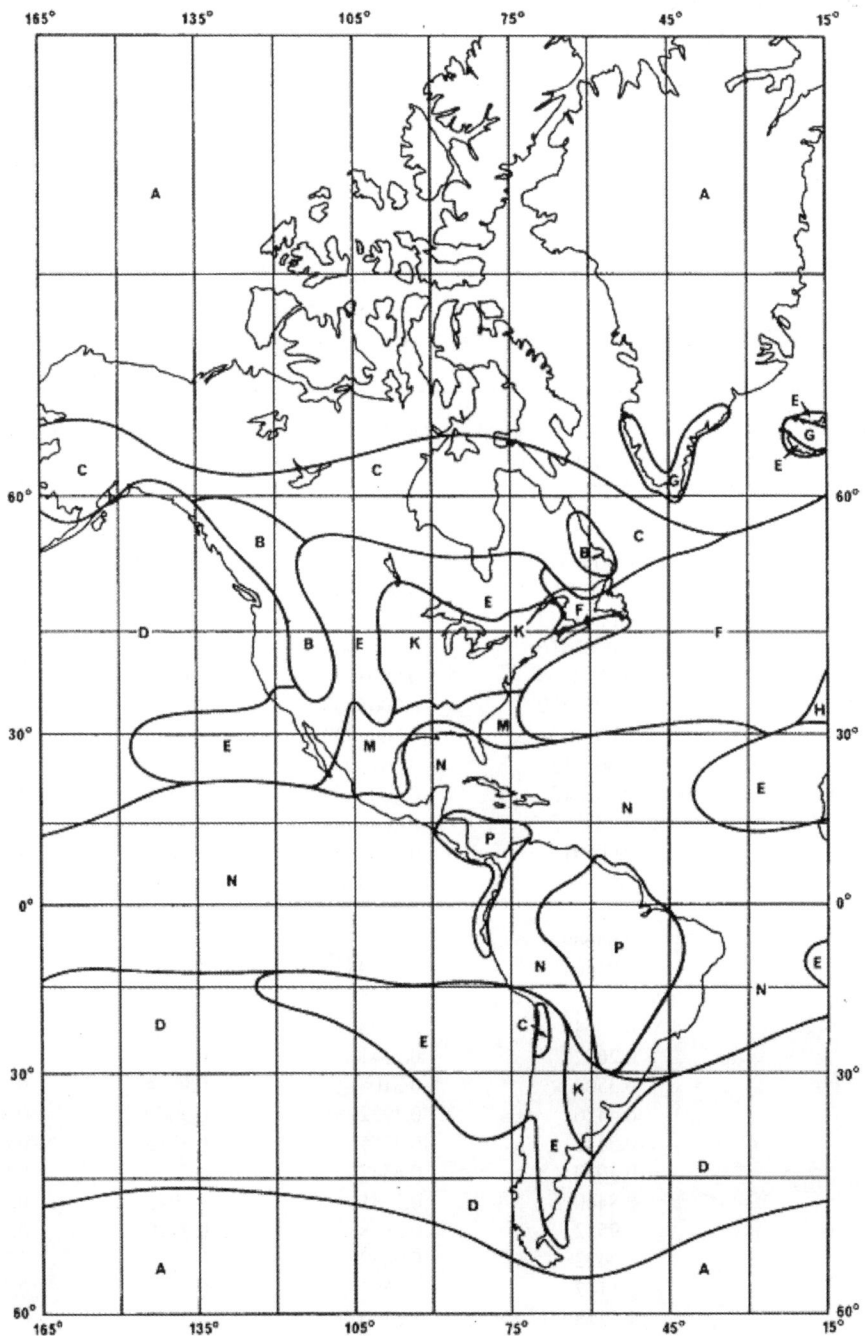

Figure 3.50 Rain charts for the Americas (Source: ITU-R PN 837-1. Reproduced with the kind permission of ITU.)

Table 3.12 Rain climatic zones showing the rainfall intensity exceeded [mm/h] (Source: ITU-R PN 837-1. Reproduced with the kind permission of ITU.)

Percentage of time [%]	A	B	C	D	E	F	G	H	J	K	L	M	N	P	Q
1.0	<0.1	0.5	0.7	2.1	0.6	01.7	3	2	8	1.5	2	4	5	12	24
0.3	0.8	2	2.8	4.5	2.4	4.5	7	4	13	4.2	7	11	15	34	49
0.1	2	3	5	8	6	8	12	10	20	12	15	22	35	65	72
0.03	5	6	9	13	12	15	20	18	28	23	33	40	65	105	96
0.01	8	12	15	19	22	28	30	32	35	42	60	63	95	145	115
0.003	14	21	26	29	41	54	45	55	45	70	105	95	140	200	142
0.001	22	32	42	42	70	78	65	83	55	100	150	120	180	250	170

Table 3.13 Values k and α for vertical and horizontal polarisations (Source: ITU Recommendation 838. Reproduced with the kind permission of ITU.)

Frequency (GHz)	k_H	k_V	α_H	α_V
1	0.0000387	0.0000352	0.9122	0.8801
1.5	0.0000868	0.0000784	0.9341	0.8905
2	0.0001543	0.0001388	0.9629	0.9230
2.5	0.0002416	0.0002169	0.9873	0.9594
3	0.0003504	0.0003145	1.0185	0.9927
4	0.0006479	0.0005807	1.1212	1.0749
5	0.001103	0.0009829	1.2338	1.1805
6	0.001813	0.001603	1.3068	1.2662
7	0.002915	0.002560	1.3334	1.3086
8	0.004567	0.003996	1.3275	1.3129
9	0.006916	0.006056	1.3044	1.2937
10	0.01006	0.008853	1.2747	1.2636
12	0.01882	0.01680	1.2168	1.1994
15	0.03689	0.03362	1.1549	1.1275
20	0.07504	0.06898	1.0995	1.0663
25	0.1237	0.1125	1.0604	1.0308
30	0.1864	0.1673	1.0202	0.9974
35	0.2632	0.2341	0.9789	0.9630
40	0.3504	0.3104	0.9394	0.9293
45	0.4426	0.3922	0.9040	0.8981
50	0.5346	0.4755	0.8735	0.8705
60	0.7039	0.6347	0.8266	0.8263
70	0.8440	0.7735	0.7943	0.7948
80	0.9552	0.8888	0.7719	0.7723
90	1.0432	0.9832	0.7557	0.7558
100	1.1142	1.0603	0.7434	0.7434
120	1.2218	1.1766	0.7255	0.7257
150	1.3293	1.2886	0.7080	0.7091
200	1.4126	1.3764	0.6930	0.6948
300	1.3737	1.3665	0.6862	0.6869
400	1.3163	1.3059	0.6840	0.6849

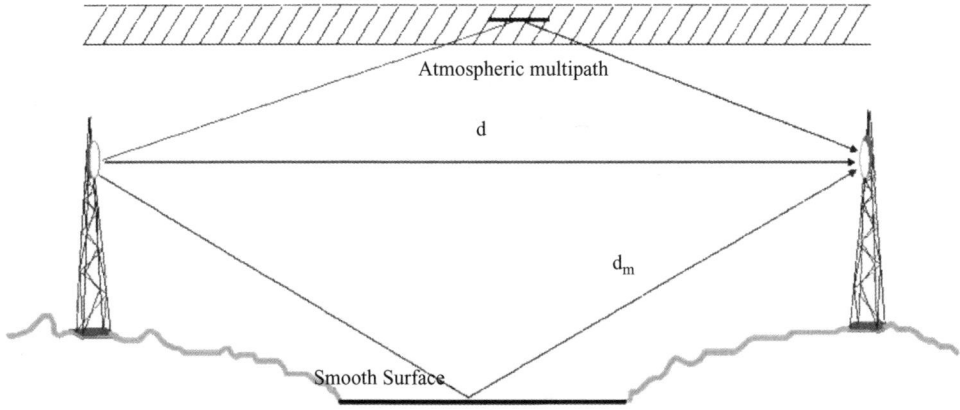

Figure 3.51 Multipath fading

signal. The received indirect signal consists of the signal reflected from the surface of the earth and atmospheric layers.

When the direct and indirect signals arrive at the receiving antenna with a half-wavelength phase difference, fading will take place. The severity of fading depends upon the amplitude and phase of the indirect signals. Thus, the probability of fading exceeding a given fade depth depends upon two factors:

(a) Amplitude of the indirect signals
(b) Percentage of time for which fading is present

Due to the interaction between the direct and indirect signals, the resulting signal will consist of constructive peaks and destructive troughs. There are two types of fading here, minimum and nonminimum delay phase fading. Assume that the direct signal level is unity and when the amplitude of the indirect signal is less than unity, e.g. b, then the 'notch' depth B (shown in Figure 3.52) would be given as

$$B = -20\log_{10}(1-b) \tag{3.50}$$

If the indirect signal has an amplitude greater than unity, then the amplitude of the peaks and troughs will be $b+1$ and $b-1$ respectively. The former is called minimal phase fading while the latter is called nonminimal phase fading. For a given notch depth and the frequency distance from it, there is a critical fade depth which, if exceeded, would result in a BER greater than the threshold level. In detailed link budget calculations, the term 'signature' is used to take in the effect of the above-mentioned phenomenon. Signatures are defined for both minimum and nonminimum phase fading. The signature is defined as the locus of notch depth required to produce a specific BER when the flat fade margin (FFM) and relative delay are fixed. The ability of the radio to withstand a notch or its slope within its bandwidth is defined as the area contained by the locus. The receiver signature is important when the bandwidth is higher (with increased capacity) and selective fading becomes important. However, the concept of selective fading becomes relevant at low frequencies, high capacity and over long hops.

The outage probability due to frequency selective fading can be calculated as

$$P_s = 2.15\eta \left(W_M \times 10^{-B_M/20} \frac{\tau_m^2}{|\tau_{r,M}|} + W_{NM} \times 10^{-B_{NM}/20} \frac{\tau_m^2}{|\tau_{r,NM}|} \right) \tag{3.51}$$

$$\tau_m = 0.7 \left(\frac{d}{50} \right)^{1.3} \text{ ns} \tag{3.52}$$

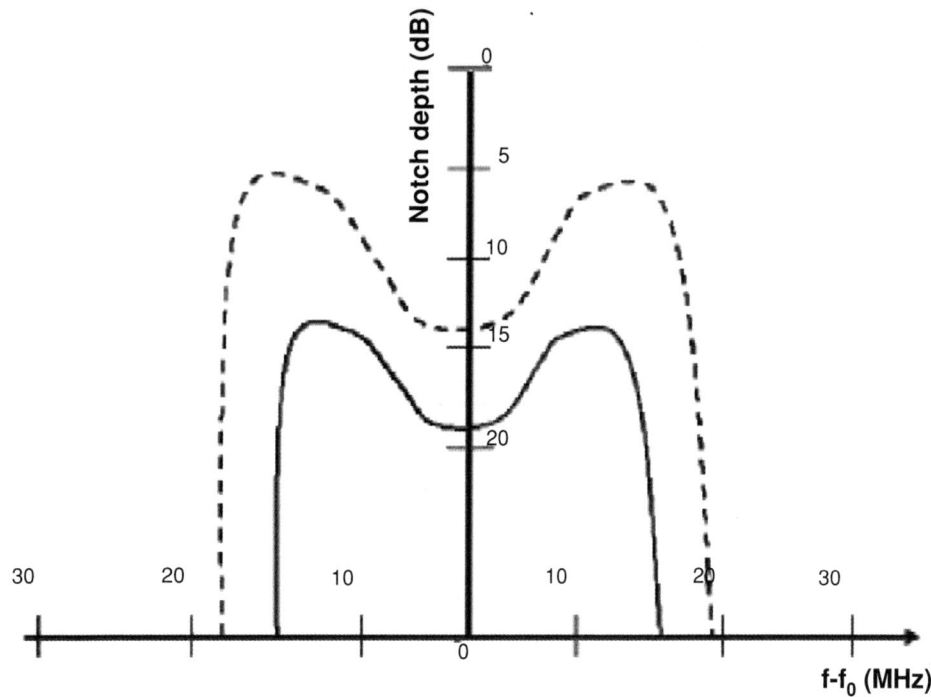

Figure 3.52 Signature results (laboratory simulated)

where

W_x = signature width (GHz)
B_x = signature depth (dB)
$\tau_{r,x}$ = reference delay (ns) used to obtain the signature, with x denoting either the minimum (M) or the nonminimum (NM) phase fading.

The amplitude and phase of the signal bening reflected from the surface is also dependent on the surface type. The smoother the surface, the more severe the effect of the indirect signal on the direct signal will be. The maximum damage is caused when the amplitude of the direct and the indirect signal is the same while they are in anti-phase. The conductivity and permittivity of the surface affect the amplitude of the reflected signal and depend on the polarisation and frequency of operation. The roughness factor of the surface determines the amount of signal that will be reflected from the surface. This is given as

$$g = 4\pi \left(\frac{S}{\lambda}\right) \sin \phi \qquad (3.53)$$

where

S = standard deviation of the surface height about a local mean (within the FFZ)
λ = wavelength of the signal
φ = grazing angle

The surfaces with $g < 0.3$ are considered to be smooth. The higher the values of g, the smoother the surface will be, thus resulting in maximum reflection of power. Hence, microwave signals passing over water bodies or a paddy field should be avoid0ed as they reflect the signals by almost 100 %.

Microwave Propagation

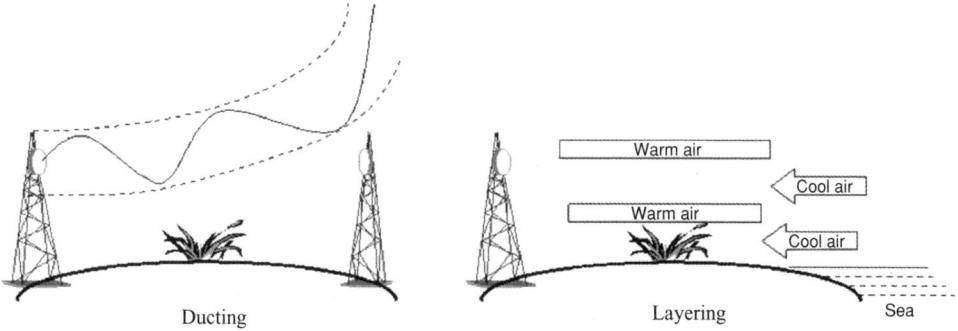

Figure 3.53 Ducting and layering phenomenon

Microwave engineers should note that the vertically polarised signals are less susceptible to the destructive interference from ground signals.

Layering

Layering is a part of the multipath phenomenon. As discussed at the beginning of this chapter that atmosphere consists of layers and it is these layers that help in the signal transmission from the transmitting to the receiving antennas. However, phenomena such as advection, frontal systems, radiation nights and subsidence can affect the refractivity lapse rate, resulting in (usually) degradation of the signal level.

Advection
This effect of temperature inversions is caused by the movement of the land based dry air over moist air over the sea by high pressure systems.

Frontal Systems
These systems are caused by changes in the signal levels observed due to the passage of warm (increase in signal level) or cold fronts (decrease in signal level).

Radiation Nights
When a cold night follows a warm day, the earth surface cools faster. This results in the air near the surface cooling faster than that above it, thereby causing temperature inversion.

Subsidence
In a high pressure system, dry air that descends becomes heated by compression and spreads over cool moist air, causing a temperature inversion.

The layering phenomenon is usually observed in an area where links are either near the large water bodies, e.g. the sea, and/or in desert areas. In the valley regions, the drainage of the cold air from the higher slopes can create multiple layers, and in the coastal areas, land sea breezes and trade winds can cause layers. Sometimes, this layer phenomenon also leads to ducting, wherein the signal becomes trapped between two layers (shown in Figure 3.53). Ducting occurs when the refractivity gradient of the atmosphere becomes less than $-157\,N$ units/km. When the refractivity gradient of the atmosphere is equal to that of $-157\,N$ units/km, the signal will trace a path similar to the curvature of the earth. However, when the refractivity gradient becomes less than this, the signal becomes trapped in the atmospheric layers, causing ducting. Fronts, subsidence, radiations nights, etc., cause the formation of fronts. However, the type duct, its thickness, height and layers within also have an impact on the received signal.

3.5.3 Overcoming Fading

As mentioned above, the fading phenomenon usually causes a reduction in the signal strength. Thus, it becomes quite important for microwave engineers to introduce the aspects that counter fading during the design phase itself. It is quite a tedious process to optimise a microwave link, especially when a link is installed.

k-fading

Once the link is 'up and running' it is very difficult to change the height, position, etc., of the microwave tower without losing the traffic (and revenues) for the period of time that the changes are taking place. With weather being the most unpredictable aspect of microwave link planning, the best a microwave engineer could do is during the link designing phase itself. Precautions against k-fading can include getting the required clearances by inputting various values of the k-factor. Also, if a choice is to be made between taller and smaller towers, then the former should be chosen. This is because a taller tower will give more flexibility of space (to move antennas) and protection against phenomena like ducting, etc. Based on the ITU recommendation and researches, there are design principles that a microwave engineer can follow to choose antenna heights.

Links under 15 km (climatic conditions: cold and temperate)

- The first fresnel zone is free of obstacles for $k = 1.33$.
- An additional check may be done with small k (e.g. $k = 0.5$) to get at least zero clearance.

Links between 15 km and 30 km (climatic conditions: cold and temperate)

- The first fresnel zone is free of obstacles for $k = 1.33$.
- An additional check may be done with small k (e.g. $k = 0.7$) to get at least zero clearance.

Links under 15 km (climatic conditions: tropic, hot and dry climates), where the following rule can be used in addition to the above:

- At least zero clearance should be obtained for $k = 0.3.6$-2

Links above 15 km (climatic conditions: tropic, hot and dry climates), where the following rule can be used in addition to the above:

- At least zero clearance should be obtained for $k = 0.5$.

While choosing antenna heights, it is better to choose higher antenna heights than lower ones. The reason is that once the tower is installed, it is easier to lower the antenna. On the contrary, if the tower is chosen to be lower and later a need to increase the antenna height arises (due to, for example, ducting), the tower would need to be taken off and a new one installed, which would result in a loss of traffic, time and the huge costs associated with replacing the tower.

Rain

Rain plays an important part in link budget calculations above frequencies of 10 GHz. At these frequencies the link lengths start to become shorter as compared to ones at lower frequencies, such as 2 GHz. There are some rules of thumb for protection against rain:

- The microwave loop should have a diameter more than 3 km.
- The angle of separation between two links should be more than $60°$.
- Links that belong to the same loop should not cross each other.

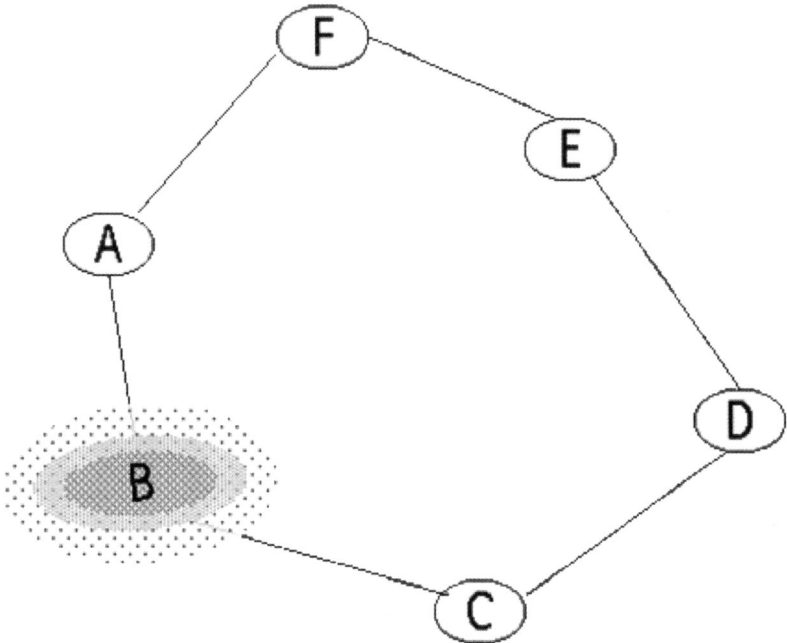

Figure 3.54 Protection against rain

Now consider a loop of microwave links. When heavy rainfall takes place near site B shown in Figure 3.54 and the angle between line BA and link BC is very small, fading due to rain may take place on both the links at the same time, resulting in loss of traffic generated at site B. However, if the links are widely separated, the probability of the traffic loss becomes less. The same is true of the loop having a wider area and links from the same loop crossing paths. Microwave planning engineers should remember that the rules of thumb given above only reduce the probability of fading, but much also depends on the atmospheric conditions.

Multipath

Diversity techniques can be used to prevent multipath fading. Space, angle and frequency diversity are the three main types of diversity techniques. Frequency diversity is recommended in cases where frequency selective fading is expected. However, as the spectrum is usually a problem, this technique is not used very much. Space diversity used in cases such as signals reflected from water/reflection surfaces are causing degradation of received power. In such cases, angle diversity, i.e. tilting of antennas, is also used. Angle diversity is generally used in conjunction with space diversity, but it can also be used alone in cases where microwave towers do not have space for a diversity antenna. The improvement due to space diversity can be calculated as

$$I = \left[1 - \exp\left(-3.34 \times 10^{-4} S^{0.87} f^{-0.12} d^{0.48} P_0^{-1.04}\right)\right] \times 10^{(A-V)/10} \qquad (3.54)^*$$

where

$$P_0 = p_w \times 10^{A/10}/100 \qquad (3.55)^*$$

$$V = |G_1 - G_2| \qquad (3.56)^*$$

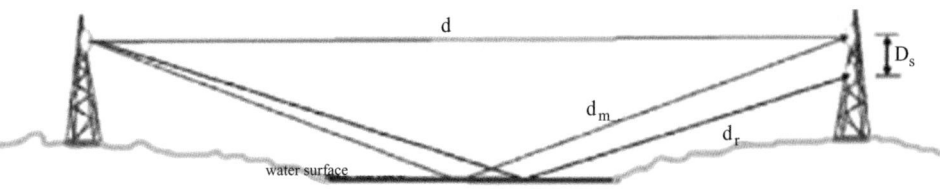

Figure 3.55 Space diversity

and

A	= fade depth (dB) for the unprotected path
p_w	= percentage of time the fade depth A is exceeded in the average worst month
P_0	= fading occurrence factor
S	= vertical separation (centre-to-centre) of receiving antennas (m)
f	= frequency (GHz)
d	= path length (km)
G_1, G_2	= gains of the two antennas (dBi)

However, microwave engineers should try to avoid a path over the water surface. If unavoidable, it is best to do a reflection calculation to find the point of reflection on the surface of the earth. During the design phase itself if the point of reflection falls onto a water body, then the transmission and receiving antenna (and tower) should be placed at the point where the reflection point falls over the water (shown in Figure 3.55). However, if the water surface is totally unavoidable at the point of reflection and the link performance is degrading, space diversity should be used. In this method, at the receiving end, another antenna is placed at a distance d from the main antenna. This is called the diversity antenna. The distance at which the antenna is placed is such that d_m–d_r is an odd multiple of the half-wavelength. On the receiving side, a combiner is used that combines the signals received from the two antennas, thereby increasing the signal level. The outage probability of such a system can be calculated as in the following steps:

1. Calculate the multipath activity factor η:

$$\eta = 1 - e^{-0.2(P_0)^{0.75}} \qquad (3.57)^*$$

where $P_0 = p_w/100$ is the multipath occurrence factor corresponding to the percentage of the time p_w (%) of exceeding $A = 0$ dB in the average worst month, as calculated in Equation (3.39).

2. Calculate the nonselective outage:

$$P_{ns} = p_w/100 \qquad (3.58)^*$$

3. Calculate the square of the nonselective correlation coefficient k_{ns} from

$$k_{ns}^2 = 1 - \frac{I_{ns} P_{uns}}{\eta} \qquad (3.59)^*$$

where the improvement I_{ns} can be evaluated from Equation (3.54) for a fade depth A (dB) corresponding to the flat fade margin F (dB) and P_{ns} from the equation above.

4. Calculate the square of the selective correlation coefficient k_s from

$$k_s^2 = \begin{cases} 0.8238 & \text{for } r_w \leq 0.5 \\ 1 - 0.195(1 - r_w)^{0.109 - 0.13\log(1 - r_w)} & \text{for } 0.5 < r_w \leq 0.9628 \\ 1 - 0.3957(1 - r_w)^{0.5136} & \text{for } r_w > 0.9628 \end{cases} \qquad (3.60)^*$$

where the correlation coefficient r_w of the relative amplitudes is given by

$$r_w = \begin{cases} 1 - 0.9746\left(1 - k_{ns}^2\right)^{2.170} & \text{for } k_{ns}^2 \leq 0.26 \\ 1 - 0.6921\left(1 - k_{ns}^2\right)^{1.034} & \text{for } k_{ns}^2 < 0.26 \end{cases} \qquad (3.61)^*$$

5. Calculate the nonselective outage probability P_{dns} from

$$P_{dns} = \frac{P_{ns}}{I_{ns}} \qquad (3.62)^*$$

where P_{ns} is the nonprotected outage given by the equation above.

6. Calculate the selective outage probability P_{ds} from

$$P_{ds} = \frac{P_s^2}{\eta\left(1 - k_s^2\right)} \qquad (3.63)^*$$

where P_s is the outage due to selective fading as calculated in Equation (3.51).

7. Calculate the total outage probability P_d as follows:

$$P_d = \left(P_{ds}^{0.75} + P_{dns}^{0.75}\right)^{1.33} \qquad (3.64)^*$$

The process to calculate the probability of outage in the frequency diversity scheme of things is similar to the above except that the improvement factor due to frequency diversity is considered. The improvement due to frequency diversity is calculated as

$$I_{ns} = \frac{0.8}{fd}\left(\frac{\Delta f}{f}\right) 10^{F/10} \qquad (3.65)^*$$

where

Δf = frequency separation (GHz)
f = carrier frequency (GHz)
F = flat fade margin

For angle diversity, the process is as follows:

1. Estimate the average angle of arrival μ_θ from

$$\mu_\theta = 2.89 \times 10^{-5} G_m d \quad \text{deg} \qquad (3.66)^*$$

where G_m is the average value of the refractivity gradient (N unit/km). When a strong ground reflection is clearly present, μ_θ can be estimated from the angle of arrival of the reflected ray in standard propagation conditions.

2. Calculate the nonselective reduction parameter r from

$$r = \begin{cases} 0.113 \, sin\,[150\,(\delta/\Omega) + 30] + 0.963 & \text{for } q > 1 \\ q & \text{for } q \leq 1 \end{cases} \qquad (3.67)^*$$

where

$$q = 2505 \times 0.0437^{(\delta/\Omega)} \times 0.593^{(\varepsilon/\delta)} \qquad (3.68)^*$$

and
- δ = angular separation between the two patterns
- ε = elevation angle of the upper antenna (positive towards the ground)
- Ω = half-power beamwidth of the antenna patterns

3. Calculate the nonselective correlation parameter Q_0 from

$$Q_0 = r \left(0.9399^{\mu_\theta} \times 10^{-24.58\mu_\theta^2} \right) \left(2.469^{1.879(\delta/\Omega)} \times 3.615^{(\delta/\Omega)^{1.978}(\varepsilon/\delta)} \times 4.601^{(\delta/\Omega)^{2.152}(\varepsilon/\delta)^2} \right) \tag{3.69}*$$

4. Calculate the multipath activity parameter η as calculated above in Equation (3.57)
5. Calculate the nonselective outage probability from

$$P_{\text{dns}} = \eta Q_0 \times 10^{-F/6.6} \tag{3.70}*$$

6. Calculate the square of the selective correlation coefficient k_s from

$$k_s^2 = 1 - \left(0.0763 \times 0.694^{\mu_\theta} \times 10^{23.3\mu_\theta^2} \right) \delta \left(0.211 - 0.188\mu_\theta - 0.638\mu_\theta^2 \right)^\Omega \tag{3.71}*$$

7. The selective outage probability P_{ds} is found from

$$P_{\text{ds}} = \frac{P_s^2}{\eta \left(1 - k_s^2 \right)} \tag{3.72}*$$

where P_s is the nonprotected outage as calculated in Equation (3.51).

8. The total outage probability P_d is calculated as

$$P_d = \left(P_{\text{ds}}^{0.75} + P_{\text{dns}}^{0.75} \right)^{1.33} \tag{3.73}*$$

Other hydrometers such as snow, hail, cloud and fog droplets have some impact on the performance of microwave links. However, attenuation due to moist snow is less than the equivalent rainfall, while wet snow has an attenuation approximately double the equivalent rainfall and watery snow has an attenuation triple the equivalent rainfall. Hail, which is a mixture of water, air and ice, has an impact at frequency of 34 GHz, with losses going as high as 6 dB/km. Cloud and fog droplets with a radius less than 0.1 mm attenuate the signal by 0.3 dB/km at a frequency of 40 GHz.

3.6 Interface Planning

Once the quality issues have been addressed in link planning, capacity planning is the next step. In the GSM networks, A_{bis} and A_{ter} interface planning helps to define the equipment required, especially microwave radios. Through this planning/dimensioning exercise, the number of 2 Mbps links can be determined, which further helps to define the type of radios.

To dimension the interfaces, the following aspects should be kept in mind:

- no blocking between the BTS and BSC;
- no blocking between the MSC and transcoders;
- minimal blocking of about 0.1 % is permitted on the A_{ter} interface.

Based on factors such as capacity, coverage and quality, radio planners need to determine calculations such as the number of TRXs required per cell. Although the number of TRXs determines the 2 Mbps links that are required from the BTS to the BSC, topology and protection also play an important role in capacity planning.

3.6.1 A_{bis} Planning

One A_{bis} contains 32 timeslots of 64 kbps capacity. Each timeslot (TS) has four traffic channels of 16 kbps capacity. TS0 is used for management. The remaining 31 timeslots can be freely to carry traffic and signalling channels (allocation can vary, as described later in the chapter). Generally TS31 is used for carrying management bits related to synchronisation, loop control, etc. This means that 30 TSs are used for user traffic and associated signalling. Each TRX needs two TSs for traffic and 0.25 TSs for signalling (if the signalling used is 16 kbps). Using this principle, it can safely be assumed that for 12 TRXs, 24 TSs are required, and depending upon the topology and the number of BTSs connected on 1E1 (1 PCM), the remaining six TSs can be used for BCF and TRX signalling.

Assume that a three sectored BTS-A site of $4 + 4 + 4$ capacity is connected to a BSC directly. This means that the 1E1 link would be enough to carry the traffic. The traffic can be carried by leased lines or a 2E1 PDH radio. However, if the same BTS-A site is followed by two more base stations (BTS-B and BTS-C) of similar capacity, then the traffic running on the BTS-BSC link would be three times the original, i.e. 3E1, thus increasing the radio requirements to that of a 4E1 PDH radio. However, if the BTS-B and BTS-C sites are of capacities $2 + 2 + 2$ each, then the number of E1s required would reduce to 2E1. This is possible by 'grooming'. This means that the traffic channels are organised so as to fill the 2 Mbit frame more efficiently. Once A_{bis} is dimensioned, the link between the BSC and TCSM (transcoder submultipleyer)/MSC needs to be planned. The following steps are performed for dimensioning the A_{ter} interface:

- Total traffic (in Erlang) is offered through the A_{bis} interface of a BSC.
- Define the blocking probability on the A_{ter} interface.
- Apply the Erlang-B formula.
- Calculate the number of traffic channels, Y.
- Find the number of traffic channels that are carried by A_{ter}, e.g. 120.
- Capacity required (in E1) = $Y/120$.
- Actual capacity planned = capacity required + spare capacity.

For interface planning in EDGE networks, A_{bis} interface planning is similar to that of GSM networks. However, if the feature of Dynamic A_{bis}* is used then A_{bis} interface planning in EDGE transmission networks is much more complicated than in GSM transmission networks.

3.6.2 Dynamic A_{bis}

Under this scheme, some of the timeslots on the A_{bis} are reserved specifically for the packet data users. The planning is based on the coding schemes that are used in the EGPRS. In EGPRS, there are nine coding schemes used, each having its own data rate, as shown the Table 3.14. Based on the bit rates, the number of TCH required on the A_{bis} to carry the traffic is calculated. These traffic channels are basically of two kinds: master and slave. First the TCH is the master and the rest are slaves, e.g. in the case of MCS-1, as there is only one TCH, it is a master; however, for MCS-7, first the TCH is master while the remaining three are slaves.

In GSM A_{bis}, almost 96 TCHs (i.e. subscribers) can be 'talking' at the same time. Due to the packet data requirements, the numbers of simultaneous users decreases in the EGPRS system and the A_{bis} capacity becomes a bottleneck. Assume that there is one site that is EDGE capable. The capacity of the site is $4 + 4 + 4$. In the case of the GSM network, the A_{bis} will be able to carry 96 users. However, if the users need to access data services and have a requirement of the highest data rate that is possible. This means that there are 24 TSs present for traffic (with the assumption that TS0 and TS31 are used for management and TS25-TS30 are used for TRX and BTS signalling). If the first user needs MCS-9, then 1.25 TS is

---

* Note that this is a vendor specific feature.

Table 3.14 TCH required on A_{bis} for EGPRS coding schemes

Coding scheme	Bit rate [bps]	TCH required on A_{bis}
MCS-1	8 800	1
MCS-2	11 200	2
MCS-3	14 800	2
MCS-4	17 600	2
MCS-5	22 400	2
MCS-6	29 600	3
MCS-7	44 800	4
MCS-8	54 400	5
MCS-9	59 200	5

required/used by him/her (refer to Table 3.14). Similarly, the second user will take another 1.25 TS. Thus, 24/1.25 will mean that only 19 users will be logged on at the same time, which is substantially lower than that of the GSM network. Not only that, these users might not be using the capacity optimally, i.e. some of them might be reading only the downloaded email. Taking this into account, it is beneficial to use a Dynamic A_{bis}. By using the Dynamic A_{bis}, the capacity is fully utilised. This means that when the user is reading the downloaded email, the TSs that are not being used by the user can be used by another user. Hence, by using the Dynamic A_{bis} concept, the probability of the A_{bis} capacity becoming a bottleneck is reduced; hence utilisation is optimised. Dynamic A_{bis} planning (DAP) is implemented as a software feature. Some aspects that may be helpful in dimensioning the dynamic A_{bis} are:

- *BSS achievable throughput.* An A_{bis} pool is a shared resource that can limit the throughput per RTSL. Hence RNP and TNP together should dimension together to achieve a BSS throughput.
- *RTSL capacity (k).* This is the amount of data transmitted through a fully utilised RTSL. This is dependent mainly on the average C/I distribution.
- *Average user throughput (N_u).* This is the average throughput perceived by the user. Therefore, if N_u TSs are allocated, then the average throughput can be approx imately (kN_u).
- *Reduction factor (RF).* This is the measure of the congestion of the EGPRS network. Its value lies between 0 and 1. In a nearly saturated network, it is near to 0 and in less saturated network, it is closer to 1. The RF is based on the number of TS N_s (total system capacity), TS N_u and resource utilisation (U).
- *Resource utilization.* This is the ratio between the PS traffic intensity and the total system capacity N_s.
- *RTSL mean throughput.* This is the product of the RTSL capacity and its utilization.
- *A_{bis} data rate reduction factor.* The reduction of the mean throughput due to a lack of the transmission resources (i.e. it is the effect of sharing a few transmission resources to transport data coming from a higher number of RTSL or system capacity, N_s).

The probability of successful trials for each call can be theoretically calculated using the following binomial distribution:

$$B(n, N, p) = \sum_{x=N+1}^{n} P(x) \qquad (3.74)$$

where

N = number of timeslots available in the pool
p = channel utilisation of the EDGE channels in the air interface

n = number of traffic channels used in the air interface
B = blocking probability of the DAP

The formula basically tries to calculate how much the blocking on the A_{bis} interface would be due to the (lack of) timeslots available in the pool. Obviously, a higher number of timeslots, i.e. a larger pool size, would mean less blocking, i.e. a lower probability of pool overflow.

Inputs from the RNP are:

- RTSL utilisation
- RTSL capacity
- EGPRS territory (dedicated + default timeslots)

While inputs from the TNP are:

- A_{bis} data rate reduction factors (usually agreed with the RF and should be low, e.g. 0.1 %)
- Available TSs that can be used for the pool.

Sharing of DAP

Consider the following example where a DAP of three TSs is created. Each of these timeslots carry data traffic of coding schemes MCS-3, MCS-9, MCS-7, MCS-5 and MCS-6. The required number of slave A_{bis} channels required is 11 (summation of the A_{bis} slave channels required by the individual coding schemes). The DAP is designed with four timeslots, which mean $4 \times 4 = 16$ A_{bis} channels. As the requirement is for 11 A_{bis} (slave) channels, all the requests would be accommodated without any problems.

MCS-3	MCS-5	MCS-6	MCS-6
MCS-7	MCS-7	MCS-7	
MCS-9	MCS-9	MCS-9	MCS-9

Now consider the following example. There are six calls MCS-3, MCS-9, MCS-7, MCS-5 and $2 \times$ MCS-6. This means that the total number of requests for the A_{bis} slave channels is 13. The pool size is three timeslots, i.e. $3 \times 4 = 12$ A_{bis} channels. This means that the resources in the pool are not sufficient to handle the calls as they are. Thus, there will be some reductions in call rates in order to make sure that all calls go through but with a lesser data rate than requested. Thus, in this case, MCS-9 and MCS-8 calls reduce to the MCS-7 coding scheme, while the MCS-7 call is reduced to the MCS-6. Thus, by reducing the coding schemes of the calls, i.e. by reducing the requested throughput, the requested calls are able to go through.

MCS-3	MCS-5	MCS-6	MCS-6
MCS-7	MCS-7	**MCS-6**	MCS-6
MCS-9	MCS-9	MCS-9	MCS-9

Observation

From the above two cases, it can be seen that the size of the DAP pool plays a very important role. If the DAP pool is not sufficiently large, then the coding schemes of the calls would be downgraded. This means that the effective throughput goes down. Thus, when designing the DAP, the average throughput

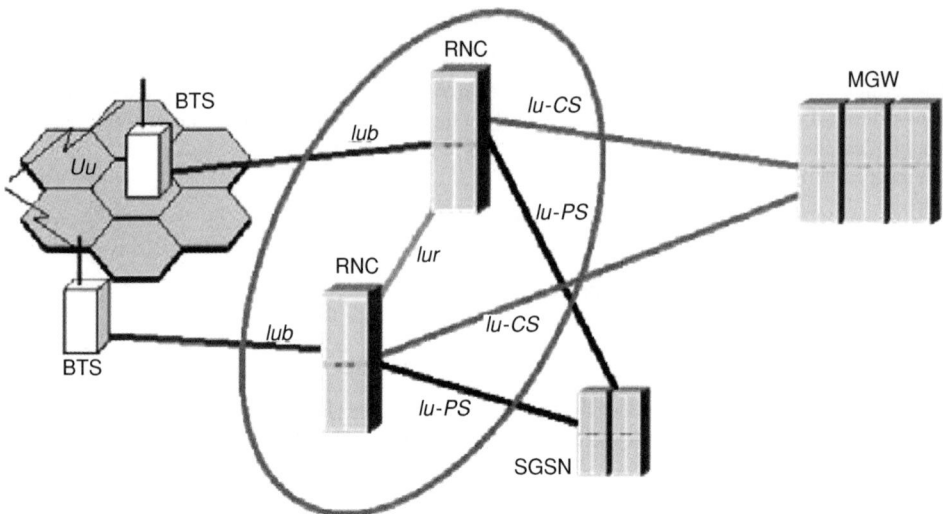

Figure 3.56 Transmission interfaces in WCDMA

reduction basically signifies that the coding scheme of the calls would be reduced, thereby reducing the effective throughput.

3.6.3 Interface Planning in the UMTS Access Transmission Network

Interface dimensioning in WCDMA networks is a key aspect of transmission network planning. It is much more complicated than the interface dimensioning that was seen in the GSM/EDGE networks. The following interfaces are needed to be dimensioned by access engineers (see Figure 3.56):

- I_{ub}: interface between the BTS and RNC
- I_{u-cs}: interface between the RNC and MGW (media gateway)
- I_{u-ps}: interface between the RNC and SGSN
- I_{ur}: interface between the RNC and RNC

In dimensioning, the capacity required is calculated in order to decide what equipment is required. Capacity calculations in WCDMA networks requires an understanding of some concepts because, apart from the user traffic, some overheads are needed (which was not the case in GSM/EDGE networks). Each of these interfaces and related dimensioning will now be discussed.

Protocol Stack

In Figures 3.57(a) and (b) the user plane protocol stacks for several interfaces are represented (where FP is the Frame Protocol, PH the physical layer, UP the user plane, SNAP the subnetwork Access Protocol, PDCP the Packet Data Convergence Protocol and UDP the user Datagram Protocol). The part that involves transmission is in the bottom part of I_{ub} and I_u interfaces. The common layers in all the interfaces are the physical layer and the ATM. The physical layer is used as the generic name in the figures since the practical solutions for a physical layer are many. The physical layer can consist of wire (e.g. coaxial cable, optical fibre) and wireless media (e.g. microwave link) and on top of them different kinds of digital hierarchies can be used (e.g. PDH, SONET, SDH), which accommodate different carriers (e.g. E1, T1,

Interface Planning

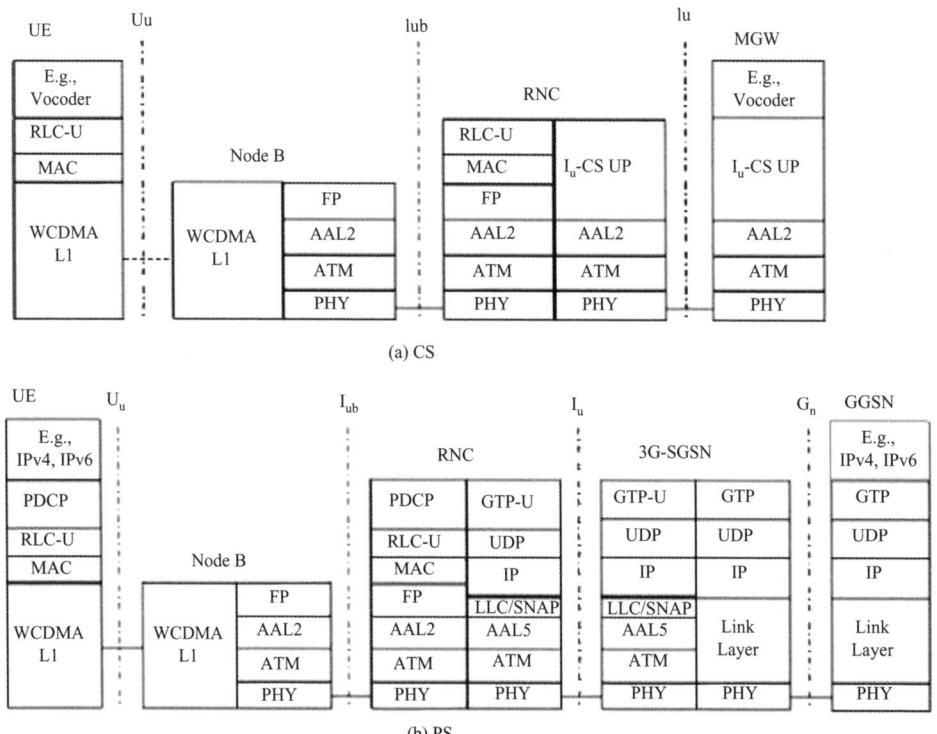

Figure 3.57 Protocol stack of user plane traffic

JT1 (JT is a japanese standared the equivalent of ANSI), STM-1). In the case of the wireless physical layer such as the microwave radio link, the physical layer will also have radio framing for the digital baseband bit stream.

The ATM is chosen as the layer 2 protocol for WCDMA transmission to pack the data into cell format and to provide the possibility of effectively gaining from differentiating the services according to their different kinds of service level and bandwidth requirements. AAL2 is chosen as the ATM adaptation layer in the I_{ub} interface for all user traffic to keep the data units small in narrowband connections, but for the I_u interface the choices are different for I_{u-cs} and I_{u-ps}. I_{u-cs} uses AAL2 as well since it fits well with its small packet size for use in real-time connections like traditional voice. In I_{u-ps}, however, the services are different and less sensitive to delays, so AAL5, which has a larger packet size, is chosen as the adaptation layer, which means, for example, less overheads.

In addition to I_{u-ps}, IP, UDP and GTP-U protocols are also used for layers 3 and 4 to provide as close as possible the full flexibility and gains of packet switched data communication networks to the WCDMA transmission network.

Signalling in the WCDMA Access Network

There are three main categories of signalling that can be distinguished in the WCDMA transmission network. These are signalling for transport, signalling for the WCDMA system and signalling between the radio network controller and the user element.

Transport Signalling

Signalling for transport means signalling to establish and release AAL2 connections for user plane connections. The signalling itself is done using AAL5 connections. This signalling is needed everywhere where AAL2 connections are used, i.e. in I_{ub}, I_{u-cs} and I_{ur} interfaces. AAL2 signalling (Access link control application protocol, or ALCAP) bandwidth dimensioning is directly proportional to the amount of connection establishments and releases, which is then further directly proportional to the average length of service usage, network bandwidth reallocation capability and policy, mobility of the users and soft handover policy.

The average use of traditional voice services is still reasonably easy to predict and project against changing call plans, but forecasting the user behaviour with new services provided by the WCDMA network is challenging. It is even more difficult bearing in mind that it might depend on adjustable or fixed system parameters when there is an automatic release time for AAL2 connections of different services. In other words, if that is the case then user behaviour for the average idle time (e.g. reading time) per service should be forecasted with some accuracy. Quick release will keep the user bandwidth operating more efficient by, but more signalling will be generated.

Network bandwidth reallocations either are or are not possible in the transmission network, but if they are possible then rules concerning how the bandwidth is being increased and/or reduced are likely to vary. In some networks these might be adjustable rules while in some other networks they might be fixed rules set by the vendor. If bandwidth reallocations are not possible or are disabled then obviously no transport signalling traffic is generated, but having these rules enabled, on the other hand, can provide great benefits on the user perception side by keeping the network accessible and providing the best possible bandwidth to the end user at any given moment. Each bandwidth reallocation, however, generates additional transport signalling.

Users moving around the coverage area will need new AAL2 connections during their service usage when they are moving from one base station coverage area to another. Therefore the more mobile users there are the more AAL2 signalling is required in order to establish new connections and release others. The mobility of the users between base stations also depends somewhat on how the radio access planning is done, i.e. how the base stations and their antenna beams/sectors are planned along the most mobile areas such as highways. A soft handover policy is just a special case of handover that is not necessarily due to mobility but to strengthening the signal by combining several radio access connections between base stations and the mobile user element. However, the number of simultaneous soft handover connections might be restricted, again with either adjustable or fixed parameters, and accordingly the created transport signalling amount will differ. Transport signalling is needed more when users are mobile in an area where the base station sectors are narrow compared to the direction of the movement and where the soft handover connections are not limited by any additional parameterisation. On the other hand, a stationary end user near a base station in a network where the soft handovers are disabled will not generate any AAL2 signalling traffic.

To summarise the paragraphs above, making a statement about how much transport signalling capacity is required for a WCDMA base station is very hard. A safe value can range practically from 20 kbps (single carrier rural base station) to 200 kbps (urban high capacity base station) and even up to 400 kbps (multi carrier urban base station) per base station according to simulations depending on the above conditions. For I_{ur} and I_{u-cs} interfaces the loading is not likely to go so high since one major load contributor, the mobility of people, has only a small effect. Also, soft handovers have a small effect and bandwidth reallocations are likely to be very rare. It all this information is put together transport signalling capacities of 100 kbps for each 2 Mbps in I_{ur} and 50 kbps for each 2 Mbps in I_{u-cs} should be reasonable.

WCDMA System Signalling

Signalling for the WCDMA system means signalling that is needed by the WCDMA network for communication between nodes (i.e. base stations, RNCs, SGSN and MSC). In the I_{ub} interface this signalling uses the Node B application part (NBAP) Signalling Protocol, in the I_{ur} interface between RNC signalling it uses Radio Network System Application Part (RNSAP) Signalling Protocol and in I_u interface signalling it uses the Radio Access Network Application Part (RANAP) Signalling Protocol.

NBAP signalling is used for procedures of logical operation and maintenance of WCDMA base stations. This includes signalling related to cell configuration, controlling of random access, forward access and paging channels, initialising measurements in the cell or base station and reporting the results and finally managing faults. The major difference in dimensioning the NBAP signalling channel is made by reporting the frequency of cell and/or base station measurements as well as the extent of those measurements. The needed signalling bandwidth is larger if there are heavy measurement reports to be frequently signalled to the RNC.

NBAP signalling is also used for adding, releasing and reconfiguring radio links, controlling dedicated and shared channels, controlling softer handover combining in base stations and initialising and reporting radio link measurement as well as managing radio link faults. Dimensions of the NBAP signalling channel depend on radio network parameters and radio access conditions, which affect the number of radio link additions, releases and reconfigurations per time unit. Controlling softer handover combining can also create significant signalling traffic into the NBAP signalling channel in certain conditions. Finally, reporting the measurements of radio links can require significant signalling bandwidth if done extensively.

NBAP capacity requirement after all these considerations is likely to be slightly more than for transport signalling. The final capacity estimate again depends on so many variables and parameters that giving one number or even a simple formula is impossible, but the allocated capacity should be according to simulations from 40 kbps (single carrier rural base station) to 250 kbps (urban high capacity base station) and even up to 500 kbps (multicarrier urban base station), depending heavily on the capacity of the base station.

RNSAP signalling can be used for signalling information about a user moving across the RNC boundaries, organising the establishment of a user connection between RNCs for soft handover and eventually releasing such a connection as well as transferring cell and base station information between RNCs. RNSAP signalling channel dimensions therefore heavily depend on how the RNC boundaries are planned. The need for RNSAP signalling between two RNCs that share a boundary where the highway goes across the boundary is obviously different compared to the case where the RNCs have a boundary in a mountainous, uninhabited area. Practically it makes sense to dimension RNSAP signalling according to user plane dimensioning and user plane dimensioning makes sense when granularity that is available for the physical layer is taken into account. As with transport signalling, if all this information is put together an allocation of 100 kbps for each 2 Mbps for RNSAP signalling should give a solid and reasonable base line.

Finally, RANAP signalling is used to change the serving RNC and for a hard handover. Other uses of RANAP signalling are, for example, paging and location reporting. Following the earlier base line of 50 kbps for each 2 Mbps should be safe but reasonable.

Signalling between the Radio Network Controller and the User Element

This signalling consists of some or all of the logical control channels, i.e. the broadcast control channel (BCCH), paging control channel (PCCH), dedicated control channel (DCCH) and common control channel (CCCH). These channels are established within the transport network user plane and are therefore within the acceptance/rejection decision scope of the call admission control. It is possible to judge the needed bandwidth for these signalling channels simply by simulating the results of the call admission control with real-life channel characteristics. Conversely it is of no significance to guess the needed bandwidth without knowing the algorithm used for the call admission control and how the algorithm sees these channels.

Interface Dimensioning

I_{ub} *Interface*

This interface connects the BTS and RNC. It carries CS voice and data and PS data traffic. Interaction between various transmission/transport layers is shown in Figure 3.58. As seen in the previous chapter,

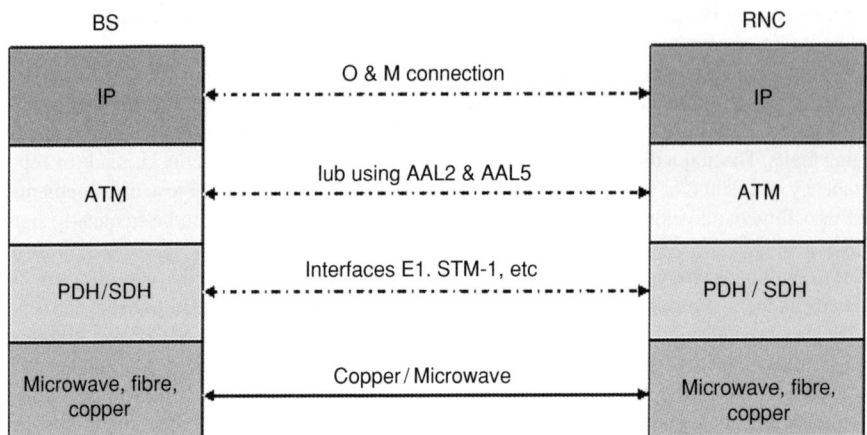

Figure 3.58 I_{ub} transport and transmission layers

various kinds of services used by the subscriber are categorised under four classes of conversational, streaming, interactive and background. Of these, conversational and streaming generate the CS traffic while the PS is generated by all four classes. Air interface data channels are carried over I_{ub} by the Frame Protocol (FP). These frame protocols are then segmented in AAL2 packets. These mini cells are then carried by ATM cells within the ATM VCCs, which are then mapped on to the physical interface, e.g. microwave/fibre optic/copper.

Frame protocols are applied to transfer user data and the necessary control information between the serving RNC (SRNC) and the BTS. They are characterised by a frame structure consisting of the payload, containing the data and a header/trailer with control information. Three different kinds of FPs are specified in the UTRAN, i.e. I_{ub} CCH, RAN DCH and I_{ur} CCH. When data pass through each of the layers, an 'overhead' is added to it. These overheads increase the required capacity. Apart from the protocol stack overheads and frame protocol overheads, there are other overheads that should be used when doing interface dimensioning.

With actual traffic, signalling is required and usually a capacity between 1 % and 10 % is reserved. Also, as voice, CS data and PS data flow on the I_{ub}, signalling capacity requirements are the highest there.

In the previous chapter, the concept of soft handover was discussed. Soft handover increases the performance of the WCDMA network, but it is a feature that increases the required capacity by almost 40 %. This means that approximately 40 % of UEs are connected to two or more Node Bs at the same time. Furthermore, it can be assumed that 35 % are in a two-way SHO and 5 % in a three-way SHO. Consequently, on average every call occupies $0.6 \times 1 + 0.35 \times 2 + 0.05 \times 3 = 1.45$ physical channels.

Retransmission and buffering also increases capacity requirements. The system retransmits the data to the subscriber's UE because the first time transmission had errors. This feature is used for PS services but not for CS services as it does not make any sense. Some capacity is also utilised for buffering. Usually 10–15 % of the overheads are utilised for each of the two features.

I_u Interface

The I_u interface is of two types: one for CS and one for PS. The I_{u-cs} interface carries the circuit switch traffic (e.g. voice and real-time data) and connects the RNC and the MGW. The I_{u-ps} interface carries the packet switch traffic (e.g. non-real-time data) and connects the RNC with the SGSN. The blocking probability is quite low (e.g. ∼0.1 %). The overheads are quite similar to those of the I_{ub} interface. However, as the calls are terminated at the RNC, there are no overheads related to soft handovers. Also, the signalling is on of the lower side of the assumption mentioned above, ∼1%–2 %. However, the I_{u-ps}

Interface Planning

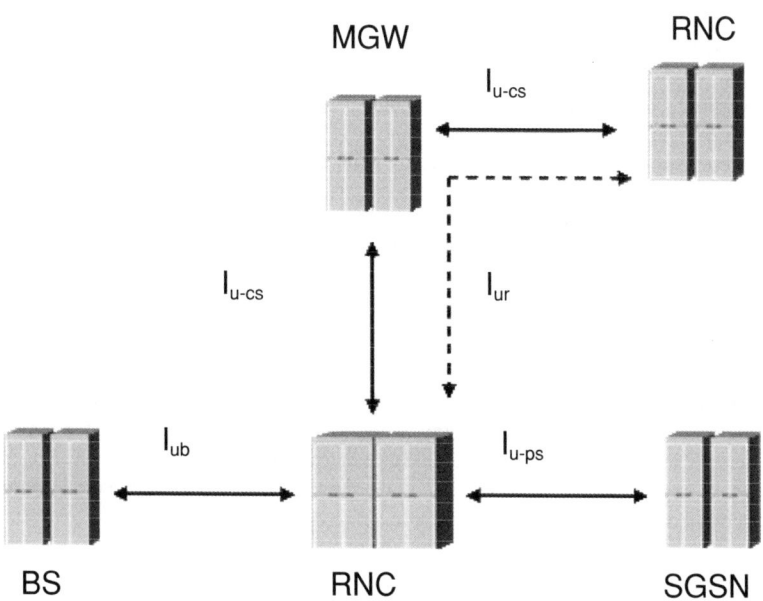

Figure 3.59 Interfaces in the UMTS transport network

interface overheads are quite different. The data are routed to the SGSN through the tunnelling protocol (GTP). Tunnelling of the user data requires the GTP protocol plus UDP and IP (encapsulated) running on AAL5 ATM transmission media. Protocols GPT and UDP contain headers that are 12 bytes and 8 bytes long respectively. The IP layer also adds a header of variable lengths with the minimum being 20 bytes long. Over the ATM adaptation layer 5, IP uses LLC encapsulation, leading to overheads of around 8 bytes per datagram. Then there are trailer bytes, which are added to the end of the packet data unit. Adding all these bytes together leads to substantial overheads on the $I_{u\text{-}ps}$ interface. Signalling on the $I_{u\text{-}ps}$ interface is similar to that of the $I_{u\text{-}cs}$ interface, i.e. ~ 1–2 %.

I_{ur} *Interface*

The I_{ur} interface connects two RNCs. The I_{ur} is not a physical interface. It is a logical interface that connects two RNCs via an MGW. For capacity calculation purposes, it is assumed that the I_{ur} carries traffic that is a small percentage of the total I_u traffic.

Example: Interface Planning

As mentioned before, access interface planning is the most important aspect of transmission network planning. Consider the following inputs for dimensioning interfaces (shown in Figure 3.59). The following assumptions are made at the beginning:

- U_u interface blocking = 2 %
  - This means that there is a modest chance left for any service request to be blocked by congestion in the air interface; 2 % is commonly seen as a value that is both acceptable by subscribers and reduces costs for the operator.
- SHO = 40 %
  - Soft handover. This consists of assuming 50 % mobile terminals without soft/softer handover, 15 % with softer handover, 25 % with soft handover, 5 % with softer–soft handover and 5 % with soft–soft

handover. Studying all the legs of these user element connections, it can be found that there are 40 % more connections in I_{ub} due to a soft handover.
- VAF = 50 %
  - Voice activity factor (VAAF). This is statistically easy to explain and the safe value is 50 %.
- Voice traffic = 12 mErl/sub, voice rate 12.2 kbps
  - This represents quite low subscriber activivity compared to current actual values in 2G networks, but this works here just as an example and could also be justified by lower voice service usage in the 3G network. (Voice services can be pushed to overlaying the 2G network.)
- CS data 10 mErl/subscrber, service class RT 64 kbps
- PS data 600 megabytes/month/user, service class NRT 128 kbps
- 200 BTSs (1 + 1 + 1 configuration)
- 1000 subscribers/BTS

The dimensioning is needed for I_{ub}, $I_{u\text{-}cs}$, $I_{u\text{-}ps}$ and I_{ur} interfaces. Dimensioning is done for both circuit switched and packet switched traffic. In the following, to give some starting point, the dimensioning is done on the generic level to find out how much end user data are loaded on the system. The results must be revised later with information about vendor specific system features that affect the dimensioning. Those features can have a big impact on the results.

I_{ub} Interface Dimensioning
For circuit switched traffic the inputs used are:

- Voice traffic: 12 mErl/subscriber
- CS data at 64 kbps: 10 mErl/subscriber
- 1000 subscribers/BTS
- Soft handover: 40 %
- Voice activity factor: 50 %
- I_{ub} signalling: 10 %*

Step 1. Calculate the voice and circuit switched data channels/BTS.

The first assumption is that due to user behaviour the use of the voice service and circuit switched data services are independent consumption decision events and as such are statistically handled separately. The second assumption is that even if in 3G no circuits are assigned for services, describing user connections as circuits and using Erlang B as the connectivity quality measure for them on average is close enough.

Voice traffic/BTS = 12 mErl/subscriber × 1000 subscribers/BTS = 12 Erl/BTS
Based on Erlang B, with 2 % blocking, the total number of channels = 19
Soft handover = 40 %
Hence, the total number of voice service user connections/legs over I_{ub} = 19 × 1.4 = 26.6 = ~27

Circuit switched data traffic/BTS = 10 mErl/subscriber × 1000 subscribers/BTS = 10 Erl/BTS
Based on Erlang B, with 2 % blocking, the total number of channels required = 17
Soft handover = 40 %
Hence, the total number of circuit switched data service user connections/legs over I_{ub} = 17 × 1.4 = 23.8 = ~24

---

* The I_{ub} signalling value is dependent upon the topology, system parameter values and vendor.

Interface Planning

Step 2. Special considerations for voice service.

Voice service user data rate in the active mode in the U_u interface = 12.2 kbps
Voice service user data rate including ATM overheads in the active mode = 18.9 kbps and in the silent mode = 4.5 kbps
Using voice activity factor 4= 50 %,
Hence, the average voice service user data rate = $18.9 \times 50\% + 4.5 \times 50\% = 11.7$ kbps

Step 3. Calculate the total circuit switched user data load in the I_{ub} interface.

As the user data rate for the circuit switched data service is 64 kbps, total circuit switched data in kbps/BTS = $64 \times 24 = 1536$ kbps/BTS
As the average voice service user data rate is 11.7 kbps, the total voice service user data rate in kbps/BTS = $27 \times 11.7 = 315.9$ kbps/BTS
Adding ATM overheads (23%) on circuit switched data services = $1536 \times 1.23 = 1889.2$ kbps/BTS
I_{ub} signalling for this example was assumed to be 10 %
Total circuit switched traffic in the I_{ub} interface = $(315.9 + 1889.2) + 10\% \; I_{ub}$ signalling = 2429.69 kbps

Step 4. Calculate the busy hour usage per user for packet switched traffic on the I_{ub} interface.

Packet switched data traffic = 600 MB/month/user

$$\text{BH usage per user} = \frac{600 \times 1024 \times 1024 \times 8 \text{ (bits / byte)} \times 10\%}{30 \times 3600} = 4.66 \, \text{kbps}$$

Step 5. Calculate the total packet switched traffic on the I_{ub} interface.

Assumptions:

- Retransmissions: 25 %
  - This is just given here as a value for this example. Estimating retransmissions is a network-specific task.
- Soft handover = 40 %
- ATM overhead = 30 %
- Packet switched data services do not have any delay requirements but are best effort

Traffic/user = $4.66 + 40\% + 25\% = 8.155$ kbps/user
Traffic/BTS = 8.15 kbps/user $\times 1000$ users = 8155 kbps
With ATM overheads
Traffic/BTS = $8155 + 30\% = 10\,601.5$ kbps/BTS
With assumed I_{ub} Signalling = 10 %
Total packet switched traffic in I_{ub}/BTS = $10\,601.5$ kbps + 10 % I_{ub} signalling = 11 661.65 kbps

Step 6. Calculate the $I_{u\text{-}cs}$ traffic.
Inputs:

- Circuit switched data service channel in $I_{u\text{-}cs}$ = 78.7 kbps (64 kbps + 23 % ATM overhead)
- Assume I_u signalling traffic = 1 %
- Assume $I_{u\text{-}cs}$ blocking = 0.1 %
- Assume 50 3G BTS/RNC

Total voice traffic = 1000 subscribers $\times 12$ mErl/subscriber $\times 50$ BTS = 6000 Erl
Total voice service user connections based on Erlang B = 549
Total Circuit switched data traffic = 1000 subscribers $\times 10$ mErl/subscriber $\times 50$ BTS = 500 Erl

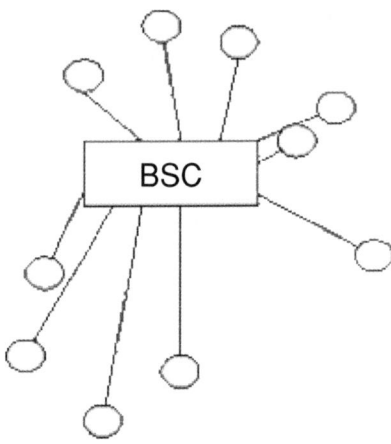

Figure 3.60 Point-to-point topology

Total Circuit switched data service user connections based on Erlang B = 458
Total $I_{u\text{-}cs}$ capacity = [11.7 kbps × 549 + (78.7 × 458)] × 1.01 = 42.89 Mbps

Step 7. Calculate $I_{u\text{-}ps}$ traffic.
Inputs:

- Assume again I_u signalling = 1 %
- Total protocol overheads = 24.2 %

Total packet switched user traffic/BTS = 4.66 kbps/subscriber ×1000 subscribers/BTS = 4660 kbps/BTS
Total packet switched user traffic/RNC = 50 BTS × 4660 kbps/BTS = 233 Mbps
Total PS traffic between RNC and SGSN = 233 + 1 % + 24.2 % = 292.28 Mbps

Step 8. Calculate the I_{ur} traffic.

The I_{ur} traffic can be assumed here to be, for example, 1–2 % of the total I_u traffic. This assumption depends heavily on the network plan and parameters, as explained later. In this case, the total I_u traffic is
$I_{u\text{-}cs} + I_{u\text{-}ps}$ = 42.9 Mbps + 292.3 Mbps = 335 Mbps
Thus, the total I_{ur} traffic = 2 % of 335.2 Mbps = 6.704 Mbps

3.7 Topology Planning

After individual link planning is done, the next step is to define the topology of the network. The topology is not only important from the perspective of defining the BOQ (bill of quantity) but also the performance of the network. Now consider the site location shown in Figure 3.60. When the sites are connected directly to the BSC, the topology is called 'point-to-point' topology. This topology is easy to plan and implement. However, the capacity is not fully utilised. If all the sites need only 0.5E1s, then the rest of the link capacity is wasted. Apart from this the protection is expensive as the usual protection that is used is equipment diversity. This topology is usually used for sites that cannot be incorporated into any of the topologies (discussed next).

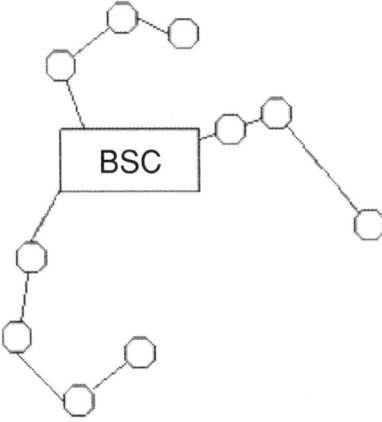

Figure 3.61 Chain topology

Chain topology (shown in Figure 3.61) is an extension of the point-to-point topology. In this topology, only some of the sites are connected directly to the BSC. These are the sites that are either very near to the BSC or do not have a possible connection to other neighbouring sites. The advantage of this topology is that the possibility of utilising the link capacity increases further. Sometimes transmission nodes are connected to the BSC using this technology. These transmission nodes basically act as digital multiplexers that groom the traffic of far-away locations and transfer it to the BSC. The major disadvantage of this topology is that if the first hop or site towards the BSC and/or the transmission node and/or the links is connected to the BSC, a large amount of traffic would be lost, resulting in loss of revenues.

In thus figure, if the last sites of the two chains are connected, it would become a loop (as shown in Figure 3.62). This means that additional equipment required to create a single link would be required. However, this would give an increased protection to the links. When loop topology is used, equipment redundancy is not used. Thus, the cost required to put up one additional link results in increased protection and increased availability.

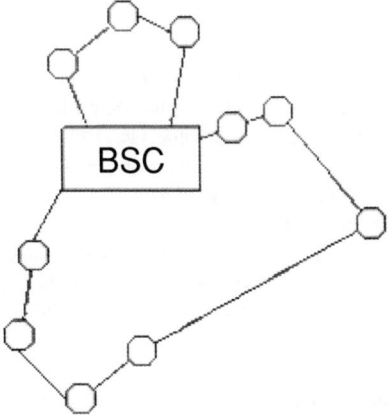

Figure 3.62 Loop topology

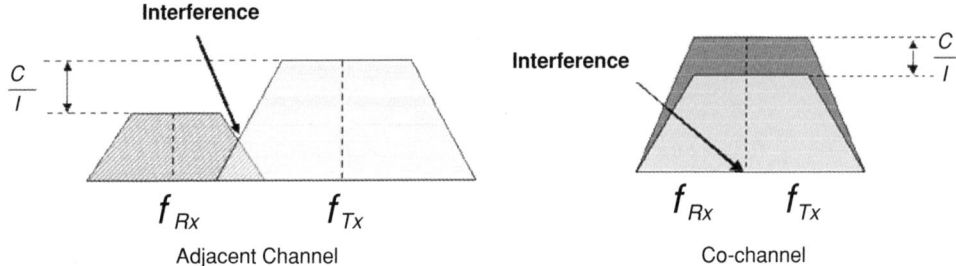

Figure 3.63 Co-channel and adjacent channel interference

3.8 Frequency Planning and Interference

Once the frequency bands are defined by international bodies such as ITU, IEEE, etc., it is up to the national regulatory bodies to determine the distribution of the frequency bands (which is usually done by paying a fee) to different bodies intending to use them for public or private applications. This has an impact on the type of equipment that would be required in the network, leading to an impact on link budget calculations, topology planning, etc. In some networks the number of bands available for microwave planning engineers might be less and hence frequency planning plays an important role. A bad 'frequency planned' network would see a drop in its quality and subsequently a drop in revenue for the operator.

The main idea behind frequency planning in the microwave network is to make sure that the wanted 'carrier' signal reaches the receiving end without getting hindered by any other 'interfering' signal. The closer the two signals are in terms of frequency and signal strength, the higher will be the level of interference. Interference may take place due to many phenomena taking place in the network, some of which are described below:

(a) Both the interfering and carrier signals have the same frequency channel but different polarisations.
(b) The interfering signal has an adjacent frequency channel to the carrier frequency channel and the same polarisation.
(c) The interfering signal has an adjacent frequency channel to the carrier frequency channel but a different polarisation.

One hop may have two frequency channels either having same the frequency channels and the same polarisations (or different polarisations) or adjacent channels having the same (or different) polarisations, as shown in Figure 3.63. Sometimes the signals (back lobes) from the antennas transmitted in different directions that are co-located or located too closely may interfere with the signals at the receiving antenna. Also, interfering signals may reach the receiving antenna due to diffraction or the over-reach effect. Over-reach is caused when the signal from a distant antenna (usually in a chain) interferes with the main signal for the initial hops (or vice versa). As shown in Figure 3.64, the signal 'I' acts as an interfering signal by over-reaching the far end antenna in the chain.

The parameter used to measure the interference levels is the carrier-to-interference (C/I) ratio. When carrying out frequency planning, the C/I ratio is optimised so that the minimum fade margin required to meet the objectives is actually available on the interfered hop. The main parameters taken into account when evaluating interference effects are:

- Modulation type and spectrum shaping
- RF channel frequency spacing
- C/I sensitivity

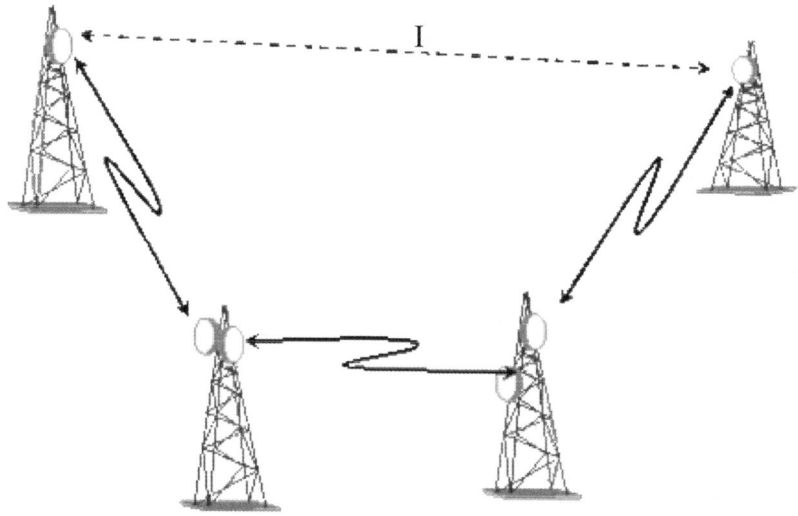

Figure 3.64 Over-reach

- Cross-polarisation discrimination
- Antenna patterns
- Filtering selectivity

Due to interfering signals, the receiver threshold level increases (at a given BER). Due to the increase in the strength of the interfering signal, the signal-to-interfering (S/I) ratio decreases, making a negative impact on the threshold degradation of the receiver. This has a direct impact on the link budget calculations. To counter the impact of this phenomenon, the input power to the receiver should be increased by certain means, such as increasing the transmitted power, etc. Thus, link budget calculations are looked into twice, once when designing individual links and again when assessing the impact of other links on the network. It is assumed that site A is being interfered by the signal from site D. The interference level at site D is a difference between the (interference) signal level at site A and the combined discrimination at both ends. The received interference signal is calculated by using the normal link budget calculation process described previously. The discrimination aspect can be calculated from the radiation patterns of the antennas at both site A and site D. S/I ratio is the ratio of the carrier signal to the interfering signal in the unfaded state. In the faded state, the carrier signal level is the sum of the threshold degradation and the carrier signal level. Hence, the threshold degradation can be calculated as

$$\text{Degradation} = -10\log_{10}\left(1 - e^{2.3026[1.3 - 0.1(S/I)_{\text{faded}}]}\right)$$

Based on these discussions, certain basic rules should be followed when performing frequency planning. A microwave link uses two frequencies, one for the TX and the other for the RX directions. The separation of two frequencies depends upon the duplex separation of the filters. This frequency separation should be provided separately in the frequency plans. Intermodulation is another aspect to be considered during frequency planning. Intermodulation is present when more than one frequency is present. It should be avoided as it creates problems for the third receiver. Consider the following example.

There are five frequency spots F1, F2, F3, F4 and F5 and five sites that need to be connected in a loop. Frequency allocation should be such that the same or adjacent channels are not next to each other or hops are placed in a way where over-reach interference possibilities exist. The polarisation also changes with every hop, with switching between vertical (V) and horizontal (H) polarisations, with the frequency

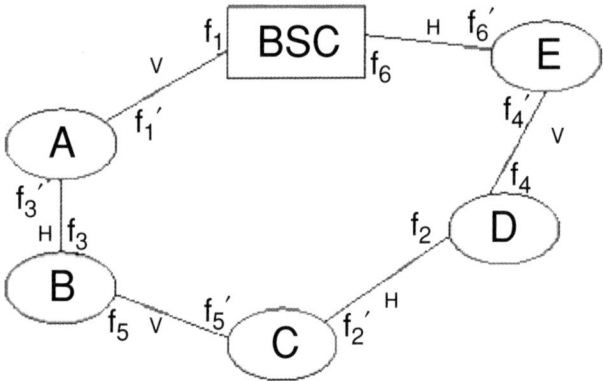

Figure 3.65 Example of the frequency plan

spots at each site being either high or low. When doing frequency planning, both high and low spots at the same site should be avoided. An example is shown in Figure 3.65.

3.8.1 Loop Protection

Loop protection is one of the most robust techniques. When one link fails in a loop, traffic can be routed through to the other side. In Figure 3.66 the link between site 1 and site 2 becomes unavailable; hence the traffic from site 1, which was being routed (to BSC) through sites 2, 3 and 4, will in the present conditions be routed directly to the BSC. The failure of the link between site 1 and site 2 can be due to a power cut, equipment failure, human errors, fading, obstacles, etc. The switching of the traffic flow is performed in such a short duration that there are no drop calls. Another major advantage of the loop is increased availability. The availability is increased by more than 10 times when compared to that of a single link.

The unavailability in a loop can be calculated as

$$P_M = \left(\sum_{i=1}^{M} p_i\right)\left(\sum_{i=M+1}^{N} p_i\right) \tag{3.75}$$

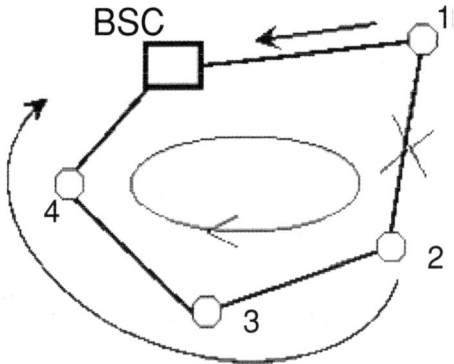

Figure 3.66 Loop protection

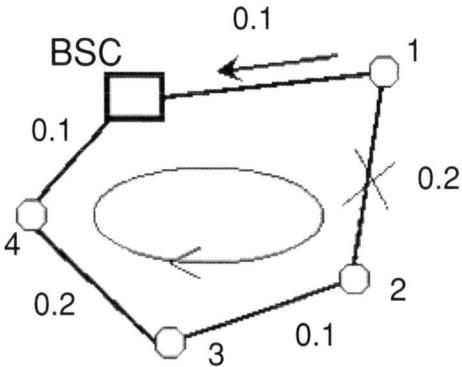

Figure 3.67 Availability in the the loop

where

P_M = site unavilability
N = number of hops in the loop
M = consecutive number of hops from the hub
p = probability of outage or unavailability of a hop in absolute terms (equipment, propagation, human errors)

Consider site 2 in Figure 3.67. The unavailability of site 2 in a condition that it was in a chain with site 1 would be:

Unavailability of the site 2 (in chain with site 1) = 0.1 % + 0.2 % = 0.3 %

Using the formula given in Equation (3.75), the unavailability of the site 2 in the present scenario = (0.1 % + 0.2 %) (0.1 % + 0.2 % + 0.1 %) = 0.3 % × 0.4 % = 0.0012 %. Thus 0.0012 % is a huge advantage over 0.3 % (or 0.4 % if connected via site 3 and site 4). This increased availability helps in designing links of longer hop lengths. Also, antenna sizes can be decreased. However, as seen in the section above, there is always an increased cost for the additional link.

3.9 Equipment Planning

3.9.1 BSC and TCSM Planning

The number of BSC required in a network depends upon the BSC features and its capacity to support:

- Number of TRXs
- Number of BTSs
- Number of PCMs
- Number of PCU cards (only in the case of GPRS/EGPRS networks)

Calculation of the TRX and BTS is simple (just add up all those connected to one BSC). The number of PCMs that are connected to the BSC can decrease using the concepts of grooming, topology planning, etc., ensuring that the PCMs connected to the BSC are fully utilised. However, as the number of BSCs increases, the cost goes up astronomically. In smaller networks where just one BSC may be sufficient, the operators may prefer using two BSCs for protection purposes. In many cases, the number of SS7 (signalling system 7) channels and LAPD channels that can be supported by the BSC are also taken into account.

PCU Planning

In EGPRS networks, the packet control unit (PCU) becomes an important part of the equipment planning exercise. The PCU is a plug-in unit that controls the EGPRS radio resources. It receives and transmits the transcoding and rate adaptation unit (TRAU) frames to the BTS and frame relay (FR) packets to the SGSN. Important aspects to remember when doing PCU equipment planning are:

- Number of A_{bis} channels supported by the PCU
- Number of radio timeslots supported by the PCU
- Processing capacity of the PCU
- Number of TRXs and BTSs supported by the PCU

Based on these factors, the number of PCU cards required can be calculated.

3.10 Timeslot Planning

Timeslot allocation planning along with the 2 Mbps planning form a part of detail transmission network planning. Understanding the following features will help in making the strategy for TS allocation on the A_{bis}. A good TS allocation plan is one that takes into account the next releases of software and hardware, BTS configurations, topologies, upgrades (due to traffic and signalling) in the near and far future, etc.

3.10.1 Linear TS Allocation

In this model, the next TS is utilised automatically. Assume that the configuration of a BTS is $1 + 0 + 0$. In this configuration, TS 1 and TS 2 are used. However, when the configuration becomes $1 + 1 + 0$, TS 3 and TS 4 are used for the next cell. When the configuration becomes $2 + 1 + 1$, the next TRX on the first cell is assigned TS 5 and TS 6 and so on. The methodology is simple to implement but difficult to manage after some time.

3.10.2 Block TS Allocation

In this model, a block is reserved for each cell. Usually there are three sectors for each BTS. Therefore, TS 1 to TS 8 are reserved for traffic in the first cell, TS 9 to TS 16 for the traffic in the second cell and TS 17 to TS 24 for the traffic in the third cell. This means that the configuration of the BTS is $1 + 0 + 0$ and the TRX will be assigned TS 1 and TS 2. However, if the configuration is $1 + 1 + 1$, then the TS assigned will be TS 1 and TS 2 for cell 1, TS 9 and TS 10 for cell 2 and TS 17 and TS 18 for cell 3. Although using this methodology causes the spare capacity to be fragmented, the upgrades are easier.

Bits are reserved for transmission management (discussed later), controlling the loop, i.e. the direction in which traffic will flow under normal conditions, and in conditions when the link breaks.

3.10.3 TS Grouping

TSs can be grouped according to TRXs, BTSs or by usage. When grouping the TS by the TRX is done, the traffic and signalling channels of the given TRX are kept together. This is valid for both the block and

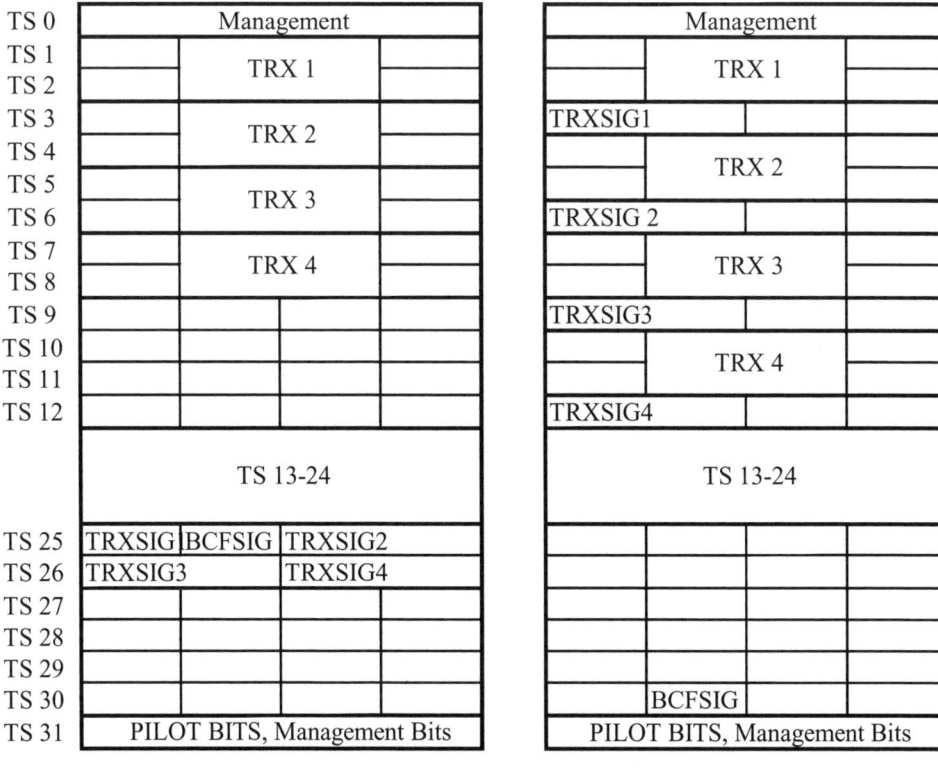

Figure 3.68 TS allocation strategies

the linear allocation. In this case, the BTS signalling channels are placed at a different location. In the case of grouping by the BTS, the traffic and signalling channels (both TRX and BCF) are placed together. In the case of TS grouping by usage, the traffic channels and signalling (SIG) are at different locations; i.e. all the traffic channels are placed together and all the signalling channels are placed together, as shown in Figure 3.68.

3.10.4 TS Planning in the EDGE Network

TS planning in EDGE networks becomes slightly more complicated if the concept of dynamic A_{bis} is used. In the GSM and GPRS transmission networks the A_{bis} is known as 'static A_{bis}' while in the EDGE network, A_{bis} becomes more dynamic (**as seen in Section 3.6.2**). Due to the transfer of the packet data, the number of A_{bis} required for a single fully loaded BTS (e.g. $4 + 4 + 4$) is usually more than one. Thus, the TS allocation in the EDGE transmission networks is different from those in the GSM and GPRS networks as some TSs are dedicated for the packet data. Thus, as seen in Figure 3.69, TSs in the pool are placed together while traffic and signalling channels are placed separately (TS grouping by usage). It is important to note that the way the TSs in the pool are assigned depends upon the equipment vendor design specifications (shown in Figure 3.68).

	1	2	3	4	5	6	7	8
0	LINK MANAGEMENT							
1	TCH.1		TCH.2		TCH.3		TCH.4	
2	TCH.5		TCH.6		TCH.7		TCH.8	
3	TCH.1		TCH.2		TCH.3		TCH.4	
4	TCH.5		TCH.6		TCH.7		TCH.8	
5	TCH.1		TCH.2		TCH.3		TCH.4	
6	TCH.5		TCH.6		TCH.7		TCH.8	
7	TCH.1		TCH.2		TCH.3		TCH.4	
8	TCH.5		TCH.6		TCH.7		TCH.8	
9	TCH.1		TCH.2		TCH.3		TCH.4	
10	TCH.5		TCH.6		TCH.7		TCH.8	
11	TCH.1		TCH.2		TCH.3		TCH.4	
12	TCH.5		TCH.6		TCH.7		TCH.8	
13	TCH.1		TCH.2		TCH.3		TCH.4	
14	TCH.5		TCH.6		TCH.7		TCH.8	
15	TCH.1		TCH.2		TCH.3		TCH.4	
16	TCH.5		TCH.6		TCH.7		TCH.8	
17	TCH.1		TCH.2		TCH.3		TCH.4	
18	TCH.5		TCH.6		TCH.7		TCH.8	
19	TCH.1		TCH.2		TCH.3		TCH.4	
20	TCH.5		TCH.6		TCH.7		TCH.8	
21	TCH.1		TCH.2		TCH.3		TCH.4	
22	TCH.5		TCH.6		TCH.7		TCH.8	
23	TCH.1		TCH.2		TCH.3		TCH.4	
24	TCH.5		TCH.6		TCH.7		TCH.8	
25	TRXSIG1		OMUSIG1		TRXSIG2		OMUSIG2	
26	TRXSIG3		OMUSIG3		TRXSIG4		OMUSIG4	
27	TRXSIG5		OMUSIG5		TRXSIG6		OMUSIG6	
28	TRXSIG7		OMUSIG7		TRXSIG8		OMUSIG8	
29	TRXSIG9		OMUSIG9		TRXSIG10		OMUSIG10	
30	TRXSIG11		OMUSIG11		TRXSIG12		OMUSIG12	

(Timeslots 17–24 → DAP Pool)

Figure 3.69 Dynamic A_{bis} (OMU, operation and maintenance unit)

3.11 Synchronisation Planning

The cellular networks generally consist of three types of equipment: asynchronous, plesiochronous and synchronous. Asynchronous equipment, e.g. PDH multiplexers with bit stuffing and line systems at higher than the primary bit rate, is used to transport data between digital exchanges. PDH networks have islands of synchronous networks that are connected to each other by the highly accurate synchronous clocks. Synchronous networks do not require any synchronisation clocks. Because of the advantages of using SDH networks, the backbones are usually made of the SDH while the access networks usually comprise a combination of plesiochronous and asynchronous equipments. For operators, the SDH is clearly the best choice for new transport networks. It is suitable for all types of networks: from core networks to regional transport networks, for city access and residential access networks, for cellular transport networks, etc. However, from the perspective of the network design engineer, a cellular network is a mixed network in terms of synchronisation. This kind of scenario, where both the PDH and SDH exist in the same network, will be there for quite some time to come. However, before the principles of synchronisation are discussed, it is important to understand why synchronisation is needed in mobile networks. Consider a cross-connect equipment having different bit rates at the incoming and the outgoing interfaces. The buffers in these equipments take care of the speed differences, but due to the difference between the incoming and outgoing traffic, there is a possibility of 'slips' happening. The bit errors due to these 'slips' result in 'clicks' taking place during the call. In the worst cases, the calls may even drop. Synchronised payloads can then easily be transferred within the network even if no synchronisation is applied, as the pointer mechanism would takes care of frequency differences. However, synchronisation is required due to the presence of the PDH interfaces. The performance of the SDH de-synchronisers and PDH access equipment may be degraded by pointer movements. With proper synchronisation the pointer events are

Synchronisation Planning

rare and do not cause bit errors in data services or clicks in speech. Also, the frequency reference with the equipments, e.g. the BTS, is not accurate enough. By providing external synchronisation, a more accurate frequency reference is provided for these equipments. ITU-T Recommendation G.803 outlines the architecture of the transport network based on the SDH and the synchronisation networks. Highlights of this recommendation are the reference clocks and clock distribution.

ITU G.803 recognises a hierarchy of clocks that can be used as a reference clock source. They are the primary reference clock (PRC), the Slave clock and the SDH equipment clock (SEC). The PRC has the highest accuracy and hence is at the top of the hierarchy level. It can be derived from an autonomous system or nonautonomous system, with the autonomous system deriving the accuracy from primary atomic standards such as caesium (using a caesium beam atomic oscillator). The PRC must maintain a long term frequency accuracy of 1×10^{-11} or better, with verification to universal time coordinated (UTC). The accuracy requirements of the PRC are given in ITU-T Recommendation G.811. These clocks are highly accurate but are also expensive.

Slave clocks such as the synchronisation supply unit (SSU) is a stand-alone synchronisation supply equipment (SASE). The accuracy of the SSU is specified in ITU-T Recommendation G.812. They are always locked to the higher level clock references via the network. These are generally used in long chains to refresh the clock signal. The traditional technology for implementing the SSU is the rubidium atomic oscillator. They have a lower cost than caesium and have excellent short time stability. However, their lifetime is limited. Modern quartz technology allows low cost oscillators to be built. Although their accuracy may not be as good as that of rubidium oscillators, they not only satisfy G.812 but are also cheaper and have a longer lifetime.

The SEC is a built-in clock and comes under the slave category (i.e. needs a master clock). Its accuracy as specified by ITU-T Recommendation G.812 is of the level $\pm 4.6 \times 10^{-6}$.

In cellular networks, the master–slave method is used to distribute the clock in the system, as shown in Figure 3.70. In this system, a higher level clock is used to synchronise a lower level clock. The lower level clock can then further distribute the timing signal in the network.

The following are synchronising elements in the cellular network:

- The MSC can be supplied by the clock through the caesium PRC or the GPS. However, some networks may also supply the clock to the MSC through the PSTN networks.
- The BSC receives the clock from the MSC. The MSC transfers the clock to the SDH and through the retiming function the clock is transported to the BSC.
- The BTS receives the clock from the BSC in a manner similar to the clock transfer from the MSC to the BSC.

There are a few things that transmission planning engineers should know when planning synchronisation in a cellular network:

- The clock always flows from the MSC to the BSC and the BSC to the BTS. The MSC gets the clock from the PRC.
- The maximum number of equipments in a chain should be limited; e.g. the number of SSUs should not be more than 10. Jitter and wander become accumulated when the synchronisation chains are long. Thus, long chains should be avoided.
- When planning synchronisation in a loop, there should not be any synchronisation loops. For this purpose, the master control bit (MCB) and the loop control bit (LCB) are used in TS planning.
- To maintain the quality of a synchronisation network, a highly accurate clock should be used as the PRC and there should always be more than one clock source available at all times for the equipment/element needing synchronisation.

An example of a typical synchronisation network is shown in Figure 3.71.

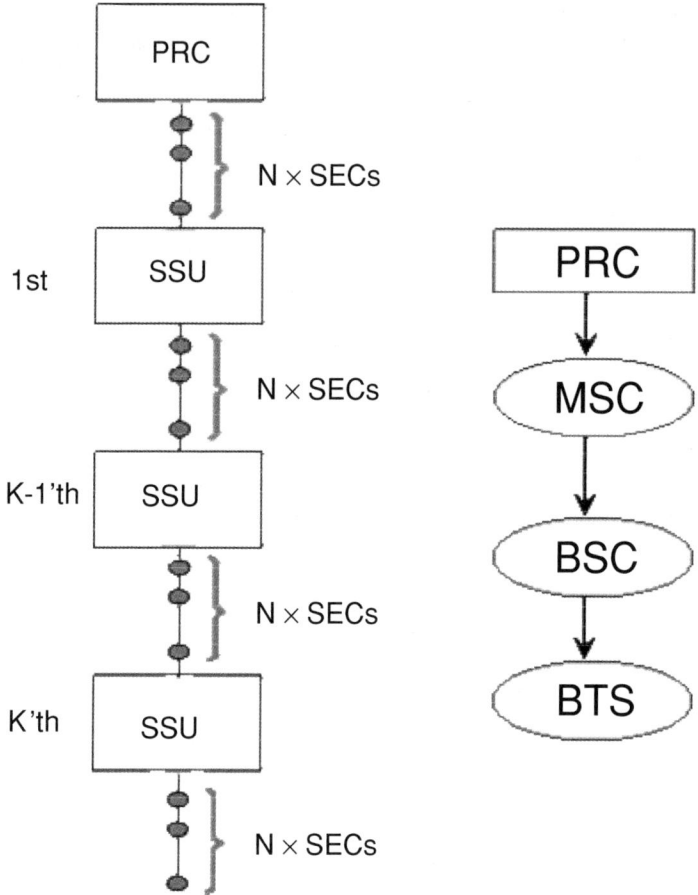

Figure 3.70 Synchronisation chain

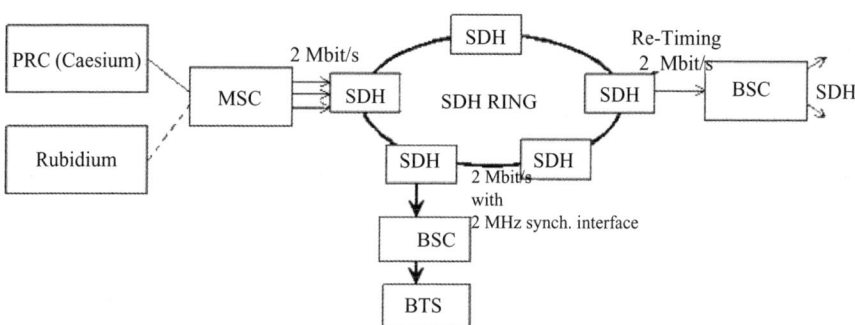

Figure 3.71 Typical synchronization network

3.12 Transmission Management

Management of the cellular network from one central location is very important for the operator, given the area size of the cellular network, where the sites may not only be located far away but also at not so easily accessible locations. Thus, a centrally located network management system as handy for the operators. An advanced management system gives the operator total management control of the cellular network from one central location. It is possible to control most of the elements, such as the BTS, BSC, MSC, TCSM, transmission system, etc., using most of the network management systems available from the equipment manufacturers. The more integrated the management system, the easier it is for the operator to control its network. Some of the advantages of the advanced management systems for operators are:

- better network availability;
- easy management of the cellular network;
- faster changes possible in the network with an integrated system;
- the number of visits required to the sites decrease considerably (although for hardware related problems, site visits will still need to be undertaken);
- core team of personnel located centrally can manage the entire network (which is a definite advantage over small teams managing various areas regions of the network);
- information flow is faster and hence the problems can be rectified quickly;
- software upgraded for new releases can be done easily;
- there is no extra payload required for the management data.

A typical management system is able to perform the following functions:

- Alarm management. This provides the information of the current status of the network element. In the case of any faults, etc., it generates an alarm.
- Security management. This feature protects the cellular network by detecting security breaches in the network. It detects and prevents the network from cases such as misuse of the network resources and service by the user, recovery from theft of services, etc.
- Configuration management. This is used for the installation of the network element, their connection to other network elements, etc. This feature also allows the NMS to receive and deliver data to the identified network element and/or to the connection between the network elements.
- Fault management. This allows the NMS to detect the failures and follow-up such as the repair scheduling, return to service, etc.
- Element management. The element or node can be configured or controlled using this feature.
- Performance management. This feature helps the operators to evaluate and report the behaviour and effectiveness of the network element.

The management of the transmission network is done through management buses. These buses deliver the message from the management system to the network elements and vice versa. The management system is planned on the master–slave protocols. Thus, planning of transmission management accounts for the following steps:

- Identify the masters for the elements to be managed.
- Identify the management buses.
- Plan the routing of the management buses.
- Set the parameters of the management buses.
- Plan for the protection of the management buses.

3.12.1 Element Master

There are two types of management schemes possible: BSC based management and BTS based management.

BSC Based Management

A set of management buses are created at the BSC, which therefore acts as a master. In this scheme, the management bus is carried from the E1 port of the BSC on any TS between TS 1 and TS 31 to the transmission equipment near the BSC. The BSC based management scheme is not used for the whole network but only for the network elements placed near the BSC. The number of elements that can be controlled by one BSC can be in the hundreds (depending upon the NMS capability) and there can be around 20 management buses.

BTS Based Management

In this scheme the BCF supervises the transmission equipments that are located near the BTS. Thus, the BTS acts as a master. A definite advantage of this scheme is that the management is faster as the buses are shorter in length. The number of elements that can be controlled by the BTS is less than 50 (this depends on the NMS capability of the vendor).

3.12.2 Management Buses

The management bus can be transferred using:

- External cables.
- Using the bits between TS 1 and TS 31. In some cases bits in TS 0 can also be used.
- Auxillary data channel in the frame overheads of microwave radios, DSLs (copper line connections), optical fibres, etc.

Based on the type of equipment, its location and distance from the master, these techniques can be used to transfer the management bus from the master to the slave.

Routing of Management Buses

Generally, management bus interfaces are of two types: maintenance interfaces and data interfaces. The cabling is planned in such a way that the connections from the master come from the maintenance interface while the outgoing signal uses the data interface. This is done because the maintenance interface is connected to the microprocessor of the equipment that needs the management instructions from the master. There are various topologies in which these interfaces can be connected for which NMS vendor instructions should be followed.

Parameter Planning

As shown in Figure 3.72, various parameters need to be defined for the management bus. This includes defining the addresses of the management bus. Along with the equipment, station and interface, the ID should be defined. The addressing scheme is based on the specifications given by the vendor. Another parameter that needs to be defined is the speed of the management bus, also called the baud rate. It should be noted that if a single management bus is connected to many equipments, then the lowest bit rate defines the speed of the whole bus. If the PCM is used to carry the management bits, then those bit numbers along with the TS carrying those bits should be defined, e.g. TS 31, bits 1 and 2.

Transmission Unit			
Group	Setting	Value/Operation	Note
Synchronization		According to defaults	
Identifications	Equipment ID	According to plan/ station identification	ID (text) of the TRUA equipment on this site.
Service options	Mgmt address	4080 or 4081 (Physical address)	Displayed with command 6,1,2,0
	Mgmt Bus baud rate	9600 bit/s	Default:9600 bit/s
	Data hybrid	OFF	Default: OFF
	O&M configuration (local master)	ON	Default: ON
	Data Channel Protection	OFF	Default: OFF
2M settings / TS(0) settings DIR 1	Use of data channel	Default: Fixed 1	Default: Fixed 1
2M settings / TS(0) settings DIR 2	Use of data channel	Default: Fixed 1	Default: Fixed 1
2M settings / TS(0) settings DIR 3	Use of data channel	Default: Fixed 1	Default: Fixed 1
2M settings / TS(0) settings DIR 3	Use of data channel	Default: Fixed 1	Default: Fixed 1
HW Database Settings	TRU control mode	BCF-mode	Default: BCF-mode (user interface: MMIDATA software)

Figure 3.72   Example of parameter planning for a transmission unit in the BTS

3.13 Parameter Planning

Parameter settings/configuration is one of the aspects of UMTS transmission networks. Unlike in GSM/EDGE transmission networks, there are many parameters that need to be set. Quite a few of the parameters are dependent upon the vendor specification/equipments. The majority of the parameters are related to the AXC (ATM cross connect) and RNC. Parameters are related to the application layers, transport layer, physical layers, interfaces, synchronisation and management. Thus, parameter definitions are made for the type of interfaces, ATM terminations and cross-connections, Internet protocols, etc.

3.13.1 BTS/AXC Parameters

Interface Equipment Parameters

Parameters related to the SDH and PDH interfaces are set. For the PDH, the setting has to be done according to the standards in use, e.g. E1, T1, JT1, etc. Automatic testing features such as loopback functions also have to be set and the number of timeslots that are used for traffic related to the ATM or GSM need to be specified. The structure of the SDH increases the number of parameters that are required to be set. Parameters related to physical, regeneration, multiplexer and container sections have to be planned, which mainly include assigning the IDs, enabling testing features (as in the PDH), defining the BER thresholds, etc.

ATM Parameters

As seen in the section above, ATM connections need quite a few parameters to be configured. Many of these parameters are related to traffic and quality of service apart from defining the parameters related to routing. Primary parameters involve defining the VCs, VPs, VCIs, VPIs, connection types such as CBR and UBR, etc. Other parameters include defining the bandwidths of connections, type of traffic carried and type of interconnection and cross-connections.

IP Parameters

The IP setting mainly consists of IP addresses both public and private. The addresses need to be defined for ATM and Ethernet interfaces. Parameters related to routing tables are also defined, especially if the communication to the indirectly connected nodes is there. Destination and gateway addresses are defined along with the masking bits (if required).

Synchronisation

As mentioned in previous sections, e.g. Section 3.11, the main parameters include defining the clock sources (primary and secondary), interface units and clock priorities.

Network Management

Some parameters are defined for the interaction between the BS and NMS. These include parameters related to equipments IDs, configuration management, registrations and alarms related to them.

3.13.2 RNC Parameters

RNC is an interface between the BS, MGW and SGSN. It is the 'brain' of the network and results in many more parameters being defined for the RNC as compared to the BTS/AXC. However, the synchronisation concept remains the same as defined in BTS/AXC.

Equipment Interface Parameters

The parameter configuration flow is shown in the Figure 3.73. As discussed in the AXC/BTS, the PDH and SDH parameters need to be configured and include the PDH exchange terminal (PET) and the SDH exchange terminal (SET) as well as the parameters related to exchange terminals, protection group, ATM access profiles, physical interface, etc. The configuration includes defining the operation mode (E1, T1, etc.), frame alignment mode, loop back testing, TS usage (i.e. the number of TSs required to carry the GSM and 3G traffic respectively), BER threshold levels, protection, etc.

ATM Parameters

These include the parameters VP, VC, VPI and VCI. The ingress and egress bandwidths are defined along with the type of traffic that is carried, e.g. CBR, UBR, etc. Cross-connections for virtual paths and virtual

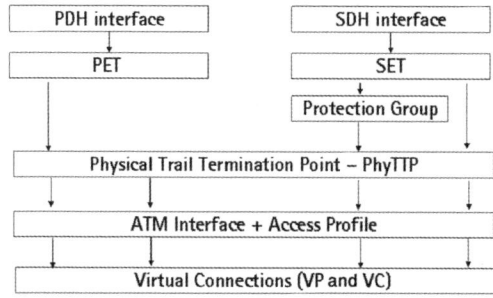

Figure 3.73 Flowchart to configure interfaces

channels are defined. With traffic comes the quality of services parameters (given in Section 3.3.3), which need to be defined.

Interface Parameters

Parameters required to configure the I_{ub}, I_u-C, I_{u-ps} and I_{ur} are defined. Configuration of the interfaces usually contains steps that include configuring the physical interface parameters, creating the radio network connection configuration parameters, creating the data connection networks, defining signalling and routing parameters, creating IP over ATM parameters, etc. Physical interface and ATM parameter creation has been dealt with in the above section. Connection configuration parameters include defining the parameters related to interface and control and the signalling unit (which is responsible for the function such as handovers, power controls, packet scheduling, etc., and is present in RNC), AAL2 user and signalling link VCIs, routing, etc. For the PS side, parameters are defined for interfaces, VCIs, VPIs, etc., apart from IP addressing, packet scheduling and routing, and forwarding/routing tables. An example is shown in Figure 3.74.

Signalling Parameters

The signalling data are carried by the SS7 signalling network. Before planning a signalling network, issues such as the type of signalling network, kind of signalling allocations scheme, restrictions related to the interconnection to other networks, application that needs to be carried, etc., should be addressed. Parameters generally include the defining type of network, identifying the signalling point codes, links and their IDs, standards used (e.g. ITU-T, JAP16, ANSI, etc.), VCs, VPs, VCIs, VPIs, etc. The routing of

ATM Interface parameters

Name	Value
ATM Interface Identifier	13
Physical Layer Trail TP Identifier	13
Interface Type	uni
Administrative State	unlocked
Maximum Bandwidth (ingress)	150000
Maximum Bandwidth (egress)	150000
Maximum VCI Length In Bits	8

Interface ID (13)
  VPI: (2)

Figure 3.74 Connection configuration (CoCo)–ATM Interface

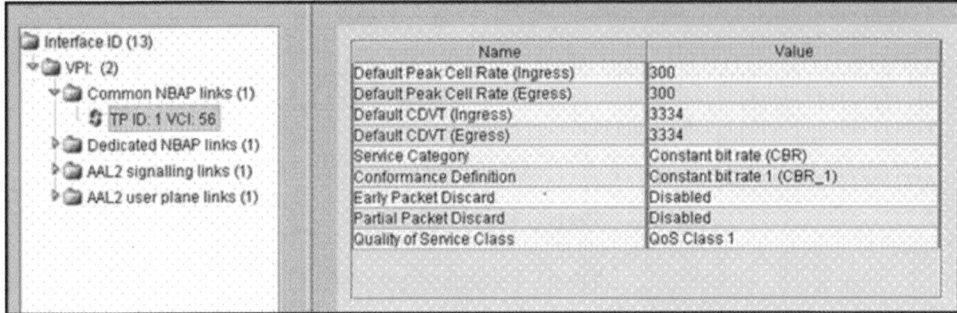

Figure 3.75 Connection configuration (CoCo)–D-NBAP links

signalling links should also be determined, including the priority. Parameters are defined for signalling links, signalling link sets, routing and also for signalling sublayers. An example is shown in Figure 3.75 (where CDVT is the cell delay variation tolerance and D-NBAP is the Dedicated Node B Application Protocol).

3.14 Transmission Network Optimisation

To start with, two things need to be specified. What is meant by transmission in this context and what is really meant by optimising it? This might sound like hair-splitting but unfortunately transmission optimisation and transmission optimisation in cellular networks are loosely used phrases that are heard often but very seldom (if ever) made clear as to what is really meant. It is evident that the purpose is to improve something and that there is a willingness to do this, but that is about as far as it gets before the situation becomes fuzzy.

3.14.1 Definition of Transmission

To get started, it is important to acknowledge that when talking about transmission optimisation in cellular networks the discussion concerns a very special kind of transmission solution and a very narrow implementation of transmission. Transmission as a word just suggests that some information is being moved from one place to another or other places. This means that in cellular networks the interface between the ME and BTS involves transmission. Obviously when talking here about transmission optimisation that type of transmission is not meant. The optimising interface between ME and BTS is discussed in other chapters. That case is just one example to highlight where transmission optimisation might not be a very good expression to use because it is so vague. Therefore an agreement is needed here to define the scope.

Let the agreement be that what is meant by transmission in this context is the use of a data transport system between access network elements and core network elements. The data transport system can be dedicated to this purpose but it can also be used for several different purposes, data transport for the cellular network being only one of them.

There is one important remark to be made concerning this agreement. It quite clearly shows that network services and data transport services should be considered as two separate systems, which are only part of a section. The section can be complete from the data transport network viewpoint if it has no other purpose than serving the 3G network, but the section can also be incomplete if the transport network is also used for some other purpose or purposes that have nothing to do with the 3G network.

The importance of this is that reference to the 3G network actually loses most of its significance in this context. The transmission optimisation in 2G networks or in some other type of network would actually have the very same principles. Of course, the 3G network has certain characteristics that specify what kind of transmission network can provide the data transport service for it, but that has relatively little effect on the main optimisation principles.

It can also be said that the 3G network is specified so that the expectation is for 'perfect enough' data transport between the network elements. In other words, the expectation is that from the 3G system viewpoint data transport is not supposed to interfere with the quality of the system itself. This is another opportunity to make the same aforementioned remark that when talking about transmission optimisation and 3G optimisation different things are meant.

This also clarifies why it is easy to get confused when the optimising interface between the ME and BTS and optimising transmission are mentioned close to each other. The fundamental difference between these two optimisation challenges is that the 3G network does not make any perfection assumptions about the interface between the ME and BTS. Rather it assumes that the interface is a long way from perfect. That is a very fair assumption since between the ME and BTS the connection is made in three-dimensional free space and the connection relies on reflections and interferences of electromagnetic fields. It is clear that the whole system has to focus on performing well in that complex environment and there can be many parameters to tune.

However, the nature of transmission is 'borrowed'. Transmission is something that is needed but where a fair assumption is that it must be able to provide such a level of service that it is transparent to the 3G network. It does not need to be 'own' network as long as it can provide needed services. This is also a fair assumption since the transmission network is usually point-to-point and has interference protection. It could be described as having fewer dimensions.

It can also be said that due to this weak interlinkage between these systems, looking at 3G system related parameters does not provide any transmission optimisation solution. This can be said even though the meaning of optimisation has not yet even been defined.

Definition of Optimisation

Optimisation, on the other hand, is generally a synonym to improvement, but there are different approaches that can be taken. A common nominator to these approaches is that they all try to achive the same goal, i.e. to increase the profit of the cellular operator. In Western capitalism this goal is as natural as staying alive and reproduction is in nature, but yet that goal can be easily forgotten when optimising some detail of the network.

However, the importance of that fundamental goal cannot be overestimated since it provides the base line where the effort of optimising some detail is not optimising the profit but actually costing more than could possibly be gained from profit. It is easy to become blinded and focus on optimising some detail, but this overall effect should always be kept in mind.

After this lengthy introduction to what really is important and worth optimising, the ways of optimising revenue can be categorised roughly into two main streams: those optimising efforts that focus on maximising the revenues and those that minimise the costs (see Figure 3.76). In addition, the optimisation effort differs in the time line, i.e. when the optimum is to be reached. Some optimisation effort focuses on achieving as optimal a status as possible immediately, while some other optimisation effort looks two years into the future. The actions are naturally very different in such example cases. They could be called tactical optimisation and strategical optimisation – or perhaps they might be called short-sighted and long-sighted. However, that might be overstated since both have situations where they are applicable.

Making decisions about when to reach the optimum falls to business management, but when talking about transmission optimisation here that can be ignored. In the following, the focus is put on two earlier mentioned main categories of optimisation and reference is made to describe only the expected effect on the time line.

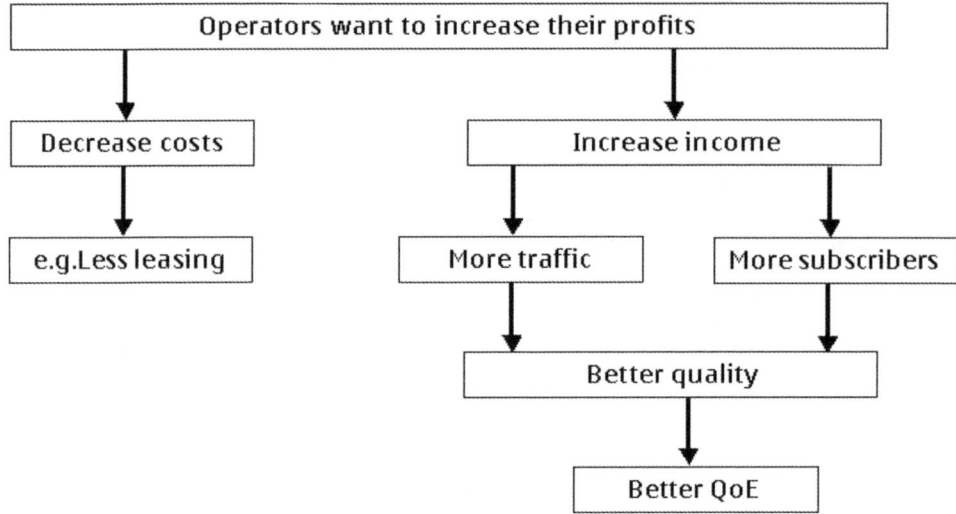

Figure 3.76 Overview of optimisation (QoE, quality of end user)

Maximising revenues

Those optimising efforts that seek to maximise revenues are based on the assumption that customers will choose the best serving, the best quality operator. It can be argued how well that thinking actually holds true and how much people are willing to sacrifice and search for cheaper plans and rates if thought is given, for example, to how things have developed recently in commercial aviation. It seems to be the case that people are more often guided in their service consumption decisions by price rather than quality. In other words, people seem to be seeking fundamentally for cheaper prices. When making the decision quality seems to be only a secondary parameter that is evaluated. Two example scenarios can illustrate this.

In the first scenario the consumer has been using service A and becomes aware of a similar service B that is cheaper. The parameters that the consumer is likely to calculate are:

1. Price difference. How much cheaper is service B compared to service A? Is the difference so big that it makes sense to disturb a current stable situation?
2. Quality difference. Is service B promising less or if the offering is the same can I trust that it is true? Have I heard or can I find any negative feedback about this service?

In other words, what is the inertia in losing an existing customer?

In the second scenario the consumer has been using service A and becomes aware of a similar service B that has the same price but claims to have better quality or has a reputation of having better quality. Now the parameters that the consumer is going through are:

1. Satisfaction difference. Will I be so much happier with service B that it makes sense to have the effort to disturb a current stable situation?
2. Quality difference. Is service B able to hold its promise and really provide better quality service?

In other words, what is the inertia in gaining a customer with quality?

When thinking about these two scenarios it can be concluded that there certainly is inertia for most consumers to make such a decision. How much depends on the individual, but it is safe to say that competing with quality and using quality as a sales argument to maximise revenues is not easy. The price of service is likely to have more of an effect but even with that there is some inertia.

Think about it yourself. How many people do you know who have changed their operator due to network quality? What about due to service pricing? How many do you know who have never changed operator at all?

As a bottom line, if the network is not bad but is able to provide reasonable service quality then it really is arguable how much improving the network quality can really maximise revenues. On the other hand, virtual operators have lately been an increasing trend, which makes forecasting the customer perception and behaviour even more difficult. The general public is not likely to be very well aware of the share of responsibilities between a virtual operator and an underlying network operator providing network services to a virtual operator.

After this discussion about how effective quality improvements actually are, an assumption will be made that it has some effect and what kind of things consumers consider to be quality will be discussed. Roughly, consumer perception of quality can be categorised into three categories:

1. Accessibility. What is the success rate of getting the service established and something through?
2. Retainability/sustainability. What is the success rate that a consumer does not get interrupted during the consumption of the service, but completes the service by himself/herself?
3. Bandwidth. How fast/slow is the service or response time?

These categories will be looked at one by one, putting extra effort into the main focus of this chapter. How can the transmission network be optimised so that the effect is in that particular aspect of quality? First aspects of GSM/EDGE transmission network optimisation and then UMTS transmission network optimisation will be discussed.

3.14.2 GSM/EDGE Transmission Network Optimisation

Performance Monitoring

Before performance monitoring begins, it is important to study the actual plans and compare them with what was implemented/commissioned. As in most of the networks the number of sites implemented are in hundreds and thousands and are set up at a very high speed, e.g. a few hundred sites in the air in a few months, there is every possibility that an antenna will not be mounted at the exact height or facing the correct direction, power output will not be at the right level, interference problems will develop from the same or external networks, etc. Capacity may also be a problem as the growth of mobile subscribers may be higher than expected. For capacity related problems in the GSM, the optimisation exercise would include studying the existing capacity plans, which means studying the site configurations, topology and transmission media. Once the radio planning engineers come up with new configurations, transmission planning engineers should be able to connect the sites with upgraded configurations. This is a usual problem in green-field networks. Hence, the best cure for such a problem is to make sure that there is a sufficient extra capacity during the planning phase itself so that new upgrades (after a few months) can take place. Assume that two BTSs each with a 2 + 2 + 2 configuration are connected to a BSC through the 1E1 link. This means that there is no extra capacity (the assumption is that the 1E1 is able to handle 12 TRXs). In few months time, the operator realises that a shopping mall in the vicinity of the two sites is to be built. This would mean that these two sites would need upgrades leading to a requirement for extra PCMs. If in the very beginning each site had been assigned one PCM each, there would be less problems at a later stage. However, such decisions (i.e. assigning one PCM to each site) are not easy in the beginning as the revenues may not have started to flow and obtaining leased lines may be costly in some countries.

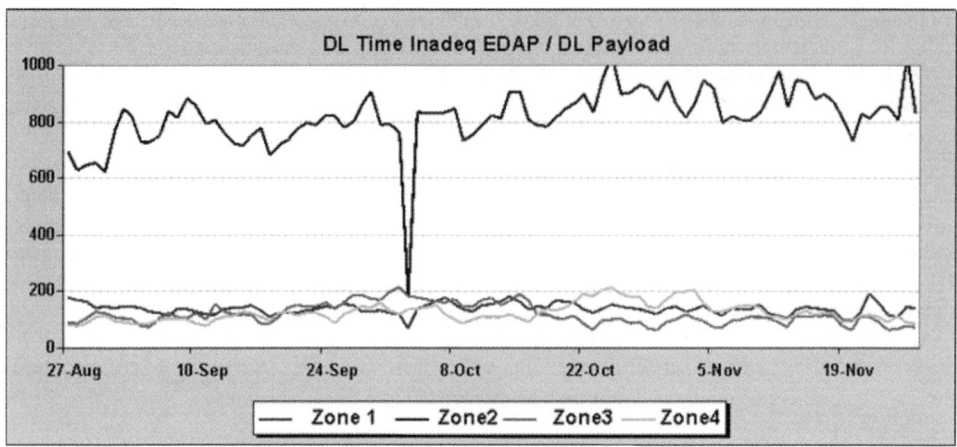

Figure 3.77 DAP performance in various regions of the network

Upgrading the GSM network to the EDGE network definitely leads to capacity problems. This is mainly due to the dynamic A_{bis} pool. As seen previously, the dynamic A_{bis} pool would need some reserved timeslots for itself. There may be some problems with the dimensioning for the DAP, which would need to rectified during the optimisation phase. The performance can be monitored using the network management system and the sites/regions with problems can be identified. Figure 3.77 shows the performance of the network in various regions. It shows that the DAP in zone 2 is the most congested. This means that the utilisation rate is more than 100 %, thereby indicating that users are not able to access the data at the maximum rate, e.g. the MCS-9 coding scheme. The reason lies in the dimensioning and hence one solution could lead to correction of the DAP size. The congestion (as shown in Figure 3.77, zone 2) may also be due to some other factors, such as the availability of the PCUs and/or reservation of the signal processing resources.

In a network upgraded from the GSM to EDGE capable, the problems may lie in the PCM lines. Due to nonavailability (or higher costs) of PCM lines, the pool size may be kept small (e.g. one or two timeslots) as the number of EGPRS users may be limited during the launch phase. In such cases the solution is to assign another PCM line to cater for the voice user of certain cells and a primary PCM can then be used for the DAP with more timeslots. This also results in a lower number of PCU units assigned in the BSC or a limit in the number of signal processing units. There are certain counters in the management systems (vendor dependent) that should be monitored before making changes. With careful assessment, the number of timeslots in the DAP and/or PCU units in the BSC should be increased.

Quality

There are few parameters that are measured for quality of microwave links. Some parameters in the ATM sections have been observed to find the quality of the network. However, these parameters define only the quality of the ATM network. To determine the performance of the access transmission network, the parameters should be defined with the network operator. For the access transmission network, especially GSM networks, there are not many parameters when compared with those in the radio network. However, as the main focus is on capacity and quality, the parameters defined, performance monitored and optimisation done revolve around these two aspects. Some basic parameters defined in ITU Recommendation G.826 on error performance and availability need to be understood. The Recommendation gives the basic error performance definition and related parameters.

Transmission Network Optimisation

Error Performance Events

Errored Block (EB)
This is a block in which one or more bits are in error.

Errored Second (ES)
This is defined as a 1 second period with one or more errored blocks or at least one defect.

Severely Errored Second (SES)
This is defined as a 1 second period containing $\geq 30\%$ errored blocks or at least one defect. SES is a subset of ES.

Error Performance Parameters

Errored Second Ratio (ESR)
This is the ratio of ES to total seconds in available time during a fixed measurement interval.

Severely Errored Second Ratio (SESR)
This is the ratio of SES to total seconds in available time during a fixed measurement interval.

Background Block Error Ratio (BBER)
This is the ratio of background block errors (BBE) to total blocks in available time during a fixed measurement interval. The count of total blocks excludes all blocks during SESs.

Block Sizes
Some examples of block sizes are:

- For 1.544 kbps PDH path, the PDH block size is 4632 bits.
- For 2.048 kbps PDH path, the PDH block size is 2048 bits.
- For 1664 kbit/s VC-11 path, the SDH block size is 832 bits.
- For 2240 kbit/s VC-12 path, the SDH block size is 1120 bits.

Availability and Unavailability

A period of unavailable time begins at the onset of 10 consecutive SES events. These 10 seconds are considered to be part of unavailable time. A new period of available time begins at the onset of 10 consecutive non-SES events. These 10 seconds are considered to be part of available time. A bidirectional path is in the unavailable state if either one or both directions are in the unavailable state.

Quality degradation in an access transmission network may be due to capacity problems, microwave link performance related issues including interference, synchronisation failures or even those due to the parameter setting (e.g. addressing scheme) in the network management system. The process can start with monitoring the PCM failures from the network management system (as shown in Figure 3.78). In this figure the two peaks on the right-hand side are a cause for concern as they degrade the quality of the system. The reason for the outage could be human error, power failures or performance degradation in terms of aspects mentioned above.

The impact of the DAP on the access transmission system and how it can be resolved was discussed in the section above. However, sometimes the transmission network is simply not capable of taking the load of the subscriber growth. This situation takes place when the sites are added and without considering the impact it will have on the transmission system. In networks that have hundreds of sites, the SDH backbone is advisable so that capacity problems do not happen in a crucial part of the network. Some networks that had a PDH backbone in the initial stage outgrew the capacity and hence faced huge amounts of drop calls. In that case, the capacity upgrade should be done immediately to prevent loss of revenue.

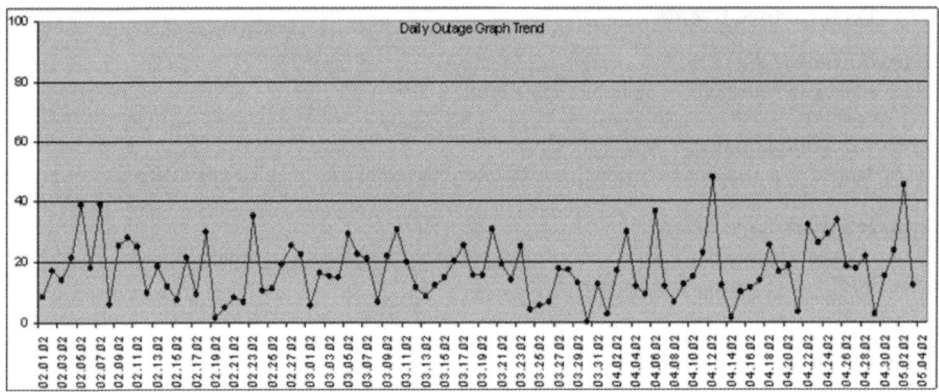

Figure 3.78 Example of recording PCM failures at the BSC

Performance problems related to the microwave link can be twofold: one is related to the performance of a single link and the other to collective performance of the links. The former may be related to errors in link design and atmospheric changes while the latter may be due to errors in system design, e.g. bad frequency planning. In either case, it is good to review both the original plans – design and implemented. Link budget calculations should be checked and compared to commissioning results in terms of transmitted power, received power, losses and availability. If possible, the performance of the 'problem link' should be recorded for a 24 h period (shown in Figure 3.79) for certain periods of time (ranging from a week to a few months). Especially in cases where phenomena like ducting or layering take place, i.e. microwave links in a valley, near large water bodies, etc., the measurement over a substantial period of time is helpful. Based on the results, changes to the microwave antennas with respect to transmitted power, antenna heights or orientation should be made.

When talking of system performance with respect to a microwave link, frequency planning and interference are the two major aspects to be looked into. Frequency planning in bigger networks with less frequency spots may be a cause for concern, especially in the roll-out phase. When re-examining the frequency plans, the rules of thumb that were seen in the sections above should be followed. Techniques

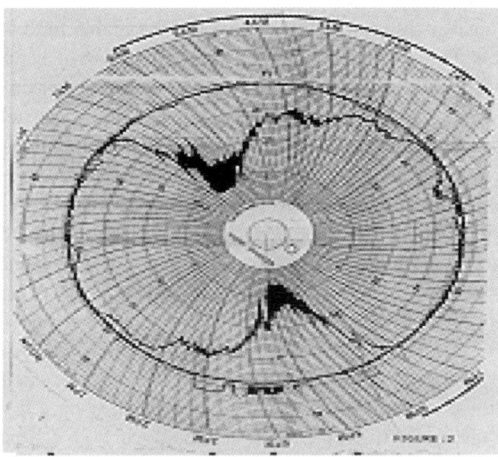

Figure 3.79 Typical performance of the link recorded on a 24 h basis

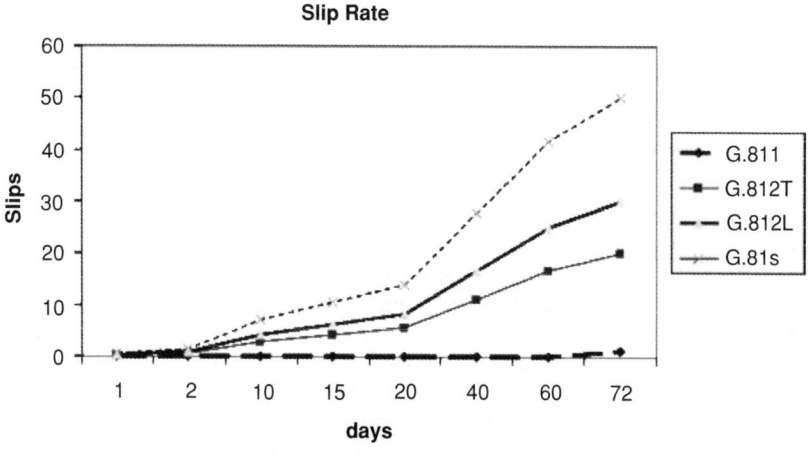

Figure 3.80 Slip rate

such as the high–low concept, radius of a loop not less than 3 km, vertical and horizontal, etc., should be applied. Even after this interference is spotted, the source should be found, whether the source is external (other than an own network) or internal (own network). To reduce interference, some techniques such as cross-polar discrimination or antenna tilt power reduction should be applied.

For issues related to synchronisation, the slip frequency limit parameter should be observed from the management system (an example is shown in Figure 3.80). The most common problem associated with a high slip rate will be handover failure within the GSM network. The mobile with the need to hand over cannot find the neighbouring BTS because the frequency has drifted too far. This contributes to the drop call rate. The BTS with high slips should be checked by tracing the synchronisation flow or by checking the timing priority bit in case a loop configuration occurs.

3.14.3 UMTS Transmission Network Optimisation

Accessibility

First consider accessibility. Poor quality in accessibility means that the consumer is not able to make a call at all or is not able to establish a data connection. The well-known cause of this is poor radio coverage and in most cases radio coverage is also the area where there is most room to improve. Radio interface optimisation is discussed in other chapter but it is good to note that poor radio coverage is something that the consumer can monitor. The cellular phone has a display for field strength, which works in two ways. It gives the consumer the ability to monitor the quality of the radio interface even when not planning to make a call. The measurement device goes with the consumer in the phone. On the other hand, it guides the behaviour of the consumer. It sets consumer expectation, since the consumer can see while planning to start consuming a network provided service what is the field strength and whether the radio field is nonexistent or weak; the consumer might then choose to try some other location later.

Whether consumers really ever or regularly monitor the field strengths of their operator while travelling around can be questioned. For most people it would be safe to assume that they do it only if they are extremely bored. Nowadays when mobile games are readily available in cellular phones monitoring the network quality in operation is quite a theoretical option. However, when people check their phones for missed calls or short messages they do subconsciously notice the field strength – at least on a level of whether there was a radio field or not. As a bottom line it can be said that the easiest and maybe the only way to improve the perception of accessibility quality is to work on improving the radio interface.

There is also the capacity side (or rather, say, the interference side in 3G) in the radio interface, which can be congested even if the field strength is appropriate and which is not able to be monitored by the end user. That will face similar end user perception as transmission problems of accessibility, but fixing those problems is discussed elsewhere in this book.

Why is this so important and why has all this time been spent in a transmission optimisation chapter to go through it? It is important to understand what a consumer really knows and appreciates to be able to make decisions about what is worthwhile to optimise in a transmission network. Remember that the aim is not to make a perfect system, but to maximise revenues!

Accessibility problems due to poor radio interface coverage and to congested transmission result in very similar end results. The point is that the consumer has gained more understanding about accessibility problems of radio interface coverage than about problems of transmission. This is because they can monitor the radio interface with their phone (which amounts to coverage), choose where and when to make the service request and, last but not least, because they, as everyone else, know that there needs to be 'field' to be able to make a connection. They cannot monitor the transmission network in the same way. This last part means that the radio interface has had a lot of publicity, partly perhaps because it is now the new 'thing' in cellular networks compared to fixed line services. It does not mean that the general public would know much about radio propagation, not at all, but since everyone carries a device capable of measuring the field strength, it is quite human for many people to act as if they were experts. On the other hand, radio propagation is so complex, three-dimensional and literally in the air that even real experts can have only an estimate about the reality. Therefore everyone is qualified to have an opinion.

However, the same does not apply to transmission. The main causes of accessibility problems in transmission are congestion in signalling links or congestion in the user traffic channel.

Transmission Signalling

Congestion in 2G signalling links has now been in focus from the very end of the 1990s because the number of different types of services that use those signalling links has exploded and transmission network planning, which was done in many cases in the mid-1990s, was definitely not done for such a scenario. As 16 kbps signalling links have been widely used they are now in danger of becoming congested. Therefore around the world operators have lately been changing signalling links from 16 kbps to 64 kbps for most critical base stations.

In 3G there is already much more uncertainty in the first place since no one yet has a firm idea of how these networks will evolve or what services will break through in addition to traditional 2G services. That is why dimensioning the signalling links is a very challenging task. Simulations have been made, but depending on the selected services very different kinds of results can be achieved. As a rough guideline 10 % of user traffic for signalling is frequently seen. That maps quite well to 16 kbps signalling link usage in 2G, which as discussed has proved to be underestimated in most loaded base stations. This does not necessarily mean anything since 2G and 3G services are different, but taking into account the new ATM layer and its signalling needs in 3G as well as the importance of signalling links to user experience it would definitely be safer to overestimate the signalling capacity.

To optimise end user experience about network quality, when it comes to 3G transmission network signalling the guideline of 10 % is fine, but to be more safe in the future 15–20 % of total link capacity for signalling should be considered. 3G is expected to have high capacities so from a quality optimisation and configuration management viewpoint allocating a bandwidth equal to one E1/T1/JT1 for each 3G base station should be considered for signalling. This has not been practically possible due to the slow start of the 3G markets and the cost optimisation of networks, but if those difficulties abate then it is definitely something to consider.

A way to split signalling capacity between NBAP signalling and AAL signalling links needs to be simulated case by case, with particular service and traffic scenarios. There might also be some differences in the arrangements for signalling between different vendors, so getting into more detail here might not be generally applicable, even if the direction is clearly to have truly open interfaces in the access networks.

An important point to note is the fact that the end user does not have insight into signalling capacity problems so understanding why access to the network cannot be achieved becomes vague if signalling capacity causes the failure. It becomes solely a 'network problem' whereas with coverage problems the end user takes part of the blame. The end user knows that he/she could be in a bad spot for the field but with patience could find a better place after trying once or twice more in a different place, but if transmission signalling fails it is the 'network' that does not work and gets the whole blame.

Transmission Capacity

It is equally possible that signalling does work properly, but there is not enough bandwidth in the transmission links for user traffic channel allocation, thus causing accessibility problems. Since the environment is so complex in the radio interface and transmission interfaces of 3G networks, admission control algorithms are used to calculate whether the agreed service quality is met for all ongoing services if a new service request is accepted. Based on that calculation the new service request is either accepted or rejected. This means that the system is calculating all the time what is the load of the network. This is the reason why TDMA and CDMA networks in radio interface and circuit switched and packet switched networks in transmission interfaces differ so fundamentally.

There are no specifications an how exactly admission control algorithms should do that calculation and it is more than likely that each vendor has his/her own way of doing it. This means that for end users to find the transmission capacity and its possible congestion they will have to use the specific algorithm from the vendor. Creating an algorithm that is fast, trouble-free, guarantees the quality and is capacity-friendly is not actually a very easy task, and depending on which optimisation approach that vendor has chosen surprisingly different results can be calculated using the same input values.

Being fast and trouble-free are fundamental requirements for the admission control algorithm. However, if guaranteeing quality is the main approach the algorithm easily becomes a glutton for capacity or if saving in capacity is the main approach then the algorithm is to a certain extent jeopardising the quality of services. Finding the golden mean is the challenge.

Looking back to quality and end user perception from the transmission capacity viewpoint, if a service request by the end user is rejected by the admission control algorithm that will translate to bad network quality in end user perception. The regular end user does not know or care about the admission control algorithm's calculations and decisions. For the end user the success of the service request is the only meaningful consideration.

As far as the admission control algorithm is concerned, the service quality should be good if a service request is accepted, so the expectancy is that, if accepted, the end user will not have any quality issues that could be improved. This, however, might not always be true if the vendor has chosen an admission control algorithm that tries to optimise the capacity usage of the network by taking some risk in quality. If there is such an algorithm in use, then performance engineering can optimise the calculation of the algorithm only if some input values are made accessible to engineering by the vendor. If the required input parameters are not made available then there is very little an operator or performance engineer can do. They can ask the vendor to make such changes to the software and provide a software change delivery version including those changes or they can change the vendor, which is quite a huge decision, but otherwise there is nothing they can do.

On the other hand, if there are admission control algorithms with input parameters that can be made to take adjustments then tweaking them can perform the optimisation. The possible parameters are defining in one way or another the acceptable levels of delays and acceptable levels of losing information packets (i.e. in transmission this means AAL units or ATM cells). The nature of the algorithm and what it is designed for suggests that tweaking them can easily mean that the input parameters go to an unauthorised zone. This becomes taking those parameters to absolute values. It should, however, be remembered that some sacrifices have to be made, probably in the accuracy of the algorithm, in order to keep it fast and trouble-free but on the safe side. Therefore, practically tweaking input parameters to an unauthorised zone means that a grey area is entered where the input parameters are indeed in an unauthorized zone but where the results of the algorithm might still be on the safe side compared to using the most accurate

Table 3.15 Example of varying loss and delay parameter

Loss probability	Delay [ms]	Bandwidth need bps
10^{-3}	10	772673
10^{-5}	5	1110068

possible admission control algorithm. This is where the optimisation risk is on the engineer who is actually performing the task.

To highlight the differences in tweaking the parameters, Table 3.15 gives an example of the same traffic mix in the same base station configuration with different loss and delay parameters settings for an AAL connection admission algorithm used some time ago. The graph in Figure 3.81 illustrates how delay (cell buffers), bandwidth (offered load) and loss (CLR) are coupled in general.

Transmission Quality Loss

Transmission quality loss is the third possible cause of quality problems in transmission accessibility. The network might have the signalling capacity and it might have the needed capacity for user traffic channels, but still the access might fail due to either occasional corruption of information in connection setup messaging or to some more frequent corruption cause of information in connection setup messaging.

If it is only some occasional corruption caused by, for example, a voltage surge then there is not much to do. No system is bulletproof to all occasional incidents, so those cases are just the statistical failures that do not pay off the effort to be optimised. However, if it is due to repeated corruption that cannot be explained then it makes sense to find out what is the root cause of that. The possible causes are, for example, poor copper cable quality, bad connectors, hardware problems microwave transmission link problems and extra-equipment problems like heat, condensation, ice, etc. The list can get long. It is a good idea to start excluding possible causes from extra-equipment candidates, since those are the easiest ones normally to fix. Then continue with connectors, plug-in units and subsequently backplanes, finally

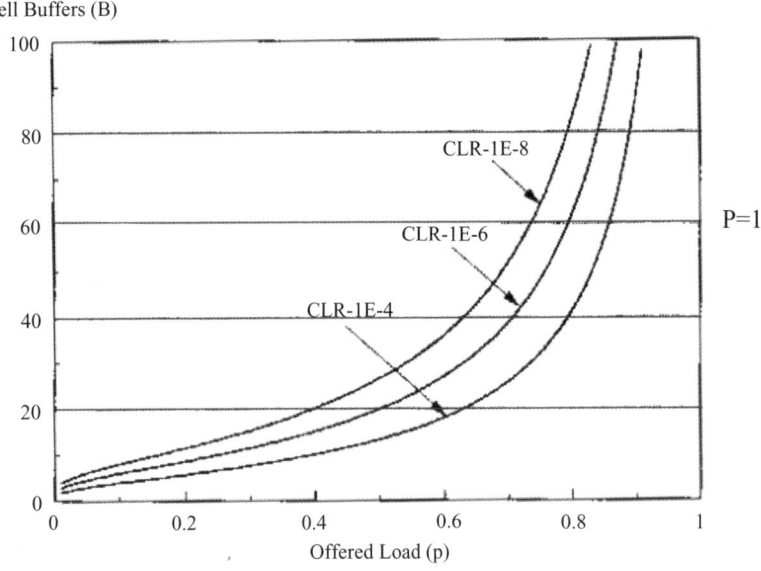

Figure 3.81 Offered load versus cell buffers

considering the possible transmission media issues and what kind of options there are available to either improve the existing media or change it.

Transmission Equipment Failure

Fourthly, and supposedly the most rare cause of accessibility problems in transmission, is a very fundamental one, caused by transmission equipment failures. These are easy to monitor, but are time-consuming and troublesome to fix.

Equipment failures are impossible to prevent completely, but effective prevention can be done to reduce the chances of negative quality effects of equipment failure by naturally first of all selecting equipment that promises a long time between failure and also by making the transmission network plan as protected as possible. This can be achieved in many ways. The most cost-efficient and very effective one is to use physical layer loops with cross-connections that monitor the signal and switch over to an other direction if one direction of the loop is failing. If nothing else, it is wise at least to avoid long branches of transmission without any secondary routes. Creating long series of single-point-of-failure routes is asking for trouble.

Trying to protect the network in case of equipment failures is again an area where there is practically no limit to how much money and effort can be used to try to get more and more incremental improvements to the network under the banner of optimisation, so it must be carefully determined from the end user perception viewpoint whether it is worth it. For example, a very safe transmission network would have a physical loop for each base station with two different media, e.g. a fibre and microwave radio link, and two different plug-in units serving these different directions. That would protect even from accidental cable cut caused by a careless excavator or an unexpected squall of rain. However, even with such a very expensive arrangement there could be the possibility of failure.

Retainability/Sustainability

The transmission network should not be vulnerable when it comes to retainability/sustainability. The radion interface due to mobility of the end user is much more likely to have these issues due to coverage differences. However, the transmission network will also face challenges to keep the ongoing service running if the user is constantly moving, i.e. moving from one base station area to another, or if the radio interface planning is done so that there is some underlay–overlay plan with, for example, 900 and 1800 networks or 2G and 3G networks. In such a case the new transmission connection will face the same accessibility issues as discussed earlier; they just seem like retainability/sustainability issues from the end user perspective.

If the end user is stationary and there is no need for handovers due to other abov-ementioned reasons, then the situation is simplier for the transmission network.

Transmission Signalling

Transmission signalling can still cause some retainability/sustainability quality problems, because some services rely on signalling links themselves to provide the service. This means that from an end user perspective a well-started service can have problems during the end user experience. From the transmission network viewpoint this results from a series of instances of signalling network use, but again the end user does not know or care about that. Refer to an earlier discussion about how to optimise the transmission signalling network.

Transmission Capacity

Transmission capacity should be protected in 3G by admission control so that once the connection(s) for the end user service is/are admitted and established then the capacity should be guaranteed according to the service level agreement as long as the service is ongoing and untill it is dissolved.

There is, however, possibility that the transmission network is planned to be able to adapt to loading of the network. In other words if the network load increases to a certain level existing services might be reallocated new narrower bandwidth to be able to accommodate and admit still new service requests. And

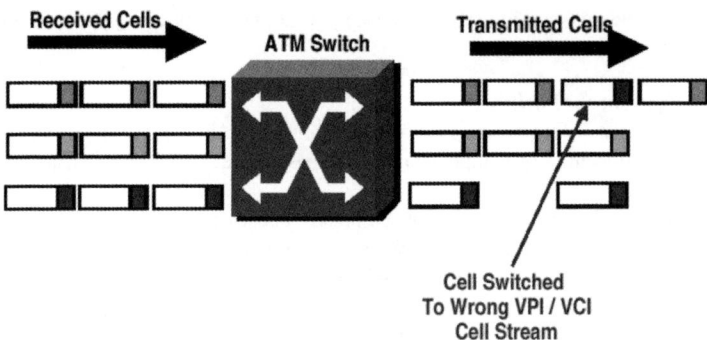

Figure 3.82 Retainability/sustainability problems

vice versa if the network load decreases existing services might be reallocated new larger bandwidth. This kind of adaptation even if it is of course very useful network feature overall, creates a sort of handover situation between two capacity allocations and like in any handover there is some chance of failure. RNCs that support such adaptation most likely always offer a chance to disable the feature, so if the quality is really main concern then one might consider disabling it. However generally the positive effects of such feature are likely to exceed the negative ones.

Transmission Quality Loss

This is the main cause of transmission retainability/sustainability quality issues. In a packet switched network the transmission network is increasingly vulnerable to bit errors. In 3G the ATM cells, AAL units and IP packets carry large amounts of data in one cell, unit or packet compared to the 2G era. Any errors in those headers can cause a loss of information and in the worst case misinsertion of information as well. Naturally these layers do provide retransmissions in case information is lost, but that helps only in cases of non-real-time services. Real-time services cannot utilise retransmissions (other than human retransmission: 'What did you say?') and so are more vulnerable in 3G transmission.

Figure 3.82 illustrates one example of retainability/sustainability problems caused by transmission quality loss. Refer to an earlier discussion about how to prevent transmission network quality loss.

Transmission Equipment Failure

This is supposedly a rare chance to experiance equipment failure while having an ongoing service, but it is possible, as discussed earlier, to have accessability problems due to the same reason. However, the causes and possible approaches needed to achieve optimisation have already been discussed above.

Bandwidth

This end user quality aspect is purely defined by transmission capacity, which can be allocated to the user, and the possible transmission channels that can be assigned to a certain service. There might also be some subscription policies related to bandwidth. This quality issue only covers non-real-time services, for obvious reasons.

The spectrum of different 3G non-real-time services is so broad (with many breakthrough services still not even foreseen) and end users so different that it is hard to tell what is exactly slow and what is a fast enough service. This amounts to psychology. However, simulations are needed to make sure that admission control, offered services, assigned transmission channels, subscriptions and average end user expectations do meet each other. Once again admission control should protect the transmission network so that no one should have a slow service if services and transmission channels are properly mapped.

Table 3.16 Physical layer counters

Description of counter	
Severely errored cells	Physical layer termination point counters for counting all errored cells with one or more errors in the cell header (containing the header error check error)

Table 3.17 IMA counters

Description of counter	
IMA group unavailable seconds	IMA group counter for counting one second intervals where the group traffic state machine is down
IMA link severely errored seconds	IMA link counter for counting one second intervals at the near end containing more than 30 % of the IMA Control Protocol (ICP) cells counted as ICP violation or one or more link defects, loss of IMA frame or link out of delay synchronisation defects except during IMA unavailable second condition
IMA link unavailable seconds	IMA link counter for counting unavailable seconds at the near end. Near end unavailability begins at the onset of 10 contiguous seconds which count severely errored IMA group seconds, including the first 10 seconds to enter the unavailable IMA group seconds condition, and ends at the onset of 10 contiguous seconds with no severely errored IMA group seconds, excluding the last 10 seconds to exit the unavailable IMA group seconds condition

Conclusion

This should conclude the discussion about what makes a difference in 3G transmission networks, what constitutes network quality and how revenues of the network operator can be maximised. The following tables give an example and insight into the kind of counters a regular 3G network should provide for monitoring the quality of transmission, as discussed above.

Physical Layer, ATM Transmission Convergence Performance and Inverse Multiplexing ATM (IMA) Performance

The counters in Table 3.16 are examples of what can be collected on the physical layer termination point (SET and PET). The physical performance indicators are the ESR, SESR and BBER, which have been defined before. The indicators using MML commands are shown in Figure 3.83. When the unit type is PET, the command is YMO: PET, 14; outputs.

Based on the IMA counters given in Table 3.17, the following key performance indicators can be set to monitor IMA performance:

$$\text{IMALink\_SeverelyErroredSeconds} = 100 \times \frac{\text{IMALinkSeverelyErroredSeconds}}{\text{ns} - \text{IMALinkUnavailableSeconds}} \quad (3.76)$$

where ns = number of seconds in the measurement period.

ATM Adaptation Layer Performance

The ratio of a successful AAL2 connection to the attempts made is very critical to measure the performance and the availability of the AAL2 user plane. Based on the results of the AAL2 KPI, if the target is not reached, a check of each counter permits identification of what is configured or set up wrongly. Some examples of AAL2 counters are given in Table 3.18.

```
PCM STATISTICS

PERMANENT MEASUREMENT

UNIT           START TIME

PET-14         2002-10-09 00:00


CRC PERFORMANCE MONITORING:

                TOTAL      AVAIL     UNVAIL
                TIME       TIME      TIME
LOCAL END       08:37:29   08:37:29  00:00:00
REMOTE END      08:37:29   08:37:29  00:00:00


(%)          EFS      ES     SES    DM     BBE
LOCAL END    100.00   0.00   0.00   0.00   0.00
REMOTE END   100.00   0.00   0.00   0.00   0.00
```

Figure 3.83 Example of PCM statistics (EFS, error-free second; DM, degraded minute)

The AAL2 successful connection ratio is as follows:

$$\text{AAL2UserPlaneSignallingSuccess} = 100 \times \frac{\text{SuccessfulConnectionEstablishment}}{\begin{array}{l}\text{SuccessfulConnectionEstablishment}\\+ \text{NetworkOutofOrder} + \text{TemporaryFailure}\\+ \text{Congestion} + \text{UnavailabelChannel}\\+ \text{Unavailable Resource}\\+ \text{Un Supported AALParameters} + \text{InvalidMessage}\\+ \text{Missing Information Element}\\+ \text{Not ImplementedMessage}\\+ \text{InvalidInformationElement} + \text{InvalidInformation}\\+ \text{ERQ\_TimerExpiry} + \text{REL\_TimerExpiry}\\+ \text{RES\_TimerExpiry} + \text{BLO\_TimerExpiry}\\+ \text{UBL\_TimerExpiry} + \text{UnrecognizedParameter}\\+ \text{OSAIAllocationFailure} + \text{ChannelUnavailable}\\+ \text{AdjacentNodeUnaviable}\end{array}} \quad (3.77)$$

Table 3.18 Examples of AAL2 counters

Description of counter	
Successful connection establishment	Signalling link counter counting for successful connection establishments. The number of connection events started in the AAL2 signalling.
Network out of order	Signalling link counter counting for network out of order messages (cause 38)
Temporary failure	Signalling link counter counting for temporary failure messages (cause 41)
Congestion	Signalling link counter counting for switching equipment congestion messages (cause 42)
Unavailable channel	Signalling link counter counting for requested circuit/channel not available occurrences (cause 44)
Unavailable resource	Signalling link counter counting for resource unavailable unspecified occurrences (cause 47)
Unsupported AAL parameters	Signalling link counter counting for AAL parameters can not support occurrences (cause 93)
Invalid message	Signalling link counter counting for invalid messages (cause 95)
Missing information element	Signalling link counter counting for mandatory information element missing occurrences (cause 96)
Not implemented message	Signalling link counter counting for message type nonexistent or not implemented occurrences (cause 97)
Invalid information element	Signalling link counter counting for information element nonexistent or not implemented occurrences (cause 99)
Invalid information	Signalling link counter counting for invalid information element contents occurrences (cause 100)
ERQ(establish request message)_timer expiry	Signalling link counter counting for recovery on ERQ_timer expiry occurrences (cause 102)
REL(release request message)_timer expiry	Signalling link counter counting for recovery on REL_timer expiry occurrences (cause 102)
RES(reset request message)_timer expiry	Signalling link counter counting for recovery on RES_timer expiry occurrences (cause 102)
BLO(block request message)_timer expiry	Signalling link counter counting for recovery on BLO_timer expiry occurrences (cause 102)
UBL(unblock request message)_timer expiry	Signalling link counter counting for recovery on UBL_timer expiry occurrences (cause 102)
Unrecognised parameter	Signalling link counter counting for message with unrecognised parameter, discarded occurrences (cause 110)
OSAI allocation failure	Signalling link counter counting for originating signalling association identifier (OSAI) allocation failures
Channel unavailable	Signalling link counter counting for requested AAL type 2 channel was not available in the destination AAL type 2 node occurrences
Adjacent node unavailable	Signalling link counter counting for connection establishment rejected since the signalling relation into the adjacent AAL type 2 node was not available

ATM Layer Performance

A list of example ATM counters is shown in Table 3.19. The ATM KPIs could then be as follows, where the ingress cells can be estimated using the following formula:

IngressCells ≈ DiscardedCLP0Cells + DiscardedCLP1Cells + EPDCells + PPDCells
    + DiscardedPolicingCells + TransmittedCells + DiscardedChecksumErrorCells
    + DiscardedUTOPIAInterfaceCells + DiscardedCRCErrorCells + DiscardedErroredCells
(3.78)

The ingress lost cells can be estimated using the following formula:

LostCells ≈ DiscardedCLP0Cells + DiscardedCLP1Cells + EPDCells + PPDCells
    + DiscardedPolicingCells + DiscardedChecksumErrorCells
    + DiscardedUTOPIAInterfaceCells + DiscardedCRCErrorCells + DiscardedErroredCells
(3.79)

A good estimation of the ingress CLR per port would then be

$$\text{CellLossRate} \approx 100 \times \frac{\text{LostCells}}{\text{IngressCells}} \tag{3.80}$$

On the egress side similar calculations can be made.

Loading

Using the formula above and the Ingress PCR of all the ports in the interface unit, we can estimate the average load per unit in ingress using the following formula:

$$\text{AverageLoadPerUnit} \approx 100 \times \left[\frac{\sum_{b=1}^{b=n} \text{IngressCells}_b}{T \left(\sum_{a=1}^{a=n} \text{IngressPCR}_a \right)} \right] \tag{3.81}$$

where

n = number of used ports per unit
T = measurement period

Minimising Costs

Two ways to minimise costs are minimising the transmission link capacities and transmission media selections.

Minimising Transmission Link Capacities

The costs of the transmission link capacity can build up from leasing costs of transmission capacity owned by some other company and/or from purchase, installation and maintanance costs of owned transmission capacity. Minimising the transmission link capacities is not a very effective method of improving network profitability if the operator owns the copper, fibre and/or microwave transmission links. Naturally when selecting the media and the bandwidth that will be bought it makes a difference to calculate carefully the foreseen service scenario to the base station level and make decisions about the media and bandwidth based on that calculation. However, when the purchase is done and the transmission links are installed then obviously transmission links should not be minimised but rather maximised within the installed bandwidth to get the most out of it.

Minimising the transmission link capacities is, however, quite a fundamental way to minimise the costs if the operator is using a leased transmission capacity. Leased capacity is normally in the form of copper or fibre, but very rarely, if ever, microwave radio links and used. Common granularity of leasing

Table 3.19 Example of ATM counters

Description of counter	
Discarded errored cells	Physical layer termination point counter counting for cells with more than one bit error. A count of the number of incoming ATM cells discarded due to a header error check (HEC) violation. HEC is used for checking and correcting an error in the ATM cell header. One bit errors are corrected. If there are more errors in the header it cannot be corrected but the cell is discarded
Discarded CLP0 cells	Computer unit counter counting for the total cell loss priority (CLP) 0 cells over the threshold on ingress. This tells how many high priority cells buffer management has discarded. There is a discard threshold for CLP0 cells. When the number of CLP0 cells in the buffer reaches the CLP0 threshold, incoming cells are discarded
Discarded CLP1 cells	Computer unit counter counting for the total cell loss priority 1 cells over threshold on ingress. This tells how many high priority cells buffer management has discarded. There is a discard threshold for CLP1 cells. When the number of CLP1 cells in the buffer reaches the CLP1 threshold, incoming cells are discarded
EPD cells	Computer unit counter counting for the total cells discarded due to early packet discard (EPD) on ingress. When congestion occurs and buffers are filling, EPD discards new packets arriving at a queue. All cells associated with a new packet are discarded. The remaining buffer space can then be used for ATM cells belonging to packets that have already entered the queue
PPD cells	Computer unit counter counting for the total cells discarded due to partial packet discard (PPD) on ingress. PPD discards all the cells associated with the packet discarded during buffer overflow
Transmitted cells	Computer unit counter counting for the total number of ingress cells transmitted to the switch fabric
Discarded policing cells	Computer unit counter counting for ingress cells discarded due to policing action. Policing discards the cells, which can affect the quality of service of established connections. Policing at the user–network interface is referred to as usage parameter control. Policing at the network–network Interface is referred to as network parameter control
Discarded checksum error cells	Computer unit counter counting for ingress cells discarded due to checksum errors
Discarded UTOPIA interface cells	Computer unit counter counting for ingress cells discarded due to parity errors at the UTOPIA interface. Count of cells when error was detected as a result of a parity check at the universal test and operations interface for ATM (UTOPIA)
Discarded CRC error cells	Computer unit counter counting for ingress cells discarded due to cyclic redundancy check (CRC) errors

on the physical layer is in terms of E1/T1/JT1 or VC-12/VC-4 in the case of fibre. On the ATM layer the bandwidth is leased in terms of kbps or cps. In one way or another, however, there is a granularity where the bandwidth is leased. Minimising the leased capacity to just meet the desired traffic scenario is the optimisation objective. The importance of this optimisation is especially large at narrowband links where a decision between leasing, for example, one or two E1 often means selecting between single and double costs. At more broadband links the effect decreases if, for example, the selection is made of whether eight or nine E1s should be leased. In the end the same maximisation of the usage principle applies as with owned capacity. When the leasing agreement is done and the transmission links are taken into use then obviously the transmission links should not be minimised but rather maximised within the leased bandwidth to get the most out of it.

So far it might seem quite a simple objective, but the wild card of this minimisation task is to consider the time line. Sooner or later the needed capacity is supposed to exceed the installed or leased capacity. The trick is to optimise that moment. There are big installation and purchase costs involved in owned capacity in case it needs to be upgraded and there are considerable lead times for purchasing new equipment and installing them. There can also be considerable lead times and extra installation costs for leasing more bandwidth. The decision about the period under optimisation should come for transmission optimisation from operator top management.

Transmission Media Selections

The selection of transmission media might not be very relevant to all operators since there are plenty of operators with some installed base of transmission capacity due to an already existing fixed line network or 2G network. However, capacity needs for transmission in 3G are likely to grow so high that some new selections and modification are likely to be needed, especially for the last mile for base stations. Even if there is not significant installed base of transmission capacity, selecting the transmission media is partly linked to a larger decision of whether to use own or leased transmission. That decision has the same parameters as any other owning versus leasing decision that individuals need to make, such as with a car, apartment, etc.

If top management decides to go for own transmission then microwave radio links provide an attractive option. Microwave radio links are normally faster to install, even if most radio links are subject to permission from authorities controlling the radio frequencies, and the failures with microwave radio links are easier to overcome. They do carry a larger risk of failure, however, compared with copper or fibre, due to radio propagation problems in rainy conditions, misalignment, etc., but conveversely, on the credit side of copper and fiber, once they are installed they are very resistant to errors. The main reason for problems with fibre and copper is human error; either someone cuts the cable or by error disconnects it. The downside of having fibre or copper is the long lead times to get it installed, with first the delay to get all the permissions to do the wiring and then to do the actual wiring itself.

If the top management decision is, however, to go for leasing then the main options are basically dark fibre, xDSL/PDH capacity, SDH/SONET capacity or ATM capacity. Dark fibre and xDSL/PDH capacity mean practically that the fibre ends are delivered to two locations to be interconnected or copper cable is delivered to digital distribution frames (DDFs) in two locations. On the other hand, SDH/SONET capacity as well as ATM capacity normally mean that some terminating equipment is delivered and the operator connects the base stations directly to that.

Summary

There is no way to guide anyone through the problems step by step, as there is such a variety of possible cases, but hopefully the information here is enough to give the reader a checklist of what are the important aspects in this optimisation task so that wise decisions can be made.

Endnote

* Source: ITU-R P 530-7. Reproduced with the kind permission of ITU.

4

Core Network Planning and Optimisation

James Mungai, Sameer Mathur, Carlos Crespo and Ajay R Mishra

PART I CIRCUIT SWITCHED CORE NETWORK PLANNING AND OPTIMISATION

4.1 Network Design Process

Due to the highly complex nature of network data design, it is imperative to follow a structured process for design of the circuit switched (CS) core network (CN). The network design process (Figure 4.1) can be done in several phases, as described below:

- Network assessment. The purpose of network assessment is to show the restrictions and advances of the network and to produce dimensioning guidelines concerning user distribution, traffic profiles and service penetration. These issues are taken into consideration in later planning steps.
- Network dimensioning. The objective of dimensioning is to produce a model of the network in accordance with the operator's forecast user and service profiles. The correct model should achieve the operator's technical objectives while optimising costs of investment. The results of network dimensioning form an important part of the network master plan, showing selected network architecture and the number of network elements required during each network roll-out phase.
- Detailed network planning produces the documents needed for implementation of the network and is based on the master plan prepared in the previous module.
- After detailed network planning, source data for each network element identified in the dimensioning phase is prepared and audited, ready for implementation.

4.1.1 Network Assessment

The main objective in network assessment is to study the existing network infrastructure and provide a feasibility report on its relevance to the network under design. During network assessment, the network growth and deployment is determined and new site interconnection requirements are given for a smooth

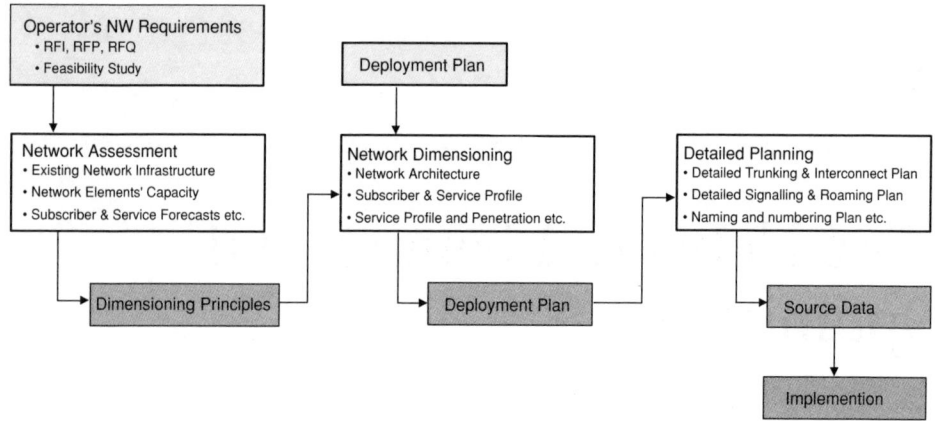

Figure 4.1 Network design process (RFI, request for information; RFP, request for proposal; RFQ, request for quote)

network implementation. The assessment phase may be performed for a new network or for expansion of an existing network. A network that is already planned or where preliminarily investments have been made, but not necessarily implemented, may also be assessed.

Many aspects of the network may be studied to ensure proper network design and planning. Some of the studies that need to be performed include:

- Subscriber forecast
- Service penetration
- Numbering plan
- Traffic routing study
- Country regulations

Tasks of the Network Assessment

The following information can be collected about the network:

- Solution required and technology
- Network elements, quantity, capacity and site location
- Site availability/spare capacity
- Transmission requirements and interfaces
- Existing transmission network
- Required services and new product roll-out schedule
- Subscriber forecasts
- Numbering plan
- Network topology
- Geographical information
- Charging plan and data used
- Traffic measurements

A careful evaluation of project requirements is needed to determine which part of this information is necessary in the process of network assessment. The requirements may vary from project to project and not all of the above information may be relevant to every project.

Table 4.1 Subscriber forecast

	Phase_1	Phase_2	Phase_3
MSC_01	130 000	210 000	350 000
MSC_02	112 000	182 000	240 000
MSC_03		190 000	260 000
MSC_04			50 000

Table 4.2 Subscriber categories

	Phase_1	mErl
Business	30 %	45
Post-paid	20 %	18
Pre-paid	50 %	10

Network Subscriber Forecast

An important input and key parameter required for network dimensioning is the number of subscribers forecasted and the subscriber growth for each of the next phases of network growth. These are usually provided by the operator's marketing department. In the absence of such data, a market service penetration forecast may be used.

If market service penetration is assumed, a population census of the main population centres may be used to determine the number of subscribers expected from these centres for each phase and each service. The geographical distribution of the subscribers will in this case be directly proportional to the population distribution. It is easier to assume switching locations in the major population centres or cities to reduce transmission costs. The number of subscribers in each switch at the end of each phase can therefore be presented in table form, enabling the network design engineer to estimate the required number of network elements (Table 4.1).

Sometimes it is possible to specify the percentages of different users, i.e. business users, post-paid users and pre-paid users (Table 4.2). Each group of users generates different amounts of traffic and a distribution of subscribers to these user types enables a more accurate estimation to be made of network traffic. While business subscribers may generate higher traffic per subscriber (e.g. 45 mErl/subscriber), nonbusiness category subscribers would usually generate lower traffic (e.g. 18 mErl/subscriber) and even less for pre-paid subscribers (e.g. 10 mErl/subscriber). It would therefore be important to make as detailed a study as possible to estimate the correct traffic amount and their sources.

Generally each subscriber group may be considered separately when dimensioning the traffic before adding all the traffic together to get the total traffic in the network. A tool for this purpose may be necessary.

If it is not possible to specify the ratio of different subscribers in the different subcategories, then it is sufficient to take an average amount of traffic per subscriber (in mErl) and assume a single subscriber category. In expansion projects of an existing network, it is possible to get the number of active subscribers from MSC/VLR measurements of active subscribers in the VLR database during the busy hour (BH).

Network Service Deployment

A careful study of planned and existing services should be done in order to assess its impact on the network. Some of the basic services are voice telephony, value-added services, Internet, intelligent network (IN)

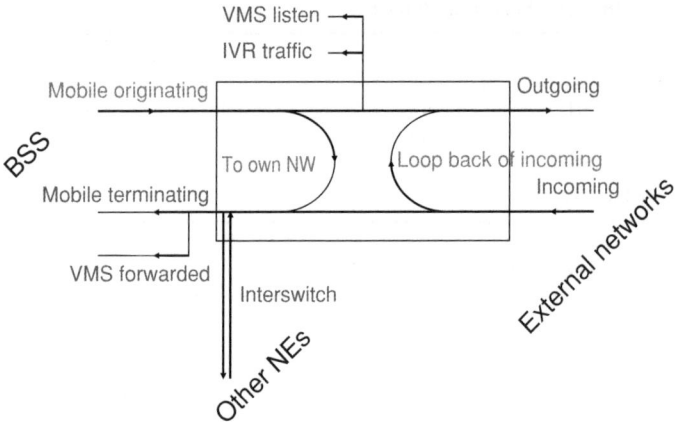

Figure 4.2 Definitions of interswitch traffic (VMS, voice mail server; IVR, interactive voice response)

services, etc. Details of the services to be supported by the network and their usage should be documented. Consider service growth and forecasts for different service types.

Service penetration can provide important input on the number of subscribers in the network using the service. This information should be available from the operator's marketing department and may be used to dimension the traffic due to a given service.

Network Naming and Numbering Plan

The network naming and numbering convention is important. It should be considered during the network assessment phase. For an existing network, a study of the existing naming and numbering principles should be done. Network element signalling point codes, engineering numbers (i.e. global title address, roaming numbers, handover numbers, etc.), allocated subscriber number ranges and IMSI number ranges for each HLR should be determined with the operator. This information is important when preparing the detailed design document for preparing network element source data.

For a completely new network an elaborate numbering plan is required from the operator. Once a number or number ranges have been allocated and used in the network, any change may cause unforeseen failures in the network. Naming and numbering schemes should leave enough room for future network elements, expansions and addressing requirements.

Traffic Model of a Switching Element

It is very important to get accurate data concerning network traffic (see Figure 4.2). The operator may provide the latest traffic data from an existing network. It is recommended, however, to generate the repartition matrix (to own network, to other networks, regional) with the help of a planning tool.

The main traffic model inputs are:

- subscriber growth per phase: phase 1, phase 2, phase 3, etc.;
- subscriber distribution per network site;
- ratio of 2G to 3G subscribers;
- traffic per VLR attached subscriber (2G, 3G);
- mean holding time of calls (voice and video);

- ratio of CS (transparent and nontransparent) data calls to total number of calls;
- traffic percentage share: M–L (mobile to land), L–M (land to mobile), M–M (mobile to mobile).

It is assumed that MOC + MTC = 100%. Thus either of the two values is needed. Also the percentages of 'subscribers to own network' and 'loop back of incoming' traffic are needed. It is important to take care not to end up with negative traffic numbers. For example: MOC = 75% (MTC = 15%). 'Subscribes to own network' is a relative value, with respect to MOC. If it is too high, say 44 %, then the absolute value, 33 % (44 % of 75 %), exceeds the value MTC = 15%!

It should be remembered that the traffic distribution is changing as the network evolves. In the initial phases of a new network the number of 'subscribers to own network' is small and grows at the same pace as the growth of the own subscriber base. In some network cases, the amount of traffic per subscriber varies from site to site, from one network region to another. This needs to be taken into consideration for a more accurate modelling of the network.

Country Specific Regulations

It is important to become familiar with country specific regulations at an early stage of the whole planning process. Regulatory issues may impose certain restrictions on the planning of networks, sometimes requiring special solutions to meet the requirements. Yet at times the regulations provide tailored solutions for a given country and all operators need to adhere these.

Results of Network Assessment

The results of the network assessment phase are presented in a detailed report. This report may be listed as a deliverable. The network assessment report should include all the inputs already discussed as well as the following points:

- Description of the network planning limitations imposed by the existing network infrastructure, e.g. the current network has several hot spots with congestion and/or the majority of all traffic is handled by a limited number of switches.
- Description of the network development considerations for the long term. With regard to the longer term growth, the planning report should point out issues that may limit network expansion. Considerations introducing a higher network layer (for transit or signalling STPs) could be discussed.
- Outline the network growth strategies in terms of costs and customer strategies.
- Dimensioning principles.

The report can be used as base information for the dimensioning process.

4.1.2 Network Dimensioning

The objective of dimensioning is to model an operator network based on subscriber and traffic forecasts to produce a technically optimum model of the network. The network model is presented in a master plan indicating the selected network architecture and the amount of nodes needed during the roll-out of the network.

Planning a network over several phases is done by starting with the last phase first. The network should be dimensioned in such a way that it can handle the traffic at the end of the phase. The pace of the subscriber growth determines the phases (in the ideal situation of unlimited resources). At the end of each phase the master plan should be reviewed and modifications for the upcoming phases can be made if necessary.

Different things will influence the design, such as

- Network architecture
- Quality and performance goals
- Subscriber traffic profile
- Service penetration and profile
- Traffic growth estimates

The planner has to consider everything that will affect the outcome of the final network design. For the dimensioning process the dimensioning principles are needed. They are collected during network assessment.

Switching Locations

The switching locations are selected based on available floor space in operator sites and traffic intensity. In general, network elements should be placed in sites with high traffic intensity. Transmission lines are needed to carry traffic from sites with low traffic intensity. These, however, come at a cost to the operator and therefore switching nodes should be placed closest to main sources of user traffic in the network.

Traffic Matrix

The total traffic originating from a given site may be estimated according to the total number of users and service usage. The distribution of that traffic depends on the traffic matrix, which can be deduced from measurements in the existing network or modelled according to subscriber distribution and calling/moving interest.

The end-to-end traffic matrix (Table 4.3) shows the distribution of calls to other switching sites in the network. Calling interest is the probability of a call terminating in a given switching site. This probability reduces with distance in an exponential way. Moving interest models the moving behaviour of subscribers in a given network. The probability that a subscriber temporarily relocates to a new location further from their home location reduces exponentially with distance. Other factors affecting the traffic matrix include subscriber distribution. Traffic moving to a given switching site will be directly proportional to the amount of subscribers in that location of the network.

Routing Plan

The routing plan would define the routing strategy for internal calls, external calls and international calls. It is also important to note the routing strategy for other types of calls, i.e. PABX (private automatic banch exchange), calls centre, service numbers, other services, etc. The routing strategy depends on the switching network structure, that is:

- the interconnect plan, i.e. switching connections by direct link, and which connections are made indirectly via transit switches;
- the number and locations of transit switches;
- the number of switching gateways to other networks;
- the number of levels of transit switches used;
- re-routing mechanisms for the congestion and failure situation, e.g. alternative routing, load sharing, dynamic routing, etc.

As a basic rule, a direct interconnect is provided between two switching sites if there is intense traffic between the two exchanges. Should the amount of traffic be low, however, it is preferable to use existing interconnect and exchanges as a transit to reduce transmission costs. The decision of how to connect

Table 4.3 End-to-end traffic matrix

Network total interswitch traffic

From	To Total	A1 350	B1 10 232	C1 350
A1	10 232	116.7	9999	116.7
B1	350	116.7	116.7	116.7
C1	350	116.7	116.7	116.7

Outgoing traffic

From	To Total	A1 130	B1 10 232	C1 130
A1	10 086	43.3	9999	43.3
B1	130	43.3	43.3	43.3
C1	130	43.3	43.3	43.3

Incoming traffic

From	To Total	A1 220	B1 147	C1 220
A1	147	73.3	0	73.3
B1	220	73.3	73.3	73.3
C1	220	73.3	73.3	73.3

Measured interswitch traffic

From	To Total	A1 346	B1 9999	C1 346
A1	9999	104.6	9999	104.6
B1	346	104.6	104.6	104.6
C1	346	104.6	104.6	104.6

the switches to each other is made by considering traffic distribution, the location of the local switching nodes and points of interconnection (POI) with other networks. This also enables a decision to be made as to whether a transit layer is required to carry transit traffic.

Signalling Plan

The signalling network architecture depends on common SS7 topology (see Figure 4.3). In TDM and ATM networks, this is dependent on the MTP topology of the network, but with Sigtran this depends very much on the underlying IP transport technology.

As a basic rule it is recommended that a minimum of two signalling links be used in a linkset or, in the case of Sigtran, two associations in an association set. Signalling routes and route sets are important in narrowband and broadband SS7 and the recommendation is to have at least two signalling routes in a route set.

The following are some of the important inputs for dimensioning the signalling network:

- Busy have call attempts (BHCAs) per subscriber
- Periodic location updates

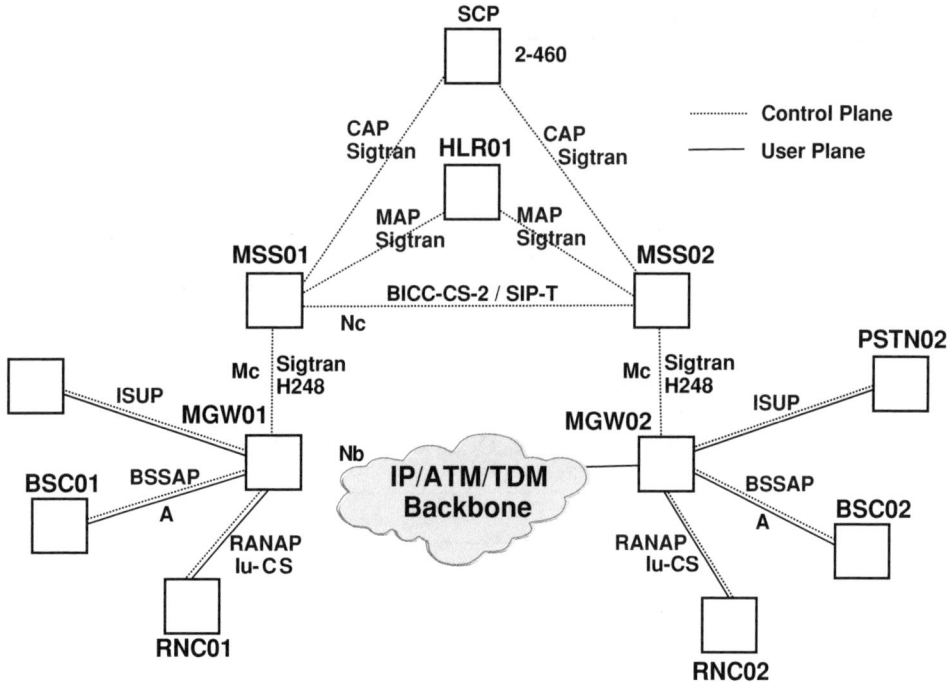

Figure 4.3 Example of signalling network architecture (SCP, service control point; CAP, Cammel application part; MAP, mobile application part; BICC, bearer independent call control; ISUP, ISDN user part; SIP, Session Initiation Protocol)

- Intra-VLR location updates
- Inter-VLR location update – incoming
- Inter-VLR location update – outgoing
- Number of attach subscribers
- Number of detach subscribers
- Authentication frequency
- EIR query frequency
- TMSI access
- Send authentication parameters and SAI responses
- Send ID and send parameters (IMSI inquiry) request
- IN calls per subscriber BH
- SMS MO during BH per VLR attached user
- SMS MT during BH per VLR attached user

The signalling interfaces may be dimensioned for each SS7 protocol using the message flows as defined in the standards.

Network Master Plan

The output of the network dimensioning phase is a network master plan with details of all required network elements, subscriber capacity, traffic capacity and transmission requirements. The master plan also provides other important information regarding the network architecture and phases of network deployment that serve as inputs for detailed network planning. The master plan should incorporate or contain information concerning network element deployment, architecture, internal traffic routing strategy, external traffic routing strategy, traffic intensity and transmission requirements for each interface.

The master plan is an important input to the next phase of planning: detailed network planning.

4.2 Detailed Network Planning

Detailed network planning is the phase where documents necessary to prepare source data for network element implementation are prepared. The implementation documents are based on the master plan prepared in the previous module and other inputs provided by the operator. After the detailed network planning documents, the other important task in detailed network planning is to produce the source data necessary to integrate for each network element.

The input information needed to complete the data sheets is to some extent collected during network assessment and network dimensioning and is contained in the network master plan. Some of the main inputs for detailed network planning are:

- ET allocation numbers and types
- Points of interconnection (POI)
- Number of PCMs towards each POI
- Signalling point codes for each network element and whether STP usage is allowed
- Signalling point names
- Signalling links and link sets: names and numbering
- Signalling route and route sets: names and numbering
- Network element naming and numbering plan
- Naming and numbering plan principals
  - Circuit groups
  - Destinations
  - Subdestinations
  - Charging cases
  - Routes
- Analysis trees and their usage
- Preanalysis
- Emergency and service call routing: routing zones, service and emergency numbers
- Detailed numbering plan for engineering, subscriber number ranges
- Announcements, etc.

Some of the important document deliverables after the detailed network planning are:

- Naming and numbering plan
- Network topology and trunking plan
- Network interconnect plan
- Roaming plan
- Routing plan
- Signalling plan
- Network synchronisation plan
- Announcements and network tones, etc.

Naming and Numbering Conventions

It is important to create a naming and numbering plan prior to preparing source data forms. Once such guidelines have been prepared, they should be strictly followed. Changes of identifiers and labels in a later phase are likely to cause errors and data inconsistency in the network.

As an example:

- Naming of circuit groups to MGWs:

    MGNNXXY
    uniquely identifies a circuit group (CGR) to specific MGWs where

    NN   identifies the MGW location
    XX   identifies the MGW X by number
    Y   identifies consecutive circuit groups the same as MGW

- Circuit group number should be unique in the network element. The CGR numbers can be dedicated for certain interfaces. For example:

    300–699   MGW CGR
    700–899   BICC circuit groups
    900–999   Interconnect to POI

- Signalling link set name. Name the signalling link set that could be linked to the destination signalling point to where the link is set up.
- Name of subdestination. The subdestination is used for routing a call inside the network. The most informative way of naming is according to the name of the circuit group used for the call. This enables easier troubleshooting during network operation.
- Naming of destination. The destination name should identify the switch to which the call is to be routed or the network and POI for the interconnect calls. The destination is determined during digit analysis or, in case of incomplete analysis, the next switch where the call would be further analysed.

Network Architecture

The detailed design documents include a set of diagrams showing the architecture and network topology at a more detailed level. The relationship between the switching nodes, the used interfaces and their main parameters should be drawn. Detailed network diagrams should contain the following information:

- Trunking plan: interswitch connections (network internal)
- Network interconnects: points of interconnections with external networks
- Signalling network plan
- Network service platforms connecting plan
- Synchronisation plan

Routing Plan

A routing plan has to be created for all new network elements and updates to existing network elements (Figure 4.4). Table 4.4 shows an example of how to create a routing table. Direct connection serves as the default route while the connections via other MSCs serve as alternative routes. If the network has a transit layer and transits are implemented in mated pair configurations, the traffic may be load-shared between the transits in a mated pair.

Detailed Signalling Plan

A detailed signalling plan contains information of the signalling points, point codes, links, link set and routes and route sets for the entire network. This serves as the basis for signalling definitions in the network elements. Figure 4.5 presents an example of a detailed signalling plan for an example network. The signalling definitions for the example network are presented in Table 4.5 and signalling linkset definitions are presented in Table 4.6.

Table 4.4 Routing plan example

Originating Point	Destination	Alternative 0	Alternative 1
MSC01	PSTN01	Direct	MSC02
	MSC02	Direct	MSC03
	MSC03	Direct	MSC02
	MSC04	MSC03	MSC02
MSC02	PSTN01	Direct	MSC01
	MSC01	Direct	MSC03
	MSC03	Direct	MSC04
	MSC04	Direct	MSC03
MSC03	PSTN01	MSC01	MSC02
	MSC01	Direct	MSC02
	MSC02	Direct	MSC01
	MSC04	Direct	MSC02
MSC04	PSTN01	MSC02	MSC03
	MSC01	MSC03	MSC02
	MSC02	Direct	MSC03
	MSC03	Direct	MSC02

Table 4.5 Example of signalling definitions for the MSC (NA, national)

Signalling point name	Signalling network	Signalling point code (SPC)	Global title CC + NDC + SN
PSTN1	NA0	1200	
MSC01	NA0/NA1	1340	254 733 000001
MSC02	NA0/NA1	1341	254 733 000002
MSC03	NA1	1342	254 733 000003
MSC04	NA1	1343	254 733 000004
HLR01	NA1	1320	254 733 000101
SMS01	NA1	1380	254 733 000201
SCP01	NA1	1360	254 733 000210

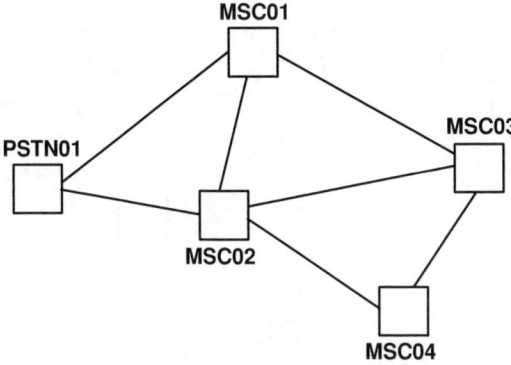

Figure 4.4 Example network for routing

Table 4.6 Example of a signalling link set definitions table

Signalling point name	Signalling network	Originating point code (OPC)	Destination point name	Destination point code (DPC)	Number of links	Signalling link bit rate
MSC01	NA0	1340	PSTN1	1200	8	64k
MSC02	NA0	1341	PSTN1	1200	8	64k
MSC01	NA1	1340	HLR01	1320	4	64k
			MSC02	1341	4	64k
			MSC03	1342	4	64k
			SMS01	1380	2	64k
MSC02	NA1	1341	MSC03	1342	4	64k
			MSC04	1343	4	64k
			SCP01	1360	2	64k
MSC03	NA1	1342	MSC04	1343	2	64k
			HLR01	1320	2	64k
			SMS01	1380	2	64k
MSC04	NA1	1343	SCP01	1360	2	64k

Detailed Numbering Plan

The detailed numbering plan is provided by the operator. It consists of an allocated numbering range for the network MSISDN numbers in the E.164 format. It also contains the allocated IMSI number ranges in the E.212 format. The operator may also divide the provided number ranges for different purposes in the network, i.e. engineering numbers and subscriber MSISDN number allocations.

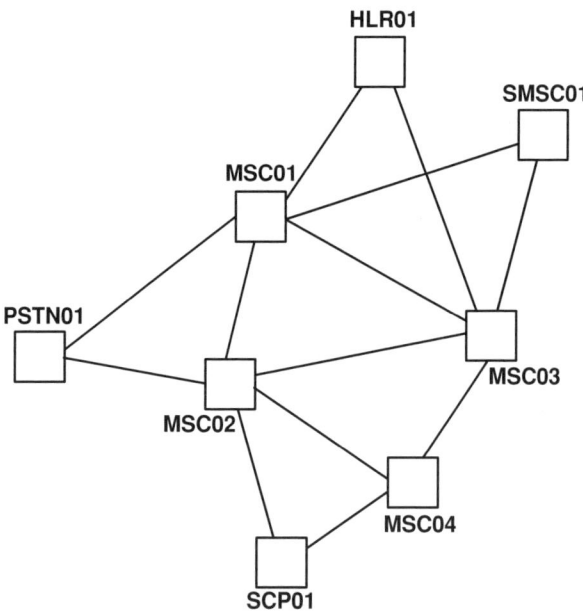

Figure 4.5 Detailed signalling plan (SMSC, short message service centre)

Table 4.7 Numbering plan for the example network

MSISDN number range	
E.164	
254 733 0000&&9	GT (global title) number range
254 733 0010&&9	MSRN numbers
254 733 0020&&9	HO numbers
254 733 1&&9	Subscriber MSISDN numbers
E.212	
639 02 1&&9	IMSI number range

The E.164 number format is

$$CC + NDC + SN$$

where

   CC = country code
   NDC = national destination code
   SN = subscriber number

The E.212 number format is given as

$$MCC + MNC + MSIN$$

where

   MCC = mobile country code
   MNC = mobile network code
   MSIN = mobile subscriber international number

The engineering numbers consist of network element global titles, MSRN numbers and HO numbers. The subscriber numbers consist of number ranges for test users, operator employees and commercial users.

Table 4.7 presents a number allocation for the example network.

4.3 Network Evolution

4.3.1 GSM Network

A 2G or GSM network consists of a base station subsystem (BSS) and a network subsystem (NSS) (Figure 4.6). The GSM network is suited to provide voice telephony services for mobile subscribers. In 2G phase 2+, the GSM network has been further enhanced with a general packet radio service (GPRS) to provide high speed data services for GSM subscribers. The GPRS uses a packet mode technique in GSM to transfer high speed and low speed data and signalling in an efficient manner. The GPRS optimises the use of network and radio resources.

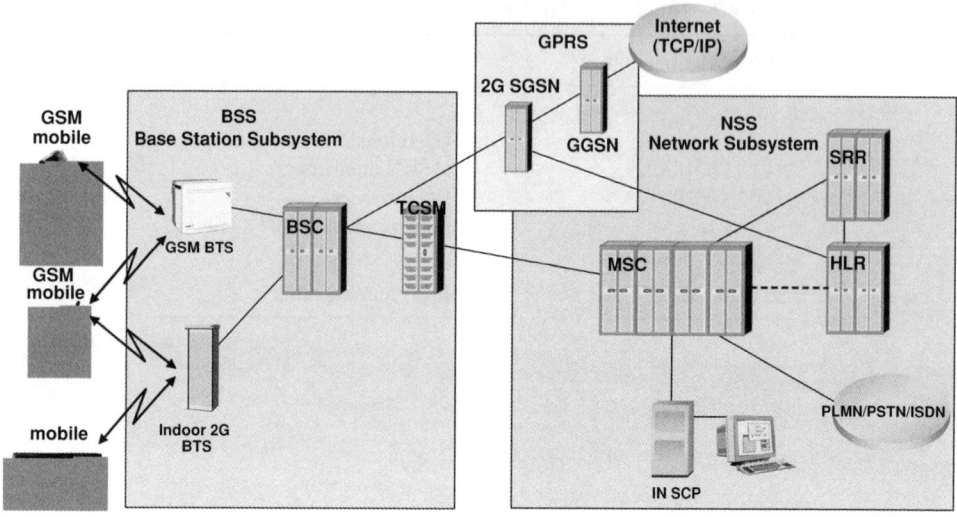

Figure 4.6 GSM network architecture (SRR, service routing register)

2G GPRS Network

The GPRS system brings the packet switched bearer services to the existing GSM system. Support for GPRS in the HLR includes GPRS subscriber data storage, a standard MAP interface to the serving GPRS support node (SGSN) and a subscriber data management interface for GPRS subscriber data.

4.3.2 3GPP Release 99 Network

Release 99 Network Architecture

The architecture of this network is shown in Figure 4.7.

3G MSC in Release 99 Architecture

The 3G MSC is a central network element for release 99 mobile networks. The release 99 3G MSC consists of two network elements, the MSC and MGW. The MSC contains the visitor location register (VLR) and service switching point (SSP) functionalities. The 3G MSC is able to serve the 3G radio access network (RAN) by supporting the Iu interface through a new network node, the ATM module (otherwise known as the MGW), and acting as a multiplexing point for the Iu interface, so that one Iu interface connection is needed from the RNC.

The A interface towards the GSM BSS is supported by the 3G MSC, as required. The MSC functionalities can be categorised into three main areas:

- Call control and switching functionality
- Visitor location register (VLR)
- Service switching point (SSP)

Network Evolution 329

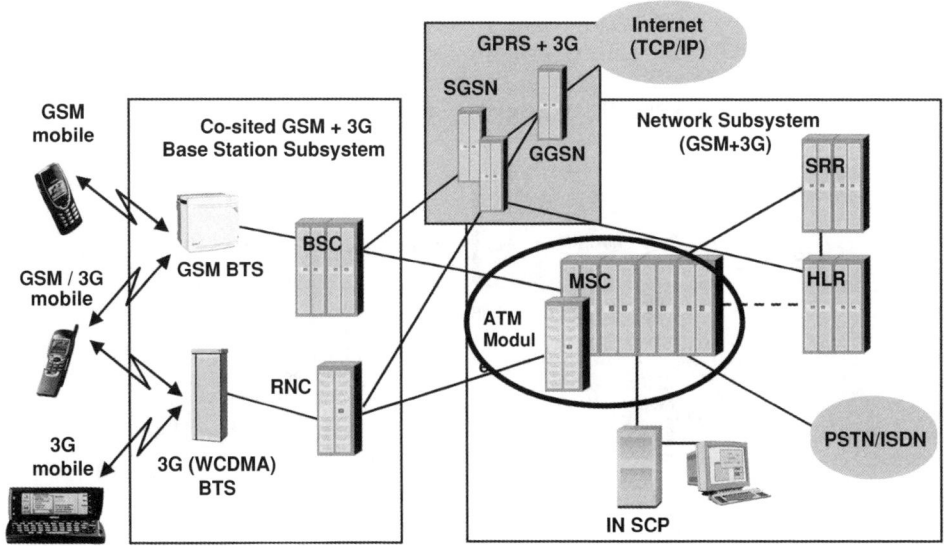

Figure 4.7 3GPP (Third Generation Partnership Project) release 99 network architecture

Release 99 MGW

The main function of the release 99 MGW (ATM module) is to provide UMTS terrestrial radio access network (UTRAN) interworking with the mobile switching centre (MSC). The release 99 MGW also provides transcoding (TC) services for calls originating in the UMTS network. The new interfaces introduced to the network are:

- the Iu interface connecting the RNC to the MGW;
- the *A* interface connecting the MGW to the MSC.

The MGW's main functionalities in 3G release 99 environments can be categorised as:

- the IWU function for the ATM to TDM transcoding;
- signalling Gateway functionality for the RANAP to BSSAP conversion;
- the ATM interface to RAN connectivity.

4.3.3 3GPP Release 4 Network

Release 4 Network Architecture

The next step of network evolution contains the 3G release 4 networks. The network architecture of a 3G release 4 network comprises three functional planes: the user plane, the control plane and the application plane (Figure 4.8). The user plane contains all network elements mainly involved in actual switching and handling of user data. The control plane network elements perform call control, subscriber mobility management, charging and MGW control functions. The 3G release 4 architecture introduces several new network elements, i.e. the application server (APSE), the MSC server (MSS), the gateway control server (GCS), the media gateway (MGW) and the circuit switched data server (CDS).

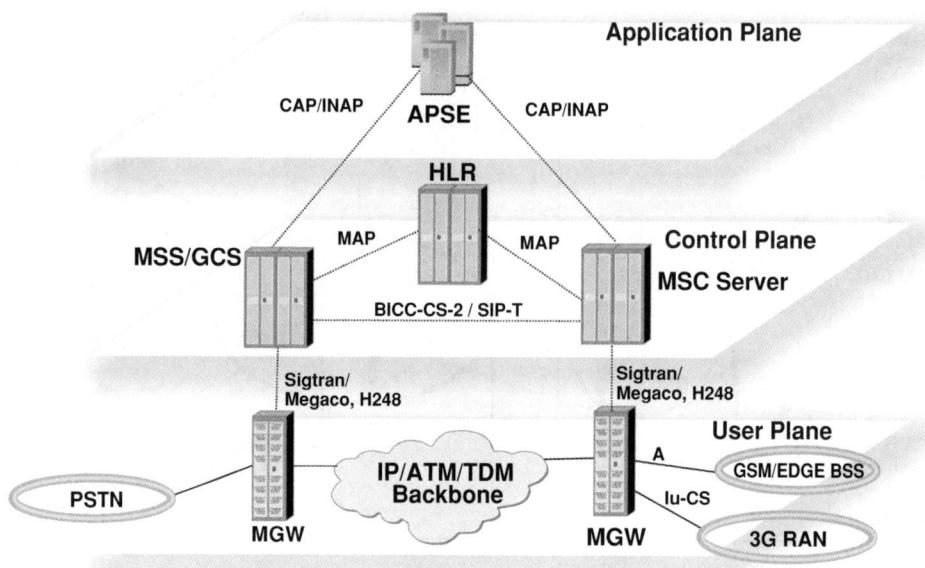

Figure 4.8 3G release 4 network architecture

MSC Server (MSS)

As mentioned earlier, in the MSC server concept, the user plane and the control plane are separated, as introduced in the 3GPP specification TS 23.205. Separation of the two planes in the core network means that 2G/3G MSC functionalities evolve so that user plane traffic switching functionality is devolved to a new user data switching element called the media gateway (MGW). The MSS and GCS elements maintain call control, mobility management and MGW control functionalities of the network. The MSS/GCS also handle the connection of media channels in the MGWs.

The MSS has several main functionalities:

- Gateway MSC (GMSC) functionality for call control
- Visited MSC (VMSC) functionality for mobility management
- Media gateway control function (MGCF) functionality for media gateway control
- Switching functionality available in an integrated MSC server solution
- Interworking function (IWF) functionality is also available for integrated MSC server solution

The GCS is a new network element in the MSS solution and provides for interconnections with PSTN and external networks. The GCS is intended for operators with high transit traffic in the network and thus require a transit layer or dedicated PSTN access for IP telephony. The main functionalities of the GCS are:

- GMSC functionality
- MGCF functionality

The standard also introduces three new interfaces, the Nc (network control) interface, the Mc (MGW control) interface and the Nb (network backbone) interface. In order to control MGWs to route user plane traffic through different bearer channels in the backbone, 3GPP release 4 circuit core standards recommend the use of BICC-CS2 as the signalling protocol on the Nc interface. BICC-CS2 is able to establish bearer channels for user plane connections both with an ATM and IP based backbone. Moreover, BICC-CS2 is independent from the signalling message transport layers, which means that BICC-CS2 can be carried on top of TDM (SS7), ATM (SAAL, on signalling ATM adaptation layer) or IP (SIGTRAN).

When the backbone network is based on IP transport, the signalling for call control between MSS/GCS elements in the release 4 core circuit network can be based on the SIP protocol. SIP messages are carried on top of the UDP/IP. An MSS/GCS may use domain name server (DNS) functionalities to get IP addresses of the destination MSS/GCS when performing call control related signalling. The use of IP routing for the signalling messages and DNS resolution to resolve domain names to IP addresses is used in the transport of signalling messages on the control plane. There is not much need for higher layer signalling transfer points when SIP is used. It has to be noted that the use of the SIP protocol in the release 4 circuit core is a proprietary solution, not included in 3GPP standards.

Media Gateway (MGW)

The MGW in the MSS system solution is an independent network element. It is, however, controlled by the MSS through an internal Media Gateway Control (MEGACO). Protocol running on IP. It has two main functions:

- Trunk switching functions
- Termination of the A, ATM and IP interfaces

Network Backbone

The 3G release 4 soft switch solution offers operators the flexibility to choose a suitable network backbone depending on availability and suitability of transmission in the local market environment. The network backbone can be based on any available transport technologies, i.e. IP, ATM or TDM. This flexibility enables operators to maintain a common backbone technology for their CS and PS services.

Packet switched (PS) traffic is expected to increase in release 4 networks and later in release 5 networks, thus ensuring a common backbone with greater synergies: IP, ATM or TDM. The separation of user and control plane architecture anticipates the evolution to an all-IP architecture, which at the same time supports the already existing networks. This enables operators to reduce operating expenditure (OPEX) with the MSS system compared to a similar network in a TDM network environment, due to high capacity compact network elements and a common network backbone. The number of required network elements and sites is expected to be lower as a result of this common circuit switched network for both GSM and UMTS networks.

The 3GPP release 4 network contains an additional element called the IP multimedia subsystem (IMS). The IMS, when fully specified, will contain a uniform way to maintain VoIP calls, thus offering the operators a way to deliver VoIP calls between UMTS networks. In addition to 'universal VoIP' the IMS offers a platform to other real-time and non-real-time IP services, such as, for instance, multimedia services.

The IMS consists of the media gateway control function (MGCF), the call state control function (CSCF) and the media resource function (MRF). These three functionalities form an extended model of connection management when compared to the 3GPP release 99 network. Basically the MGCF controls the MGW used in the connections, whether when using conversions, echo cancelling, etc. The CSCF and MRF together form the logic of how the transaction using IMS is treated.

The MGCF–CSCF–MRF model brings a big and new aspect to the system, which is related to the services, their creation and treatment. The 3GPP release 4 network contains a service platform, which is taken into use through the MGCF–CSCF–MRF chain. The basic concept exists in the 3GPP release 99 network and actually the service capabilities are presented there, but in the 3GPP release 4 network the core network structure supports the effective use of service capabilities.

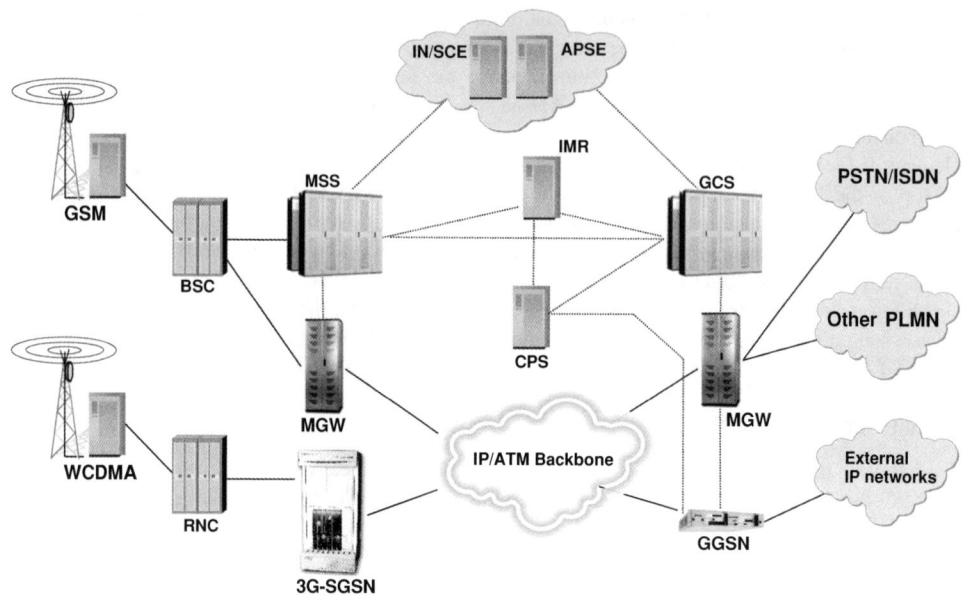

Figure 4.9 3G release 5 network architecture (including GSM) (SCE, service creation environment; IMR, IP multimedia register; CPS, connection processing server; GGSN, GPRS gateway service node)

The 3GPP release 4 network contains all the possibilities of traffic treatment. If the transaction coming from the access network is packet switched, it may be relayed to the external network either in a packet switched or a circuit switched manner. Also, if the transaction coming from the access network is circuit switched, it may go to the external network either in a circuit switched or a packet switched manner. In the 3GPP release 99 network the nature of the connection (circuit/packet) remains the same through the network.

The 3G release 4 network elements also ensure future proof investment as all the new network elements introduced can be upgraded to support future releases. The MGW for a legacy PSTN interconnect may also be used in an all-IP network. This will also provide operators with an early experience for voice transport over the IP.

Circuit Switched Data Server (CDS)

Mobile circuit switched data requires a suitable interworking function (IWF) to be provided by the core network. In the 3GPP release 4 architecture this functionality is located in the media gateway (MGW), but in the pre-release 4 architecture it is located in the MSC. Some vendors do not provide integrated interworking functions for circuit switched data calls in the MGW but rather in the dedicated network node for nontransparent data calls. The IWF functionality is therefore provided in an external circuit switched data server (CDS).

4.3.4 3GPP Release 5 and 6 Networks

The 3GPP release 5 network aims to introduce a UMTS network where the transport network utilises IP networking as much as possible (Figure 4.9). Therefore this goal has been given the name of an 'all-IP' network. IP will be used in both the network control and user data-flows. The mobile network implemented according to the 3GPP release 5 specifications will be an end-to-end packet switched cellular network using IP as the transport protocol instead of CCS7 (common channel signalling 7), which holds a major position in existing circuit switched networks. The IP based network will still support circuit switched services. The 3GPP release 4/5 networks will also start to utilise the possibility of using new radio access techniques. In the 3GPP release 99 network the basis for the UMTS terrestrial access network (UTRAN) is WCDMA radio access. In the 3GPP release 4/5 networks another radio access technology derived from GSM with enhanced data for GSM evolution (EDGE) will be specified to create the GSM/EDGE radio access network (GERAN) as an alternative to the UMTS mobile network.

In the 3GPP release 5 network the access network will experience more changes, with the changes done in the core network being minor by nature. In the 3GPP release 5 network the traffic is always packet switched and may be either real-time or non-real-time. In the development of release 5 the focus has shifted to the PS domain, which has been extended with the IMS functionality. The services, their accessibility and creation are emphasised in the 3GPP release 4/5 core network implementation. The service capability layer has already been introduced in the 3GPP release 99 network but in further implementations its role will be increased by open service architecture (OSA) based solutions. The OSA provides mechanisms for universal service creation and management.

The changes performed between release 4 and release 5 should not be visible to the end users. The UTRAN radio path still works in the same way and the terminals being used are also still working. Within the access network, however, the transport technology could be IP instead of ATM.

Besides UTRAN the evolved GSM BSS named GERAN can be connected to the core network with the Iu interface. Thus the traffic coming from GERAN gets the same treatment as the traffic coming from UTRAN in the sense of interfaces. If the operator has the IMS in use, the CN CS domain is not basically needed any more; the main step (besides the new radio access alternatives) between the 3GPP release 4 and release 5 networks is whether to quit the CN CS domain or not. This will depend significantly on the direction and maturation of the VoIP development.

Release 6 is an enhancement of earlier 3GPP releases, aiming to bring mobile users a complete 3G experience. The 3GPP release 6 network includes numerous new features, among them being high speed uplink packet access (HSUPA), the second phase of the IP multimedia subsystem (IMS), interworking with wireless local area networks (WLAN), the multimedia broadcast multicast service (MBMS) and enablers for push-to-talk (PoC). Release 6 brings many notable new services to mobile device users. Symmetrical data services enabling two-way high speed data communications for services such as video conferencing and mobile email are specified. High speed uplink packet access (HSUPA) will complement high speed downlink packet access (HSDPA) for a rich, two-way, interactive wireless broadband user experience. HSUPA and HSDPA together will enable symmetrical data communications at a high speed, supporting multimedia, VoIP, etc.

The second phase of the IMS comprises all of the core network elements offering multimedia services. The IMS makes it possible for operators to offer mobile users multimedia services using the Session Initiation Protocol (SIP). The SIP enables mobile users to use services based on Internet applications. In release 6, the IMS is developed to support interworking with circuit switched networks, non-IMS networks and 3GPP2 based CDMA systems.

Release 6 also defines interworking with wireless local area networks (WLANs). The interworking is defined in a very flexible way, enabling different multiradio scenarios. The multimedia broadcast multicast service (MBMS) makes it possible to distribute multimedia content efficiently to multiple recipients. Such content could be, for example, video or music clips. Conversational services such as push-to-talk (PoC) are also specified in release 6, together with open mobile alliance (OMA).

The main products are the connection processing server (CPS) and the IP multimedia register (IMR). The CPS provides a centralised point of registration, session control and charging for IP multimedia services. The IP multimedia register complements the existing HLR with additional subscriber and service information for SIP subscribers as well as a service deployment mechanism. The IP multimedia core fits easily into the existing network environment and complements existing investments.

Connection Processing Server (CPS)

The connection processing server (CPS) is a centralised entity providing control of end-to-end IP multimedia services using the Session Initiation Protocol (SIP). The CPS generates charging data and sends it to the charging gateway.

Ready-made incoming and outgoing traffic treatment services are provided for the end-user, e.g. personal lists for session forwarding including redirection to personal contact pages. The CPS also supports the ISC interface towards external SIP based application servers that can offer extended services to subscribers.

IP Multimedia Register (IMR)

The interface between the 3GPP HSS (HLR server system) and call state control function (CSCF) is known as the Cx interface. The user mobility server (UMS) in the IMR implements the standard Cx interface. The Cx interface is based on the Internet Engineering Task Force (IETF) Diameter Base Protocol and 3GPP Cx application extensions. The diameter layers are run on top of the SCTP (Stream Control Transport Protocol)/IP. The Cx interface supports the transfer of data between the UMS and the CSCF functionality in the CPS. The main procedures that require information transfer through the Cx interface are:

- procedures related to registration and the serving CSCF (S-CSCF) assignment;
- procedures related to routing information retrieval from UMS to CSCF, as in a mobile terminating session;
- procedures related to updating the user's data in the S-CSCF.

There is also a proprietary interface between the service execution environment (SEE) of the CPS and the service and subscription repository (SSR). This interface is called Xc and is also based on the Diameter Protocol with application extensions. The Xc interface is used to deliver service profiles, service logic and service data from the data repository in order to service execution.

IP Multimedia Service

What is a common feature of the successful mobile services of today, namely mobile voice and short message service (SMS)? They are person-to-person services and satisfy a basic human need: the need to communicate. Operators worldwide are currently deploying networks with IP capabilities (GPRS, EDGE and WCDMA). These capabilities can today be used to fill some human needs, such as the need for information (content downloading and browsing), the need for basic entertainment (content downloading, streaming, downloadable games), etc. However, in order to increase revenue further, operators need to find other human needs and to fulfil existing needs better. One such need is the need to share experiences: share things that are seen, done together and emotions.

IP has been designed to carry multimedia traffic. Today's networks offer IP capabilities for person-to-content services, but lack a mechanism to establish IP connections between terminals. The IP multimedia core enables IP based person-to-person services over mobile networks as well as to/from alternative access networks. IP connections between people enable a range of new multimedia communication services as well as the ability to create interactive applications like interactive games. Communication services no longer use just one media, but the end users can add and drop different media within one session. These new services will enable the users to communicate and interact in innovative ways, to save money, time, trouble and have fun. For the operator, the key issue is to launch a wide variety of services cost-efficiently by using a single connectivity mechanism that also provides centralised charging. This is possible with the IP multimedia core.

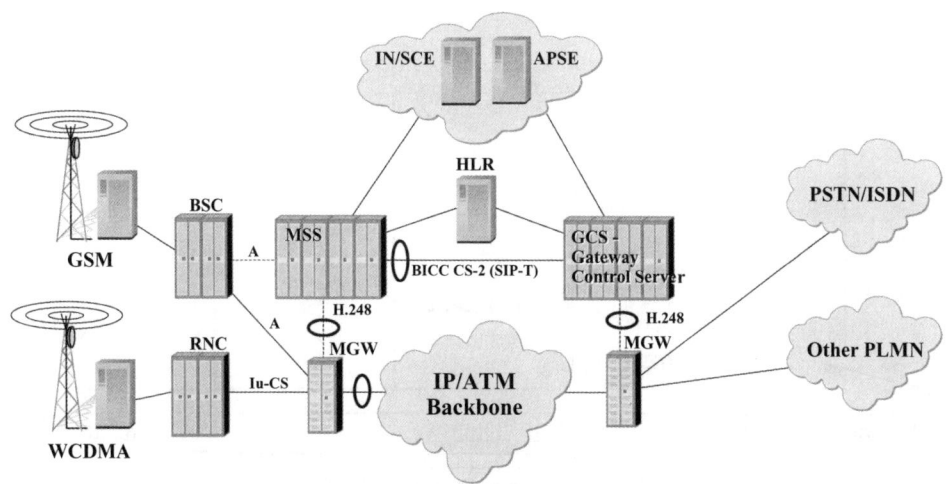

Figure 4.10 MSC server network architecture

4.4 3GPP Release 4 Circuit Core Network

4.4.1 Release 4 Core Network Architecture

Figure 4.10 presents one soft-switch solution implementation of the 3GPP release 4 architecture. The user plane and the control plane are separated in the MSC server concept. This reflects the split architecture as defined in the 3GPP specification TS 23.205. The separation of the two planes in the core network means that 2G/3G MSC functionalities evolve such that user plane traffic switching functions are performed by the media gateway network element, while call control and mobility management functions are performed by MSS/GCS network elements. The MSS/GCS network elements also take care of the connection of media channels in the MGW.

3GPP release 4 circuit core standards recommend BICC-CS2 as the signalling protocol in the Nc interface to be used to establish bearer channels for user plane connections both with an ATM and IP based backbone. Furthermore, BICC-CS2 is independent from the signalling message transport layers, which means that it may be carried on top of the TDM (SS7), ATM (SAAL) or IP (SIGTRAN). The H.248 (MEGACO) Protocol is used by the MSS to control the MGWs when routing user plane traffic through different bearer channels in the backbone.

When the backbone network is based on IP transport, the signalling for call control between MSS/GCS elements in the release 4 core circuit network can be based on the SIP. SIP messages are carried on top of the UDP/IP. If domain names are used the MSS/GCS may use a DNS to change domain names to IP addresses that may be used to route call control related signalling to the target MSS/GCS. The use of IP routing for signalling messages and of DNS to get those IP addresses eases tasks of configuration of signalling networks and is not limited by the number of MSS/GCS in the network. The need for a signalling transfer point is no longer necessary when using SIP. It has to be noted, however, that the use of the SIP in the release 4 circuit core is a vendor proprietary solution, and is not included in the 3GPP standards.

4.4.2 CS Network Dimensioning

Dimensioning Inputs

The actual process of network planning has already been discussed. The CS core network planning for 3G release 4 networks would follow the same planning process. It is important to remember that network

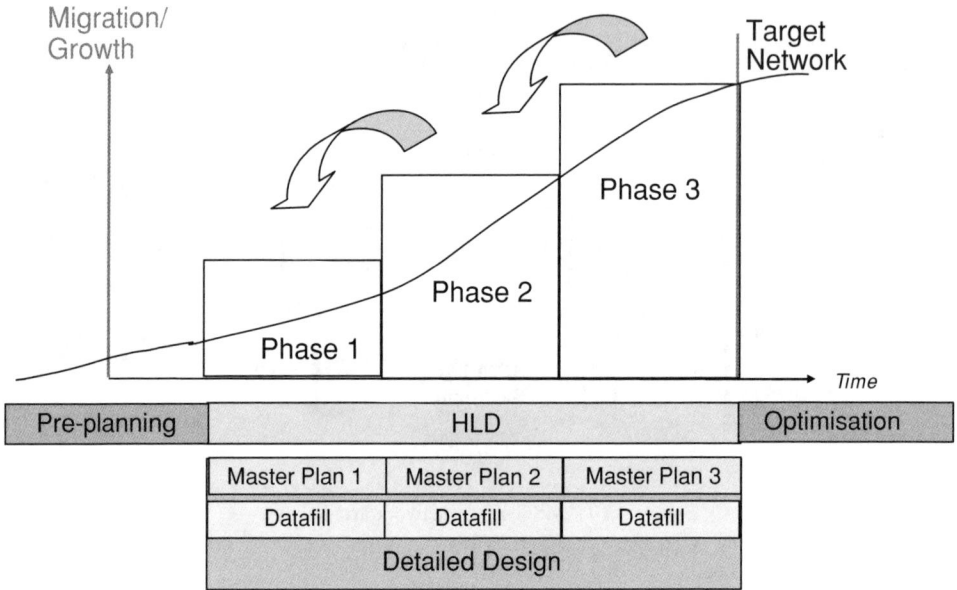

Figure 4.11 Phased planning of the network (HLD, highlevel design)

assessment is still a fundamental part of the process. If not explicitly carried out as a separate module, information gathered during network assessment is an important input for network dimensioning module.

Prior to dimensioning the network, the following information should be agreed with the operator before performing network dimensioning:

- existing sites, floor space available, and transmission equipment installed and available capacity, existing equipment technology, quantity and location;
- planned services to be provided by the release 4 network;
- subscriber forecasts, traffic profile, forecasts or measurements;
- country specific regulations, customer specific requirements;
- network element capacity and functionality roadmaps.

Network Dimensioning

The purpose of network dimensioning is to find the most cost-effective and technically optimised design of the network identifying selected network architecture and the required number of sites and equipment needed during roll-out of the network. The network design can be achieved through several planning phases (Figure 4.11). The designed network should be able to carry the total traffic at the end of each phase. The pace of the subscriber growth determines the phases. At the end of each phase the master plan should be reviewed and where necessary modifications for the upcoming phases can be made. When several planning phases are required the last phase is optimised first. The network should then be dimensioned such that there is no unnecessary capacity introduced that is not needed in later phases.

Planning is generally an iterative process. The network architecture and topology can be selected with the inputs collected, based on the coverage area of each element. Each network element is dimensioned according to site location and subscriber/transmission capacity per road map. In the release 4 split architecture, the user plane and control plane are calculated separately. The end-to-end signalling traffic is calculated between the network elements. The traffic can then be routed based on the interconnection of the elements. It is important to check that the load on each element in terms of processing capacity and number of interfaces does not exceed its nominal capacity. Calculate the transmission needs to implement the chosen configuration. If the network element capacity or interface is exceeded, then a repeat of the iterative process is required until an optimal solution is found.

User Plane Planning

As one of the innovative aspects of the 3GPP release 4 network is the separation of the user plane and control plane, the location and coverage areas can be defined for both the MGW and MSS/GCS. In brief, the user plane planning process can be summarised as follows:

- Determine the number of MGWs, site locations and coverage areas.
- Calculate end-to-end traffic matrices in Erlangs.
- Determine the architecture for MGW interconnections.
- Route the traffic in Erlangs.
- Ensure that the network element processing load and interface capacity are not exceeded.
- If no modifications are needed, from the routed traffic determine the bandwidth requirements in the backbone links.

Number of MGWs, Site Locations and Coverage Areas

The MGW location and coverage area are important factors affecting the total number of MGWs required for a network.

MGW Site Location
The location of an MGW site depends on several factors. Some of the factors are:

- location of the RNC and BSC;
- total traffic from the MGW coverage area (both RNC and BSC traffic);
- subscriber distribution;
- call profiles;
- location of the point of interconnection with the PSTN/PLMN;
- transmission availability and cost.

Though the MGW does not contain a subscriber database, it is necessary to consider the required traffic handling capacity to determine the required number of subscribers. The required capacity depends on traffic generated by subscribers in the MGW coverage area. The coverage area of an MGW may be considered as the area served by the RNCs/BSCs parented to a given MGW. The number of MGWs with radio access interfaces (Iu or A interfaces) can be determined by:

(a) the total traffic generated by the RNC/BSC in the MGW coverage area in terms of the BHCA;
(b) total interfaces required to carry traffic from the coverage (RNC/BSC) area;
(c) call processing capacity of the MGW, while depends on the amount of resources required to handle a given type of call (i.e. UE–UE, UE–MS, MS–MS, UE–PSTN, MS–PSTN, etc.).

The number of MGWs dedicated to external interconnects (i.e. other PLMNs and PSTNs) is determined by:

(a) the total traffic of the interconnect in terms of the BHCA;
(b) the location of the POI location and size;

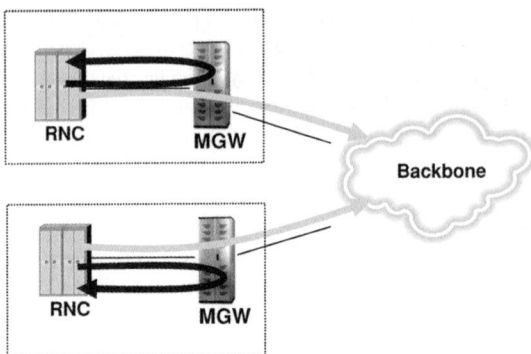

Figure 4.12 MGW distributed architecture

(c) the number calculated from the routing of the end-to-end traffic matrices;
(d) the call processing capacity of MGW, which depends on the amount of resources required to handle a given type of call (i.e. UE–UE, UE–MS, MS–MS, UE–PSTN, MS–PSTN, etc.).

The definition of the coverage areas for MGWs is quite flexible as the constraints are on the maximum size but not the optimal size. There are therefore two possible solutions, i.e. distributed and centralised architectures.

Distributed Architecture of MGWs

In a distributed architecture, the MGWs are located close to the RNC/BSC, which is the major source of traffic in the network (Figure 4.12). This enables switching of user traffic close to the traffic source, thus reducing transmission costs. Traffic generated by the RNCs/BSCs is connected directly to a given MGW and terminated in the RNCs/BSCs connected to the same MGW and can be switched directly close to the originating site with no need to carry it to the core site. In order to maximise the transmission savings, traffic terminating in other networks (POIs) should be switched locally at the MGWs. Therefore the operator should consider providing a POI to the external network (i.e. PSTN/PLMN) at the host MGW. The interconnect design for the outgoing traffic should allow local PSTN/PLMN traffic to be emptied directly at the closest POI.

The distributed architecture of the MGW reduces the required transmission costs but at a cost because it increases the number of MGWs as more MGWs are required to collocate with the RNC/BSC. In sites with high user traffic, this solution is cost-effective but in sites with low user traffic the transmission savings are too low to justify the increased number of MGWs.

Centralised Architecture of MGWs

The centralised architecture of MGWs is more suitable for access sites with low user traffic (Figure 4.13). The MGWs can be located in a centralised core site while the traffic from the access site is carried to the core site using transmission lines. The result is a reduced number of MGW elements in a few selected core sites, where traffic is collected and switched to/from all RNCs/BSCs in the network and to external networks. Fewer MGWs makes easier management of the network configuration, interconnections and traffic routing.

In a real project situation a pure decentralised or centralised architecture of the MGW is rare but a hybrid solution with some sites with heavy access traffic with a decentralised solution and some remote sites with low user traffic in a centralised solution is common. The selection of an architecture solution is therefore the prerogative of the network designer to ensure an optimum solution. As a general rule the distributed solution can be used when there is heavy 'local' traffic at a site. This often takes place when

3GPP Release 4 Circuit Core Network

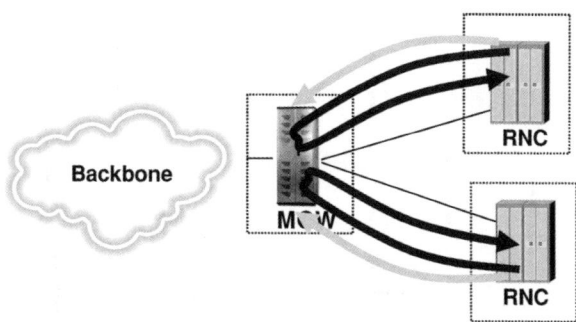

Figure 4.13 MGW centralised architecture

the RNC/BSC site covers a large geographical area (rural area) with low user density. The POI for MOC to PSTN should be located close to the RNC/BSC sites and the transmission costs are high compared to costs of deploying and operating an MGW.

Another situation where a distributed solution could be selected is in the replacement of outdated old switching equipment (e.g. a low capacity subrack MSC) with an A interface re-homing to MGW elements. In this case, placing one or more MGWs of the same capacity as the replaced MSCs in the same site allows minimal reconfiguration and enables reuse of existing transmission links to the BSCs. A centralised solution is the preferred solution when the inter-MGW traffic is high.

The total number of MGW to meet the operator's requirement therefore depends on the selected user plane network architecture. Finally, the capacity statement for the MGW in certain SW (software) releases enables the total number of required MGW elements to be estimated. The capacity of the MGW depends on the following factors:

- call handling capacity, i.e. the total number of BH call attempts and simultaneous calls;
- port capacity, i.e. E1/T1 ports, ATM ports, IP ports;
- hardware configuration, i.e. signalling units, transcending units, etc.

It is also important to take into account any resiliency requirements imposed by the operator. Should there be a resiliency requirement to introduce some redundant capacity margin, then it should be taken into account when dimensioning the MGW elements.

End-to-End Traffic Matrices

The initial number and location of the MGW and MSS/GCS is determined by the number of users served by each MGW element. The number and location of MGW elements together with other network information gathered during network assessment (e.g. service supported, number of users, traffic profiles per user) are used to calculate end-to-end traffic matrices.

The end-to-end traffic matrices show the traffic intensity between any two MGWs in Erlangs. As the user plane traffic is routed by MGW elements, end-to-end traffic matrices are therefore a valid support in deciding the interconnection architecture and represent the first step in network element dimensioning.

The end-to-end traffic matrix may be calculated using specialised tools, e.g. the interswitch module of the NetAct Transmission planner. It is also possible to use measured traffic values if available. The calculations are based on CS network dimensioning principles. The traffic matrix should be studied carefully and compared with the available data in order to start the dimensioning process using the conect assumptions.

340   Core Network Planning and Optimisation

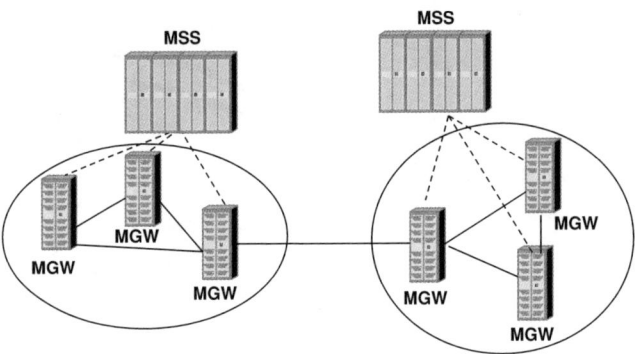

Figure 4.14   MGW transit layer concept

MGW Interconnection Architecture and Transit Layer Concept

All the MGWs controlled by the same MSS/GCS are part of the same MSS area. In order to minimise the number of elements on the user plane path for a call generated and terminated in the same MSS area, the MGW in one MSS area should have direct links. The architecture of the MGW under the same MSS area should therefore present a full meshed structure. In large networks with several MSS areas, there is a possibility of having a fully meshed architecture between all the MGWs or of selecting transit MGWs for inter-MSS area traffic. Figure 4.14 presents an example of MGW transit layer topology. The transit topology should be selected carefully.

When the Nb interface between MGWs is implemented on top of an IP backbone there is no need for a transit layer. In fact, during the call establishment the MSSs controlling the origin and destination MGWs decide which origin and destination IP addresses will be used by the origin and destination MGW for the call. As the IP backbone offers routing functionalities, this is all that is needed by the MGW to set up the call. All the MGWs in the network can be considered to have direct connections and the architecture to be fully meshed. This allows a maximum of two MGWs to be used in each call scenario (i.e. one MGW on the mobile originating and the other on the mobile terminating part of the call).

When the Nb interface is implemented on top of an ATM backbone there could be a need for a transit layer. As mentioned above, during call establishment, the origin and destination MSSs decide which MGW to use for the user plane. The user plane is transported on AAL2 links and the establishment of a channel for the call between the MGWs is handled by the MGWs using the AAL2 signalling. The AAL2 signalling uses dedicated AAL2 signalling links that must be configured and maintained.

In large networks the configuration and maintenance of fully meshed links (both for AAL2 signalling and the user plane) between MGWs can become quite complex. In such cases a transit layer should be considered. The transit MGWs will interconnect several groups of fully meshed MGWs. Each group of fully meshed MGWs may include one or several MSS areas.

There are delay considerations to take into account when planning a layer of transit MGWs. Each MGW element on the user plane path adds some delay to the transmission of user traffic. The main sources of delay in the network are transmission delay, packetisation/buffering delay, codec delay and propagation delay. As a general rule, delay is a critical parameter for real-time traffic and should be kept at a minimum. The maximum delay for voice traffic as defined by ITU-T is 400 ms. Greater delays than this are unacceptable as the voice quality becomes highly distorted. Delay due to transmission is very low compared to delay introduced by packetisation/buffering and codec delays. These are introduced in the user equipments, radio access network and MGW elements. It is therefore important to keep the number of MGWs taking part in the call establishment to a minimum.

As a partial solution to the problem, in the MGW and following, the AAL2 nodal function is extended to the Nb interface. The AAL 2 nodal function makes it possible for originating and destination MGWs to establish a user plane channel through transit MGWs without involving the controlling MSS. Moreover, the internal handling of the traffic in the transit MGWs would be significantly different and that would reduce some delay.

MGW Termination of the A interface

Implementation of the MGW for a 3GPP release 4 network offers the possibility of terminating an *A* interface of the GSM. This allows a common backbone network to be used for both GSM and UMTS networks. When an *A* interface is terminated in the MGWs, the BSSAP signalling is terminated in the controlling MSS. MGW becomes a signalling gateway and terminates the SS7 MTP transport layer while relaying application part messages on top of Sigtran to the MSS.

The MGW performs the needed adaptations of the user plane traffic from the G.711 codec received on the *A* interface to the codec used in the Nb or MGW–MGW interface. The bandwidth needs for the GSM user plane traffic (voice) in the Nb interface depends on the codec supported in the MGW and on the capabilities of the transcoder (TC) element used between the A_{ter} and *A* interfaces.

There are several methods used to enhance voice quality for GSM traffic in the release 4 network.

Tandem Free Operation (TFO)

TFO is achieved by the TCs using in-band signalling. The transcoders are physically present along the communication path but the transcoding is partially or fully disabled. In this way the codec used on the A_{ter} interfaces (e.g. EFR or AMR) can be forwarded over the backbone, improving voice quality. No transmission savings would be obtained in this way as the codec on an interface is G.711 and the transcoders will override part of the G.711 frame with the EFR codec and fill the rest with padding. Some MGWs have the feature to recognise the use of TFO between the TCs. TFO is activated through in-band signalling during the call establishment procedure.

Forced Payload Compression over the Backbone

When TFO is not supported by the TCs or when the MGWs operate fixed bandwidth reservation at the call establishment phase and it is not possible to detect the use of TFO, the MGW may be configured to use a default codec (e.g. EFR or AMR) in the Nb interface or backbone, introducing supplementary transcoding operations. This enables transmission savings but it reduces the voice quality perceived by the end user as up to four transcoding operations can take place in tandem for each communication path.

G. 711 Codec Directly Transmitted Using AAL2 Bearer Channels

This results in a 64 kbps transparent channel for voice, even in the UMTS packet backbone, and can be used when TCs do not support TFO or when the MGW has a fixed reservation for GSM channel bandwidth and the voice quality cannot be reduced using forced compression. In fact, if the MGW is reserving bandwidth at the call setup and is not able to modify dynamically the reservation made, then detection of the TFO mode used by the transcoders has no benefit as the MGW will consider a 64 kbps channel as permanently allocated to the call even if only a fraction of the capacity is needed during the call.

User Plane Dimensioning

User data are carried over AAL2/ATM on the Iu-CS interface and over RTP/UDP/IP or AAL2/ATM on the Nb interface. The protocol stacks for user data transport over Iu-CS and Nb interfaces are depicted in Figure 4.15. The MGW supports several voice codecs, among them UMTS AMR, GSM EFR, GSM FR and G.711 *A* and μ law codecs.

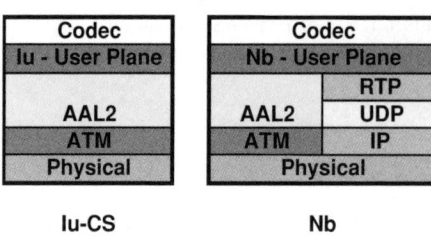

Figure 4.15 User plane protocol stack

Voice Codec on the User Plane
Adaptive Multirate (AMR) Speech Codec

The AMR codec was developed by the European Telecommunications Standards Institute (ETSI) and standardised for GSM; the codec has also been chosen by 3GPP as the mandatory codec for 3G systems. It is a multimode codec with eight narrow band modes with bit rates between 4.75 and 12.2 kbps. The sampling frequency is 8000 Hz and processing is done on 20 ms frames, i.e. 160 samples per frame. The codec was developed to preserve high speech quality under a wide range of transmission conditions.

The AMR modes are closely related to each other and use the same coding framework. The codec is designed with a voice activity detector (VAD) and generation of comfort noise (CN) parameters during silence periods. The AMR codec, therefore, can reduce the number of transmitted bits and packets during silence periods to a minimum. The operation to send CN parameters at regular intervals during silence periods is usually called discontinuous transmission (DTX) or source controlled rate (SCR) operation.

The format for packetisation of AMR encoded speech signals into the real-time transport protocol (RTP) is specified by RFC3267. The payload format supports transmission of multiple channels, multiple frames per payload, use of fast codec mode adaptation, robustness against packet loss and bit errors, and the inter-operation with existing AMR transport formats on non-IP networks. The speech bits encoded in each AMR frame have different perceptual sensitivities to bit errors. This property has been used in cellular systems to achieve better voice quality using unequal error protection and detection (UEP and UED) mechanisms. The UEP/UED mechanisms focus the protection and detection of corrupted bits onto the perceptually most sensitive bits in an AMR frame. In particular, speech bits in an AMR frame are divided into classes A, B and C, where bits in class A are most sensitive and bits in class C least sensitive. A frame is only declared damaged if there are bit errors found in the most sensitive bits, i.e. class A bits. On the other hand, it is acceptable to have some bit errors in other classes, i.e. class B and C bits. Moreover, a damaged frame is still useful for error concealment at the decoder since some of the less sensitive bits can still be used. This approach can improve the speech quality compared to discarding the damaged frame.

The AMR codec offers eight different modes, each mode offering a different bit rate (see Table 4.8). For dimensioning purposes it is assumed that AMR mode 7 corresponds to the GSM EFR mode. This mode provides the best voice quality and consequently the highest bit rate. For example, the number of bits in the mode 7 frame is 244 bits, or 31 bytes (30.5 to be exact). A frame is produced every 20 ms, producing a bit rate of 12.2kbit/s.

AMR specifies silence suppression, which is aimed at reducing the effective bandwidth to about 7.5 kbps for each connection. This is based on the following assumptions:

- The voice/silence ratio is 60/40.
- The SID (silence detection) is sent every 8 frames (0.16 s).
- The SID frame is 39 bits long.

In one second, with 60 % activity, 1000 ms $\times$ 0.6 / 20 ms = 30 speech samples are generated. There SIDs are generated. Thus the resulting bandwidth (in one direction) is 30 samples $\times$ 31 bytes + 3 samples $\times$

Table 4.8 Modes of the AMR codec

Index	Mode	Class A bits	Speech bits total
8	AMR SID	39	39
7	AMR 12.2	81	244
6	AMR 10.2	65	204
5	AMR 7.95	75	159
4	AMR 7.4	61	148
3	AMR 6.7	58	134
2	AMR 5.9	55	118
1	AMR 5.15	49	103
0	AMR 4.75	42	95

5 bytes = 7560 bit/s. For comparison, G.711 A-law and μ-law produce a constant bit rate of 64 kbps and no silence suppression is used.

G.711 Codec

G.711 is the international standard for encoding telephone audio on a 64 kbps channel. It is a pulse code modulation (PCM) scheme operating at an 8 kHz sample rate, with 8 bits per sample, fully meeting the ITU-T recommendations. Assuming that G.711 coding is used with a constant bit rate of 64 kbps and one sample is sent every 20 ms, the overhead consists of Nb_UP/RTP/UDP/IP, which totals 204 bytes every 20 ms, or 80 kbps (without the link level overhead; see Figure 4.15).

Protocol Stack for the User Plane

The Nb-UP Protocol is specified by 3GPP in specification 29.415. It is used to convey data between MGWs over the Nb interface. The Nb-UP Protocol can be either in the support mode or transparent mode.

The transparent mode may be used when the bearer does not need any services of the Nb-UP Protocol except data transfer. Therefore Nb-UP frames are not generated in the transparent mode. The support mode is intended for use when the bearer requires other services of the Nb-UP Protocol besides data transfer. So far, the only support mode defined is the support mode for predefined SDU size. For example, an AMR codec might use this support mode because it requires some procedure control function, e.g. initialisation and rate change.

The Nb-UP framing is identical to the Iu-UP framing defined in 3GPP standards TS 25.415. In the support mode, the Nb-UP overhead is 3 or 4 bytes, depending on whether the CRC is used or not. In the transparent mode, the Nb-UP header is absent, and thus the overhead is 0 bytes.

IP Backbone RTP

The Real-Time Transport Protocol (RTP) is a simple protocol used to encapsulate voice and compressed video codec information for end-to-end delivery. The purpose of the RTP is to offer a remedy for the delays and packet losses inherent in IP based networks. The RTP uses the UDP for transport. The services provided by the RTP, and its companion protocol, the Real-Time Control Protocol (RTCP), include:

- Payload type identification
- Sequence numbering
- Time stamping
- Delivery monitoring

Table 4.9 Data bit rates in the IP backbone

	G.711 (64 kbps)		AMR (12.2 kbps)		AMR VAD	
			Nb-UP (3-4 bytes)			
			RTP (12 bytes)			
			UDP (8 bytes)			
	IPv4 (20)	IPv6 (40)	IPv4 (20)	IPv6 (40)	IPv4 (20)	IPv6 (40)
Bit rate [kbps]	80	88	30	38	19.2	24.5
			Ethernet (42 bytes)			
Bit rate [kbps]	96.8	104.8	45.2	53.2	29.2	34.5

An RTP packet consists of a fixed RTP header, a possibly empty list of contributing sources and the payload data. Up to five speech frames can be packed into a single RTP packet. In the MGW, RTCP packets are generated every 5 seconds, and therefore their effect on the bandwidth requirement is negligible.

The contributing source of the RTP container (CSRC) identifiers are present in the header if the payload consists of a mix of traffic from various contributing sources. An example could be, for example, audio conferencing streams, which are multiplexed in an RTP mixer, which would then include the stream source of the RTP container (SSRC) identifiers contributing to this stream in the CSRC field. Thus the resulting RTP overhead is 12 bytes.

Table 4.9 presents the overhead introduced by a user plane, as well as the data rate for each used codec with or without the Ethernet layer. In the case of IPv6 the header is 40 bytes, compared to 20 bytes for IPv4. In the table AMR mode 7 (12.2 kbps) has been used. If a different mode is used, refer to the AMR modes to get the bit rates. If the G.711 codec is used, then the Nb user plane does not introduce any header, while for the AMR a header of 3–4 bytes is added.

Ethernet overhead bytes:

$$12 \text{ gap} + 8 \text{ preamble} + 14 \text{ header} + 4 \text{ trailer} = 38 \text{ bytes/packet w/o 802.1q}$$
$$12 \text{ gap} + 8 \text{ preamble} + 14 \text{ header} + 4 \text{ trailer} = 38 \text{ bytes/packet w/o 802.1q}$$

Example 1. If the G.711 codec is used on top of the IPv4 backbone, then the data bit rate for one call is calculated as:

$$\text{Data bit rate (kbps)} = 64 \text{ kbps} + [20 \text{ bytes (IPv4)} + 8 \text{ bytes (UDP)} + 12 \text{ bytes (RTP)}] \times 8 \text{ bits/20 ms}$$
$$= 80 \text{ kbps}$$

If an Ethernet header is added, then

$$\text{Data bit rate (kbps)} = 64 \text{ kbps} + [42 \text{ bytes (Ethernet)} + 20 \text{ bytes (IPv4)}$$
$$+ 8 \text{ bytes (UDP)} + 12 \text{ bytes (RTP)} \times 8 \text{ bits/20 ms} = 96.8 \text{ kbps}$$

Example 2. If an AMR mode 7 (12.2 kbps) codec is used on top of an IPv4 backbone, then the data bit rate for one call is calculated as:

$$\text{Data bit rate (kbps)} = 12.2 \text{ kbps} + [20 \text{bytes (IPv4)} + 8 \text{bytes (UDP)} + 12 \text{bytes (RTP)}$$
$$+ 4 \text{ bytes(Nb–UP)}) \times 8 \text{ bits/20 ms} = 29.8 \text{ kbps (approximately 30 kbps)}$$

Table 4.10 Data bit rates in the ATM backbone

	G.711	AMR	AMR VAD
		Nb-UP	
		AAL2	
		ATM	
Bit rate [kbps]	75.3	17.1	11
Cells/s	178	45	26

If an Ethernet header is added, then

Data bit rate (kbps) = 12.2 kbps + [42 bytes (Ethernet) + 20 bytes (IPv4) + 8 bytes (UDP) + 12 bytes (RTP) + 4 bytes(Nb−UP)] × 8 bits/20 ms = 46.6 kbps

ATM Backbone (AAL2)
The AAL2 specifies bandwidth-effective transmission of variable length packets over a single ATM VC. Up to 248 AAL2 connections can be multiplexed on a single VC. The AAL2 adaptation layer consists of two parts: a common part sublayer (CPS) (ITUT specifications I.363.2) and a service specific convergence sublayer (SSCS) (ITUT specifications I.366.1). The CPS specifies a format for CPS-PDU, which consists of a CPS-PDU start field and a CPS packet.

The CPS packet has a 3 byte CPS packet header followed by 1, ..., 45 octets payload. The start field is one octet long, followed by a CPS-PDU payload. The CPS-PDU payload may carry zero, one or more complete or partial CPS packets. The CPS packet may overlap one or two ATM cells, and the boundary can be anywhere, including the CPS packet header.

The SSCS consists of three parts:

- SSSAR (service specific segmentation and reassembly), which supports payloads up to 65 536 bytes long;
- SSTED (service specific transmission error detection);
- SSADT (service specific assured data transfer).

Only the SSSAR is used and implemented on the Iu-CS and Nb interfaces. The SSSAR does not add any overhead to the SSCS-SDU.

Backbone Bandwidth Dimensioning
The backbone bandwidth dimensioning depends on the type of codec used (G.711, AMR, etc.) as well as on the total number of voice calls. In the previous paragraphs it has been shown how to calculate the bit rate required for a voice call in the backbone depending on the different codecs, IP backbones and physical layers (e.g. Ethernet) used.

Example 1. Given the traffic in Erlangs, calculate the total number of voice calls by applying the Erlang B formula and the required bandwidth for a given codec on the IP or ATM backbone:

X = traffic (Erl)
Y = blocking probability (%)
Z = voice calls = number of traffic channels = Erlang B calculator (X, Y)
D = data bit rate
B = bandwidth (kbps) = ZD

Example 2. Calculate the required bandwidth for:

- Traffic = 6540 Erl
- Blocking probability = 0.1%
- IP backbone using the AMR 12.2 codec, without including the Ethernet header and IPv4

Solution:

- Voice calls = Erlang B calculator (6540, 0.1%)
- $Z = 6687$ calls
- Bandwidth (kbps) = $6687 \times 45.2 = 295.1684$ Mbps

Example 3. Calculate the required bandwidth for

- Traffic = 6540 Erl
- Blocking probability = 0.1%
- On an ATM backbone using the AMR 12.2 codec

Solution:

- Voice calls = Erlang B calculator (6540, 0.1%)
- $Z = 6687$ calls
- Bandwidth (kbps) = $6687 \times 17.1 = 111.6677$ Mbps
- ATM cells per second = $6687 \times 45 = 300.915$ kcells per second

4.4.3 Control Plane Planning

The control plane planning process follows similar steps:

- Determine the number of MSS/GCS and parenting of MGWs (i.e. MGWs controlled by MSS/GCS).
- Determine the number of calls between nodes from the end-to-end traffic matrices between MGWs and the information about which MSS/GCS controls which MGW.
- Determine signalling between MSSs/GCSs.
- Determine amount of signalling between MSS/GCS and controlled MGWs.
- Determine the architecture for the MSS/GCS interconnections.
- Route the signalling traffic.
- Ensure that the network element processor load and signalling transfer load is not exceeded.
- If no modifications are needed, from the routed traffic determine the bandwidth requirements in the backbone links.

In the following paragraphs, single issues affecting the planning process are addressed. The approach is to show all the options when planning a 3GPP release 4 circuit core network with solutions from the most generic to the most detailed, bearing in mind that each practical case will have a unique combination of the presented scenarios and it should be carefully considered how to combine all those issues.

Number of Signalling Points and Sites

Dimensioning of the MSS/GCS needed in the network is based at first on the subscriber number (as each MSS has a limited VLR capacity) and BHCA (as each MSS/GCS has a limited processing capacity).

3GPP Release 4 Circuit Core Network

The number of MSSs is calculated by determining coverage areas based on the subscriber distribution and traffic profile. Each coverage area should contain a number of estimated subscribers that:

- fits into the VLR capacity of the element (that should not be fully utilised to allow user mobility);
- generates total traffic in terms of BHCA (as well as other processing capacity constraints, e.g. SMS) within the processing capacity of the MSS.

The coverage areas of MGWs and MSSs are not in a fixed relation: one physical MGW can be logically divided into several virtual MGWs under the control of different MSSs, thus allowing a great degree of flexibility. It has to be noticed that one RNC can be connected to several virtual MGWs but only to one MSS; therefore the division of subscribers among MSSs is based on the RNC's size. The initial plan should consider the coverage area of one MSS as the union of the coverage areas of the controlled MGWs and consider the effect of virtual MGWs only when consistent unused capacity is available in a given MSS. For instance, three MGWs with a capacity of 400k BHCAs can be controlled by three MSSs partially loaded ($3 \times$ 400k BHCAs) or by two MSSs at the maximum capacity ($2 \times$ 600k BHCAs), if this is allowed by the subscriber distribution between the RNCs.

The number of GCSs is calculated based on the MGWs dedicated to the PSTN interconnection and their traffic load in terms of BHCAs (usually calculated after the user plane is routed). In the case of new delivery stand-alone MSSs/GCSs, as they are handling only control plane traffic, the only constraint on the physical location of the elements is the availability of IP connectivity for the signalling traffic. In the case of integrated or upgraded MSSs/GCSs, their location is determined by the fact that the element provides interworking functionality for CS data or not.

Signalling Network Dimensioning

In mobile networks, signalling network dimensioning can be considered in two major areas:

- Call-related signalling. In call-related signalling, all SS7 signalling messaged required to establish, maintain and tear down a call would be considered.
- Non-call-related signalling. In non-call-related signalling other signalling services for mobility management, SMS, IN, etc., would be considered.

The message flows for SS7 signalling are defined by the 3GPP standards. Average message sizes depend on the message signal unit (MSU) payload sizes and OH introduced by underlying SS7 protocol layers. The signalling messages can be analysed per service and signalling interface.

SIGTRAN – SS7 Transport over IP

SIGTRAN is a standardised (IETF) method of transporting CCS7 signalling over IP with the option of using one of three solutions of user adaptation layers, i.e. SUA signalling connection control part ((SCCP) user adaptation) layer, M3UA (MTP3 user adaptation) layer and M2UA (MTP2 user adaptation) layer. SIGTRAN is mandatory in 3G release 4 networks (see Figure 4.16).

The overhead introduced by SIGTARN depends on the application part data, user adaptation layer used and OH introduced by underlying layers. Typically ISUP messages are relatively short and in the order of a few tens of bytes, whereas CAP and MAP messages can be significantly larger. The overhead also depends on the number of data units included in one SIGTRAN message. Figure 4.17 presents an example of one of SIGTARN's implementations.

The SCTP is standardised by RFC 2960 in IETF and may be carried on an IPv4 or IPv6 layer. The SCTP is carried directly over IP. The SCTP common header is 28 bytes and each chunk has a 4 bytes header. The SCTP header should contain a common header, chunk description and payload identifier.

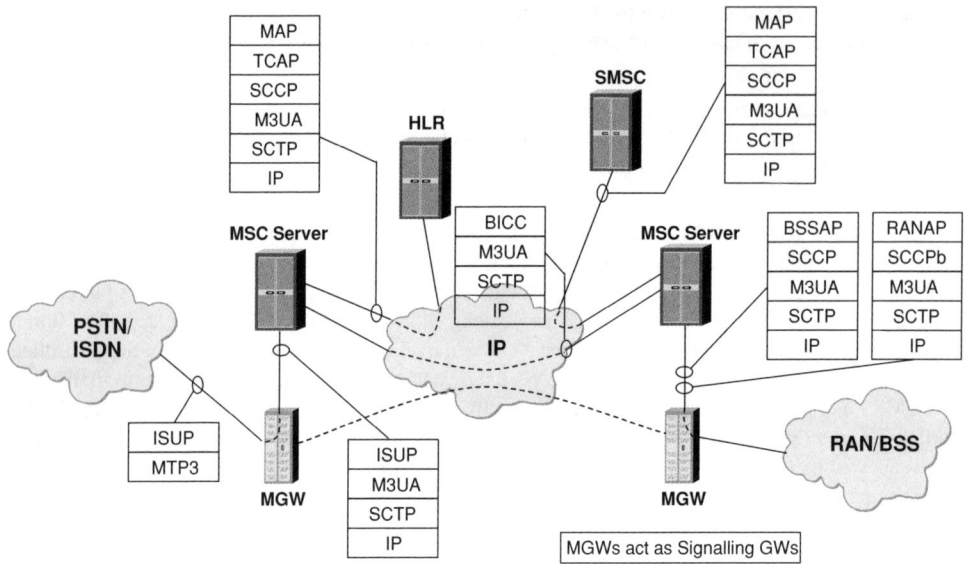

Figure 4.16 SIGTRAN implementation in a release 4 network (TCAP, transactions capability application part; GW, gateway)

The length of the common header is 3 dword. The length of the chunk description is 1 dword. The length of the payload identifier is 4 dword. The total SCTP header size is 28 bytes.

The size of the common message header of the M3UA is 2 dword. The common message header exists in every M3UA message. The main part of the bit rate is generated from the payload data messages. Therefore, an example describes only the structure of the payload data. The optional network appearance parameter is used in M3UA; its size is 2 dword. The routing label of the MTP3 message is encoded in separate fields. The OPC and DPC fields are both 1 dword long and the rest of the routing label parameters, like SI (a transaction capability application part), NI (a gateway), message priority and a signalling link set (SLS), are encoded in 1 dword. The total length of the M3UA message header is 24 bytes in the payload data message.

The size of the IP header depends on what version of the IP stack is used, IPv4 or IPv6. The length of the IPv4 header can vary from between 20 and 60 bytes; 20 bytes is the length of the basic IP address in IPv4 while the remaining 40 bytes is the length of the option field. Options are not commonly used

BSSAP	RANAP	BICC	
SCCP			ISUP
M3UA (24 bytes)			
SCTP (28 bytes)			
IP (20 IPv4, 40 IPv6)			
Ethernet (38 bytes)			

Figure 4.17 SIGTRAN generic protocol stack

Figure 4.18 MEGACO protocol stack

in IPv4. The length of the IPv6 header is 40 bytes + option fields. Usually some of those option fields are included in the message when it is using IPv6. Including the IPv4 protocol overhead, transferring an upper layer payload message would introduce 72 bytes of overhead.

H.248 (MEGACO) Protocol – Mc Interface

H.248 or MEGACO is the protocol used by the MSS/GCS to instruct the MGW about the actions to perform in relation to a call (see Figure 4.18). H.248 or MEGACO has been developed by IETF MEGACO workgroup in cooperation with the ITU-T. The MEGACO protocol was standardised by the ITU and is used to perform the following tasks:

- reserve and connect terminations;
- connect or release echo canceller to terminations;
- connect or release tones and announcements to terminations;
- send/receive DTMF tones.

The dimensioning of the bandwidth needs between a given MGW and its controller (MSS or GCS) is based on the total number of calls, at the MGW, during the busy hour, the estimated signalling messages needed per call and average message size. (The BHCA load of the element is obtained from the NetAct tool. In case a MGW is handling transit traffic by means of the AAL2 nodal function, this transit traffic should not be considered for the MEGACO calculation as it does not require any interaction between the MSS and MGW.) The H.248 messaging load per call is difficult to estimate as it depends partly on the implementation (i.e. how transactions are packed into the H.248 message) and partly on the call case.

As an example, one call consisting of context and termination reservations, tone connection, speech path through the connection, speech path disconnection and termination releases is expected to require 800–2500 bytes in one direction with ABNF (ASNI binary codes or ASCII) coding. Such a call requires eight transactions and each transaction is packed into one message. There will therefore be $8 \times (28 + 20) = 384$ bytes of SCTP and IPv4 overhead. For example, if an MGW handles 600 000 BHCA, the total amount of traffic in one direction including the overheads (no L1 overhead) is 1.6–3.9 Mbps.

A Interface

When the A interface is connected to the MGW, BSSAP signalling traffic is routed through the MGW to MSS using SIGTRAN. The MGW acts as a Signalling gateway for the BSSAP traffic from normal SS over TDM to SIGTRAN. The signalling protocol layers are presented in Figure 4.19.

Iu-CS Interface

Circuit switched traffic between UTRAN and the core network is carried in Iu-CS interface. This interface is ATM based and carries a user plane using an AAL2 adaptation layer and a control plane using an AAL5 adaptation layer. The control plane consists of the RANAP signalling protocol. In the core network Iu-CS is connected to the MGW. Signalling gateway functionality in MGW relays the control plane to the MSS using the SIGTRAN protocol (see Figure 4.20).

The calculation of the RANAP message load depends on the number of subscribers and on the traffic and mobility profile of each subscriber. If no better knowledge or tool is available, the following estimation of one particular user profile can be used. The average number of RANAP signalling messages sent and

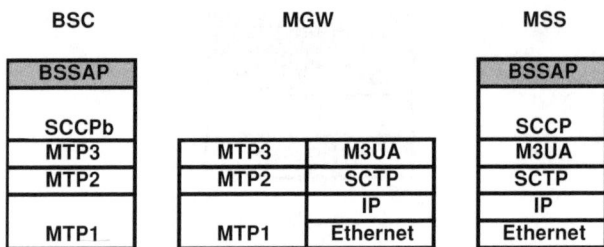

Figure 4.19 MGW as the signalling gateway for BSSAP

received by one user per hour is estimated to be 70, or 35 in one direction (1 call, 1.5 SMS, 2 location updates). Assuming an average message length of 70 bytes, two ATM cells are needed to be transmitted; i.e. the bandwidth required for signalling by 50k users, including the ATM layer, is

$$50\,000 \text{ subscribers} \times 35/3600 = 486 \text{ messages/s}$$

and

$$486 \times 2 \times 53 \text{ bytes} \times 8 \text{ bits} = 412 \text{ kbps}.$$

The RANAP signalling is carried from the MGW to the MSS/GCS using SIGTRAN. The calculation of the bandwidth needs is done considering the same number of messages and message size as in the Iu-CS interface and applying the SIGTRAN overhead to each message, i.e. again assuming a 70 byte average message length, 50k users and that each message is carried in its own SCTP message, get $(70 + 12 + 28 + 20) \times 8 \text{ bits} \times 486 \text{ messages/s} = 505$ kbps in one direction between the MGW and the MSC server. This includes the M3UA (12 bytes), SCTP (28 bytes) and IP (Ipv4, 20 bytes) overheads.

Nc Interface

The use of the BICC CS-2 is defined in 3GPP release 4. The BICC CS-2 is a bearer-independent call control capability set 2 signalling protocol on the Nc interface for the IP/ATM backbone. It supports a user plane based on the ATM or IP backbone in the Nb interface. Figure 4.21 presents implementation of the control plane protocol stack for the Nc interface. In MSS and GCS, the SIP can be used on the Nc interface as an alternative to the BICC if an IP backbone is used on the Nb interface. In the MSC server, the SIP is used as a tunnelling method for ISUP messages. The use of the SIP is not required in 3GPP release 4 but the implementation can be done in accordance with IETF specifications.

SIP requires the domain name server (DNS). The result of a number analysis in the MSC server will be a domain name that represents an address on the target point of call control signalling. The MSC server then sends a query to the DNS to resolve the domain name to an IP address that will be used to route the signalling message through the IP network. Unlike SIP, BICC call control signalling does not require DNS.

	RNC		MGW		MSS
	RANAP				RANAP
	SCCPb	SCCPb	SCCP	SCCP
	MTP3b	MTP3b	M3UA	M3UA
	SAAL-NNI	SAAL-NNI	SCTP	SCTP
	AAL5	AAL5	IP	IP
	ATM	ATM	Ethernet	Ethernet

Figure 4.20 MGW as the signalling gateway for RANAP (NNI, network-to-network interface)

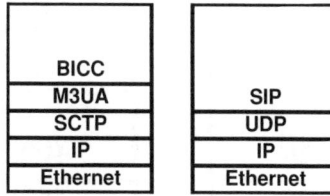

Figure 4.21 Call Control Protocol stacks for the Nc interface

Dimensioning BICC and SIP interfaces follows a general procedure as outlined below. When BICC signalling is used, it is assumed that one signalling message size is 40 bytes (also used for ISUP). A total of six BICC messages are required per call. Two of the messages (IAM (initial address message) and APM (application transport mechanism)) contain the SDP (session description protocol) payload of approximately 150 bytes. This results in $4 \times (40 + 12 + 28 + 20) + 2 \times (40 + 150 + 12 + 28 + 20) = 900$ bytes/call or 450 bytes/call/direction. As an example, assuming 50k users, each initiating 1 BHCA, then 450 bytes/call/direction $\times$ 8 bits $\times 50\,000$ calls/h/3600 = 50 kbps are made including the M3UA (12 bytes), SCTP (28 bytes) and IPv4 (20 bytes) overheads.

When SIP is used, the calculation of number of messages and size is more complex as it depends on the call case, the lists of codecs supported and even the length of domain names selected by the operator for signalling addresses of network elements. When no better estimation from measurements is available, then it is assumed that one signalling message is 500 bytes. A total of 14 messages are required per call in both directions. This results in $14 \times (500 + 8 + 20) = 7392$ bytes/call or 3696 bytes/call/direction. Assuming 50k users, with 1 BHCA per subscriber then 3696 bytes/call/direction $\times$ 8 bits $\times$ 50 000 calls/h/3600 = 410 kbps are made including the UDP (8 bytes) and IPv4 (20 bytes) overheads.

MAP Interfaces C, D, E, F and G

The MAP-C interface connects the MSS/GCS to the HLR. The MSC server fetches the subscriber location information from the HLR via the MAP-C interface. The 3GPP release 4 circuit core standards recommend the use of BICC-CS2 as the signalling protocol in the Nc interface. The BICC-CS2 is able to establish bearer channels for user plane connections both with the ATM and IP based backbone. Moreover, the BICC-CS2 is independent from the signalling message transport layers, which means that the BICC-CS2 can be carried on top of the TDM (SS7), ATM (SAAL) or IP (SIGTRAN). A default implementation is to carry the BICC-CS2 messages on top of IP, using the SIGTRAN protocol stack.

As SIGTRAN uses permanent 'association' between signalling units in the counterpart element to deliver signalling messages through an IP network, in a full mesh topology all the signalling links need to be configured and maintained, making it challenging (but not impossible) to implement a full mesh topology in large networks. One solution is to split large networks (i.e. with more than 10/20 MSSs/GCSs) into regions (on a geographical base or administrative base), with all the MSS/GCS in one region fully meshed for signalling, and then add some more elements (GCSs) dedicated to STP functionality between regions. Figures are not currently available for estimating the signalling transfer capability of the GCS used as the STP and the value of 1M BHCA for GCS can be used even if conservative (as the GCS used as the STP will not have GMSC or SCP functionalities, the signalling processing capacity will probably be higher).

When the backbone network is based on IP transport, the signalling for call control between MSS/GCS elements in the release 4 core circuit network can be based on SIP protocol. SIP messages are carried on top of UDP/IP (or TCP/IP) and an MSS/GCS uses DNS functionalities to get IP addresses of the counterpart MSS/GCS when performing call control related signalling. Therefore, the use of IP routing for the messages and of DNS to get those IP addresses makes the configuration of signalling networks fairly easy and not limited by the number of MSSs/GCSs in the network. When using SIP, there is no need for a transit signalling layer.

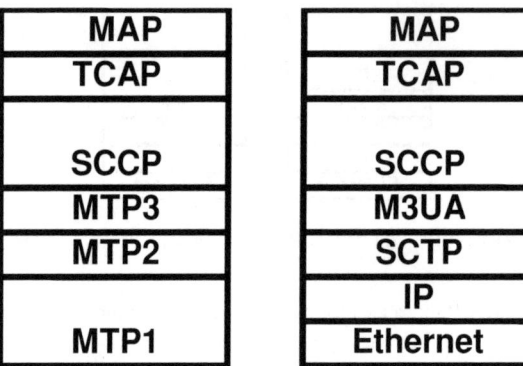

Figure 4.22 Alternative protocol stacks available in MAP interfaces C, D, E, F and G in the MSS/GCS

The MAP-D interface between the visited MSC server and HLR is used for keeping the mobile subscriber location information in the HLR updated. The MAP-E interface between visited MSC servers is used in handovers where the serving MSC server is also changed. The MAP-E interface is used for delivering short messages between the visited MSC server and the SMS-GMSC server for mobile terminated short messages and the visited MSC server and the SMS-IWMSC (interworking MSC) server for mobile originated short messages. Alternatively, the visited MSC server can be connected directly to the SMSC (short message service centre) if it is part of the SS7 network and the SMSC can make the necessary HLR queries itself. The MAP-F interface is located between the MSC server and the EIR and the MAP-G interface is located between the VLR and the VLR (see Figure 4.22 for the available protocol stacks).

Protocols supported are the CAP, INAP and MAP, which can use SIGTRAN or the PCM based SS7 as the transportation layer.

4.5 CS Core Detailed Network Planning

As discussed before, the 3GPP release 4 standards introduce the separation of the user plane from the control plane. In release 4 network architecture, the user data are carried on the user plane, while the signalling traffic is carried on the control plane. The user plane is therefore carried independently of the control plane on the Nb, Iu-CS and A interfaces and the external interface to the PSTN. On the Iu-CS and Nb interfaces, the user plane traffic is carried over the IP/ATM networks.

4.5.1 Control Plane Detailed Planning

Despite being separate entities, the user plane and the control plane are linked in an indirect and flexible way via the user plane destination (UPD) and user plane destination reference (UPDR) parameters.

Signalling Interfaces and Protocols (Figure 4.23)

Detailed planning of the control plane requires:

- signalling point code allocations for MSS, MGW, HLR, RNC, BSC and PSTN POI;
- GT allocations for MSS, HLR, RNC and MGW.

CS Core Detailed Network Planning

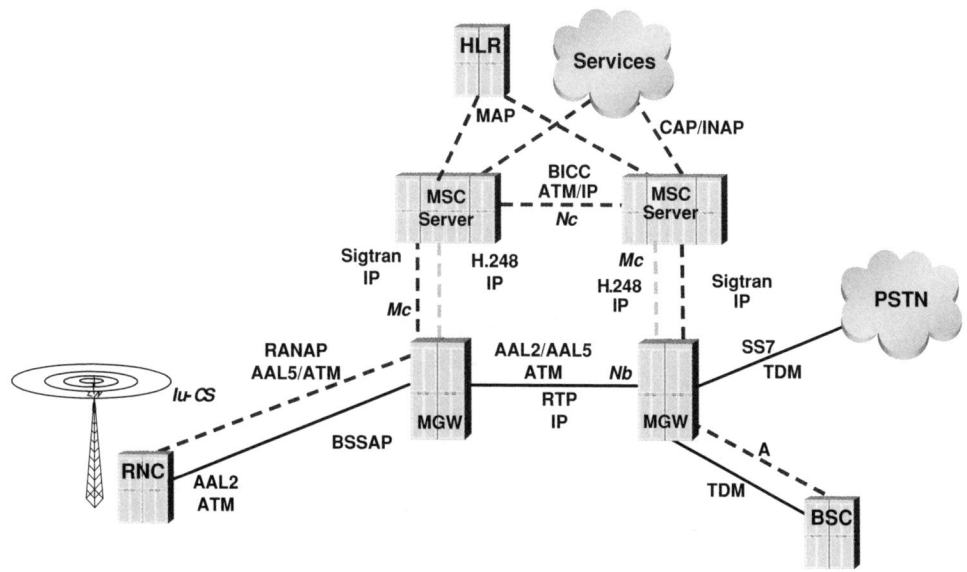

Figure 4.23 Signalling interfaces and protocols

To create a control plane for each interface requires the creation of:

- signalling association sets for Sigtran
- narrowband signalling linksets
- broadband signalling linksets for Iu-CS interface

Signalling Transport over IP (SIGTRAN)

SCTP Associations and Association Sets

The Stream control transport protocol (SCTP) was developed as the basic lower layer protocol to serve different signalling adaptation layers. The SCTP runs directly over the IP. It is similar to the TCP but has some enhancements. It features:

- reliable data transfer (continuous monitoring on reachability);
- connection oriented with multiple streams;
- ordered or unordered delivery;
- congestion and flow control;
- protection against known attacks (in IP networks).

SCTP associations are logical 'point-to-point' signalling channels between two nodes (see Figure 4.24). There is no physical resource allocation per association. Associations are created between a pair of signalling units.

SCTP Multihoming

Each signalling unit has a pair of interfaces each with a logical IP address. During the establishment of an SCTP association the two peer signalling units exchange their IP addresses. Either IP address may be used as source or destination addresses but one of the IP addresses is the primary address and is normally used while the other is the secondary IP address and is only used when the primary is unavailable. An

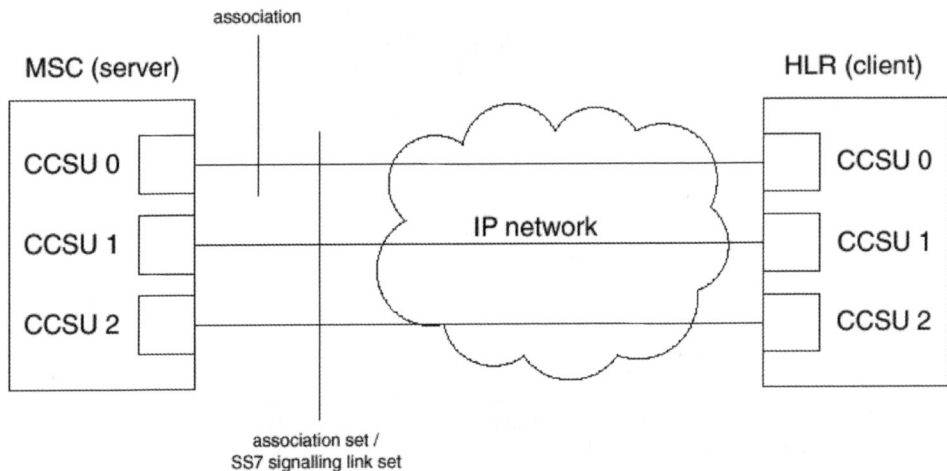

Figure 4.24 SCTP association/association set (CCSU, common channel signalling unit)

SCTP association is said to be multihomed when more than one address/path to the peer signalling unit exists for use in failure situations.

An association set is a set of SCTP associations connecting two signalling points. The bit rate of signalling traffic in an association is not limited but it is recommended that as many associations as there are signalling units are created in order to spread the signalling load on the signalling units. In the current implementation there can be up to 16 SCTP associations in an association set.

IP Signalling Link Sets

An IP signalling link contains exactly one association set. An IP signalling link set contains exactly one IP signalling link.

IP Signalling Route Set

The M3UA is used to transport SS7 signalling traffic over IP interfaces. The MSS allows the use of M3UA with the same signalling user parts as in TDM signalling links. Configuration of these IP signalling links (Figure 4.25) is similar to the configuration of PCM based signalling links.

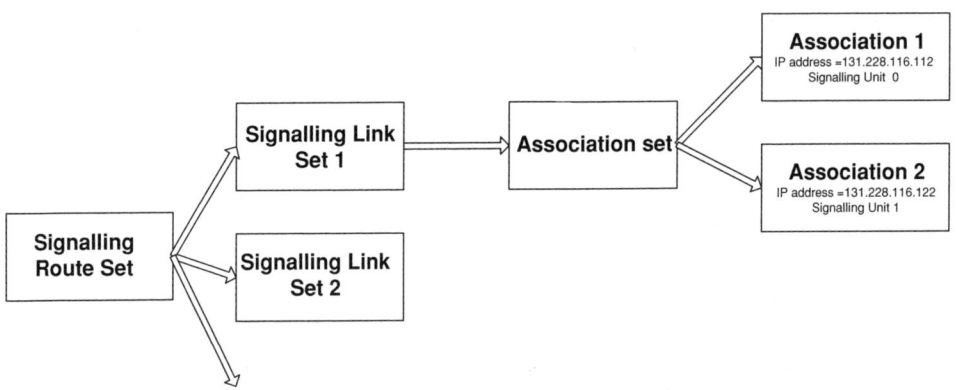

Figure 4.25 Signalling components

CS Core Detailed Network Planning

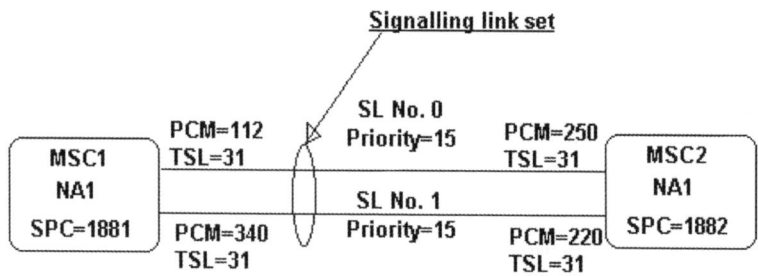

Figure 4.26 Signalling link set

Narrowband SS7 Signalling

Signalling points are connected to each other by a signalling link. A signalling link can contain a minimum of one signalling channel created in one PCM–TSL (PCM and timeslot). A 2 Mbps PCM link has 32 channels of 64 kbps timeslots. An SS7 signalling link can, however, occupy any one of the timeslots, increasing the bit rate for each link.

A signalling link set connects two signalling points in a given signalling network (Figure 4.26). A link set consists of a number of signalling links connecting two signalling points. In SS7 the maximum number of links in a link set is 16.

The signalling route is a predetermined path between two signalling points. The signalling link set can be seen as a 'physical' resource, like a pool of channels that can be used, while the route is the logical association between one destination and the link set used to reach it. A signalling route contains one signalling link set.

A signalling route set contains up to eight signalling routes. Each signalling route in a route set is given a priority. The signalling route with the highest priority is used while signalling routes with a lower priority will be used only when the primary route is out of use. The signalling load may be shared between several routes with the same priority.

Narrowband SS7 components are defined and used on TDM interfaces (i.e. A interface and PLMN/PSTN interfaces).

Broadband SS7 Signalling

The broadband SS7 protocol stack is used on the Iu-CS interface, i.e. between RNC and MGW. Broadband SS7 is needed to enable transmission of mobile specific protocols like RANAP over the ATM. The ATM based signalling links are based on the ATM adaptation layer 5 (AAL5) VPI–VCI channels. The user and application part protocols of the SS7 protocol stack (e.g. TCAP, SCCP, MAP, INAP/CAP, RANAP, etc.) require broadband adaptations to be carried over the ATM network.

The main difference to the upper layers is the longer 4 kbytes maximum message length. The SCCP needs small modifications to SCCP level message segmentation. MTP level 3 functionality changes are also minor. These are specified in ITU-T Recommendation Q.2210. The MTP level 2 will be replaced by two new protocols, the Service Specific Connection Oriented Protocol (SSCOP) and the Service Specific Coordination Function (SSCF)-NNI. The SSCOP is a new fast data transfer protocol designed for high speed networks. The SSCF-NNI provides adaptation between the SSCOP and MTP-3.

The broadband SS7 link consists of a single ATM link between two signalling points on the same signalling network. The signalling link is based on the AAL5 channel created on a VPI-VCI. To create an ATM signalling link set:

- create ATM termination points;
- create a signalling channel VPI-VCI;

- create signalling links using a created signalling channel;
- create an SS7 signalling link set;
- create a signalling route set;
- activate the MTP configuration;
- create the SCCP configuration;
- activate the SCCP configuration.

Signalling User and Application Parts

When the lower SS7 layer components (links, link sets, route and route sets) are created, it is possible to define the higher layer SS7 user and application parts. The following issues should be taken into consideration when planning the SS7 signalling network.

Use of Signalling Gateways

The MSS without any non-IP signalling connections and with M3UA signalling connections requires the presence of the signalling gateway network element. In the MSS concept such functionality can be realised by the MGW (alternatively in an STP or SRRi, or signalling routing register).

Signalling gateways terminate narrowband and broadband signalling link sets and act as STPs for traffic towards the MSS network elements. The presence of a signalling gateway in the path between two signalling points in the MSS system is perceived as a normal STP. The signalling routing configurations are therefore similar to routes with a signalling gateway as the STP.

Use of M3UA (MTP3 User Adaptation)

M3UA is used to transport SS7 signalling traffic over SCTP/IP interfaces. The MSS allows the use of M3UA with the same signalling user parts as in TDM signalling links. Configuration of these IP signalling routes is similar to the configuration of PCM based signalling links.

Configuring SCCP in MSS

The MSS can transport the SCCP user part over either IP or TDM based signalling links. The configuration of both MTP3 and M3UA layers is very similar because both provide a similar interface to the SCCP. The SCCP is configured in the MSS for the following application parts:

- MAP traffic
- BSSAP traffic
- RANAP traffic
- IN traffic (INAP/CAP)
- ISDN supplementary services (SSAP)
- SCCP level STP traffic towards other signalling points

In the example presented in Figure 4.27, a possible configuration of the SCCP between the RNC is connected through the Iu-CS interface to MGW1. In this example, MGW1 acts as an STP for RANAP signalling which terminates in the MSS. An IP signalling link set is created between the MSS and MGW1. A signalling link set is also created between the RNC and MGW, but this link set is based on the SAAL (signalling ATM adaptation layer), used as transport for MTP3b messages.

For RANAP, the MSS has a signalling route towards the RNC destination. Similarly, the RNC needs a signalling route towards the MSS. In these route sets, the MGW1 is defined as an STP and also plays the role of a signalling gateway between broadband SS7 signalling on the Iu-CS interface and SIGTRAN on the Mc interface.

Once the MTP layer components have been created and the SCCP has been created as the MTP level service in the MSS, then SCCP level configuration data can be created.

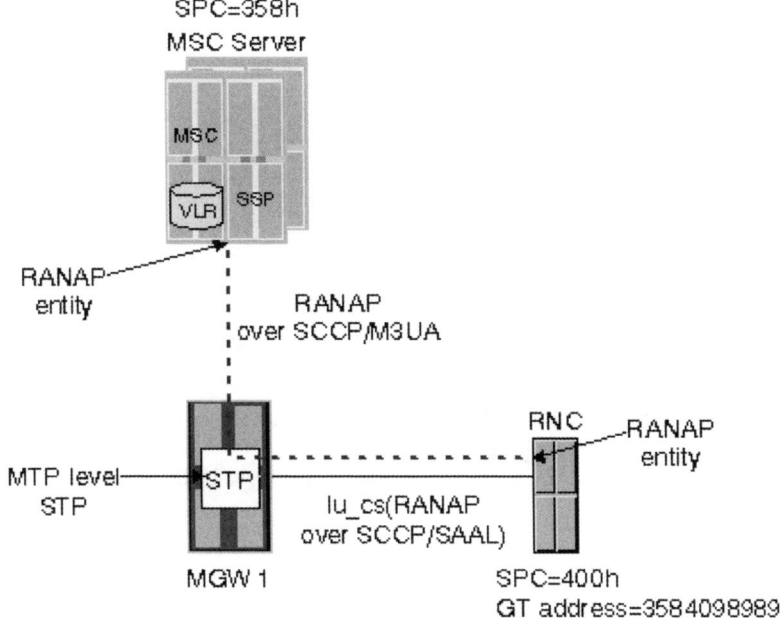

Figure 4.27 SCCP configuration

BICC and SIP Signalling

The UPDR parameter is used as a link between the control plane and the user plane. For incoming calls, the operator configures the UPDR parameter to the circuit group data, while for outgoing calls, the operator configures the UPDR parameter to the route data (Figure 4.28). On a per call basis, the value of the UPDR is delivered to a user plane control application to be used as an input attribute in several phases of the user plane analysis along with many other input attributes. Both BICC and SIP signalling have a similar behaviour in this respect.

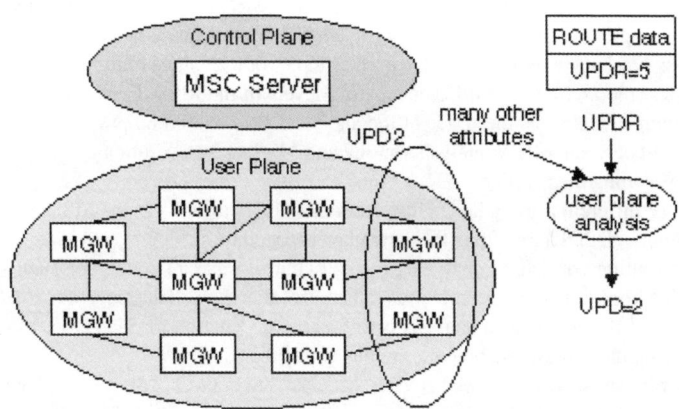

Figure 4.28 UPDR parameter on the outgoing side

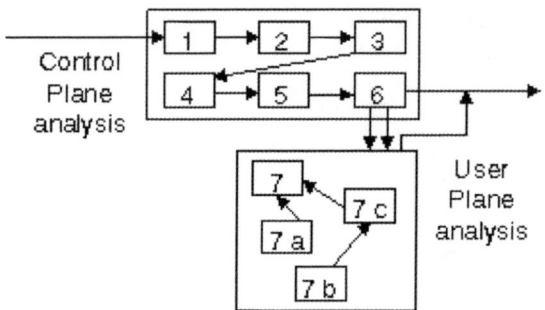

Figure 4.29 Logical hierarchy of routing levels in the MSS

In Figure 4.28 the UPDR value 5 is read from the outgoing ROUTE data. It is used as an input attribute for the user plane analysis with many other attributes as defined in the user plane analysis attributes. Analysis results for UPD = 2 means that two MGWs belonging to the UPD2 are allowed to be used for a call.

4.5.2 Control Plane Routing

Implementation of the MSS routing configuration is split into two separate parts as a result of the split architecture of the MSS and MGW. These routing parts need to be configured before any user traffic can be carried in the release 4 network. Section user plane routing and analysis planning, described above, give the configurations used in user plane routing.

The following sections will therefore describe the configuration of the control plane routing, including the following:

- configuring analyses;
- configuring UPD and UPDR;
- configuring SS7 signalling;
- configuring charging.

The control plane routing is the highest level of routing configuration present in the MSS. Routing is done by analysing the dialled MSISDN of the subscriber, MSRN or other routable E.164 number.

User plane routing is executed after finding the correct control plane destination for the call. Some results from the control plane routing level are used as input for user plane routing. Figure 4.29 represents an example logical hierarchy of the different routing levels in the MSS. Each box represents a certain analysis and a number indicates the order of the analysis. This should be presented earlier in order to understand the split between MGW analysis/routing and MSS analysis/routing before going into each of these in detail, as commented earlier.

As there are two different routing layers instead of one layer currently in the MSC, it is the user plane destination reference (UPDR) that links these two layers together.

Control plane routing consists of analysing the call control-related parameters, such as the different attributes available from signalling, the previous call control analysis and the received called party number. The main purpose of the call control analysis is to find the destination for the call, set the charging and execute tasks during the call establishment or at the end of the call.

The UPDR parameter serves as a link between the separated control plane and user plane procedures. By configuring the UPDR to the circuit groups/routes, the control plane is able to assign the call to the right MGW. The UPDR values can be configured in the incoming circuit group with the commands of the circuit group handling and in the outgoing routes using the commands of the route handling, but only if the signalling is BICC or SIP/SIP-T. Thus the following configurations are performed:

- configuring the UPDR to the incoming circuit;
- configuring the UPDR to the outgoing route;
- configuring the UPD for RANAP signalling in the RNC.

4.6 User Plane Detailed Planning

4.6.1 Configuring Analyses in the MSS

MSS functionality affects only the end-of-selection analysis, attribute analyses and digit analysis. The following examples show the capabilities offered by these analyses when configuring the control plane routing in the MSS.

Digits Analysis

The linkage between the control and the user plane layers is handled via the UPDR parameter, which can be set in the route resulting from the digit analysis. The planners can attach the UPDR parameter to route or circuit group level parameters before using it in user plane analysis. The succeeding UPDR (SUPDR) is attached to the outgoing route. The preceding UPDR (PUPDR) is attached to the incoming circuit group. The user plane analysis using P/SUPDR, which is not attached to the control plane level, will not be executed properly or not at all.

The PUPDR is one of the attributes of the following subanalysis:

- preceding UPD determination (PUPD);
- succeeding BNC characteristic determination (SBNC);
- CMN determination (CMN);
- succeeding UPD determination (SUPD);
- succeeding action indicator determination (SAI);
- interconnecting BNC characteristics determination (ICBNC).

The SUPDR is one of the attributes of the following subanalysis:

- succeeding BNC characteristics determination (SBNC);
- CMN determination (CMN);
- succeeding UPD determination (SUPD);
- succeeding action indicator determination (SAI);
- interconnecting BNC characteristics determination (ICBNC).

EOS Analysis

MSC server functionality uses the internal clear codes in the end of selection analysis to actuate control plane routing.

Routing Attribute Analysis

Routing based on cell-dependent routing category functionality provides a tool to manage the local routing of the calls within one MGW. The cell-dependent routing category can be used in each different attribute

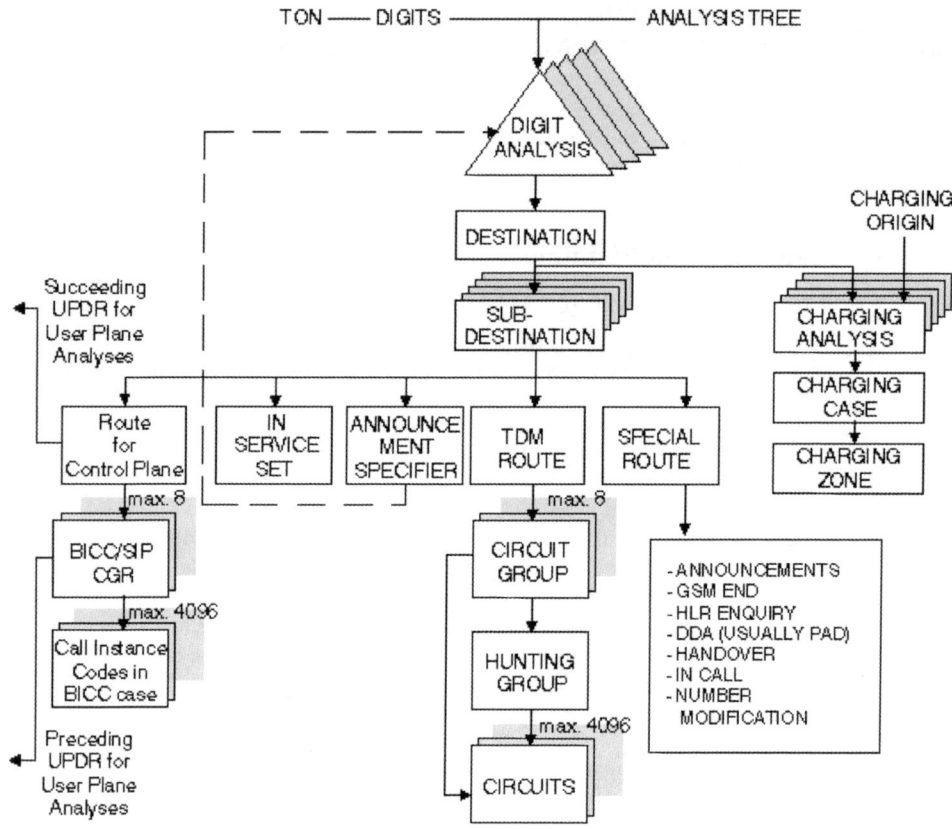

Figure 4.30 Digit analyses in MSS (DDA, direct data access; PAD, packet assembler/disassembler)

analysis as an attribute. For example, all calls originating from certain cells to the PSTN may always need be routed to the nearest PSTN interconnection, but with no restrictions applying to calls towards other mobile subscribers. In such a configuration, the suitable cell-dependent routing (CDR) category attribute can be configured to the cells belonging to the BSC/RNC, which are physically connected to MGW1. The same category value is then analsed in the routing attribute analysis which gives the suitable tree as a result, e.g. 91. This tree 91 has analyses for both roaming numbers as well as PSTN numbers. Digit analysis of the PSTN numbers will lead to the route (and circuit group) containing the circuits that are present in the same MGW1.

4.6.2 Routing Components of the MSC Server

The routing framework in the MSS is shown in Figure 4.30.

4.6.3 User Plane Routing

User plane routing is responsible for controlling the user plane transmission in the MSS in a network where media processing is distributed to several MGWs (see Figure 4.31). In general, the user plane routing scheme can be summarised as follows:

User Plane Detailed Planning

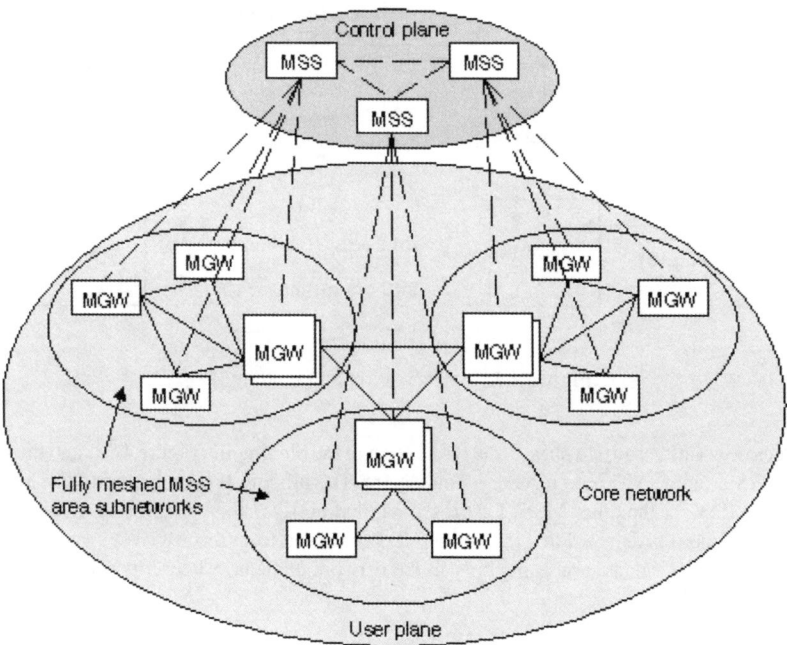

Figure 4.31 Distributed network architecture

- call initiation from the user equipment via the originating RNC to the MSS and setup confirmation from the terminating RNC;
- topology database query and user plane analysis;
- MGW selection and resource reservation request sending;
- bearer connection establishment;
- RAB assignment request and access side bearer establishment between the MGW and the RNC;
- alerting and call connection.

The user plane can be routed via the ATM and IP core network. ATM resources can be reserved by BICC signalling. IP resource reservation is handled by BICC or SIP.

User Plane Signalling

Bearer establishment can be achieved with the following signalling protocols:

- BICC
- SIP
- RANAP

With the Iu-CS interface, the user plane is always established using the RNC. Therefore, the BIWF (bearer interworking function) address (RANAP: transport layer address) and binding ID (RANAP: Iu transport association) are reserved from the chosen MGW and delivered to the RNC using the RANAP assignment request message. Currently only the AAL2 type of user plane connection is allowed in the Iu-CS interface. The binding ID is the connector factor of RANAP/BICC signalling at the control plane and AAL signalling at the user plane. The binding ID is needed in AAL2 PVC (permanent virtual connection).

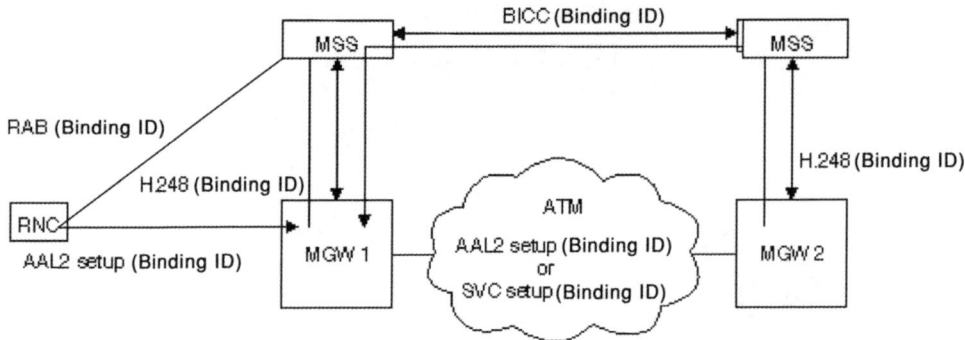

Figure 4.32 Reservation of a binding ID

Resources reservation has two phases, the reservation of the binding ID (Figure 4.32) and the signalling part. The MSS asks the MGW to reserve a binding ID. The binding ID is returned to the MSS, which sends it to the RNC or the other MSS. That reserved binding ID is then received in the AAL2 or UNI signalling setup and it is detected that the binding ID is reserved from this MGW.

The binding ID has valuable meaning only in the network element, where it is reserved. It has to be unique inside the MGW.

User Plane Analysis in the MSS

A user plane analysis is a new analysis introduced in the MSS system. It consists of several subanalyses, which can be chained and linked to different kinds of results. The structure of one analysis is presented in Figure 4.33. A user plane analysis, like the extended preanalysis and attribute analysis, has attributes to be analysed. Each attribute can be analysed in one or more subanalyses, and the subanalyses can be chained.

An attribute is a call-related variable. The value of the attribute is the value of the variable. In one subanalysis, there can only be one attribute, but the handling of the different values of this attribute may differ. For example, the analysis may continue from the next subanalysis with one value of the attribute, but with some other value the analysis goes to the final result.

The following describes the relationship between the six phases of user plane analysis and its results:

- *Call mediation node*. This indicates whether the MSS should act as a transit switching node (TSN) or as a call mediation node (CMN).
- *Interconnecting backbone network connection characteristics*. This indicates what type of bearer is used between two MGWs controlled by one MSS.
- *Preceding user plane destination*. This identifies the user plane destination for the incoming side.
- *Succeeding action indicator*. This indicates what bearer establishment method is used at the outgoing side.
- *Succeeding backbone network connection (BNC) characteristics*. This indicates what type of bearer is used at the outgoing side.
- *Succeeding user plane destination*. This identifies the user plane destination for the outgoing side.

Other Analyses in the MSS

The other analysis services provided by the MSS are actually no different from the services supported in the MSC (GSM environment). Those basic routing function analyses (mainly call control related) include:

User Plane Detailed Planning

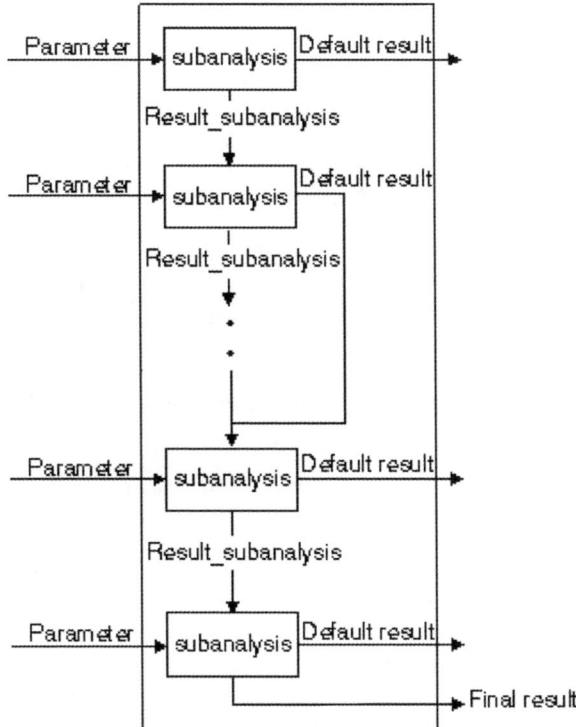

Figure 4.33 User plane analysis structure

- origin analysis;
- preanalysis/extended preanalysis;
- end of selection analysis/end of selection attribute analysis;
- routing attribute analysis;
- digits analysis;
- charging analysis, etc.

Configuring User Plane Resources in the MSS

Creating external TDM resources is only applicable in an integrated MSS. ATM and IP resources are not directly configured in the MSS. The related source data templates are circuit groups, routes and special routes.

Creating an MGW Database in the MSS

The MGW database is a common storage place in the MSS for MGW specific parameters that are used for user plane routing purposes. In the routing configuration the physical resources (i.e. the TDM) are hunted in the MSS, while the ephemeral resources (i.e. IP and AAL2) are hunted in the MGW. This requires that the MSS must have knowledge of the physical resources connected to the MGW.

After successful registration of the MGW, the MSS requests the capabilities of the MGW via the auditing process (such capabilities are, for instance, termination Ids of ephemeral resources, supported

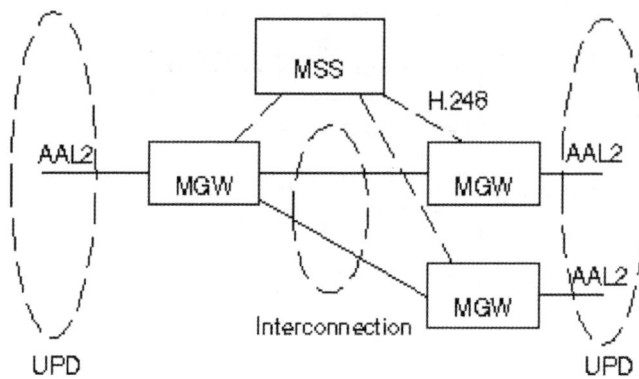

Figure 4.34 User plane topology information

packages and supported codecs which are stored in MGW database). The MGW can be registered to different MSSs, but only to one MSS at a time. However, it is possible to divide the MGW into several virtual MGWs that can be connected to a single MSS. This is recommended in the case where the network load is high.

The MSS controls the resources in the MGW by using the H.248 MEGACO protocol. Therefore creating the MGW database will give the details of the MGWs that are connected to the MSS. The related source data template is the MGW creation.

Configuring User Plane Routing and Topology in the MSS

The user plane topology database is a separate structural element in the MSS. Its main purpose is to store user plane topology information (see Figure 4.34) and, when requested, to deliver this information to the user plane control application. The user plane control application uses topology data to route the user plane to the proper destination. There are two kinds of data in the topology database.

- data records for user plane destinations (UPDs);
- data records for interconnections.

The operator can enter the actual network configuration to the database. The related source data templates are UPD creation, MGW full-meshed subnetwork and MGW interconnection.

A group of MGWs that is controlled by the MSS must be filled in the UPD creation form. In the case of a meshed network interconnection between those MGWs, the MGW identity (i.e. MGW name and ID) has to be properly defined in the MGW full-meshed subnetwork and MGW interconnection forms.

Creating User Plane Analysis in the MSS

The user plane analysis and its components are created in the MSS. The analysis consists of several subanalyses, which can be linked to the chain and to different kinds of results. There are six phases in the user plane analysis and as every phase has its own attributes, every phase has its own results too. For example, when the originating MSS has determined which terminating MGW to connect its MGW (originating) to, the succeeding user plane destination (SUPD) parameters of the user plane analysis in the originating MSS will identify the user plane destination for the outgoing route.

User Plane Detailed Planning

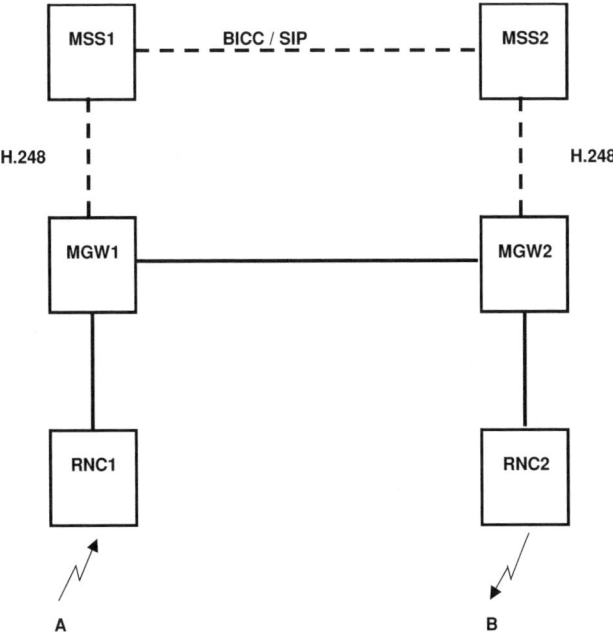

Figure 4.35 Digit analysis and routing in the MSS

Creating other Analyses in the MSS

The related source data templates and the MML commands required for creating other analyses are shown as follows:

- origin analysis;
- preanalysis and results;
- extended preanalysis and results;
- end of selection analysis;
- end of selection attribute analysis and results;
- routing attribute analysis and results.
- digits analysis;
- charging analysis and results.

Digit Analysis and Routing in the MSS System

The primary function of digit analysis and routing in MSS system network elements is to find free resources in the network to direct user traffic, voice and data connections to the desired destination. Digit analysis determines the destination of a call and selects an external route to the destination. It analyses the digit sequences that it receives. Typically the received sequence is the address of the receiving end (E.164 number, AESA address).

Digit analysis is used to find the correct route to a destination that is requested by an MSS. Digit analysis gets the address of the other network element and finds the route for the destination by analysing the address. In brief, the digit analysis process in the MSS system can be described in a call example, as shown in Figure 4.35.

Considering a call from subscriber A in MSS1 to subscriber B in MSS2, then MGW1 and MGW2 are controlled by MSS1 and MSS2 respectively.

Digit Analysis in MSS1
- The analysis of the incoming call (in tree 2) results in a special route of the HLR enquiry.
- This results in the subscriber B roaming number being returned to MSS1.
- The analysis of roaming number (in tree 50) results in an outgoing route to MSS2.
- Besides the roaming number of subscriber B, the BNC-ID (bearer network connection identifier) and BIWF address of MGW1 are also sent to MSS2.

Digit Analysis in MSS2
- The analysis of received roaming number (in tree 50) results in a special route of paging subscriber B.
- MSS2 sends the BNC-ID and BIWF address of MGW1 to MGW2.

Digit Analysis in MGW2
- The analysis of the received BIWF address of MGW1 from MSS2 (after the digit analysis process in MSS1 and MSS2) results in the establishment of a bearer network connection with MGW1.

Digit Analysis in RNC1
- The analysis of the received BIWF address of MGW1 from MSS1 (during the RAB assignment request) results in the establishment of a bearer network connection with MGW1.

Digit Analysis in RNC2
- The analysis of the received BIWF address of MGW2 from MSS2 (during the RAB assignment request) results in the establishment of a bearer network connection with MGW2.

Creating Routing Objects and Digit Analysis in the MGW

This section describes the procedure for creating routing objects and digit analyses with subdestinations and routing policy for the Nb interface with MML commands. Digit analysis is needed in the Nb interface, which connects two MGWs (MGW release 4).

There are two different approaches to creating digit analysis for the Nb interface:

- creating (basic) digit analysis, where each destination has only one subdestination;
- creating digit analysis, where each destination can have more than one subdestination.

Creating subdestinations for a destination and defining routing policy (the latter approach above) are optional features. Generally speaking, creating a basic digit analysis is sufficient, and it is recommended that the latter approach be used only if there is a definite need for several subdestinations and routing policy measures. The routing policy function allows the percentage call distribution (also known as load sharing) to be utilised. The percentage call distribution traffic to a destination can be distributed among two or more subdestinations in predefined proportions.

Creating subdestinations for a destination and defining routing policy are optional when creating the ATM backbone in the MGW release 4 network. In general, creating basic routing and digit analysis are sufficient. Subdestinations and subdestination routing policy should be used only if there is a definite need (e.g. load sharing) for several subdestinations and routing policy measures.

Digit analysis and routing are closely connected to each other. Their relationship and objects used in routing and analysis in the MGW are shown in Figure 4.36 and are further described in the following section.

User Plane Detailed Planning 367

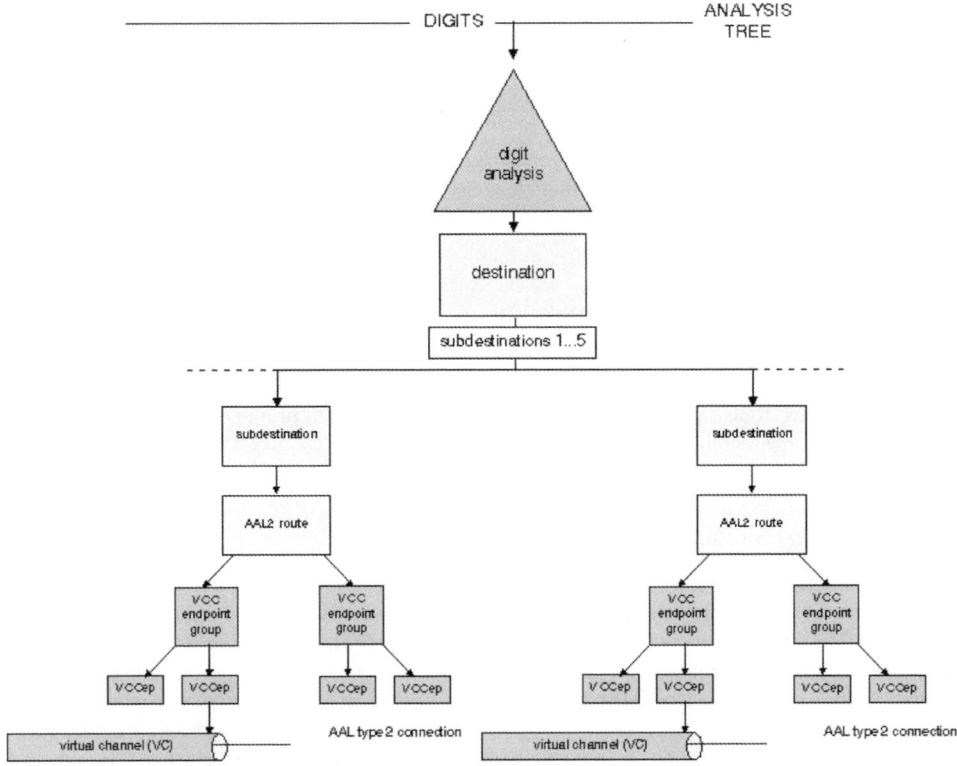

Figure 4.36 Digit analysis and routing components in the MGW (AAL2)

Configuring Charging in the MSS

Charging of the calls in the MSS is similar to charging in the MSC. In principle, the same call detail record (CDR) is created from events related or not related to calls, such as mobile originated calls (MOCs) or supplementary service execution (SUPS). A detailed description of the changes in the CDR content is available from a CDR field description.

However, the account principles of interswitch traffic of the calls using the IP or ATM transport are divided into two different levels:

- user plane accounting;
- control plane accounting.

The accounting of traffic using TDM circuits is not changed. It is recommended that multiple UPDRs should be defined in such a way as to differentiate between control plane accounting of the different MSSs within the release 4 network. At least one UPDR should be configured towards each MSS. Similarly, if more detailed accounting of different user plane resources between own MSC servers is needed, multiple UPDs should be defined towards the MGWs controlled by other MSSs.

Defining the basis of charging involves the following steps:

- *Origin analysis.* The origin analysis examines the cell tariff, the cell category and the classmark of the mobile station. The result of the analysis is the charging origin (CORG).

- *Digit analysis.* The digit analysis examines the origin data and the dialled digits. As a result of the analysis, the system receives information on routing and charging control.
- *Charging analysis.* The charging analysis receives the digit analysis result data and the call origin, and uses these to define a charging case containing the charging characteristics for the call.

The related source data templates for configuring charging in the MSS are charging day classes, change group, charging zones, charging case definition, charging attribute subanalysis and results, as well as origin analysis and digit analysis.

4.7 CS Core Network Optimisation

The CS core network optimisation is typically a part of the network operation process. It consists of a follow-up of how the planned services are actually provided to ensure efficient utilisation of existing network resources. It also provides a follow-up of the business plan in order to meet the operator's projected service quality, subscriber and service penetration forecast.

4.7.1 Key Performance Indicators

The key performance indicators (KPIs) provide information on how the network resources are utilised, whether there is underutilised capacity or bottlenecks and when resources get used up. An initial assessment of the CS core network's KPIs is done during the network audit phase. The initial values of the KPIs are determined before the optimisation process begins. Other activities include determining available optimisation tools, available measurements, measurement periods, access to network elements for the optimisation team and reporting procedures to be used during the project. The KPIs to be used in the project are defined during the project definition phase. The performance of the network is evaluated based on the agreed KPIs. The KPIs may comprise a network element's measurement counter information, external test equipment measurements or other easily accessible data that give relevant information on the quality of service and performance of the network. Some of the measured values may be used directly as KPIs but others require postprocessing to derive the desired performance indicators. The existing documents are reviewed in this module.

A follow-up procedure to be included in the project plan is defined for the selection of measurements needed for checking the success of the optimisation project and to define a regular audit plan to be followed in the regular network operation.

The main KPIs for the CS core include:

- average circuit group (CGR) utilisation;
- total successful BHCA per MSC;
- call success rate;
- signalling link utilisation;
- signalling link load;
- signalling link availability;
- VLR database fill ratio;
- HLR database fill ratio;
- computer unit load rate;
- MB load rate;
- average call setup time;
- inter-MSC handover success rate.

These KPIs may be used to evaluate the performance of the CS core network and for benchmarking.

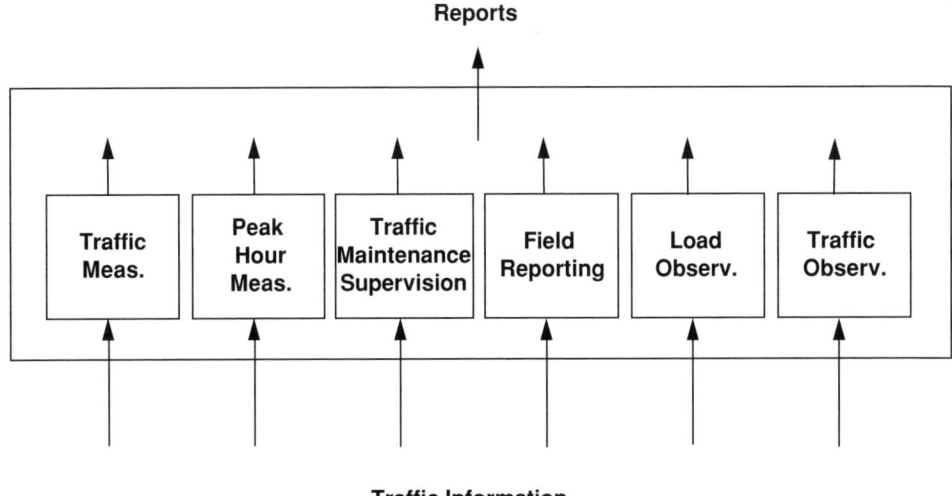

Figure 4.37 Traffic administration functions of network elements

4.7.2 Network Measurements

The traffic measurements can be obtained from measurement servers or from a network element's own performance management counters. The network element's own traffic administration capabilities provide measurement counters that enable a printout of more accurate measurement reports. In some cases external measurement tools provide independent verification of network traffic intensity. Alarm monitoring enables users to find out whether this equipment is available and working to expectations.

Traffic Administration

Traffic administration functions of network elements (Figure 4.37) cover procedures for collecting information on the switched calls, finding out the peak busy hour, determining the load of the various units in the exchange and, finding out what the availability of the exchange is. Traffic administration also contains functions for monitoring the system performance, the ratio of failed calls, the load on computer units, the circuit seizure, occupation and release times, for a given circuit group or computer unit.

Selected sets of the data used in the configuration of the switches are recovered from the network elements as input for the network parameter build upgrade and tuning.

Traffic Measurements

Traffic measurements provide information on traffic duration, intensity and distribution of the traffic in the exchange. Such information is important in analysing the traffic flow model of the network, which may be used for:

- dimensioning the exchange and the telecommunications network;
- checking the system condition;
- finding system faults.

The most important data produced by traffic measurements is the traffic intensity in the measured object.

Calculations in Erlangs are used in traffic measurements. The Erlang measurement is based on the true reservation time of the resource and the length of the results accumulation period. All of the reservation time is updated in the appropriate counter at the end of the call. For this reason remarkably large values in the Erlangs field are possible in a case in which there are many calls that have started before the results accumulation period but end in the results accumulation period.

Traffic Observations

Traffic observations produce real-time observation reports on individual calls and call attempts. Conditions for the traffic observation can be defined that, when fulfilled, cause an observation report to be generated. Some of the observation conditions are included in the exchange-specific settings that cannot be changed; e.g. reports on emergency calls are produced automatically.

The data produced by the traffic observation are important when:

- analysing an exceptional operation of calls on a given circuit group;
- finding out reasons for failed calls (e.g. on a given circuit group);
- testing connections.

Observations include:

- Circuit observation
- Number observation
- Clear code observation
- Combined observation
- Statistical observation

All the observation types can be defined to be active simultaneously. If at least one observation condition is fulfilled, the system outputs an observation report on an individual call. Even if more than one observation condition is fulfilled for a call, only one observation report is generated on the call.

4.7.3 CS Core Network Audit

A switching network audit is performed on the basis of measurements collected from a network element's own performance management counters. The collected measurements and observation counters provide the network design engineer with important information regarding the network's performance. The tasks of measurement data collection may be performed by the operator using the network element's own measurement counters or specialised performance data collection tools like Metrica and Traffica. The network element's own measurement counters can be obtained through direct Man-Machine language (MML) sessions to the network element. The operator also provides other necessary information, i.e. signalling network data, a traffic routing strategy and a current network synchronisation plan.

The main task during the network audit phase is to analyse and interpret measurement data received from the operator. The initial switching network performance is analysed in view of the assumed KPIs. The activities also include finding the network busy hour, identification of bottlenecks, modelling the actual network on Direct+, building an upgrade and tuning of the network parameters and making a comparison of actual network usage and penetration forecast.

Network Performance Survey

The main objectives of a network performance survey are to:

- analyse active measurements in the network;
- list or activate important measurements for the project;

CS Core Network Optimisation

- analyse initial values of KPIs;
- evaluate the call answer rate.

Evaluation of Preoptimisation KPIs

At the beginning of the project, the initial values of KPIs should be taken and tabulated. This can be done for each module separately or generally for the whole project. Some of the main CS core KPIs are:

- Call success ratio
- Call answer rate
- Drop call ratio
- Paging success rate
- LU success rate
- SS7 link utilisation
- HO success rate
- Total number of call attempts
- Call setup time
- Circuit group utilisation
- Circuit blocking

A detailed list of KPIs and their definitions is contained in the KPI guide. As an example the evaluation and results of the LU success rate and paging success rate analysis are presented hereafter.

Location Update Analysis

Location updates can be seen in NSS statistics mainly in the MSC/VLR. The main reports on location updates are provided by VLR measurement (Figure 4.38) and location update observation. All measurement data on location updates and IMSI attach/detach operations are collected in the VLR register units.

The procedure used in VLR measurement (the register unit is VLRU) is the following. At the beginning of the measurement period, statistics process sends the start information to each active register unit. Then every three minutes, statistics process queries the data needed for averaging the number of registered subscribers from each active register unit. At the end of the measurement period, statistics process queries the full measurement data from each active register unit and saves the data to a file. When all data from each active register unit is received or the time supervision limit is reached, statistics process generates the report and sends it to the report forming service. If statistics process does not get measurement data from each active VLRU, this is mentioned in the VLR measurement report.

In a similar way the HLR location update success rate can be evaluated and prepared for presentation. The results show the success rate of location updates from the HLR point of view. While these results testify to the performance of the cellular network, they may, however, provide valuable inputs for evaluation of improvement in the network performance.

Other KPIs listed above are evaluated further in this document. A list of the most important KPIs is given at the beginning of each module. The values may be evaluated in the network performance survey module or at initial phases of each module. The paging success rate may be evaluated as presented in the following subsection.

Paging Success Rate

VLR measurement provides the number of paging attempts through the GSM BSS network and Gs interface (SGSN) for GPRS attached mobile stations or mobile subscribers. In addition to a terminating call, these paging attempts are also updated for a terminating SMS (short message service) and USSD (unstructured supplementary service data). Observations are an effective tool for testing and solving error situations. As an example the paging success rate for MSC-WAR01 is presented in the Figure 4.39.

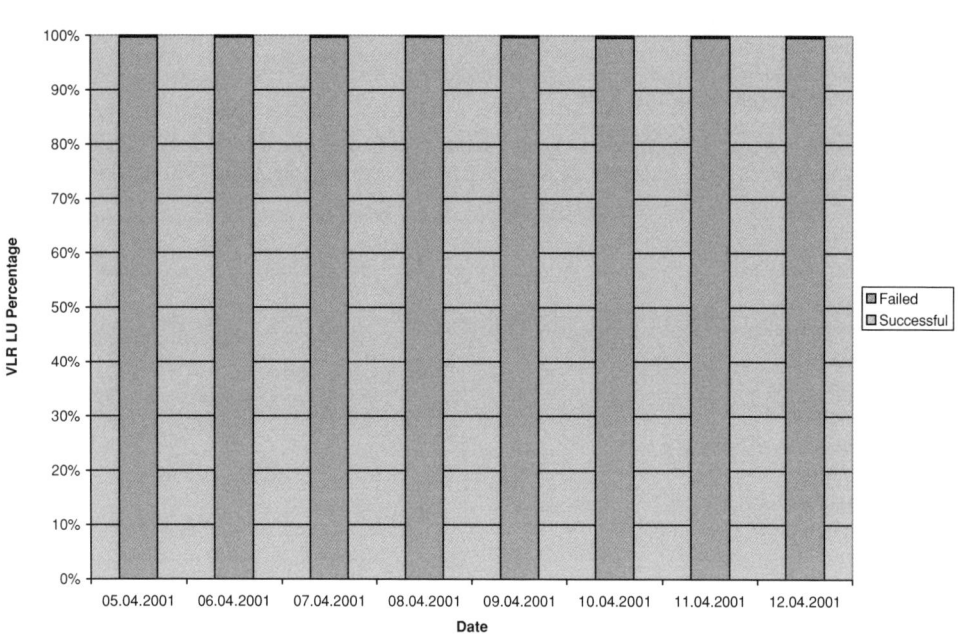

Figure 4.38 VLR location updates

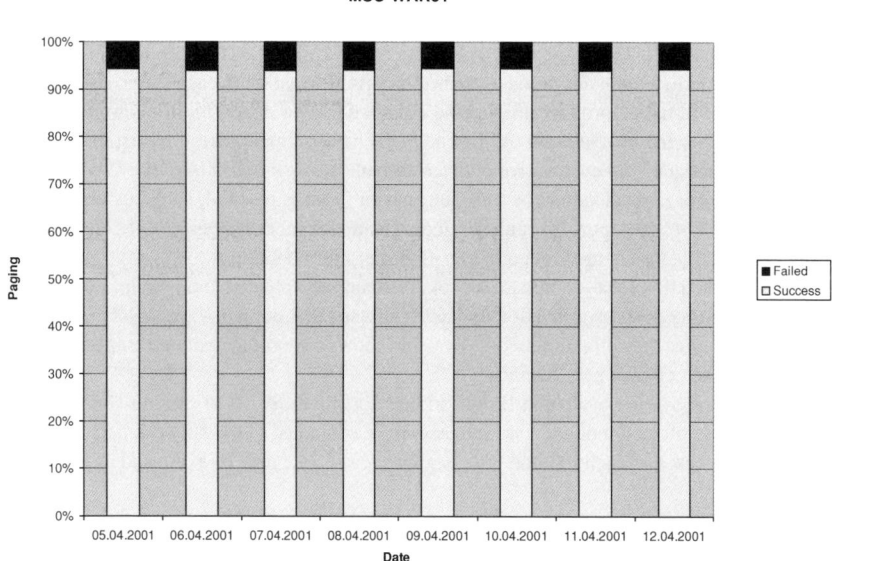

Figure 4.39 Example of the paging success rate analysis

CS Core Network Optimisation

The paging success rate in the example network is average. The situation can be improved by using greater indoor network coverage on the radio network. The data collected here come from daily measurements. It is recommended that hourly measurements should be collected in order to evaluate the need for more resources on the air interface.

Call Answer Rate Analysis

A call answer rate analysis requires a detailed study of the network clear codes. The main objectives of the call answer rate analysis are to evaluate the call answer rate and identify technical areas for network improvement.

To perform a call answer rate there is a need to change the default clear code (CC) groups in the network element. The new grouping of CCs is done by splitting the CC group for normally ended calls into answered calls and technically successful calls and then to introduce a new CC group for radio IF errors. This enables identification of areas with a new grouping of clear codes:

- Group 1: answered calls (clean code number, e.g. 000'H, normal end of call)
- Group 2: technical successful calls (clean codes, e.g. 000'H-3FF'H ex. 205H)
- Group 3: internal congestion (clean code group 400'H-7FF'H)
- Group 4: external congestion (clean code group 800'H-BFF'H minus radio IF error)
- Group 5: subscriber error (clean code group C00'H-FFF'H minus radio IF error)
- Group 6: radio interface error (clean codes 205'H, B13'H–B14'H, B16'H, B1A'H–B1F'H, D00'H–D01'H, D04'H, D07'H–D08'H)

In the new grouping, clean codes related to failures in the radio interface are grouped into group 6. The following is a list of CCs for group 6:

205'H: radio if congestion
B13'H: radio interface failure
B14'H: handover failure
B16'H: remote equipment failure
B1A'H: BSSMAP protocol error
B1B'H: radio interface message failure
B1C'H: radio interface failure reversion to old channel
B1D'H: O&M released inside network
B1E'H: radio resource unavailable
B1F'H: requested terrestrial resource unavailable
D00'H: subscriber signalling message content error
D01'H: subscriber signalling message format error
D04'H: subscriber signalling protocol error
D07'H: no response to call establishment; alerting or connect
D08'H: GSM ERH 4.08 protocol timer expires

The result of the call answer rate analysis is a report on the amount of calls successfully completed by either subscriber A or subscriber B and the share of technically successful calls. As an example, Figure 4.40 presents the results of the call answer rate analysis for nine months in one MSC. These results show an increase in technically successful calls during the analysis period.

Network Configuration Assessment

The main objectives of network configuration assessment are:

- review the existing voice network topology;
- review the existing signalling network topology;

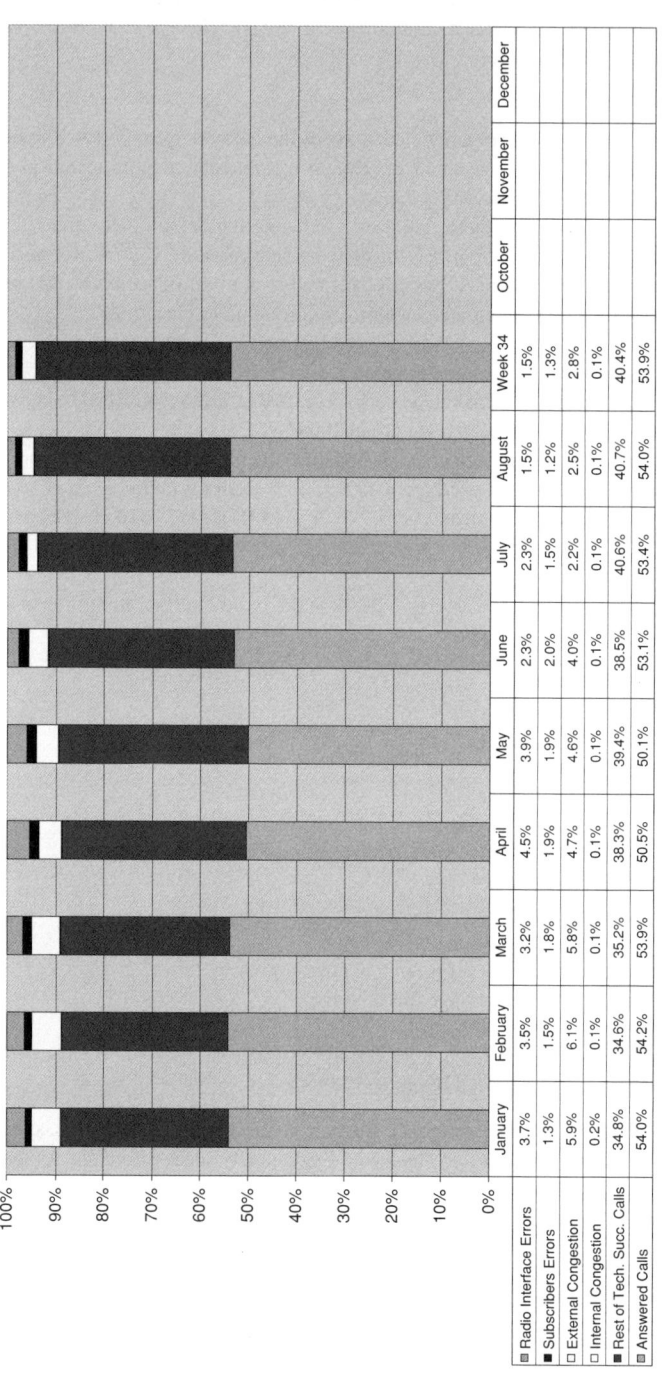

Figure 4.40 Call answer rate analysis

- analyse the network naming convention;
- VLR parameter review;
- HLR parameter review;
- proposals and suggestions for improvement.

4.7.4 Audit Results Analysis

The reports generated during the network audit serve as a vital input in the audit results analysis module. This module comprises separate tasks that may be performed independently. If the project has more available planning resources the tasks are shared and performed concurrently, speeding up project completion. The tasks performed in this module are network blocking analysis, signalling network analysis, traffic routing analysis, network load observation, alarm and fault analysis, and the network synchronisation review. A model of the present state of the network is made to identify bottlenecks, waste capacity, load-distribution problems, protection issues and underperforming parts of the network. The improved network is also modelled with Direct+ and the results are presented in the final report.

Optimisation measures are identified and outlined in a report for each separate task. A report on other optimisation measures such as network parameter build upgrade and tuning, capacity upgrade or reshuffling, and technological upgrade performed during the network audit should also be updated and prepared for implementation.

Voice Network Plan Review

The existing network architecture should be analysed for resilience in failure situation. It is important when determining the number of CS core network elements and their physical interconnections. The number of circuits and circuit groups should be determined. This is important information before collecting circuit group traffic measurements. The destinations, subdestinations and defined routes should be analysed to determine the internal routing strategy.

The amount of voice traffic per route depends on the end-to-end routing strategy used in the network. In load-sharing the amount of traffic is shared in proportions defined by the user. If there are two routes to a given destination, the amount of traffic per route may be defined as 50 % and 50 %. This also means that the two routes should have the same capacity. In alternative routing, the traffic to a given destination is routed through the primary route and only in the failure situation is the secondary route used. The primary route may therefore be dimensioned to carry the total load (i.e. 100 %) but the secondary route should also be dimensioned to carry some of the load during failures. For example, the secondary route may be dimensioned for 20 % spare capacity. It is therefore important to determine the number of internal routes and the routing strategy.

The applied routing strategy should be established. Capacity calculations may vary considerably depending on whether the operator applies load sharing or alternative routing. Generally in load-sharing of calls, voice traffic is shared equally among the routes while the alternative routing allows for disproportional distribution of the traffic load. Often, for load-sharing, a 50–50 % ratio is applied for primary and secondary routes. For example, alternative routing of 100–20 % sharing of traffic is allowed. This implies that an extra 20 % capacity is required on the alternative route and in the case of link failure 20 % capacity will be supported on the additional route.

The network environment in relation to other telecommunication networks existing in service area of the operator should be considered. A study of the percentage of outgoing traffic destined for each of the existing networks should be established in order to determine the most suitable external routing strategy. Increasing traffic between two operators presents the need to open direct POI to avoid routing failures and additional transit costs, but if the amount of traffic from an MSC area to another operator is relatively small then the traffic may be routed to the nearest POI with the operator or when it is even smaller to the nearest PSTN connection. The last option should, however, only be considered if there are no formal restrictions for transit traffic through the PSTN.

Signalling Network Plan Review

The signalling system can be used with different types of signalling network structures. The choice between different types of signalling network structures may be influenced by factors such as administrative aspects and the structure of the telecommunication network to be served by the signalling system.

In the case where provision of the signalling system is planned purely on a per signalling relation basis, the likely result is a signalling network largely based on associated signalling, typically supplemented by a limited degree of quasi-associated signalling for low volume signalling relations. The structure of such a signalling network is mainly determined by the patterns of the signalling relations.

Another approach is to consider the signalling network as a common resource that should be planned according to the total needs for common channel signalling. The high capacity of digital signalling links, in combination with the needs for redundancy and reliability, typically leads to a signalling network based on a high degree of quasi-associated signalling with some provision for associated signalling for high volume signalling relations. The latter approach to signalling network planning is more likely to allow exploitation of the potential of common channel signalling to support network features that require communication for purposes other than the switching of connections.

In order to take account of signalling message delay considerations when structuring a particular signalling network, regard should be given to the overall number of signalling links (where there are a number of signalling relations in tandem) related to a particular user transaction (e.g. to a specific call in the telephone application). In other words, there should be as few signalling transfer points as possible in the signalling network.

The optimisation engineer should collect information concerning the signalling network structure. The topology should be presented in a clear structure with all signalling elements and signalling information included. The report is an important input to the signalling network analysis module.

VLR-Specific Parameters Audit

The VLR-specific parameters are general parameters of the VLR, meaning that they do not depend on the subscriber's HPLMN (home PLMN). The VLR-specific parameters can handle:

- general VLR operations (e.g. VLR cleaning, triplet record, deregistration);
- security operations (e.g. the use of authentication and IMEI checking);
- the use of TMSI paging and searching;
- the support of supplementary services, teleservices and bearer services.

The network planner should be familiar with the most important VLR parameters and their effects on network performance, as this constitutes part of the optimisation process.

In the example above, the use of TMSI for paging is disabled. This means that the operator is using IMSI for paging purposes. It is not recommended that IMSI is broadcast on the air interface as this could lead to eavesdropping. According to product recommendations, one repaging attempt should be used if TMSI is used for paging but in the case where only IMSI is used, it is recommended that three repaging attempts should be made. The number of repaging attempts may be checked. As an example, unsuccessful paging may cause a high number of clear codes of 0005H, B number busy, and 0012H, no answered call.

A thorough study of the VLR and PLMN parameters should be done to ensure proper functioning and better results with paging, location update and the general call answer rate.

PLMN-Specific Parameters Audit

The PLMN-specific parameters control VLR-specific functions that depend on the subscriber's HPLMN. The PLMN-specific parameters can handle:

- roaming status;
- IMEI checking parameters;
- TMSI allocation parameters;
- authentication and ciphering parameters;
- advice of charge parameters;
- equal access parameters.

The PLMN parameters enable subscriber and roaming profile management. When a roaming profile is being set for the subscriber, the profile data have to be checked. Whenever creating a profile, it is recommended that a name for the profile be determined so that profiles can be distinguished from those that have not been used in subscriber data. When analysing PLMN parameters, check that no inconsistencies remain in the roaming profile data. The most reliable way to ensure this is to format all profile data relative to the PLMN analysed. The counter values mean that, for example, when the location update counter is 5, the operation (IMEI checking, TMSI allocation or authentication) is done with every fifth location update registered in the VLRU. If the counter value is 1, this means that the operation in question is done every time it is registered in the VLR. If the counter value is 0, this means that the counter is not used.

HLR Configuration Parameters Audit

The HLR-specific parameters are general parameters of the HLR (i.e. they do not depend on the PLMN, for example). These various parameters define the HLR actions in more detail and affect all subscribers in the HLR. In PLMN records the operation of the HLR with other networks is determined. The PLMN record specifies whether:

- a subscriber is removed from the VLR when the update fails in that PLMN;
- roaming is not allowed to a PLMN for subscribers with certain services;
- there are supplementary or basic services that must not be transferred to the PLMN where a subscriber is roaming.

The HLR parameters need to be studied well to determine if there are any inconsistencies in parameter selection that could lead to failures in the network. A good example is the maximum number of call forwarding that is set in the HLR. This parameter should be the same in all HLRs in the network.

When the operator makes changes in the subscriber data, the changes need to be transferred from the HLR to the VLR. The parameter number of insert attempts and the interval between insert attempts define how many update attempts are made and how long a delay there is between the attempts to update the subscriber data.

HLR PLMN Parameters Audit

The PLMN-specific parameters control HLR functions that depend on the PLMN (or even on the VLR) where the subscriber is roaming. This shows how the HLR operates with other networks. The PLMN-specific parameters can handle the following operations:

- the transferrence of subscriber data when updating the subscriber data fails;
- the defining of denied services and roaming limitations (i.e. not allowed basic or supplementary services and roaming not allowed for a subscriber with a certain service).

With these parameters the operator can define whether a subscriber with a certain service set is allowed/not allowed to roam in a given PLMN or whether the subscriber is removed from the VLR when the location update fails, as well as the supported CAMEL phase. It is important that these parameters are the same in all MSSs of the network.

Naming and Numbering Analysis

The network naming convention should be studied to ensure naming consistency within the network. Inconsistent naming of NSS elements and components makes the work of network analysis and maintenance very difficult. It should therefore be avoided where possible. A proper naming convention should be agreed with the operator and implemented consistently in all network elements.

Naming of CS Core Network Elements

The naming convention applied in the network should be examined for uniformity and consistency in the network. The operator should make maximum use of the available possibility for naming of MSCs, HLRs, VLRs, PABXs, POI (other networks), prepaid nodes, etc.

Circuit Group Naming Conventions

The operator should also make maximum use of the available possibility for naming the circuit group. For example, the naming of the circuit group to the Helsinki MSC could be MS1HE01B, which means that it is a circuit group to MSC in zone 1 (or G for the GSM network) located in Helsinki and it is the first MSC and is a bidirectional circuit group. Circuit group numbering should also be harmonised in the network and should present a consequent reasoning and consistency.

Naming of Destinations

The destination name should identify the destination MSC, network element or POI to which the traffic is routed in the network or, in the case of an incomplete analysis, the MSC in which the call will be further analysed. These can be letters, digits or a combination of both. To make troubleshooting and network maintenance easier it is recommended that a more descriptive name for destinations be used. It is therefore an operator's responsibility to select the most suitable scheme of naming destinations in the network. The network planner should make a study of the existing naming scheme and make recommendations to enhance or initiate one where such a scheme does not exist.

Naming of Subdestinations

The most informative way of naming subdestinations used for actual routing of calls is by naming them according to the circuit group they are using. Please note that in creating a route in newer MSC SW builds, subdestination and destination names not required, but the system automatically allocates numbers as names of the subdestination and destination and this complicates analysis of routing. This should therefore be avoided to ensure ease of network maintenance.

Naming of Signalling Points and Point Codes

A signalling point is identified by a signalling point name and a signalling point code. The signalling point name contains one to five ASCII characters that uniquely identify a signalling point. The name should be descriptive of the signalling point, preferably enabling easy identification with the MSC name. The signalling point naming should be consistent within the network.

Signalling point codes and a signalling network uniquely identify a signalling point at the MTP level. Some signalling point codes are issued to the operator by the telecommunication authority, while in other cases this may be decided by each operator individually. It is the responsibility of the planning engineer to understand the signalling networks in use and the point codes. The purpose here is to understand the signalling network better and identify how it is intended to work.

Introduction to the PS Core Network 379

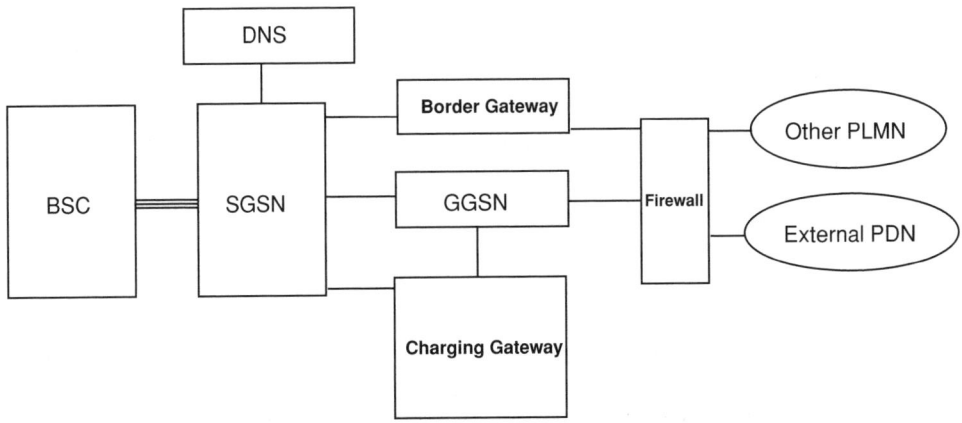

Figure 4.41 GPRS core network elements

4.7.5 Network Optimisation Results

A network optimisation project report is made in this module according to the network optimisation plan. An executive summary presents recommended changes and decisions to be made for optimisation of the network and future network performance monitoring. The full report provides the results of the project in a more detailed and explanatory format. An updated follow-up procedure is included.

It is a good practice at this stage to organise a workshop to discuss the most important findings of the optimisation project and suggestions for a regular KPI measurement reporting procedure. It is also a good forum in which to discuss any future development paths for the operator.

PART II PACKET SWITCHED CORE NETWORK PLANNING AND OPTIMISATION

4.8 Introduction to the PS Core Network

The packet switched service is a communication system in which data is divided and transmitted in packets of a predefined size. Its special feature is that communication between terminals with differing speeds and formats is possible since transmission/reception is performed after the data has been stored at the exchange. In contrast to circuit switching where a circuit is occupied until all data transfer has been completed, packet switching improves efficiency through the common use of circuits.

In a 2.5 generation GSM network, the GPRS was introduced to handle data in a more efficient manner. Due to the above-mentioned benefits of packet switching, a packet switched domain was introduced in the GPRS as CSD and HSCSD (high speed CSD) could not fulfil the requirements of handling bursty data traffic.

A GPRS network can be broadly divided into two domains, the RF domain and the packet core domain. The RF domain extends from the handset till the BSC including BTS and associated interfaces. The packet switched domain is beyond the Gb interface towards the Internet, comprising SGSN, GGSN, DNS, border gateway, charging gateway, firewalls, their interfaces and Internet access. These will be discussed in more detail later in Section 4.8.1, Basic MPC Concepts. Figure 4.41 shows the essential network elements in a GPRS core network.

Packet core remains the same for the EDGE network. An EDGE user is just another high speed user for a packet core. Hence throughput dimensioning should take care of EDGE subscribers and their required data rate. PS core planning deals with designing, planning, dimensioning and optimising of the packet core domain of the GPRS network and other IP based solutions in a mobile phone network. It involves planning of interconnectivity, dimensioning of GPRS packet core network elements like SGSN, GGSN, DNS, DHCP (Dynamic Host Control Protocol), border gateway, legal interception gateway, etc., throughput and bandwidth calculations at various interfaces. Various techniques of IP planning like VLAN (Virtual LAN), VRRP (Virtual Router Redundancy Protocol), subnetting, etc., are used to design most bandwidth efficient, cost-effective, robust and future-proof solutions. Latest technologies like ATM, SDH, PDH, SIGTRAN and MPLS (multiprotocol label switching) are judiciously and gainfully deployed.

4.8.1 Basic MPC Concepts

The packet domain allows for efficient transfer of high speed and low speed data. The network subsystem and the radio subsystem are kept totally separate so that the network subsystem can be efficiently re-used by other radio access technologies. 2G and 3G networks use a common packet core network for transfer of data packets. The quality of service (QoS) is maintained for both real-time and non-real-time services.

Packet core network functionality is logically implemented on two network nodes, the serving GPRS support node (SGSN) and the gateway GPRS support node (GGSN). The SGSN takes care of session and mobility management. It keeps track of the location of an individual mobile station and performs security functions and access control. The SGSN generates call data records (S-CDRs) for each session, which are sent to the billing centre for charging. The 2G-SGSN is connected to the BSC via the Gb interface while the 3G-SGSN is connected to the RNC via the Iu-PS interface. The GGSN provides interworking with external packet switched networks and is connected to the SGSN using an IP based packet domain backbone network. The GGSN generates call data records (G-CDRs) for each session, which are sent to the billing centre for charging.

The roles of the SGSN and GGSN will be discussed in more detail later in the chapter.

A charging gateway collects CDRs (call data records) from the SGSN and GGSN, stores them and forwards them to the billing/mediation server. A mobile station performs a GPRS attach to the SGSN as the first step in order to be known to the packet core network. The SGSN will be the home SGSN for a local subscriber and the visitor SGSN for a roamer. After a successful GPRS attach, the mobile station is available to the packet switched services, like receiving an SMS over the GPRS, paging by the SGSN or receiving notification of incoming data. In order for the MS to be able to send or receive data, it has to create a Packet Data Protocol (PDP) context with a GGSN as the next step. The GGSN may be a home GGSN or a visitor GGSN (for a roamer) depending on the policy of the service operator. If operators want roamers to use the visitor GGSN, it has to be defined in the HLR. Once a PDP context is established, an MS has access to the external packet data networks like the Internet. Thus a GPRS network does not need to understand/interpret a host of Internet protocols, thereby making things simpler.

The GPRS packet core uses the Internet protocol (IP) as specified in RFC 791 for communication with external data networks. Figure 4.42 mentions the names of important interfaces used in the GPRS to connect various network elements in the packet core.

GPRS Support Nodes (GSN)

GPRS functionality in a GSM network is implemented with the help of GPRS support nodes (GSNs). There are two types of GSN, serving GSN (SGSN) and gateway GSN (GGSN). There may be more than one of these in a single network.

The SGSN interfaces with the GSM network using the Gb or Iu-PS interface. A mobile station (MS) first attaches with the SGSN supporting its region of location. The SGSN gets subscriber data from the HLR over the Gr interface. It manages information related to mobility, location, security and authentication of the subscriber. The SGSN is further connected to the GGSN over the Gn interface through the packet

Core Network Planning and Optimisation

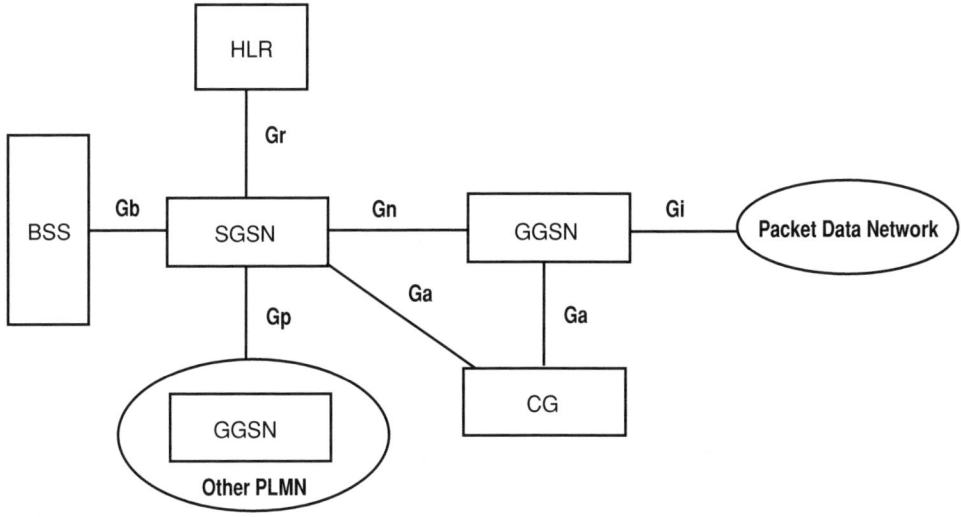

Figure 4.42 GPRS core interfaces

domain, which is IP based. The SGSN is connected to the GGSN of other PLMNs (in the case of roamers) using the Gp interface through the BG (border gateway). An SGSN helps in establishing the PDP context between the MS and GGSN. It may be optionally connected to the MSC/VLR using the Gs interface to send location information. The SGSN capacity is measured in terms of the total number of subscribers that can simultaneously attach to it, simultaneous PDP contexts and data throughput.

The GGSN provides access to packet data networks (Internet/Intranet) based on the PDP address. Packet data units (PDUs) coming from a PDN are routed by the GGSN to the SGSN to which the subscriber is attached. The SGSN and GGSN may reside in the same hardware although they are mainly separate. The GGSN is essentially a router while the SGSN may be a switching platform. The GGSN connects to the SGSN over the Gn interface while it uses the Gi interface to connect to the Internet via the firewall. In a layer 2 core switch separate VLANS are created for Gn and Gi interfaces to segregate traffic. For an external packet data network, the GGSN is the first point of contact. The capacity of the GGSN is measured in terms of the total number of simultaneous PDP contexts and data throughput.

Access Point Name (APN)

The access point name (APN) is the reference to an external packet data network to which the mobile user wants to connect. In the GPRS backbone, the APN is a reference to the GGSN, to be used in case different GGSNs are connected to different external networks. The access point is the logical connection that the GGSN provides, allowing mobiles to connect to external networks. Typically the subscriber might use one access point to connect to the corporate Intranet, another access point to access ISP and yet another for WAP (Wireless Application Protocol) or MMS (multimedia messaging service). The access point is configured on the GGSN. A single GGSN can provide several access points, even on the same physical interface. A sample APN is my.isp.com.myoperator.com.gprs.

Example of GPRS Roaming

When a subscriber of operator A is roaming in another PLMN (operator B) and tries to establish a PDP context, the APN defined in the handset (e.g. wap.operatorA.gprs) is mapped to the complete APN

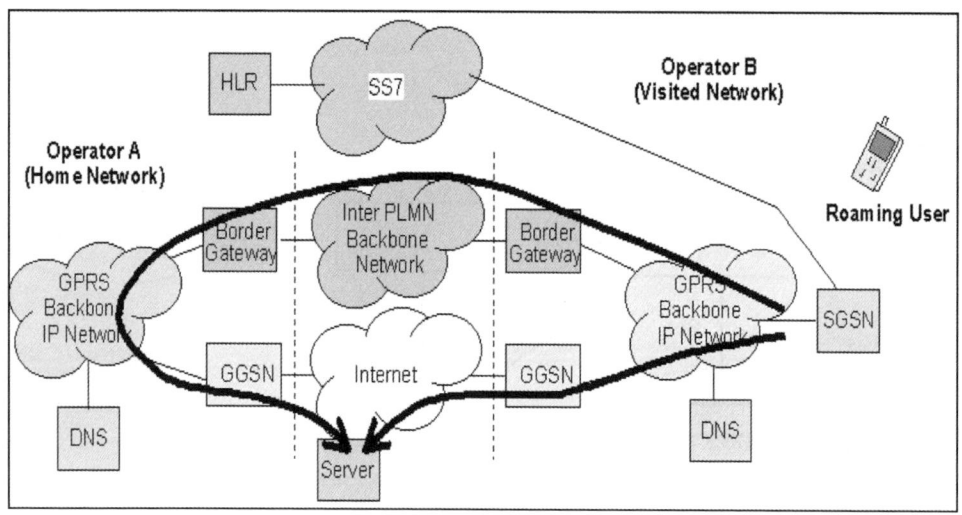

Figure 4.43 GPRS roaming

name (wap.operatorA.mnc001.mcc002.gprs) in the visitor SGSN using the IMSI information. The visitor SGSN asks its DNS for the GGSN IP address using the APN as the key. The DNS asks a root DNS whether it knows a DNS that can resolve this APN. Once the DNS of operator A is found, the IP address of the GGSN serving this APN is received. After receiving the GGSN IP address, the visitor SGSN creates a PDP context with the home GGSN serving the APN and a GTP tunnel is created over Gp interface. If allowed in the HLR and both operators have the same APNs, a visitor can use the visitor GGSN, as shown in Figure 4.43.

Tunnel End Point Identifier

The GPRS uses the Tunnelling Protocol (GTP) to transfer data between the SGSN and GGSN. A tunnel end point identifier (TEID) is a unique identifier for the receiving end of a GTP tunnel. It also uniquely identifies a PDP context. The receiving end side of a GTP user tunnel locally assigns the TEID value that the transmitting side has to use. The TEID values are exchanged between tunnel end points using control messages (GTP-C). The TEID has meaning only within the GTP protocol. In the GTP-U user plane, each PDP context has a one-to-one relationship between the TEID on the one hand and IMSI and NSAPI (network service application port identifier) on the other hand. The algorithm to compute the TEID is implementation-dependent and may vary from manufacturer to manufacturer. When a PDP context is established, the TEID is forwarded to the GGSN and is used in subsequent tunnelling of user data between the GGSN and SGSN for that particular PDP context.

4.8.2 Packet Routing (PDP Context)

The Packet Data Protocol (PDP) is used for data transfer and routing. For every GPRS subscription there is a corresponding PDP address associated with it. Each PDP address is described by one or more PDP contexts defined in the MS, the SGSN and the GGSN. The size of a PDP unit is a maximum 1500 octets. If the packet received by the MS or GGSN is bigger than this, it may be segmented or discarded, depending on the PDP type and the implementation. The packet data protocol in the MS may limit the maximum size of the PDP PDUs that are routed and transferred, for example, due to MS memory limitations. The MS is responsible for creating or modifying PDP contexts and their QoS.

Between the SGSN and the GGSN, PDP PDUs are routed and transferred with the UDP (User Datagram Protocol)/IP protocols. The GPRS tunnelling protocol (GTP) transfers data through tunnels. The tunnel end point identifier (TEID) and GSN address identify a tunnel. To support roaming subscribers, and for forward compatibility, the SGSN is not required to know the tunnelled PDP. The SGSN has the capability to transfer PDUs belonging to PDPs not supported in its own PLMN.

The GGSN could also optionally support IP multicast: this allows the MS to join multicast groups and start receiving multicast packets. The GGSN duplicates the incoming multicast packets and relays them to the already active TEIDs. These TEIDs are those of MSs that have joined a multicast group.

There are two states in which each PDP context can exist: the active state and inactive state. The PDP state indicates whether data transfer is enabled for that PDP address or not. In case all PDP contexts associated with the same PDP address are deactivated, data transfer for that PDP address is disabled.

Inactive PDP Context

In this state, data service for a certain PDP address of the subscriber is not activated. The PDP context contains no routing or mapping information to process PDP PDUs related to that PDP address. No data can be transferred. A changing location of a subscriber causes no update for the PDP context in the inactive state, even if the subscriber is GPRS attached. If a GGSN receives a data packet destined for an MS in the inactive state, it will generate error and the packet will be discarded. An Internet Control Message Protocol (ICMP) packet (error notification) will be sent to the originator. The PDP context activation procedure is always initiated by the mobile station for it to move from the inactive to the active state.

Active PDP Context

In this state, the PDP context for the PDP address is activated in the MS, SGSN and GGSN. The PDP context contains mapping and routing information for transferring PDP Units for that particular PDP address between the MS and the GGSN. An active PDP context for an MS is moved to the inactive state when the deactivation procedure is initiated. All active PDP contexts for an MS are moved to the inactive state when the handset moves to the IDLE state. A GPRS-attached mobile station can at any time activate and de-activate the PDP context in the MS, SGSN and GGSN. A GGSN can de-activate a PDP context. When the SGSN receives an activate PDP context request message, it initiates a procedure to set up the PDP context. It performs subscription checking and APN selection. If the request is to activate a secondary PDP context (a primary PDP context already exists), the SGSN does not have to perform all these functions. Instead it re-uses PDP context parameters including the PDP address but excluding QoS parameters. Once activated, all PDP contexts that share the same PDP address and APN are managed equally. At least one PDP context should be activated before a secondary PDP context can be activated for the same PDP address.

PDP Address

The PDP address is the IP address allocated to a mobile station when it activates a PDP context. It may be an Ipv4 or Ipv6 IP address. The PDP address is associated with the IMSI and it may be permanent or temporarily assigned by the GGSN, DHCP or radius, or by an external public data network (PDN).

PDP addresses can be allocated to an MS in four different ways:

- The home PLMN (HPLMN) operator assigns a PDP address permanently to the MS (static PDP address).
- The HPLMN operator assigns a PDP address to the MS when a PDP context is activated (dynamic HPLMN PDP address).

- The visitor PLMN (VPLMN) operator assigns a PDP address to the MS when a PDP context is activated (dynamic VPLMN PDP address).
- The PDN operator or administrator (e.g. an Internet service provider or corporate Intranet administrator) assigns a permanent or dynamic IP address to the MS (external PDN address allocation).

It is the HPLMN operator that defines in the subscription whether a dynamic HPLMN or VPLMN PDP address can be used.

For every IMSI, zero, one or more dynamic PDP addresses per PDP type can be assigned. For every IMSI, zero, one or more static PDP addresses per PDP type can be subscribed to. When dynamic addressing from the HPLMN or the VPLMN is used, it is the responsibility of the GGSN to allocate and release the dynamic PDP address. When an external PDN address allocation is used, the PLMN may obtain a PDP address from the PDN and provide it to the MS during PDP context activation, or the MS may directly negotiate a PDP address with the PDN after the PDP context activation procedure is executed. If the PLMN provides the address during PDP context activation in the case of external PDN address allocation, then it is the responsibility of the GGSN and PDN to allocate and release the dynamic PDP address by means of protocols such as DHCP or RADIUS. If the DHCP is used, the GGSN provides the function of a DHCP client. If RADIUS is used, the GGSN provides the function of a RADIUS (remote authentication dial-in user server) client. If the MS negotiates a PDP address with the PDN after PDP context activation in the case of external PDN address allocation, it is the responsibility of the MS and the PDN to allocate and release the PDP address by means of protocols such as the DHCP. Only static PDP addressing is applicable in the network-requested PDP context activation case.

Encapsulation

The packet domain PLMN backbone network encapsulates a PDP PDU with a GPRS tunneling protocol header, and inserts this GTP PDU in a UDP PDU that again is inserted in an IP PDU. The IP and GTP PDU headers contain the GSN addresses and tunnel end point identifier necessary to uniquely address a GSN PDP context. Between a 2G SGSN and an MS, the SGSN or MS PDP context is uniquely addressed with a temporary logical link identity and a network layer service access point identifier pair. TLLI (temporary logical link identifier) is derived from the PTMSI. NSAPI is assigned when the MS initiates the PDP context activation function.

4.8.3 Interface of the GPRS with the 2G GSM Network

In a 2G network the Gb interface connects the GPRS with the GSM. It is between the BSC and the SGSN. It may consist of E1 PCMs with a data rate of 2 Mbps, i.e. 32 timeslots of 64 kbps each. Out of these, 30 timeslots may be used for traffic (user data) and 2 for signalling. The Gb interface allows many users to be multiplexed over the same physical resource. When a user needs to send or receive data, the timeslots are allocated temporarily and are reallocated after completion of the activity. In this way, the Gb interface is more efficient than the A interface, where timeslots are dedicated to a user throughout the lifetime of the call irrespective of activity. In the Gb interface, the same user plane is used to send signalling and user data. Separate physical E1s are not required to be dedicated for signalling purposes. Access rates per user may vary without restriction from zero data to the maximum possible line rate, which is $64 \times 32 = 2048$ kbps in the case of an E1 trunk. The Gb interface can also be implemented over the IP.

Routing Area

The routing area (RA) in the GPRS is analogous to the location area (LA) in the GSM. A GPRS network is divided into several routing areas. An RA may consist of one or more cells as planned by a Gb interface planner. The location of an MS in the standby state is known to the SGSN only at a routing area level and is paged in the whole RA. A routing area identifier (RAI) identifies an RA. Non-GPRS cells are grouped in a null RA. When a mobile-terminated packet arrives at the SGSN, the MS is paged in the RA where it is located. For circuit switched services it is also paged in null RA. An RA is a subset of LA and cannot span across two LAs. An RA is served by a single SGSN. A combination of RAC and LAC should be unique:

$$RAI = MCC + MNC + LAC + RAC$$

Charging

A charging gateway (CG) is responsible for collection of CDRs (call data records) from the GGSN and SGSN in a GPRS network. It is connected using a Ga interface. The operator may define whether charging will be done on an individual MS basis and/or a PDP context basis. The SGSN CDRs contain charging information for each MS related to the radio network usage.

SGSN CDRs at least contain the following information:

- the amount of data sent/received by a MS on radio interface;
- usage of PDP addresses: how long an MS has used PDP addresses assigned to it;
- location of MS: HPLMN, VPLMN, etc.

The GGSN CDRs contain charging information related to the external data network usage. The GGSN CDRs at least contain the following information:

- destination and source addresses;
- the amount of data sent/received to/from external data networks;
- usage of PDP addresses: how long an MS has used PDP addresses assigned to it.

4.9 IP Addressing

4.9.1 Types of Network

An IP address, which is made up of 32 bits, is divided into two parts, a network address and a host address. Several machines in the same network will have the same network address prefix but a unique host address. There are three types of networks, class A, class B and class C, based on the number of bits belonging to the network or host address part.

Class A Networks (/8 Prefixes)

Each class A network address has an 8-bit network prefix with the highest order bit set to 0 and a 7-bit network number, followed by a 24-bit host number. Class A networks are also referred to as '/8s' (pronounced 'slash eight' or just 'eights') since they have an 8-bit network prefix. A maximum of 126 ((2 to the power of 7) − 2) number of '/8' networks can be defined. The calculation requires that 2 is subtracted because network 0.0.0.0 is reserved for use as the default route and the '/8' network 127.0.0.0 (also written 127/8 or 127.0.0.0/8) has been reserved for the 'loopback' function. Each '/8' supports a maximum of 16 777 214 hosts per network. The host calculation requires that 2 is subtracted from 2 to the power of 24 because all-0s ('this network') and all-1s ('broadcast') host numbers may not be assigned to individual hosts.

Class B Networks (/16 Prefixes)

Each class B network address has a 16-bit network prefix with the two highest order bits set to 1–0 and a 14-bit network number, followed by a 16-bit host number. Class B networks are now referred to as '/16s'. A maximum of 16 384/16 networks can be defined with up to 65 534 hosts per network.

Class C Networks (/24 Prefixes)

Each class C network address has a 24-bit network prefix with the three highest order bits set to 1-1-0 and a 21-bit network number, followed by an 8-bit host number. Class C networks are now referred to as '/24s' since they have a 24-bit network prefix. A maximum of 2 097 152/24 networks can be defined with up to 254 hosts per network.

Other Classes

In addition to the three most popular classes, there are two additional classes. Class D addresses have their leading four bits set to 1–1–1–0 and are used to support IP multicasting. Class E addresses have their leading four bits set to 1–1–1–1 and are reserved for experimental use.

4.9.2 Dotted-Decimal Notation

To make Internet addresses easier for human users to read and write, IP addresses are often expressed as four decimal numbers, each separated by a dot. This format is called 'dotted-decimal notation.' Dotted-decimal notation divides the 32-bit Internet address into four 8-bit (byte) fields and specifies the value of each field independently as a decimal number with the fields separated by dots. Below is shown how a typical /16 (class B) Internet address can be expressed in dotted decimal notation:

$$10010001000010100010001000000011$$
$$145.10.34.3$$

The dotted-decimal values shown below indicate how, each of the three principle classes can be assigned, where the administrator assigns Xxxx values to the hosts:

Class A: 1.xxx.xxx.xxx to 126.xxx.xxx.xxx
Class B: 128.0.xxx.xxx to 191.255.xxx.xxx
Class C: 192.0.0.xxx to 223.255.255.xxx

This is knwon as classful addressing and has several inherent limitations. A /16 network consists of 65 536 nodes while a /24 network can have only 256 nodes. Most of the medium sized organisations have requirements somewhere in between. Such organisations were allotted one /16 network. Hence /16 networks have now become scarce. Multiple /24 subnets are now allocated to an organisation that has greater than 256 nodes, but this results in a new problem of growing sizes of Internet routing tables, increasing processing capacity requirements for routers and hence increasing cost. Classless addressing, in which there is more flexibility in subnetting, takes care of this wastage in present and future networks but past mistakes cannot be corrected.

When the Internet was designed the developers never imagined the kind of growth it has witnessed. 32-bit IP addressing puts a limit of 2 to the power of 32 addresses, i.e. 4 294 967 296. With the way the Internet is growing, this will soon deplete. If more bits had been used for addressing, the availability of IP addresses would have increased exponentially. IP 6 addressing uses 128 bits, which is virtually limitless,

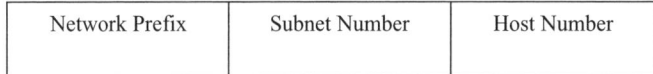

Figure 4.44 Example of subnetting

but implementing IPv6 would require organisations worldwide to patch their systems and modify existing IP addresses. As of now, organisations are simply delaying this due to the enormity of the task.

4.9.3 Subnetting

This is a technique that allows three-level hierarchies in addressing rather than two-level hierarchies, which was used in classful addressing. The network prefix remains the same but the host number is further subdivided into a subnet number and a host number (Figure 4.44).

When faced with the requirement of installing a new network in an organisation, instead of applying to the Internet authority for yet another network, an administrator can instead divide the present network into various subnetworks or subnets. These subnets are internal to the organisation. Hence there can be as many nodes as required behind only a few Internet visible public IP addresses.

Private subnetting has several advantages. Firstly, an administrator no longer has to make a request for a network number from the Internet before a new network is installed. This also accomplishes a more efficient use of the scarce IP address space. Secondly, since the subnets are not Internet visible, Internet routing tables would not grow with increasing numbers of subnets. Just one route from the Internet to the network is good enough to take care of all the machines belonging to different subnets inside it. Thirdly, every time there is a change in routing in an organisation's private network, routing tables of Internet routers need not be updated. The last but most important advantage is the efficient use of scarce address space as there can be a large number of nodes behind only a few Internet visible IP addresses.

Internet routers use the network prefix to route traffic to the correct network. Routers within the subnetted environment use the network prefix plus the subnet number to route traffic to the correct subnet. The network prefix plus the subnet number is called the extended network prefix. Therefore, if an IP address belongs to the class B network but its extended network prefix is 24 bits long, then its subnet mask will be 255.255.255.0 and not 255.255.0.0. The bits in the subnet mask whose value is 1 correspond to the extended network prefix and bits with 0 value correspond to the host number. For example, the IP address 140.1.1.1 traditionally belongs to the class B network, i.e a /16 network with the net mask 255.255.0.0. However, if its first 3 octets represent the extended network prefix and only the last octet represents the host number, then it will be represented as 140.1.1.1/24 and the net mask as 255.255.255.0. Here it would be good to mention that although /<prefix length> notation is compact and easier to understand, many modern routing protocols still expect the subnet to be in its traditional dotted-decimal format.

Subnet Planning

While designing a network, an IP planner must keep in mind present and future requirements of the maximum number of subnets and the maximum number of hosts in the biggest subnet in an organisation. When a solution comprises several solutions and sites, each solution and/or site should have at least one separate subnet that can be further subnetted as per the solution requirements.

If there is a need for six subnets to be made out of a /24 network, then 3 bits will have to be used for the subnet number, giving 2 to the power of 3 , i.e. 8 subnets, with two subnets remaining spare. Then 5 bits will be used to represent the host addresses and thus there will be a maximum of 2 to the power of $5 = 32 - 2 = 30$ hosts in each of the 3 subnets.

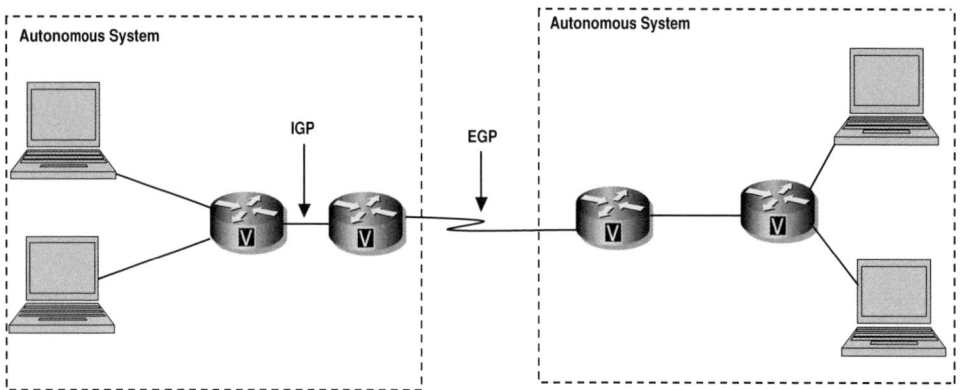

Figure 4.45 Interior and Exterior Gateway Routing Protocols

If a planner sees that the subnet requirement is going to increase in the future, then he/she should use 4 bits for the subnet number instead of 3, giving 16 subnets that can be used in the future. The same philosophy also holds good for a number of hosts. It is better and advisable to have a buffer for the future from the beginning as increasing a subnet size later on in a live network will be very difficult and undesirable.

It is to be noted that these days routers and protocols follow classless addressing but all-0s and all-1s subnets can also be used. This means that the router and its protocol are able to differentiate between a 10.10.1.0/24 and a 10.10.1.0/27 subnet. The BGP (Border Gateway Protocol) is one such protocol that has this intelligence while RIP-1 (Routing Information Protocol 1) is unable to do so. RIP-1 follows classful addressing, where it is forbidden that all bits of a subnet number be zeros or ones.

In classless addressing, host addresses are made up of remaining bits after the network prefix. Therefore a /28 network has 4 bits for the host address while a /26 has 6 bits. Now considering a /28 network, the maximum number of host addresses can be 2 to the power of 4, which is 16. Out of these, the first address consisting of all-0s represents the base address of the subnet and the last address consisting of all-1s is the broadcast address. Hence these two cannot be assigned to any host machine. Thus, the total number of usable host addresses be comes 14. In general, any network with a prefix /n can have a maximum of 2 to the power of $(32 - n) - 2$ hosts.

4.10 IP Routing Protocols

IP packets move from the source to destination based on 'routes' decided by routing protocols running on devices called 'routers'. The Internet comprises multiple networks worldwide interconnected with routers. A group of networks under the same administrative authority and control is called an 'autonomous system'. A routing protocol that is used to move information within an autonomous system is called an Interior Gateway Protocol (IGP). A protocol that is used for exchange of information between different autonomous systems is called an Exterior Gateway Protocol (EGP). Figure 4.45 depicts the use of Interior and Exterior Gateway Routing Protocols.

Various examples of the IGP are the RIP, OSPF and IS-IS (intermediate system-to-intermediate system) whereas the BGP is an EGP. These protocols work on the basis of a common philosophy of forwarding the packet to the next router, which is considered as the next hop. A route consists of a final destination network address and the IP address of the next hop to reach it. For example, for a destination network of 134.1.1.0, the next hop may be 10.132.1.1. Routing tables are defined in each of these routers. Routes are of two kinds, static and dynamic. Static routes are defined by the administrator. In dynamic routing, the protocol calculates routes. Dynamic routing algorithms adapt to changes in the network and automatically select the best route. At each node, the destination address defined in the datagram is compared with the route defined in the router, the best next hop is decided and the packet is forwarded. Each node is responsible for forwarding the packet one hop at a time. The Internet protocol does not take care of error detection and/or correction. The ICMP (Internet Control Message Protocol), which is an integral part of any IP network, is responsible for this.

Routing Information Protocol (RIP)

The RIP is an Interior Routing Protocol developed by Xerox. The RIP works well in relatively small networks. Routing decisions are based on the hop count. A route is considered as the best route if it has the least number of hops to reach a final destination. However, it ignores other important factors like line utilisation and line speed. The RIP has serious limitations in large networks. It has a limit of allowing a maximum of only 16 hops between two consecutive nodes and it is slow to converge, meaning that it takes a longer time for route changes to be propagated to all nodes. Nowadays there are more sophisticated protocols that are more efficient.

Open Shortest Path First (OSPF)

The OSPF Protocol was developed to fulfil the need for a high functionality nonproprietory protocol that was superior to the RIP and good for big networks. It is also an Interior Routing Protocol. It also takes the link state into account before taking a routing decision. It is capable of fast re-routing (convergence) and supports variable length subnet masks. In OSPF, an autonomous system is divided into routing areas. A routing area consists of closely related subnets. Link state updates are distributed only inside a routing area and not in the complete network. All routing areas must connect to the backbone area. OSPF follows hierarchical routing in an autonomous system. With OSPF, there is no limitation of hop count, unlike the RIP. OSPF uses IP multicast to send link state updates. Updates are sent only when routing changes occur and not periodically. This results in more efficient bandwidth utilisation. OSPF allows for transfer and tagging of external routes injected into the autonomous system. This keeps track of external routes injected by exterior protocols like BGP.

Interior Gateway Routing Protocol (IGRP)

The IGRP determines the best path by considering the factors of bandwidth and delay in the networks. The IGRP is fast in distributing routing changes to the nodes, thereby avoiding routing loops caused by disagreement over the next hop to be taken. It does not have hop count limitation. The enhanced IGRP is a new version of the IGRP which combines ease of use of traditional distance vector routing protocols with fast re-routing capabilities of new link state routing protocols. The enhanced IGRP requires less bandwidth as it exchanges only information related to changes in the network.

Border Gateway Protocol (BGP)

This is an Exterior Gateway Protocol used to interconnect more than one autonomous system. It is used by Internet core routers and is designed to prevent routing loops in arbitrary topologies and to allow policy based route selection. The primary function of a BGP speaking system is to exchange network reachability information with other BGP systems. Reachability information includes a list of autonomous systems (AS) traversed. The BGP supports any policy that confirms a 'hop-by-hop' routing paradigm. The latest version, BGP4, is designed to handle scaling problems of a growing Internet. BGP4 provides a new set of mechanisms for supporting classless interdomain routing. The BGP uses the TCP as its transport protocol. It uses TCP port 179 for establishing connections. The TCP meets the BGP's transport requirements and

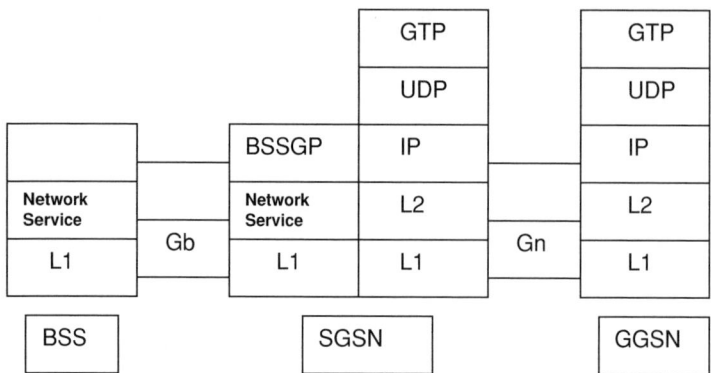

Figure 4.46 GPRS Packet Core Protocol stack

is present in almost all commercial routers and hosts. Since the BGP is running on the TCP, which is a reliable transport protocol, there is no need for the BGP to handle fragmentation, retransmission, acknowledgement and sequencing. Any authentication scheme used by the transport protocol can be used in addition to the BGP's own authentication mechanism. Any two systems that follow BGP exchange complete BGP routing tables initially. Consequently they exchange only incremental changes. Keep alive messages and error notifications are also exchanged between two BGP hosts.

4.11 Dimensioning

Dimensioning involves defining recommendations for a proposed backbone network, calculation of intersite and intrasite traffic figures for planning, calculation of the transmission capacity required across multiple core network sites and defining the appropriate number, capacity and type of recommended core equipment. Dimensioning is useful in presales support and consultancy type of work requiring a ready definition of alternative network solutions/scenarios. It helps in core network architecture design development and conceptual network architecture planning. It is also useful in the definition of network development, roll-out, migration and network re-use strategies. Dimensioning results can be used to work out alternative network scenarios and to give technical support in pricing exercises. The number of core network elements and traffic per site can be two prerequisites for input to a dimensioning tool.

4.11.1 GPRS Protocol Stacks and Overheads

Various protocols used at various interfaces add/remove data in the form of IP packets. These are called overheads. These overheads are considerable and have to be taken into account. They cannot be neglected during dimensioning. To enable the various overheads that have to be taken into account in the dimensioning of the packet core network to be understood, first the protocol stacks in the GPRS network need to be explained. Each layer (L) in the protocol stack will add additional bits to the original payload. In addition, depending on the interface, this will result in different overheads being used for network dimensioning for different nodes. Figure 4.46 shows the different protocol stacks at the different interfaces for the GPRS network.

Here is a brief description of each of the protocols shown in Figure 4.46:

- GPRS Tunnelling Protocol for the user plane (GTPU). This protocol tunnels user data between GPRS support nodes in the backbone network. The GTP-U header size is 16 bytes.

Dimensioning

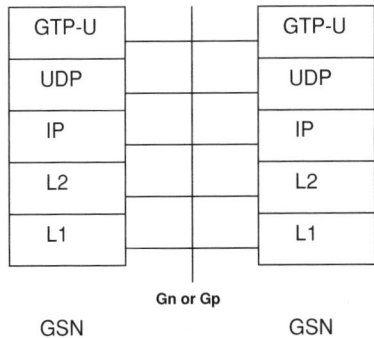

Figure 4.47 Protocol stack between two GSNs

- The UDP carries GTP packet data units for protocols that do not need a reliable transport link and provides protection against corrupted GTP packet data units. The UDP header size is 8 bytes.
- IP. This is the backbone network protocol used for routing user data and control signalling. Presently IPv 4 is being used in networks worldwide. The IP header size is 20 bytes.
- Base Station Subsystem GPRS Protocol (BSSGP). The BSSGP is used over the Gb interface between the BSC and the SGSN. BSSGP does not perform error correction. The BSSGP header size is 30 bytes.
- Network service (NS). This layer transports BSSGP PDUs. The NS is based on the frame relay connection between the BSS and the SGSN, and can multihop and traverse a network of frame relay switching nodes. The NS layer header size is 4 bytes while the frame relay header size is 6 bytes.

For GSN-to-GSN traffic, i.e. traffic between SGSN and GGSN (own or other PLMN's GGSN), the protocol stack shown in Figure 4.47 is used. GTP, UDP and IP overheads must be considered in the Gn interface between 3G SGSN and the GGSN. In the case of the 2G SGSN, the Gb overhead will be considered in the dimensioning instead as it is higher than the Gn overhead in order to cater for the worst-case scenario.

A proposed GPRS packet core network is dimensioned to support the total estimated or projected GPRS data traffic expected during the busy hour, including an appropriate overhead traffic level in addition to the actual data payload. Different applications and protocols generate packets of variable lengths. It is mainly the length of packet headers that drives the overhead size and it remains constant while the packet size varies depending on the application. Overheads as a percentage of total traffic therefore vary quite significantly for every packet length change and may be calculated as a function of the payload based on the average packet size. For example, for a packet size of 512 bytes, a 14.25 % protocol overhead is expected at the Gb interface while an 8.59 % protocol overhead is expected at the Gn interface. These overhead figures will be input to the GPRS data traffic to take into account the impact these overheads will have on the network dimensioning results.

Dimensioning Inputs and Assumptions

There are several parameters that can be taken as input from the operator based on their marketing requirements, e.g. the subscriber and traffic forecasts and the number of core sites. In case data are not provided by the operator, then some suitable figures can be assumed based on field experience of some running networks.

Some dimensioning parameters are as listed below:

- Percentage of simultaneous attached subscribers
- Percentage of simultaneous active data subscribers

- Average number of PDP context activations per subscriber during the BH (busy hour)
- Average number of PDP context deactivations per subscriber during the BH
- Average traffic demand (at the IP layer) per active data subscriber at the BH and downlink + uplink
- Ratio of downlink/(downlink + uplink)
- Number of intra-SGSN routing area updates
- Number of inter-SGSN routing area updates
- Mode of connection to the Internet/Intranet/ISP
- Ratio of transparent/(transparent + nontransparent)
- Percentage of class A, B and C mobile stations
- Number of DNS requests per subscriber during the BH
- Number of DHCP requests per subscriber during the BH
- Number of CDRs per subscriber per day
- Inter-PLMN roaming traffic: percentage of total Gn traffic
- Average packet size (bytes)
- Gn interface overhead
- Gb interface overhead
- GPRS data traffic per subscriber during the BH in bps
- EDGE data traffic per subscriber during the BH in bps

SGSN Dimensioning

The process will start again from the BSS/RAN side, the ultimate location of both sources of input, traffic and amount of subscribers. It is important to be familiar beforehand with the main characteristics and features of the type of SGSN to be deployed. The main process can be defined as finding out how many SGSNs are needed in a given core network site at a given point of the roll-out timeline to serve the previously calculated traffic from both sides, BSS/RAN and backbone, and, of course, for the expected amount of subscribers at that point of time.

The most obvious limiting factors affecting the capacity of a generic SGSN node are the number of GPRS attached subscribers to the served, simultaneous active PDP contexts and the data throughput. Data throughput is obtained as a result of the interface dimensioning exercise, whereas the amount of subscribers to be served is an input from the roll-out plan.

However, there are additional limitations, which may not be so obvious at a first glance, directly related to the traffic profile. Mobility management and session management procedures are two of the most important functionalities provided by the SGSN. Session management limitations normally arise from the amount of simultaneous PDP context active in the SGSN. In terms of mobility management, the most obvious limitation comes from the maximum amount of subscribers simultaneously attached to the given SGSN. Therefore, the two main capacity related constraints in terms of subscribers are the maximum amount of simultaneous PDP contexts and the maximum amount of subscribers simultaneously attached. Thus, it is important to obtain from the vendor product documentation of those limits, so as to ensure that the expected subscribers and their traffic can have a proper fit. In particular, the possibility should be considered of a subscriber having more than one PDP context active simultaneously. Additional possible constraints may come in the form of routing area update (RAU) procedures. The SGSN may have additional limitations from the maximum amount of procedures that can be handled in a given time, and they should be checked, especially if the input provided also specifies this type of requirement.

Having all the limiting factors affecting the dimensioning of SGSN worked out and having decided upon the geographical strategy for core site deployment, it is the moment to bring up the final calculation for the number – and configuration type, in its case – of SGSN nodes to be deployed in a given core site in a given year. The procedure is actually quite straightforward, as described in the next steps.

Firstly, consideration is given to all the expected Iu-PS interface traffic flows previously calculated that originate from all the RNCs expected to be homed into the core network site under study. The total amount of traffic, in megabits per second, is divided by the maximum throughput that one single SGSN can handle. The result, rounded up to the nearest integer number, will give the first estimation of the total amount of SGSN nodes needed in that given site for the calculated traffic in that concrete year. There are some important aspects to consider, however. The first aspect it is important to verify is that the throughput should be considered at the same protocol layer. It should not be a problem if the Iu-PS throughput is considered as the IP payload throughput for the ATM transport; otherwise the different overheads in Iu-PS with respect to Gn should be taken into account, since the maximum throughput of the SGSN is normally given in terms of the IP payload throughput. The second aspect refers to the type of Iu-PS interface cards and its maximum amount per SGSN node. Normally there should be STM-1 and E1/T1 interfaces available towards Iu-PS, the E1/T1 as components of IMA groups. Since the requirements of each Iu-PS interface are known, the game consists in fitting the Iu-PS traffic flows into the estimated amount of SGSN in a sensible way, trying not to exceed the first estimation of the number of SGSN nodes needed. The issue here is the high granularity of the STM-1 interfaces, which may cause more additional SGSNs to be needed than initially estimated, even though the throughput capacity limit is not reached because the limit of the maximum number of interfaces was first reached in a given SGSN node. For this reason, a careful check should be made after the first rough estimation of the amount of required SGSNs. The third aspect is actually brought up by the second. It may also be possible to have a mixed scenario, where not all the SGSNs are fully equipped at the given time, although this case should be contemplated in a much wider time perspective regarding expansion forecasts in the forthcoming years.

Secondly, a new rough estimation of the maximum amount of SGSN nodes in the same given core site will be made using the limit of the maximum amount of attached subscribers. As in the previous case, this should be a simple matter of dividing the total amount of subscribers expected to be served by the core site under review in a given year by the maximum attached subscribers capacity of an SGSN node, rounded up to the next integer number. The same approach as taken previously applies in this context. In this case, the point is to evaluate whether all the subscribers under any given SGSN of the core site under review, as defined by the Iu-PS homing and connectivity plan previously considered, can be served. This should be correct under the assumption that all the subscribers generate traffic uniformly, but on some occasions there might be different subscriber categories given as input – residential versus business users – that may generate different types of traffic in different areas. Therefore, this aspect should be checked in those cases.

Finally, the last calculation will be based upon the maximum number of simultaneous PDP contexts. Similar to the previous situations, the point is firstly to have a rough estimate of the maximum amount of SGSN nodes. The procedure consists of dividing the total amount of simultaneous PDP contexts expected inside the coverage area of the core site under review by the maximum amount of simultaneous PDP contexts per SGSN node, rounding up to the nearest integer. As in the previous step, in the case where all the subscribers uniformly generate PDP contexts (for instance, one PDP context per user during the busy hour), the calculation should suffice. However, in the case of having different types of subscribers with different amounts of PDP context generated by different types of subscribers, the same check as in the previous step should be carried out, to ensure that all the PDP context under a given SGSN coverage area can be served by the SGSN. It should be noted that with the introduction of new types of services and mobile phones with a multiple PDP context capability, this aspect may gain importance, although for the current situation this limiting factor is seldom the most restrictive one.

After these three aspects have been verified and the partial estimations of the number of SGSN nodes needed per core site are ready, the result of the SGSN dimensioning exercise will consist of simply taking the maximum of the values previously calculated as the amount of SGSN nodes required for the given core site. This will ensure that the most restrictive conditions are fulfilled. This process has been described generically for the number of SGSN nodes per core site, since the core site is the basic infrastructure entity in this context, but no major difficulties to expand this procedure should arise for all the foreseen core sites.

It is worth noting that different SGSN nodes from different vendors may have additional limitations, which are not described here, since only the principal aspects according to the most general types of inputs have been tackled here. The procedure should be to deal with those limitations in the same manner, considering the existing input situation and evaluating it against the network node maximum capability in terms of that given constraint.

Although the process may appear complex at a first glance, the use of a spreadsheet with simple formulas will facilitate the job considerably. In addition to that, a spreadsheet is very useful for providing partial year-based dimensioning results in a tabular format that is easy to visualise.

GGSN Dimensioning

The case of GGSN dimensioning is actually similar to that for the SGSN. Backbone design aspects will not be dealt with at this stage, only node dimensioning. There is also a component of strategic planning in this case, since there could be a GGSN node in all the core sites, or only in one or two core sites. There could also be cases of GGSN nodes with dedicated access points placed in corporate premises, for instance for ISPs or corporate clients. The actual deployment plan will be based on a points of presence strategy and transmission costs for the ISP interconnections.

As mentioned, the procedure is similar to the SGSN procedure. One difference in this case would be that, in the most general case, the physical interfaces of the GGSN will be the Ethernet for both Gn and Gi sides (possibly either a fast Ethernet or gigabit Ethernet, with a capacity 100 Mbps or 1 Gbps respectvely). This means that in most cases the interface capacity will not be effectively a limiting factor, but perhaps the amount of interfaces in some cases. From that perspective, the most important limitations will be the GGSN throughput and maximum amount of simultaneous PDP contexts. As stated earlier, should the input provide further information in terms of subscribers' dynamic behaviour, then additional constraints, like the maximum number of PDP context activations, deactivations or modifications per second, could play a role as a limiting factor in concrete cases.

Following the previous structure, the total amount of GGSN nodes required per core site will be calculated in two steps. Firstly, in terms of traffic flow, it is calculated by considering the total amount of Gn traffic that will be served in the given core site. This total amount of Gn traffic will be divided by the maximum throughput that a single GGSN can handle. This result, rounded up to the next integer, will be a rough estimation of the number of GGSNs needed for a given core site in a given year. One important aspect to consider in this case – and perhaps also in the SGSN dimensioning case – is the way to specify the traffic. Traditionally, for interface dimensioning, the uplink and downlink components of a given data traffic flow are considered separately, and the link size is then estimated as the highest of the components. This approach makes sense in traditional bidirectional symmetric transport media, like E1/T1, which can still be used in the Iu-PS interface, for instance. In the case of network nodes handling IP traffic, it is more common to specify the throughput as combined uplink plus downlink components. This fact may also apply to the case of the SGSN, depending on the infrastructure vendor, although this is not explicitly mentioned. In that case, the interface traffic input should be properly modified to add the uplink and downlink components separately, if so required.

Secondly, the maximum number of simultaneous active PDP contexts is taken into account. This is calculated by multiplying the total amount of subscribers by the total number of simultaneous PDP contexts that are expected to be active during the busy hour. Once the total number of simultaneous PDP contexts needed is obtained, the total amount of GGSN nodes required according to this criterion would be calculated as the total amount of simultaneous PDP contexts divided by the maximum amount that can be handled in a single GGSN.

In principle, session management and mobility management restrictions should also be taken into account, since the GGSN node may have dynamic limitations in terms of the maximum number of PDP context activation requests or PDP context modification requests. However, these limitations are rather difficult to characterise beforehand without having performance management measurements from a live network. Therefore, unless the node vendor provides explicit limitations that can be correlated with the traffic profiles of the dimensioning input, these aspects should be considered as part of the optimisation activities, to be dealt with later on.

Similarly to the previous case, after these two aspects have been verified and the partial estimations of the number of GGSN nodes needed per core site are ready, the result of the GGSN dimensioning exercise will consist of simply taking the maximum of the values previously calculated. This will ensure that the most restrictive conditions are fulfilled. This process has been described generically for the number of GGSN nodes per core site, since the core site is the basic infrastructure entity in this context, but there should be no major difficulties in expanding this procedure for all foreseen core sites.

It is worth noting that different GGSN nodes from different vendors may have additional limitations that are not described here, since only the principal aspects according to the most general types of inputs have been described here. The procedure should be to deal with those limitations in the same manner, considering the existing input situation and evaluating it against the network node maximum capability in terms of that given constraint.

Dimensioning for Inter-GSN Traffic

SGSN and GGSN are connected using the Gn interface. The amount of traffic that flows between SGSN and GGSN depends on the APN configuration. For example, a network has 2 GGSNs and 10 APNs. The first option is that the two GGSNs are configured in the $n + 1$ hot-standby redundancy mode. Therefore 100 % traffic flows through GGSN1 and, when it fails, the total traffic is routed via GGSN2. The second option is to configure them in a load sharing mode. In this mode, 5 APNs out of 10 may be configured on GGSN1 and 5 on GGSN2. The corresponding traffic is routed via respective GGSNs. If GGSN1 goes down, five APNs only will be affected. The third option would be to divide APNs among GGSNs based on traffic. It may be such that only three of the most popular APNs constitute 50 % of the traffic and are configured on GGSN1 while the remaining seven APNs, which constitute the remaining 50 % of the traffic, are routed through GGSN2.

In a roaming scenario, roamers normally use the visitor SGSN and home GGSN so that roaming traffic goes from the SGSN to the border gateway (BG) directly. The operators may have a tie-up such that they use common APNs and define roamers in the HLR so that they can use the visitor GGSN. In this case, inter-PLMN traffic is reduced. Therefore it can be said that roaming traffic also depends on operator policy.

The physical locations of the SGSN and GGSN also affect traffic dimensioning. If the SGSN and GGSN are co-located, the traffic will be localised. They may be connected using 100 Mbps of the Ethernet. If it is assumed that an active subscriber generates 300 bps of traffic in a busy hour, then 100 Mbps bandwidth of fast Ethernet can support around 350k subscribers. If the SGSN and GGSN are remote, then this traffic will go over the backbone. This will require IP/ATM backbone planning, which is discussed later in detail. ATM or MPLS backbone capacities, number of routers, link and topologies will have to be dimensioned accordingly. It is best as far as possible to localise the traffic coming out of the SGSNs and direct them to the GGSNs located on the same site, for minimum IP traffic across the backbone. This will save cost on building a high bandwidth backbone. However, this may not be applicable in all situations. For example, in some countries operators have to provide a service over large geographical areas and the subscriber density in each region is less and does not justify a local GGSN. Hence they opt for one centralised packet core with GGSN, DNS, BG, Internet access, etc., and remote SGSNs in each region. If roaming or DNS traffic is considerable then the BG and DNS may be co-located with the SGSN. The cost of routers, reliability of backbone transmission links and operation and maintenance issues are some other factors that should be considered while deciding upon the location of equipment.

A few other aspects which need to be taken care of during dimensioning are:

- calculation of the required number of fast ethernet or gigabit Ethernet ports on the IP routers;
- calculation of the required number of STM-1 ATM interface cards required on the ATM switches;
- account of the impact of QoS requirements in the calculations;
- introduction of router switch modules (RSMs) on the existing catalyst switches, thereby eliminating the need for the introduction of additional IP routers.

DNS Dimensioning

The SGSN caches DNS responses. This fact must be taken into account while dimensioning DNS. The time to live of a cached entry depends on how static or dynamic a network environment is. Once a day is a good option since the only time the DNS will be updated is when new APNs or routing areas (RAs) are added, removed or modified:

N number of DNS queries = name of caches per SGSN × (number of APNs + number of RA updates)

In each 24 hour period only the first APN and RA update name resolution request will be directed to the DNS. The minimum number of DNSs is two (primary and secondary).

DHCP Dimensioning

The DHCP is dimensioned according to the total number of DHCP requests during the busy hour with respect to its handling capacity in terms of requests per second. An '$n + 1$' redundancy scheme is recommended.

Firewall Dimensioning

Real-life planning scenarios and network topologies hugely impact firewall (FW) dimensioning. The firewall can be dimensioned based on the total required data traffic during the busy hour with respect to the equipment processing capacity. It is also recommended that the FWs should be implemented on an '$n + 1$' configuration running the Virtual Router Redundancy Protocol (VRRP), for equipment redundancy and load sharing functionality.

Firewalls are needed at the Gi interface for Internet access. The firewall may also be needed at the Ga interface for charging and the O&M interface for connectivity to the OMC (operations and maintenance centre). Separate traffic capacity dimensioning and port requirement calculations should be done for each of these. Traffic for access points on the Intranet may not go through the firewall; hence it should be excluded.

Border Gateway Dimensioning

The border gateway (BG) is mainly dimensioned according to the processing capacity. Just like Gi firewall dimensioning, real-life planning scenarios will impact on the dimensioning results. The BG is dimensioned according to the estimated inter-PLMN (Gp) roaming data traffic during the busy hour compared to the equipment processing capacity. The Gp traffic can be assumed to be around 10 % of the total data traffic but it may vary from operator to operator and also depends on the network topology.

The impact of implementing IPSec VPN (Internet Protocol secured, virtual private network) functionality, assumed packet size and equipment tolerance are some factors that should be taken into consideration. An equipment loading factor of 90 % can also be applied for BG dimensioning. It is also recommended that the BGs should be implemented on an '$n + 1$' configuration running the VRRP, for equipment redundancy and load sharing functionality.

Charging Gateway Dimensioning

The required number of charging gateways (CGs) is dimensioned according to the total number of CDRs generated per day. The following formula can be used to calculate the CG requirements:

$$\text{number of CGs} = \frac{\text{number of subscribers} \times (\text{CDRs/busy hour})}{\text{number of CDRs/day}/24}$$

A 90 % equipment load factor can be taken into consideration in the CG dimensioning. An '$n + 1$' node redundancy scheme is recommended.

Core LAN Switch Dimensioning

Each packet core site requires an LAN switch to provide interconnection between different network elements. The number of core LAN switches will depend on the number of packet core sites and the number of packet core network elements in each of the core sites. In addition, for circuit switched network elements that are co-located in the same site as the packet core nodes, the LAN port requirement for the IP interface for SIGTRAN is also taken into consideration when dimensioning is carried out.

For security and traffic management purposes, different VLAN configurations should be adopted for IP backbone connections. Thus, the number of ports required and the type of switch will vary from case to case during actual implementation. A more in-depth dimensioning and planning exercise should be done during the project planning phases, immediately after the successful conduct of the tendering activity.

For the core switch, there are two redundancy options: duplication on the module level or duplication on the network element level. The required redundancy option would greatly depend on the operator's policy. A minimum card requirement (in terms of 48 ports per card) is calculated. Two ports will be required per equipment connected to the switch per VLAN if LAN interface redundancy is required. Each packet processing unit, charging units and operations and Maintenance Units in the SGSN are connected separately to ports under separate VLANs created in the core switch.

4.12 IP backbone Planning and Dimensioning

In traditional 2G GSM networks, CCS7 (common channel signalling 7) traffic is carried over PCM links. SIGTRAN and SIP make it possible to carry this over the IP backbone. SIGTRAN is a set of protocols that defines a framework to transmit signalling like the SCCP over the IP. It can be applied to all mobile applications like MAP, RANAP, INAP and CAP, which use the SCCP. When SIGTRAN is implemented in a network, in the beginning there is no need to make any changes in the applications. Later on, however, applications can be optimised to take full advantage of higher efficiencies provided by SIGTRAN. SIGTRAN provides a much wider bandwidth as compared to 64 kbps per link in CCS7. Messages with a length longer than 272 octets are supported. Hence there is no need for segmentation. Load sharing and automatic re-routing of traffic is also supported. The IP and SCTP add overhead to the messages. The SIP is used to transfer ISUP signalling between mobile switching centres (MSCs).

SS7 links are not as efficient as the IP backbone in terms of capacity. The IP provides a much higher capacity and bandwidth than PCM links. Recently telecom operators worldwide have been switching over to the IP backbone. This means huge savings in investments for them. The aim of IP trunking and SIGTRAN planning is to convert an operator's network into an optimal IP based network solution, taking into account the current signalling network and future requirements.

IP trunk planning consists of three main activities:

- assessment of the current signalling network;
- dimensioning of the IP based network;
- detailed planning of the IP based network.

4.12.1 Current Network Assessment

The objective of assessment is to analyse current CCS 7 network architecture, topology, signalling plan, routing and traffic requirements in the short, medium and long term. The following factors need to be considered in the current network:

- Geographical constraints
- Network architecture and topology
- Network capacity and technology
- Technical constraints

Future demand requirements of services and capacity also need to be considered. The following inputs are required for Network Assessment:

Network architecture

- Locations and naming conventions of network elements
- Network topology (both user plane and signalling plane)
- User plane traffic routing criteria (inter-MSC, interconnect to other networks, PLMN, PSTN)
- Control plane traffic routing criteria

Signalling plan (CCS 7)

- SPC allocation
- Signalling link sets plan
- Country-specific regulations

Traffic requirement

- Timeframe definition for the study
- Services (basic voice, value added services, Internet, etc.) penetration and forecast
- Subscribers
- Traffic/subscriber service
- Grade of service

Once the above inputs are collected, the following tasks are performed:

- Network architecture study
- Signalling plan study
- Traffic requirement study

4.12.2 Dimensioning of the IP Backbone

The objectives of dimensioning are to define the technical solution in the IP perspective, to apply dimensioning rules, to evaluate bandwidth requirements and to produce an 'IP signalling master plan'. The following are required for dimensioning the IP backbone:

- The total signalling traffic should be calculated using busy hour traffic, mobile-originated calls/ subscriber, mobile-terminated calls/subscriber, SMS traffic and number of signalling messages per call.
- The total signalling traffic can be divided into three categories:
  - SIGTRAN based traffic, which constitutes traffic generated by SCCP applications;
  - SIP based traffic, which is inter-MSC traffic and has to flow over the IP;
  - TDM traffic, which is the traffic generated due to emergency calls, data/fax and handovers and has to be routed, normally using CCS7/TDM.

4.12.3 Bandwidth Calculations

- The bandwidth is required for transfer of one voice call. It is known that 13 kbps is required per full rate timeslot. It is assumed that the Ethernet overhead is 38 bytes, the IP header is 20 bytes and the UDP header is 8 bytes. Assuming no voice activity detection and no header compression, the total of these comes to 13 kbps + (66 bytes × 8 bits × 50 samples/5)/1000 = 39.4 kbps. Hence it can be stated that the bit rate requirement from the IP backbone is 39.4 kbps per call in the one-way direction:

$$\text{Total bit rate voice} = \text{number of voice channels} \times 39.4 \text{ kbps}$$

- The bandwidth required on the IP backbone for SIP messages (ISUP signalling between MSCs) can be calculated as follows:

    Average size of each SIP message = 600 bytes
    Number of SIP messages flowing per call = 8
    Calls per second = Erlangs/mean holding time
    Bit rate required = calls per second × 8(messages) × 600(bytes) × 8(bits/byte)

- The bandwidth required for SIGTRAN (SCCP based) messages can be calculated in the same way as for the voice call. The following overheads are added for each signalling message:
    - M3UA: 24 bytes
    - SCTP: 28 bytes
    - IPv4: 20 bytes
    - Ethernet: 38 bytes

Note that for each of above three categories, a calculation should be made of how much traffic would flow over the LAN (between network elements in the same site) and how much over the IP backbone (between network elements in different sites).

The next step in bandwidth calculations is detailed planning, which requires source data to be defined, implementing a technical solution in the IP perspective. The following activities are normally performed as part of detailed planning:

- IP physical connection schema towards the IP backbone;
- IP addressing plan for signalling over the IP;
- IP signalling links characteristics definition;
    - IP signalling link/link set property;
    - circuit group (CGR) mapping into the IP (IP trunk);
    - signalling route in IP.

4.13 Mobile Packet Core Architecture Planning

The Mobile Packet Core (MPC) architecture plan consists of a logical and physical network plan. It will also contain a local area network design, management system connectivity plan, switching and routing plans and charging and billing plans.

Transport Network Planning

This contains the logical and physical network architecture details, load balancing schemes, network redundancy, WAN addressing schemes, IP addressing plans, QoS parameter planning, network migration plans and routing and access control configuration. This involves the planning of the Gi and Gp interfaces.

Gi is the interface between the GGSN and the external PDN while Gp is the interface between the BG and the inter-PLMN. The plans include the connectivity to the Internet service providers, corporate clients, corporate Intranet and other services/content providers.

Mobility Management

Mobility management (MIM) is the core of the signalling plans. As discussed in Chapter 2, MM deals with the location and state of the UE in the network. Chapter 2 describes how the BSC/RNC take care of the MM functionality; however, the SGSN is responsible for the MM within the PS service domain as the UE has the MM context for both the CS and PS core networks, which work independently of each other. There are two main features that are supported by MM: GPRS attach and GPRS detach. These two are responsible for making the UE reachable and unreachable respectively; detach informs the HLR that 3G SGSN has deleted the MM and PDP contexts of the detached UE. Chapter 2 gave the definition of the location area/routing area. However, the routing area update in the context of the PS core means the inter- or intra-SGSN area update. The former one means that the mobile is within the same SGSN area but has changed the BS, while the latter one means that the mobile is now attached to a different SGSN. Until the time that the subscriber is within one routing area, data will pass through the connected SGSN to the GGSN over the Gn interface. However, when the subscriber changes the location area, i.e. attaches itself to the new SGSN, a query is sent to the HLR based on the subscriber IMSI asking for the routing area identity. The SGSN receives the routing data (RD) from the HLR, based on which it makes a query to the DNS to get the IP address of the previous SGSN. Once the IP address is received, the present SGSN contacts the old SGSN to get information that includes the GGSN tunnel ID and encryption key. Now the SGSN contacts the GGSN and establishes the session using the information received. User data information is received by the new SGSN from the previous SGSN. The new SGSN now updates the HLR with RA/LA IDs. Finally, the UE disconnects from the previous SGSN and connects itself completely to the new SGSN.

Session Management

A packet data user connection in the network is called a Packet Data Protocol (PDP) context. The PDP context is basically a set of configuration data in the UE, SGSN and GGSN, which define and manage the tunnelling of data packets between the UE and GGSN. The function of the PDP context is to transport IP packets from the UE to the GGSN and vice versa while having no knowledge of the application inside the IP packets that it is transporting. Activation of the PDP context is initiated by the UE. Controlling the actual data connections of the UE is called session management. This has two states: inactive and active. Inactive applies to attached UE with no PDP context activated while active is the opposite, i.e. with the PDP context activated. Activation of PDP contexts are of two types: primary and secondary. When the UE opens the PDP context to any access point, it is called the primary context while any subsequent context to the same access point but with a different traffic class is called the secondary context. However, both contexts share the same IP address.

The process of PDP context activation is as follows. Once the UE initiates the request for the PDP context activation, the request is received by the 3G SGSN. The request may include configuration options such as the access point name and/or the QoS profile, etc. The subscriber records are checked using the HLR by the SGSN. Once approved, the SGSN performs a DNS query which is responded to by one or more IP addresses. The IP address with the lowest cost routing is used. The radio access bearer (RAB) assignment is then completed by the RNC by establishing the RAB access using the UE. The SGSN then sends a PDP activation request to the GGSN. This request can be rejected or the QoS profile may be re-negotiated. Thus, after checking resources, the PDP context activation request is accepted.

4.13.1 VLAN

Virtual LAN is a group of workstations and network devices that are grouped together to communicate irrespective of the physical locations. This concept uses switches to create isolated domains. The main advantage of VLAN is that it reduces unnecessary congestion in the network; e.g. in an LAN, when the host broadcasts a message, all the computers receive the message, which creates congestion in the network. Thus, when using the concept of VLAN, logical segmentation is done in the network, which reduces congestion as the traffic is reduced significantly in the network (the broadcast domain is confined), increases the ability to manage the network and, increases security (as traffic is separated). The grouping can be done by using MAC addresses, protocol type, IP addressing, application type, etc. The VLANs are proposed on the basis of traffic type, e.g. O&M data are on the O&M VLAN, while charging data are on the Ga VLAN, user data on the Gn VLAN and external data on the Gi VLAN.

4.13.2 Iu-PS Interface

The Iu-PS interface (between the RNC and SGSN) has both the user plane and control planes. The signalling procedures between the SGSN and RNC are carried by the signalling plane, while the tunnelled subscriber data from the RAN to the SGSN are carried by the user plane. The user plane of the Iu-PS interface consists of the PDP context data, which are tunnelled over the ATM using the GTP Protocol. Both types of traffic are carried over the ATM interface between the SGSN and RNC. However, many of the signalling messages related to mobility and session management, SMS transfer, etc., are exchanged between the SGSN and the UE without any processing by the RNC. RANAP and SS7 are used for signalling. RANAP contains messages used for the mobility and session management procedures, such as relocation (of RNC), RAB management, location and paging, etc., while SS7 is used to carry the signalling over the ATM connections. SS7 is a protocol stack that fills the functionality between the RANAP and the low level ATM adaptations that fill the ATM cells. The protocol stack contains the SCCP (signalling connection control part), MTP-3b (message transfer part layer 3b), NNI (network-to-network interface) SSCF (service specific coordination function) and SSCOP (Service Specific Connection Oriented Protocol).

Protection of the Iu-PS Interface

This is achieved by the appropriate configuration of the tunnelling units, which depends upon the ATM transport between the 3G SGSN and the RNC. This means that if the PDP context is lost due to failure of the tunnelling units, the subscriber should be able to reconnect to the network using different link and tunnelling units. To achieve this, it is important to limit the rate of data that arrives at the tunnelling unit.

4.13.3 Gn Interface Planning

Gn Interface and Protocols

The Gn interface carries both the user data and signalling data using the GTP-U and GTP-C Protocols, where U and C denote user and control respectively. The main signalling procedures are network requested context activation, tunnel management and path management. Network request context activation messages relate to the PDU notification requests and responses, and sending the routing information for GPRS requests and responses. Tunnel management relates to the creating, updating, deleting the PDP context requests/response, etc. Path management relates to the echo requests and responses.

Redundancy of the Gn

When one active interface unit fails, the passive/spare interface unit assumes the tasks of the failed unit. The active interface unit performs this by sending gratuitous ARP messages for the failed unit. In doing so, it updates the ARP caches of the other elements in the same subnet. When the Gn interface IP addresses are not used, then the units used for tunnelling in 3G SGSN should be configured into the same routing protocols that are running on the Gn interface. Gn interfaces should be in different subnets as a router cannot have two interfaces in the same physical segment or it becomes confused when receiving the same routing protocol messages from neighbouring routers over multiple interfaces.

Domain Name Server

The domain name server (DNS) is required to translate the logical domain names into the IP addresses of the domains. DNS servers should be placed at accessible locations from the SGSN. These servers should be protected by introducing an additional DNS as a backup. Both the APN and the logical name of the routing area are translated to the IP addresses. An APN network identifier is mandatory though the APN operator identifier is optional. The APN network identifier is a label or set of labels that are separated by dots, e.g. network.com, which guarantees uniqueness to the domain name. The APN network identifiers should not end with .gprs but operator identifiers should end in .gprs. The routing area identity is given as

$$RACxxxx.LACyyyy.MNCzzzz.MCCwwww.GPRS$$

where x,y,z and w are decimal coded digits. RAC, LAC, MNC and MCC are coded as decimal numbers with leading zeros if necessary. The length is always 4 digits.

Both hardware and software are required for DNS operation. The software used in DNS equipment is called BIND, or Berkeley Internet name domain.

Gn Interface in WAN

Though the WAN interface is used for the Gi and Gp interfaces, it can also be used for the Gn interface in cases where the SGSN and GGSN are remote and not co-located. Planning of the Gn interface for WAN is based on topology, connectivity and traffic estimates. The Gn network can be extended between sites by using the routed or bridged WAN, with the major difference being the layer 2 Gn LAN at each site, with a discrete routed subnetwork in the former while in the latter it is extended to multiple sites.

Routing in Gn

The OSPF is the routing protocol in the Gn network. Routing can be both static and dynamic. Static routes are unsuitable for fallible networks while dynamic routes are determined by groups of routers sharing information. Routes can be manually defined in the router (static routes) and can be changed only by the administrator. This scheme of things is used where there are small numbers of subnets and routes or the topology is unchanged. Dynamic routing is used where there are large numbers of subnets and routes or the topology changes frequently.

4.13.4 Gi Interface Planning

The packet core network is connected to the external IP networks via the Gi interface. The external networks are mainly of three types: Internet, Intranet and Extranet. The Internet is a global network of computers, which operates using a common set of communications protocols. The Intranet is an interconnected network within one organisation that uses Web technologies to share information internally, not worldwide. An Extranet can be viewed as part of a company's Intranet but is extended to users outside the company. The reference point between the GGSN and external networks is called a Gi interface, where the access is provided to external networks via logical interfaces called access points. A single GGSN interface can provide several access points from the same physical interface. The parameters that are required to configure the access points include static and dynamic IP address ranges for subscribers, server addresses, router information, IP address allocation method, etc. When planning for the Gi interface, aspects such as required bandwidth, security, reliability, scalability, network address translations, etc., should be taken into account. Connectivity to the external networks can be done through leased lines, frame relay/ATM network, etc.

The process having access to external networks is initiated by the UE. The PDP context activation request along with the APN and PDP address is sent to the SGSN by the UE. After checking the resources, security, etc., the SGSN contacts the RNC for an RAB assignment and subsequently sends the PDP context creation request to the GGSN. If everything is all right, the GGSN returns its response message, which includes the assigned PDP (IP) address. This is passed to the UE by the SGSN. Once the IP address is received by the UE, data sessions are started with the server that it wants to connect on the external network. However, every UE that needs to access the external network needs to be assigned an IP address. These addresses can be statically assigned by using the HLR or dynamically assigned by using the GGSN, DHCP, etc. The address can be public or private. Since public addresses are limited, private addresses are normally used and the firewalls/edge routers are used for network address translation (from private to public IP addresses).

In cases where there is a need to connect two networks together across another network, tunnelling is used. This is done by encapsulating one protocol into another protocol. The protocols used are either the IP, GRE (generic routing encapsulation) or the Layer 2 Tunnelling Protocol. The payload is carried in a tunnel that exists between two end points. Tunnelling provides communication between two subnetworks that have discontinuous IP addresses. It allows subscribers to connect to remote PDNs without publishing the user IP addresses on all intervening routers, etc.

OSPF or RIP on the Gi interface are used for routing. The VRRP can be used to activate several physical routers and is seen as a single logical router, thereby eliminating a single point of failure for the router. It is transparent to the attached devices and hence no reconfiguration of the end devices are required.

4.13.5 Gp Interface Planning

The interface of the mobile packet core (MPC) network to other PLMN networks is called the Gp interface. The Gp interface facilitates the roaming feature of the UE. This helps subscribers to use their home network services irrespective of the network coverage they are roaming. For this to happen, the two network operators either need to be connected directly or they should be connected via a centralised GPRS roaming exchange (GRX). Direct connectivity is possible through the leased lines or using tunnelling protocols, while centralised connectivity is done using a separate inter-PLMN backbone network with a root DNS. For roaming to take place, the contractual agreements between different network operators should be in place. If there are many operators, the complexity and the cost related to roaming will be high. However, the GRX concept can be used whereby global coverage through one connection is possible. This is better than the former option where all operators need to be connected to each other, as the number of links required to be maintained would be less. Planning engineers can start planning using the direct connectivity approach and then, as the network builds up, should move to a more centralised approach and end with a complete centralised solution. The IP address of packet core network elements must be unique between different PLMNs. Hence public IP addresses should be used in the GP interface.

4.14 Packet Core Network Optimisation

4.14.1 Packet Core Optimisation Approaches

The actual content of the optimisation activities for the packet core network and its motivations can be a matter of discussion. There are different possibilities, and all can be argued in different real network contexts. The next sections present a few approaches to packet core optimisation, based on different motivations and viewpoints.

Concept of Optimisation as Improvement and Troubleshooting

From a more traditional perspective, there are two main approaches in terms of optimisation:

- The first is related to troubleshooting the observed problems, a practice that in some cases is directly undertaken under the network operations common denominator.
- The second refers to improving the network performance by means of improving the efficiency of the mobility management and session management procedures.

Both motivations are perfectly valid, although they trigger packet core optimisation projects in two very different scenarios, the first being related to identify, characterise and solve a concrete problem in the packet core network and the second concerning an attitude of constant improvement, even in the absence of problems. They can also be considered as short term and long term respectively.

Bearer Optimisation and Application Optimisation

There is also another way to approach the concept of packet core optimisation that has been gaining greater recognition lately, following the success of data mobile services, which is based on a layered view of mobile communications:

- bearer optimisation, which concentrates on activities related to the improvement of the bearer performance, i.e. the lower protocol layers on the multilayered protocol model;
- application optimisation, focusing on the end user perception of the actual performance of the mobile services, which are based on top of the bearer.

It is being accepted as fact that having a properly tuned network bearer is a prerequisite for having a healthy application performance. This view also supports the sequential approach of first having the bearer optimised, and then the application performance is tuned from an end-to-end perspective, in which the packet core network also plays a major role. This approach can be considered as the vertical approach, because differentiation of the activities is done vertically with respect to the protocol layer model.

System Optimisation and Subsystem Based Optimisation

There is yet another view to packet core network optimisation that can be justified in view of the actual complexity of the current mobile networks, where new services and applications are meant to be added next to the packet core network through the Gi interface:

- Subsystem based optimisation, which relies on the fact that the network subsystems (in our case, BSS or RAN and packet switched) are complex enough to be considered separately if all the potential is to be unravelled. In this context, the focus would be on packet core optimisation *per se*, when there

Packet Core Network Optimisation

is a clear need to improve performance aspects, configuration aspects or even capacity aspects of the packet core network. There is an understated link to bearer optimisation in this approach.
- Packet core optimisation as part of the concept of system optimisation, which simply assumes that the subsystem based optimisation is simply not enough to achieve the expected performance. Tuning subsystem parameters will not provide the full potential if those parameters are not aligned with the possible counterparts on the other subsystem. Under this approach, there is also an understated link towards bearer or application optimisation, having the flavour of an end-to-end approach, which has been very much under the spotlight lately.

This last approach can be considered as a horizontal approach, due to the fact that it is based on the inclusion or exclusion of network subsystems that are considered horizontally.

4.14.2 Packet Core Optimisation – Main Aspects

After being presented with different options and approaches, now is the moment to start defining the main issues in packet core optimisation.

Network Equipment-Dependent Aspects

The different network gear vendors provide different solutions for the packet core network nodes and their interfaces. Therefore, there is great optimisation potential in extracting the most out of the nominal capacity of the network nodes. This requires a thorough knowledge of the equipment. It is not enough just to know the nominal or static capacities of the network nodes, as stated in the technical sheets of the vendors, used mostly during the sales process. It is essential to understand the actual behaviour of the network nodes in the life network, with the existing traffic profiles.

This understanding should be built on a knowledge of the network performance and dynamic traffic profile, which is characterised through a series of measurements, mostly obtained through the operations support system (or network management system), which will provide the basis on which to calculate a given set of KPIs (key performance indicators). These effectively characterise the current situation of the network, and eventually will help to identify optimisation possibilities. The concept of the KPI will be dealt with in more detail in the following sections.

Thoroughly understanding the actual dynamic situation of the network will provide the tools to evaluate different scenarios for improvement of the network. A possible scenario will now be presented to illustrate this concept better. In a mobile network of a middle sized country in Western Europe, a given SGSN can be observed under heavy traffic volumes offered by a few corporate users located under three RNCs serving the business centre in a major capital. At the same time, another co-sited SGSN is connected to five RNCs serving the suburban area of the same capital, where most of the traffic is offered by commuters mainly using messaging and browsing applications, not so much volume intensive, but rather demanding in terms of mobility management due to commuting at given commuting peak hours.

A study of the OSS data indicates that the first SGSN is approaching its maximum nominal throughput during the busy hour for corporate users, somewhere around noon due to the heavy downloading activity of emails, documents, etc., while the signalling load offered for those users is rather low due to very limited mobility. In essence, a few users produce a heavy data volume, putting pressure on the throughput capability of the SGSN.

The same OSS data also reveals that the second SGSN is actually serving more subscribers, whose busy hour is between 8 and 9 am, and later in the afternoon, around 6 pm, while commuting to and from their jobs, schools, etc. The traffic volumes created for those subscribers is not so high, since they are mostly using applications such as instant messaging, multimedia messaging and browsing news. In this case, there are many more users, eventually close to the maximum nominal value for the simultaneous attached subscribers, while the throughput limits are far from the traffic volumes currently measured. Additionally, some reports also indicate that the signalling load is rather high, due to the amount of routing area updates caused by the commuters in the trains. In this case, there are many more users, eventually close to the maximum nominal value for the simultaneous attached subscribers and, in addition, some early warnings of increasing signalling load. Meanwhile, the throughput limits are far from the traffic volumes currently measured.

Based on the observations described above, an optimisation activity can be launched, with the purpose of redistributing the eight RNCs between the two existing SGSNs, with a target of distributing the current load offered by those RNCs more evenly with regard to three criteria:

- The number of subscribers
- The throughput offered per subscriber
- Mobility management of those subscribers

In cooperation with the radio access optimisation teams, the traffic profiles should also be characterised at the RNC level, with the target of achieving the correct mix of each and every one of the eight RNCs for every one of the SGSNs, in terms of optimal distributed connectivity of the Iu-PS interface links, assuming that the actual radio access topology should be kept unchanged. Hence there should be no modifications for Iu-b and Iu-r connectivity since it has already been optimised. Since both SGSNs are co-sited, the actual implementation of the optimisation exercise proposal can be undertaken as a simple re-homing exercise for the RNCs, normally carried out during the maintenance window overnight, with little – if any – disruption of the service.

Topology Aspects

It is also interesting to consider the optimisation possibilities arising from network topology aspects. Topology plays a self-evident role in radio access optimisation, but its role in packet core optimisation may not appear so obvious at first glance. Not every network offers this kind of possibility. It is, in fact, related to the relative network size, with the larger networks offering more potential in this field.

The most obvious topology optimisation subject for a packet core network can be seen as the strategic approach for deployment of the network nodes, be it a distributed architecture or a concentrated one. Obviously, the size of the network is the first aspect determining the architectural deployment strategy, since a network covering a small area such as a city state with a few tens of square kilometres, no matter the total amount of subscribers, will present a natural trend for a concentrated architecture, with all the packet core network nodes co-sited. It can be noted that during the early stages of deployment and growth of the network, it is easiest to have the nodes co-sited, since that greatly facilitates the implementation and operation of the network. However, as the subscriber base and traffic volumes grow, there is a choice to be made:

- Continue integrating network nodes in the same site, maintaining the ease of implementation and operations, at the expense of having a crowded site, not so easy to work on, besides an increasing exposure to any kind of service outage risk.
- Deploy the new network elements in a newly created site, lowering the exposure to total network service outage risks, enabling a more comfortable deployment and operation, at the expense of the increased costs and time constraints of having another operational site for core network operations.

Ultimately, the final decision will also be an optimisation outcome, in this case not so much related to technical-only aspects but rather to technical–operational–economic aspects. In any case, it is sensible to have the optimisation team participating in this type of exercise.

In the case of networks covering large geographical areas, in extensive countries, no matter what the coverage density is, the problematic is different, although the topological optimisation opportunities are actually more evident. Topology aspects to be considered for an optimised configuration include proper connectivity design from the serving network nodes to the main data centres. Also, the access to delay sensitive types of services has to be taken into account, particularly if the latency of the backbone network starts to become an obstacle to delivery of certain data services. A careful analysis of centralised and distributed types of architecture, including also the nontechnical factor, and proving the technical feasibility of both options will be the main tool for design optimisation.

In this case, physical topology has been considered. However, there are also some optimisation opportunities arising from logical topology, specifically referring to connectivity optimisation, that might well follow into the category of backbone optimisation, closer to the IT world as such. Additional aspects subject to this type of optimisation also include VLAN design optimisation and routing protocol optimisation. Up to recent times the backbones of mobile operators have not been suffering from excessive traffic loads that may trigger this type of process. However, the new IP traffic levels generated by the new telecommunications architectures, such as release 4 based soft-switch solutions, will produce a significant increase in the IP traffic levels served by the operator's backbones. Although this type of traffic is not impacting directly on the packet core network elements, the fact of sharing fixed network resources places more stress than earlier on the complete mobile network solution.

Procedure-Dependent Aspects

This type of optimisation can be described as the search for improvement in the performance of a given type of procedure. Normally, this activity arises as a result of an observed low performance of the procedure under study, noted through some of the OSS counters and KPIs, e.g. a low GPRS attach success rate. If the OSS KPIs do not give any indication, e.g. the GPRS attach average delay, this low performance of the procedure is normally detected through subscriber complaints and it is further characterised through controlled end-to-end performance tests with interface tracing. This is a technique normally associated with troubleshooting practices, and can be very useful for this type of case.

Once the source of the problem – it can also be simply an optimisation opportunity – is identified, it can generally be associated with a given network node, network subsystem or network interface. Following the previous example of the GPRS attach procedure, if the OSS data have been recording an exceptionally low GPRS attach success rate and while carrying out an interface tracing exercise an abnormally long delay is detected in the Gb interface between the request and response messages of the procedures, the optimisation activity can focus on tuning the timers that control the BSSGP associated processes in the SGSN. In a different scenario, if the delay between those messages is observed in the signalling links with the HLR for the Gd/Gf/Gr interfaces, the topology and proper dimensioning of those signalling links should also be verified and corrected. Another possible scenario could be the case where only visiting roaming subscribers experience this perceived low performance, which may lead to a verification of the service provisioning policies and roaming agreements and its correct implementation in SGSNs and HLR nodes.

This type of optimisation activity built around packet switch procedures can also be used for the PDP context activation procedure, which normally leads to backbone issues, GTP tunnelling issues or GGSN issues, especially with the newer generation of GGSNs with increased intelligence for handling PDP context requests for different services, and the extensive use in the field of new generation release 4 mobile phones with multiple PDP context capabilities.

Another type of procedure that can be the subject of procedure optimisation is the routing area update (RAU) procedure. There could be several reasons for needing an RAU procedure, but normally the OSS counters provide the lead for low performance in this aspect, presenting lower than normal success rate values. If the values can be clearly identified with a given geographical area, this could well be a system optimisation scenario, where the definition of the routing areas at the BSS side might be redesigned in order to improve the success rate of the basic mobility management procedure. Another possibility can be built around tuning the timers that control this procedure. This may be a BSS only procedure (intra-PCU RAU) or a BSS and core network procedure (inter-PCU/intra-SGSN RAU and inter-PCU/inter-SGSN RAU, the later involving also the packet core backbone and the Gn interface). This categorisation will help to identify the trouble, as it can be found only in one of the types of procedure described, and to find the correct timer values for tuning the relevant procedure. It must be noted, however, that those subtypes of the mentioned procedure are also strongly linked to the actual network topology and size, and even to the actual architecture of the network nodes implemented by the concrete vendor.

4.14.3 Key Performance Indicators

Since every vendor has a unique type of network node and unique capacities, it is important to have a set of well-defined KPIs that help the performance to be understood in a consistent manner, even in the case of a multivendor network or even for a mobile network operator having several operations in different countries. The next sections will present briefly some of the most relevant KPIs to be considered during packet core network optimisation. It should be noticed that most of them are actually generic, and they are also used in BSS or RAN optimisation. In fact, these are the same KPIs calculated in different parts of the network, and therefore their values can be different. Nevertheless, a proper study and correlation of the different values of a given KPI, as measured in different parts of the network, will provide invaluable information on where to concentrate optimisation efforts. It should be noted also that the area of performance management is in constant evolution, following the developments of the new capabilities enabled by the mobile networks. Therefore, a limited set of KPIs will be presented, which should be enough to illustrate the concepts.

Network Accessibility

With regards to accessibility the KPIs actually measure the capability of the network to provide the user with an access to the service as fast as possible. Firstly, a service-independent network access, or network accessibility, needs to be considered, i.e. the capability of the mobile station to attach to the expected GPRS network. More precisely, it can be defined as the probability that mobile data services are offered to a given subscriber, as displayed in the network indicator available on the mobile equipment. In more concrete terms, service-independent network accessibility is evaluated in terms of the GPRS attach success ratio (sometimes it is also used as the complementary failure ratio), defined as the ratio between the successful GPRS attach attempts and the total GPRS attach attempts in a given measurement period.

Service Accessibility

The second level of accessibility is that of the service access. In this case, the user has been already successfully accessing the generic network service by means of a successful registration, or GPRS attach, and now the service accessibility will provide an indication of actual accessibility for a given data service. Service based accessibility is defined separately for the different types of mobile services. However, when the service accessibility requires a PDP context activation, the PDP context activation success ratio gives a good indication of the service accessibility. In a more general case, since the PDP activation can be a prerequisite for a certain type of service provisioning (e.g. push email services), there is a need for separated definitions for the different types of mobile services. In the case of SMS over EGPRS, the indicator of service accessibility would be the ratio of successful SMS attempts to the total amount of SMS attempts in a given period. In the case of MMS, separate ratios can be defined in similar terms for mobile originated and mobile terminated MMS respectively. Likewise, for streaming, the accessibility can also be defined as the ratio of the successful stream request attempts to the total stream request attempts during a given period. For browsing services, similar ratios can be established with the content download attempts.

Integrity

The KPIs describing integrity are related to the actual quality of service, for the given service under consideration, during its use. In the case of mobile data services, this type of KPI is very closely related to those defined as retainability KPIs, but in the case of more traditional voice services it is possible to evaluate the actual speech quality, and hence the integrity aspects, from its own category in this type of approach.

Retainability

The concept of retainability refers to the ability to retain use of the service until the session is released by the user or it is disrupted. Different types of ratios have been defined for the different services from an end-to-end perspective, such as the streaming reproduction cut-off ratio, which is described as the probability that a successfully started stream reproduction can be terminated by any reason not involving the user's own will. Other types of similar KPIs defining a service measuring retainability are, for instance, the MMS end-to-end failure ratio or the MMS notification failure ratio.

4.14.4 KPI Monitoring

Monitoring KPIs is at the core of the optimisation processes. It consists of the processes and methodologies accounting, in a meaningful way, the values of the KPIs in different dimensions or aspects, usually having time as the horizontal axis reference. Time can be defined for different levels of granularity, which may make a significant difference, depending on the context. A trade-off between a fine time resolution and a significant amount of data is normally one of the dilemmas when setting up the monitoring process. Some of the next sections provide a follow-up in monitoring.

Tools and Methodologies

The basic tools utilised in packet core optimisation can be classified into two main categories, OSS tools and tracing tools. They are closely related to the fields of operational monitoring and troubleshooting. From the point of view of performance management as the basic competence in optimisation, monitoring is a daily routine task, which will eventually trigger the need for troubleshooting.

Monitoring

Monitoring is usually one of the main requirements for an OSS, and most of them will provide enough monitoring capabilities to be utilised in the packet core network environment. Normally, the optimisation engineer will be a user of those monitoring tools. The target is to detect any kind of abnormal situation having a negative impact on the network performance. The inheritance from radio access is very strong in this aspect, although from the packet core point of view it is difficult to achieve its level of efficiency, mostly due to the fact that the packet core network nodes are essentially more stable and predictable in their dynamic behaviour.

There is another variant of monitoring, not necessarily related to OSS monitoring, that enables new possibilities for core network, i.e. real-time monitoring. It is possible to define real-time monitoring at two different levels, network node based real-time monitoring and network interface based real-time monitoring. This process is essentially complex and might be intrusive for the network infrastructure, which is not a desirable feature. Existing systems are based on network probes, which are difficult to manage, or on network nodes, which are essentially easier to handle. In addition to those, it is interesting from a packet core point of view to note the approach taken by some IP based monitoring tools, which are capable of presenting a monitoring view of different application level performance aspects in real time. Although the real time feature is not essential for the optimisation approach, its instantaneous capabilities provide an additional degree of efficiency in the process.

Troubleshooting

Troubleshooting is actually a technique that is being utilised in different processes and events throughout the network lifecycle. A great deal of intuition is required for an engineer to be successful during a troubleshooting exercise, although with the flavour of optimisation, the troubleshooting techniques can be marginally better when systematically approached.

The tools utilised mostly in this area are tracing tools. The basic tools are usually protocol analysers, especially used in Gb and Iu-PS interfaces, and of course IP protocol analysers in the Gn backbone and Gi interfaces, which are used to take traces in their respective interfaces. The Gb and Iu-PS protocol analysers are essentially more complex, since they are decoding protocols such as BSSGP and RANAP, which are far more complex, cryptical and also efficient than the IP based protocols, which can be easily visualised in powerful tools, even available as freeware.

The processes of troubleshooting that are used in optimisation normally focus on the following:

- Verify normal execution of the standard procedures, which is especially important in multivendor environments with poor interoperability verifications.
- Analyse the consistency and integrity of the packets, according to the expected behaviour based on the protocol characteristics. Issues such as fragmentation, packet losses, packets out of order and jitter at the IP level can be observed using tracing tools.

After this type of analysis, it is sometimes possible to find some abnormal or suboptimal issues, which can be corrected through a normal process of implementing change requests for the originating parameters, if those are clearly identified. The worst-case scenario is when no abnormal behaviour is detected, and specific test cases should be defined in order to repeat an observed problem that appears randomly. This might be a cumbersome exercise in some cases, well beyond the scope of a normal optimisation exercise.

4.15 Security

In recent times security has become one of the biggest concerns for most mobile operators. Most of the threats are essentially similar to those known to have existed for a long time in a normal service provider environment. However, it might be useful to review some basics of security, from the optimisation point of view.

4.15.1 Planning for Security

During the planning phase, all traditional network security procedures should be a design constant and a major concern at this stage, where all the possibilities are still open. Most of the application servers can be protected by implementing appropriate firewall solutions, generally part of the design recommendations provided by the equipment vendor.

It is essentially difficult to provide precise recommendations for firewall rules in this environment, since they should be compliant with the security policy of each company. The basic advice is to ensure that the operational requirements of the mobile users are still properly allowed and the standard practices in IT security are also followed. For special features and other specific traffic flows, such as VPN based connectivity to corporate networks, a joint task force with the mobile packet core designers and the administrators of the servers in question is usually a practical approach.

It is also important to block traffic directly from mobile users to mobile users by creating appropriate access control lists in GGSNs or a similar type of access restriction mechanism. Traffic separation is also a good design practice in the core network backbone and is a common method used to create different VLANs for separated traffic flows. Another aspect to be especially careful with is the IP addressing plan, which should cover the actual requirements of the service. Only the relevant parts of the Gn VLAN should be provided with public IP addressing.

4.15.2 Operational Security

Once the mobile network is operational, one of the most important operational procedures is to verify the security from different aspects. Intrusion detection systems can be effectively deployed to detect and log intrusion attempts.

Other tools can also be utilised to detect user traffic containing some known viruses. While this type of traffic might not disturb the packet core infrastructure by itself, it is capable of creating some problems, with some potential mobile users not being aware of data sessions initiated by their mobile stations and thus being billed for traffic generated without their conscious knowledge. These types of tools are normally scanning the traffic at the Gn or Gi interface and are able to detect known data sequences that match their internal list of known viruses. The downside is that they need to be updated constantly, and the viruses that can be detected are supposed to be fully identified beforehand. The mutation of some viruses could eventually make the detection more difficult.

The rest of the aspects of operational security are defined according to the operator's specific IT routines, in which the mobile packet core network elements should also be included. This is most probably a normal practice in many mobile networks, since many operators, or their parent companies, have ISP operations.

4.15.3 Additional Security Aspects

Some additional aspects should also be considered. Physical security, for instance, also plays a role in the overall security picture, since most of the known data security attacks are carried out inside the companies, often by internal or external employees. A careful policy in premises access control and identification for the main data centres is, therefore, a normal measure.

In addition to physical security issues, personal security must be considered. There is also a need for careful selection of partners and contractors dealing with packet core equipment, which nowadays is a common practice in the industry.

It is also worth mentioning in this section the need for integrity and confidentiality of network configuration data. In normal planning and optimisation tasks, the engineers are usually dealing with confidential data regarding configuration and also internal data policies. Extreme care must be exercised when dealing with this type of sensitive data. Configuration data maintenance is one of the traditional operational routines. Additionally, live network traces, commonly used in optimisation, should be regarded as extremely sensitive data, avoiding wherever possible unprotected media for their handling. Every aspect of the optimisation process should be carried out according to the relevant policies in place for the operator in question.

4.16 Quality of Service

4.16.1 Introduction to QoS

Quality of service (QoS) is possibly the single most important concept in the field of optimisation for packet switched networks. The circuit switched networks also have a mechanism to manage the quality of the service, which is well known and well established a long time ago. The dimensioning of the number of available circuits or channels in a given network node or associated interface is done according to the criteria known as blocking probability, which can be used in the Erlang formula. However, this approach no longer serves the purpose of the packet switched networks, where multiple services are meant to use the same network resources while having essentially different characteristics and requirements. This fact was acknowledged a long time ago, and the basic IP protocol includes a field to mark every packet according to the type of service it is conveying, known as the differentiated services code point, which is a 6-bit field in the IP header that determines the per hop behaviour. Thus, the field can be used by the switches and routers in the network in order to prioritise traffic flows according to their tolerance to delay, jitter and packet losses. These were the known mechanisms, and the initial phase of deployment of radio access mobile networks, when GPRS appeared in the late 1990s and, soon after, UMTS, already had a more refined specification providing QoS mechanisms from its creation.

4.16.2 QoS Environment

It is clearly understood that in a mobile network the major constraints can be easily located in the radio access network, which is especially true in terms of QoS, where maximising the utilisation of the limited radio resources and prioritising the different traffic flows according to their nature is a complex task. However, the QoS concept in a mobile packet core network is, in principle, easier to understand and build, at least conceptually.

Nevertheless, current traffic volumes in most operators are such that the benefits of a QoS mechanism are not clearly perceived from a core network point of view. With current levels of utilisation of the network nodes in terms of traffic, the most likely scenario is that the mobility management and session management procedures offer the most limiting load to the system, hitting the capacity roof of the network nodes before the traffic volumes reach such levels that priorisation of traffic according to their nature can take place. This fact will have a major impact on the optimisation process.

4.16.3 QoS Process

Specifications for 3G define four traffic classes (conversational, streaming, interactive and background) and also specify in more detail other QoS attributes such as the maximum bit rate, guaranteed bit rate, maximum transfer delay, maximum delay variation and bit error rate. The specifications have actually been gathered in a table where the QoS requirements are defined in detail. This QoS detailed parameter table is the starting point of the QoS network implementation design and is closely related to the concept of service creation. Moreover, during the creation of a new mobile application, the QoS aspects are the major concern from the network point of view, since there is a need for an end-to-end approach for proper QoS deployment that cannot be limited to a given network subsystem.

During the network planning phase, the QoS is one of the aspects to be considered, but from a rather general approach, ensuring its deployment and functionality from a network design perspective. Only when new mobile applications are to be introduced does the QoS concept trigger attention, especially with those applications that generate more demanding traffic flows, such as video streaming. The sensible process consists of verifying the capability of the packet core network nodes to deal with the traffic created with the introduction of a new application, which may end in a new dimensioning exercise. Once the verification and eventual re-dimensioning takes place, the packet core network is ready for the end-to-end QoS implementation for the given new application. In this process, the packet core network takes a secondary role, since the most restrictive subsystem is the radio access network. However, there is a need to consider the process from an end-to-end perspective; otherwise there is a substantial risk of missing a link in the complex chain.

It can be argued that the QoS is either an essential optimisation process or a powerful optimisation tool, but it seems clear that the QoS is an end-to-end matter and can significantly contribute to increasing the network and end user performance in given situations.

Role of the Packet Core Network in the QoS Process

Normally, the main packet core network nodes, SGSN and GGSN, provide different treatments for the different traffic classes defined by 3GPP by means of implementing in those nodes some of the known functions in this area, such as queuing and shaping. The complexity is mostly in the network node design in order to deal with the expected traffic flows. Normally, there are few possibilities of tuning network node QoS parameters in order to increase the performance. Configuration of queues is one possibility, tuning the relative sizes of the queues for the different traffic classes according to the current traffic mix in terms of traffic classes. An other possibility is to tune or redefine the marking of the packets in the differentiated services code point (DSCP) field of the IP header in order to optimise the IP routing within a large and complex packet core backbone, perhaps also having multiprotocol techniques in use such as the MPLS for a more efficient handling of backbone traffic. However, it is difficult to provide concrete advice on this matter, since it is very dependent on the concrete network node implementation characteristics (some of which may not leave a margin for optimisation) and the concrete backbone design (which in some cases may not be subjected to optimisation).

Role of the HLR in the QoS Process

There is another network node that is not usually considered under the scope of a mobile packet core network but performs some of the fundamental functions and hosts some of the fundamental data for operation of the network. The HLR, one of the core network nodes, definitely plays a major role in the QoS process. It hosts all the QoS relevant parameters from a subscription point of view, at two levels: firstly, on a subscriber basis (which are determined combinations of QoS parameters), and, secondly, on an application basis (which are the QoS parameters required for a given application), normally defined per APN. The QoS parameters stored in the HLR are used during activation of the PDP context and are thus utilised for the duration of it, unless a PDP context modification takes place.

When designing new services, the service subscription data stored in the HLR has to be carefully analysed. A simplified structure of the QoS parameters is required in order to deal with the complexity of the possible combinations. Normally, a given set of parameters that are meaningful for that given application will be created as a QoS template or QoS profile, and that template will be re-used for a group of subscribers (e.g. premium, normal or prepaid) when using the given application (e.g. browsing, messaging or streaming). Thus, when a premium subscriber invokes a streaming service, he/she will be granted a PDP context with a certain set of parameters, which can be slightly different from those granted for a prepaid subscriber when using that service if the operator has so defined the access differentiation. Alternatively, the data can be structured so that the differentiation is performed only based on the application. Since the only application data available in the HLR is the APN, it can be problematic to introduce QoS differentiation when more than one application shares a common APN. The problem can obviously be solved by allocating a dedicated APN per application. This solution, in turn, leaves the subscriber with a provisioning problem, giving him/her the responsibility of selecting the correct APN when using different mobile applications. This situation does not help the adoption of new services. In order to overcome this limitation, service awareness can be introduced, both in the mobile terminal and in the GGSN hosting the APN. In the latter case, new additional aspects should be considered during the planning phase of this type of GGSN node.

4.16.4 QoS Performance Management

The previous section gives a good understanding of the complexity of handling QoS in the context of different users and different services, based on a simplified approach of utilising templates or profiles for predetermined sets of QoS parameters. This complexity has been an inherent part of the concept from the standardisation phase, a notable example being the different sets of parameters standardised in release 97/98 of the 3GPP specifications and the parameters standardised in release 99 of the same 3GPP specifications. Therefore, the different sets require a proper mapping between the parameters.

However, the amount of parameters and their possible values, originally proposed for providing great flexibility during the service design, create such a large amount of possible combinations of parameters that it would potentially make the performance management of a given service impossible. In turn, without an adequate level of performance management capabilities, the task of ensuring consistent levels of service quality for a given application to subscribers is close to impossible. Moreover, without that kind of quality assurance, the ultimate goal of ensuring the quality of experience (QoE) of the subscriber is uncontrollable from the network perspective. Hence there is a need for an adequate setup for proper QoS performance management in the network.

QoS performance management alone will not be sufficient to satisfy the QoE goals, since the QoE can only be quantified from the mobile station point of view, typically through end-to-end testing campaigns or mobile probes deployed in significant places of the network coverage. However, without QoS performance management, the perceived QoE is close to useless in the context of optimisation, since no correlation can be studied between the QoS and QoE.

The logic connection among those concepts can be expressed in the following process flow:

- The ultimate goal is to ensure a satisfactory QoE, measured through so-called QoE KPIs.
- The operator tool to ensure that goal is the mobile network.
- The closest perception of service performance in the mobile network is provided by the performance of the QoS, which can be quantified as QoS KPIs.
- The operator needs a simple mechanism to collect service based QoS KPIs.
- Once those service based QoS KPIs are collected from network elements and/or interfaces, they can be analysed.
- The closest perception of service performance in the terminal side is the QoE, which can be quantified as QoE KPIs.
- The operator needs a simple mechanism to collect service based QoE KPIs. Currently the possibilities are field test campaigns and field probes.
- Once those service based QoE KPIs are collected from field probes or end-to-end testing campaigns, they can be analysed accordingly.
- Eventually, from both parallel analyses, correlations can be found and improvement actions can be decided within the reach of current possibilities.

The main difficulties are two:

- to provide a simplified mechanism to collect those service based KPIs;
- to build meaningful correlations between service based QoS KPIs and their corresponding service based QoE KPIs.

The first difficulty can only be partially solved through a thorough study and definition of the meaningful QoS attributes that are relevant from an end user perspective and deep alignment of the OSS concept with the network nodes. This will provide a simple context of a few meaningful KPIs that can be uniquely associated to a given service or application. Some vendors in the industry are already exploring this area, with promising results.

The second one is certainly uncharted territory, where only experience can provide enough of a knowledge base for successful correlations. As an example, it is worth formulating the problem in concrete terms, as follows:

- QoE KPIs. An end-to-end measurement campaign for streaming with 100 service access requests during the busy hour of a weekday gives the result of a service accessibility of 97 %.
- QoS KPIs. During the same measurement period, the OSS registers for the aggregated streaming traffic a service accessibility of 99.6 %, out of a total of 5873 service requests.

What kind of conclusions can be extracted from those KPIs? Is access to the streaming service performing well or badly? Can the fault be associated with a given fact, geographical location, type of terminal or even type of content accessed within the given service? Answers to these questions can only be given as a result of the correlation of QoS KPIs and QoE KPIs, which will be greatly influenced by local factors and may not easily be generalised. Certainly, as time goes by and operators accumulate experience, these questions will be at least partially solved. In this context, it is worth noting that by measuring QoE KPIs and evaluating them, the problem of improving the QoE, which is the ultimate goal, will not be solved. It is relatively easy in this complex environment to confuse the process with the goal and with the risk of finishing the process short of achieving the goal. Giving a pointer towards a generic cause of bad performance is part of the process (e.g. 97 % of service accessibility is not acceptable as the cause might be in the congested packet core network node). Fully identifying the cause and enabling the solution is certainly closer to achieving the goal (e.g. while the value of 97 % of service accessibility was registered in a given radio access area, the overall aggregated value of 99.6 % of service accessibility strongly suggests that this is closely related to the geographical area where the tests were performed, further investigation showing that the RNC under consideration presented some misconfigurations affecting the handling of QoS parameters that were under investigation). Additionally, this example also illustrates the necessity for an end-to-end approach to the QoS issues.

5

Fourth Generation Mobile Networks

Ajay R Mishra

5.1 Beyond 3G

In previous chapters it was shown that all of the three previous generations of mobile network had their merits and demerits but none of them had the ability to completely replace the other. Even IMT-2000, a worldwide standard, was not able to break the bottleneck of high data rate and capacity, and so this led to the formation of a new generation, referred to as the beyond 3G generation or 4G, but there is no clear definition for it. However, it was thought that instead of developing new radio interfaces and new technology it would be better to integrate existing and newly developed wireless systems like the GPRS, EDGE, Bluetooth, WLAN and Hiper-LAN. Hence, in the eighteenth TG-8/1, a new working group WP8F was established in 1999 to look into efforts to develop the systems beyond the IMT-2000. It was felt that the next generation networks would offer some exciting opportunities, such as:

- *Performance*. The 4G systems are intended to provide high quality video services providing data transfer speeds of about 100 Mbps.
- *Bandwidth*. The 4G technology offers transmission speeds of more than 20 Mbps and is capable of offering high bandwidth services within the reach of local area network (LAN) hotspots, installed in airports, homes and offices.
- *Interoperability*. The existence of multiple standards for 3G made it difficult to roam and interoperate across networks. There is therefore a need for a global standard providing global mobility and service portability so that the single-system vendors of proprietary equipment do not bind the customers.
- *Technology*. Rather than being an entirely new standard, 4G basically resembles a conglomeration of existing technologies and is a convergence of more than one technology.

5.2 4G Network Architecture

The exact network architecture of the 4G networks is not yet defined, but some industry experts expect the system to be a layered one as shown in Figure 5.1. As seen in the figure, the next generation network is expected to be a combination of existing and advanced technologies with seamless handovers between

---

Advanced Cellular Network Planning and Optimisation   Edited by Ajay R Mishra
© 2007 John Wiley & Sons, Ltd

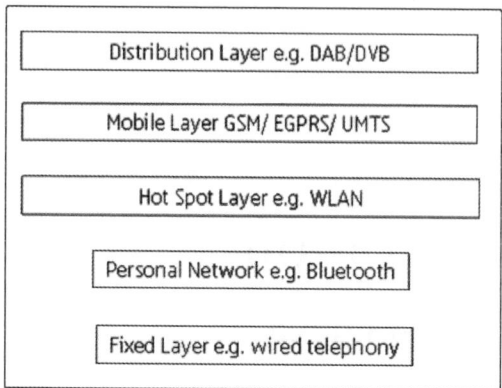

Figure 5.1 Layered structure of a next generation network (DAB, digital audio broadcasting; DVB, digital video broadcasting)

technologies irrespective of the geographical location, having high data rates, high capacity, high quality of service and low costs.

The 4G network hierarchy is expected to consist of four broad levels of networks: the personal networks, the local networks, the cellular networks and the satellite based networks (distribution networks). Above the personal networks are the local networks, which may consist of LANs using wireless LAN technology. These LANs normally have a greater coverage area than personal networks and may be used in hotspots like cafes, hotels and airports. The cellular network level comes next, which will consist of existing 2G and 3G cellular networks, as well as the enhanced cellular networks, and at the highest level will be the satellite based mobile networks, having the largest of the coverage areas of all the network levels.

As a result of this multilevel hierarchy, the 4G user devices will be expected to perform vertical handovers as well as horizontal handovers. Horizontal handovers will take place within one network level and may include soft handovers and hard handovers. On the other hand, vertical handovers will be performed between different network levels. Here is an example. User devices, e.g. a laptop that is using WLAN to communicate with servers within office premises, would perform a handover to the cellular network (e.g. EDGE, CDMA) when moving out of the office complex. Thus, 4G user devices would be intelligent devices that would be able to work with different wireless technologies and also would be able to select the appropriate technology to make use of a particular service by performing handovers.

5.3 Feature Framework in a 4G Network

The feature framework in the 4G network can be defined by using one simple word, integration, i.e. seamless integration of terminals, networks and applications (together with users):

- Apart from users, the domain includes targets such as terminals, networks and applications.
- Convergence of the above mentioned targets.
- The adaptability of the features between different targets make integration seamless.

5.3.1 Diversity in a 4G Network

The need for diversity in these networks is caused by external targets and is fulfilled by internal targets. Thus, there are two kinds of diversity in 4G networks.

External Diversity

This brings demand for adaptability features within targets. As it lies outside the targets it is therefore called external diversity. External diversity of users refers to people in different situations, e.g. background, personal preference, etc. External diversity of terminals refers to the differences of terminals in terms of both the static and mobile attributes, where the static attributes include functionality, weight, size, battery, cost, etc., and the mobile attributes include the dynamic attributes of both temporal and spatial features. The external diversity of networks can basically be defined just by assuming the large size of networks around the globe today, e.g. the Internet. Air interfaces can integrate all kinds of standards and work in different frequencies. Moreover, multiple operators deploy networks with multiple standards and protocols.

Internal Diversity

This gives the solution for adaptability. As it lies within the target it is therefore called internal diversity. Internal diversity of users are people with different interfaces, e.g. hearing, speech, etc., while internal diversity of terminals means that one terminal may integrate multiple functions, modes, interfaces, flexibilities, etc. The internal diversity of networks, on the other hand, refers to the interconnection between the various networks and the sharing of load between them. For applications, internal diversity means that one application can be tailored or drawn into multiple levels of quality, various styles, different kinds of release shape, etc.

5.4 Planning Overview for 4G Networks

Planning 4G networks would not be as straightforward as seen in the previous chapters. Each of the individual networks GSM, EGPRS and UMTS have been planned in their respective domains, e.g. radio, transmission or core. Network planning in 4G networks will require an understanding of more technologies, of which a few are covered in this chapter, and their interworking with other technologies. As the 4G network will be an integrated wireless system that enables seamless roaming between technologies, a user can be operating in a cellular technology network and then be handed over to the satellite based network and back to a fixed wireless network, depending upon the network coverage and the user's preference of charging.

Planning a network basically involves an initial layout of the system structure, which includes the spectrum, cell radius and hierarchical service area.

Spectrum

The 4G systems are expected to provide bandwidths higher than 20 Mbps and to accommodate a significantly increased amount of traffic, so sufficient frequency resources will be required. Since the lower frequency band considered suitable for mobile communications is heavily used, a frequency band for the 4G communications is to be proposed in the 3G to 5G bands.

Cell Radius

As planned, the bandwidth to be offered in 4G systems is three orders of magnitude greater than that of 2G systems. The cell radius covered by a base station (BS) generally decreases if, assuming all other conditions to be the same, radio signals are transmitted at higher bit rates than at a lower transmission bit rate in order to compensate for the increased noise level. Moreover, 4G systems may be operated in a higher frequency band so that propagation loss of the wireless signal is higher than that of 2G and 3G systems.

Using a complex set of equations, the increase in propagation loss caused by the operating frequency and channel speed can be converted into a decrease in cell radius (assuming the antenna height to be 23 m) if other transmission conditions are assumed to be the same as those of 3G systems. This means that by covering the same area as in 3G systems, the 4G systems will require four times the number of BSs. The antenna height of the BS in an urban area tends to be lower when the cell size is smaller. As a result, there may be more outage areas, even within the calculated cell radius, and it is observed that one solution to enhance the capacity is to decrease the cell radius of the BS.

Hierarchical Service Area

Although all objects will be connected to a network through wireless links, it may be difficult for small devices to be directly connected to 4G systems due to power consumption and antenna size. Compact devices will be able to access the 4G systems through a miniature base that will act as a mobile terminal (MT) for 4G systems. Employing such a configuration results in service areas consisting of multiple overlapping cells.

5.4.1 Technologies in Support of 4G

Some of the technologies that will be a part of the next generation networks are as follows.

OFDM

OFDM is a digital modulation technique in which one time symbol waveform and thousands of orthogonal waves are multiplexed, which is good for high bandwidth digital data transmission.

W-OFDM

Wideband OFDM (W-OFDM) enables data to be encoded on multiple high speed radio frequencies concurrently allowing greater security, an increased amount of data being sent and the most efficient use of bandwidth. It enables implementation of low multipoint RF networks that minimise interferences with adjacent networks. This enables independent channels to operate within the same band, allowing the multipoint networks and point-to-point backbone systems to be overlaid in the same frequency band.

The next step in the development towards 4G is the separation between uplink and downlink technologies for the transmission path. The modulation technique that holds the most potential for 4G applications is wideband OFDM. Its potential bandwidth is estimated to be as high as 100 Mbps for the capacity of one cell.

Wideband OFDM is chosen as a single carrier solution due to the lower complexity of equalisers for high delay spread channels or high data rates. In this, a broadband signal is broken down into multiple narrowband carriers (tones), where each carrier is more robust to the multipath. In order to maintain orthogonality among tones, a cyclic prefix is added. With proper coding and interleaving across frequencies, multipath fading turns into an OFDM system advantage by yielding frequency diversity.

OFDM can be implemented efficiently by using fast fourier transforms (FFTs) at the transmitter and receiver. With MIMO, the channel response becomes a matrix. Since each tone can be equalised independently, the complexity of space–time equalisers is avoided. Multipath propagation remains an advantage for a MIMO-OFDM system since frequency selectivity caused by a multipath improves the rank distribution of the channel matrices across frequency tones, thereby increasing the capacity.

Increasing demand for a high performance 4G broadband wireless mobile calls for the use of multiple antennas at both the base station and subscriber ends. These multiple antennas technologies provide high capacities suited for multimedia services and also increase the range and reliability. Multiple antennas at the transmitter and receiver also provide diversity in a fading environment.

MC-CDMA

MC-CDMA is actually an OFDM with a CDMA overlay. Here the users are multiplexed with orthogonal codes to distinguish users in MC-CDMA. In MC-CDMA each user can be allocated several codes where the data are spread in time or frequency.

Local Multipoint Distribution System (LMDS)

The local multipoint distribution system (LMDS) is the broadband wireless technology used to deliver voice, Internet and video services in the 25 GHz and higher spectra.

5.4.2 Network Architectures in 4G

The three possible architectures that can be used in the future are as follows.

Multimode Device

This configuration basically uses a single physical terminal with multiple interfaces to access different wireless networks like AMPS and the CDMA dual-function cell phone. This multimode device architecture is advantageous as it may improve call completion and expand the effective coverage area. It should also provide a reliable wireless in the case of network, link or switch failure, and the user, device or network can initiate a handoff between networks. The handling of quality-of-service (QoS) is the main issue in this case.

Overlay Network

In this case a user accesses an overlay network consisting of several universal access points (UAPs). These UAPs in turn select a wireless network based on availability, QoS specifications and user-defined choices. A UAP performs protocol and frequency translation, content adaptation and QoS negotiation–renegotiation on behalf of users. The overlay network, rather than the user or device, performs handoffs as the user moves from one UAP to another. A UAP stores user, network and device information, capabilities and preferences. Because the UAPs can keep track of the various resources a caller uses, this architecture supports single billing and subscription.

Common Access Protocol

This becomes viable if wireless networks can support one or two standard access protocols.

5.4.3 Network Planning in 4G Networks

Traffic in 3G and 4G networks is going to expand considerably and the growing data/multimedia traffic may lead to an increase in the total load on the PS core network (CN) domain elements, especially when the IP multimedia subsystem (IMS) is involved because this subsystem contains a uniform way of maintaining voice over IP (VoIP) calls and offers a platform to multimedia services. The classical teletraffic theory for the performance calculation of the PS CN domain elements can be faulty because of the high variability of burstiness in the traffic; as a result the network parameters can be underestimated.

Problems with 3G/4G network planning may arise, in particular:

- the estimation problem of the potential number of 3G/4G users;
- the prediction problem of data traffic characteristics;
- the problem of the performance estimation of PS CN domain elements taking into account the self-similar nature of the multiservice traffic.

Another important aspect of an evolution in wireless communications concerns changes in the RAN of 4G systems (4G RAN). There will also be a need to deal with the enormous amount of traffic, the base station radius needs to be small and the 4G RAN must comprise more base stations and more frequent handovers, resulting in a heavy load on the links between the elements of RAN, the remote network controller (RNC) and BSs, thereby suggesting changes to the existing 3G networks.

Taking into account the above considerations it can be seen that on the CN side of 4G systems the main purpose is to minimise the changes and utilise the existing 3G CN elements and functionality as much as possible. Also, the 4G RAN will experience major changes, in particular on a physical transmission layer. The network physical links configuration is one of the major problems of mobile communications network planning because it determines the long-term performance and service quality of networks.

The solutions for the 3G/4G network planning problems are:

- a method for estimating the number of prospective 4G users;
- an estimation method for finding data traffic generated by 4G users.

The following parameters of the data traffic are to be estimated:

- the specific (per user) rate of transactions in a busy hour;
- the average duration of 3G/4G calls;
- the specific (per user) traffic intensity created by 3G calls in a busy hour;
- the total traffic intensity on nodes of the IMS;
- the performance evaluation methods for the IMS nodes, taking into account the self-similar nature of the multiservice traffic.

The following probabilistic and time characteristics of the nodes are determined:

- the upper bound and lower bound of the server service rates;
- the average queue length in the buffers;
- the average service time of information units;
- server utilisation.

The solution methods for solving the 4G RAN problems can be supposed to be:

- The method for a quantitative estimation of reliability and cost parameters of different RAN topologies. The parameters of ring, multiring and radial topologies can be considered and compared with each other.
- The method of a physical links ring configuration between 4G RAN elements with a minimal deployment cost.

It is therefore observed that for network planning the following requirements are needed:

- the requirements to equipment internal IP networks, such as the throughput of channels and the performance of nodes and routers in the 3G/4G systems;
- the requirements to equipment gateways providing the interaction between 3G/4G systems and external networks.

However, as mentioned below, when planning a complete 4G network, an understanding of all the technologies would be required. Some technologies have been discussed in previous chapters. Here the focus will be on other technologies.

5.5 OFDM

It has been stated in the previous section that wideband OFDM is the probable candidate for the radio access technique in 4G systems. The factors that make OFDM much better than its competitions will now be discussed in detail. The two main advantages that make OFDM preferred over other multiple access techniques are:

- OFDM can sustain a high level of frequency selective fading in mobile communications. This robustness is necessary for high speed data transmission.
- By combining OFDM with CDMA, the power of combating frequency selective fading increases and there is a high scalability in possible data transmission rates.

5.5.1 What Is OFDM?

OFDM is a special form of multicarrier modulation (MCM) scheme, where a single data stream is transmitted over a number of lower data rate subcarriers. OFDM is actually a combination of both modulation and multiplexing. The question of multiplexing is applied to independent signals but these independent signals are themselves a subset of one main signal. In OFDM the signal itself is first split into independent channels, modulated by data and then re-multiplexed to create the OFDM carrier.

This can be understood by making an analogy for using a shipment by truck. There are two options: one is to hire one large truck and the other a bunch of four smaller ones. Both methods carry exactly the same amount of data. However, in the case of an accident, only a quarter of the data on the OFDM trucking will suffer. Here the four smaller trucks when seen as signals are called the subcarriers in an OFDM system and they must be orthogonal for this idea to work. The independent subchannels can be multiplexed by using frequency division multiplexing (FDM), called multicarrier transmission, or can be based on code division multiplexing (CDM), called multicode transmission.

Need for Orthogonal Signals

The main concept in OFDM is the orthogonality of the subcarriers. Since the carriers are all sine/cosine wave, the area under one period of the sine/cosine wave is zero. Even if a sinusoid of frequency n is multiplied by a sinusoid of frequency m/n, the area under the product is also zero. In general, for all integers m and n, $\sin mx$, $\cos mx$, $\sin nx$ and $\cos nx$ are all orthogonal to each other and these frequencies are called harmonics. The orthogonality allows simultaneous transmission of many subcarriers in a tight frequency space without interference from each other.

Assume a signal and divide it into four groups for the four subcarriers. To transmit them individually and at the receiver end these four modulated subcarriers are added to create the OFDM symbol by using a block called the IFFT (inverse fast fourier transform) block, as can be seen in the following steps (shown in Figure 5.2). Forward FFT takes a random signal, multiplies it successively with complex exponentials

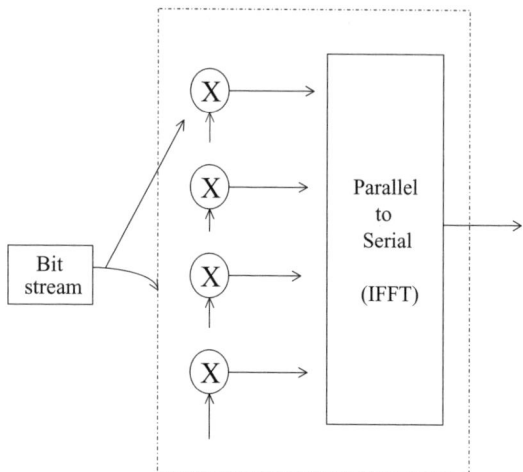

Figure 5.2 Orthogonality in OFDM

over the range of frequencies, sums each product and plots the result as a coefficient of that frequency. The coefficients are called a spectrum and represent how much of that frequency is present in that signal. The results of the FFT are basically a frequency domain signal, as shown in the Figure 5.2.

The inverse FFT (IFFT) takes this spectrum and converts all of it back to a time domain signal by again successively multiplying it by a range of sinusoids. The use of IFFT/FFT totally eliminates the bank of subcarrier oscillators at the transmitter/receiver.

Intersymbol Interference

The signal that reaches the receiving end faces attenuation during transmission. In the fading environments the signal throws a splash backwards, which needs to be avoided. When the distance between the transmitted and receiving points increases, the spread of the signal is the same as the delay spread. Thus, to mitigate the effect at the front of the symbol, the symbol is moved further away from the region of delay spread. As the blank spaces cannot be left in the signal without a guard interval between successive OFDM symbols, intersymbol interference from the $(i-1)$th symbol gives a distortion to the ith symbol. Therefore a guard interval (no signal transmission) with length $\Delta g > \tau_{max}$ should be applied, where τ_{max} is the maximum time delay due to multipath fading. This completely eliminates any intersymbol interference that may arise, but results in a sudden change of waveform containing higher spectral components, thereby leading to intersubcarrier interference. However, to counter intersubcarrier interference, cyclic prefixing is performed where the original signal is extended in the guard time (by approximately 1.25 times the original signal). Here, the end part of the symbol is copied and joined to the start of the signal until the signal length becomes 1.25 times the original length. The symbol source becomes continuous and it would appear that the symbol period is becoming longer.

Given the channel frequency selectivity, the channel time selectivity and the symbol transmission rate, the transmission performance becomes more sensitive to the time selectivity as the number of subcarriers increases because the wider symbol duration is less robust to random FM noise, whereas it becomes poor as the number of subcarriers decreases because the wider power spectrum of each subcarrier is less robust to the frequency selectivity. On the other hand, the transmission performance becomes poor as the length of the guard interval increases because the signal transmission in the guard interval introduces a power loss, whereas it becomes more sensitive to the frequency selectivity as the length of the guard interval decreases because the shorter guard interval is less robust to the delay spread.

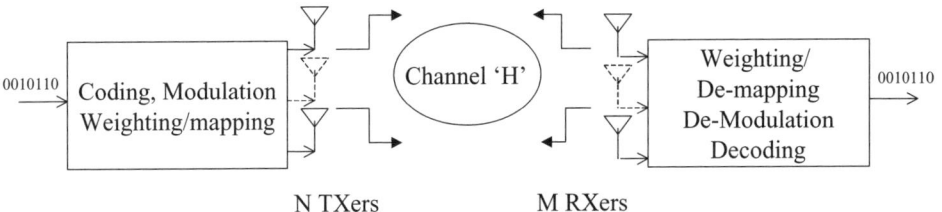

Figure 5.3 MIMO system

5.5.2 MIMO Systems

The MIMO systems need to form an integral part of the 4G system architecture as the wireless systems continue to strive for even higher data rates. Digital communication using multiple-input multiple-output (MIMO), also called a 'volume-to-volume' wireless link, has emerged as one of the most important breakthroughs in modern communications. The MIMO systems can be defined as a general wireless communication link where a link for the transmission and receiver is equipped with multiple antenna elements, as shown Figure 5.3.

The MIMO systems are chosen ahead of the single antenna systems because of their ability to turn multipath propagation, a pitfall of wireless communication, into a benefit for the user. They take advantage of random fading and multipath delay spread for multiplying transfer rates.

Principles of MIMO Systems

The idea behind MIMO is that the signals on the transmit (TX) antennas on one end and the receiver (RX) antennas on the other end are combined in such a way that the quality (bit error rate, or BER) or the data rate (bits per second) of the communication for each MIMO user will be improved. The main idea is that of space–time signal processing, in which time is complemented with the spatial dimension inherent in the use of multiple spatially distributed antennas.

As shown in Figure 5.2, a compressed digital source in the form of a binary data stream is fed to a simplified transmitting block encompassing the functions of error control coding and mapping to complex modulation symbols. The transmitter produces several separate symbol streams, which range from independent to partially redundant to fully redundant. Each is then mapped on to one of the multiple TX antennas. Mapping may include linear spatial weighting of the antenna elements or linear antenna space–time precoding. After upward frequency conversion, filtering and amplification, the signals are then launched into the wireless channels. At the receiver the signals are captured by possibly multiple antennas and demodulation and demapping operations are performed to recover the message.

The intelligence of a multiantenna system is located in the weight selection algorithm rather than in the coding side. A simple linear antenna array combination can offer a more reliable communications link in the presence of adverse propagation conditions such as multipath fading and interference. A key concept here is that of beam forming, by which the signal-to-noise ratio can be increased by focusing energy into desired directions, in either a transmitter or a receiver. If the response of each antenna element to a given desired signal, and possibly to interference signals, is estimated, the elements with weights selected as a function of each element response can be optimally combined. The average desired signal level can then be maximised or the level of other components, whether noise or co-channel interference, can be minimised.

The prospect of many orders of improvement in wireless communication performance at no cost of extra spectra is largely responsible for the success of MIMO in high rate WLANs and 3G networks and

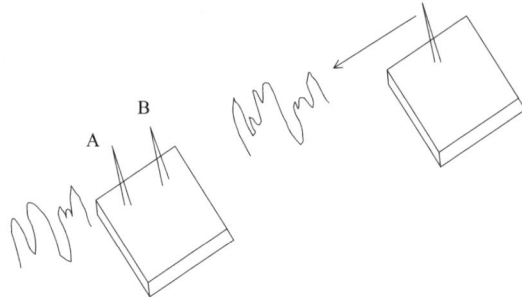

Figure 5.4   Diversity mode

is being planned in 4G networks. MIMO is basically the first wireless communications technology that can be used to multiply the link capacity. Multiplying the link capacity increases:

- Capacity
- Throughput
- Signal quality
- Low power consumption
- High data rates

The MIMO systems multiply spectral efficiency and capacity by encoding, transmitting, receiving and decoding multipath signals.

MIMO can be defined as an antenna technology for wireless communication in which multiple antennas are used both at the transmitter and the receiver ends. The antennas at both ends of transmission are combined to minimise errors and optimise data speed. MIMO operates basically in two modes:

- Diversity mode
- Spatial multiplexing

Diversity Mode

Diversity basically refers to the use of multiple antennas to increase the probability of a high quality signal path between the sender and the receiver and can be implemented at the receiver end or the transmitter end, or both ends of the wireless link.

Simple receive diversity, as shown in Figure 5.4, involves the use of two antennas that are placed sufficiently apart such that they can receive signals from independent signals paths. A basic way to select the optimal receive antenna output is switched diversity, where the receiver simply switches antennas whenever it detects weak signals or a high noise level from the current receiving level. More sophisticated diversity techniques such as maximum radio combining (MRC) receive on the multiple antennas simultaneously and apply the advanced signal processing algorithm to combine the different versions of the received signals to maximise the SNR. Switched diversity and MRC can be implemented on just the receive side of the link. The need for prior knowledge of the receiver in order to optimise the transmit paths makes transmit diversity more complicated. The simplest technique is to use the antenna from which information signals have arrived successfully from the target receiver. Hence multiple copies of the same information scheme are sent to the antenna for added redundancy. It can be seen that the same information signals must first be transformed into different RF signals to avoid interference with each other.

All-IP Network 427

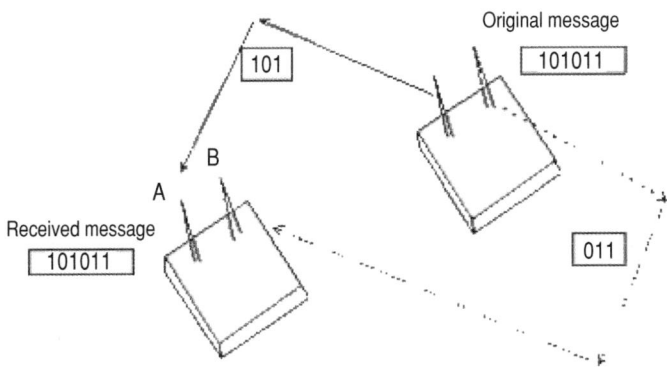

Figure 5.5 Spatial multiplexing

Diversity maximises the wireless range and coverage. It also increases the network throughput by finding quality signal paths to enable the devices to communicate using the highest data rates and avoiding the signal paths that are likely to produce packet errors and retransmissions.

Spatial Multiplexing Mode

Spatial multiplexing (SM) prospers in multipath environments as it allows the sender to transmit different portions of the user data on multiple paths in parallel in order to increase the system capacity, as shown in Figure 5.5. The target receiver must also implement a corresponding de-multiplexing algorithm to recover the original information stream from its receiver antenna. SM requires uncorrelated multipaths. Since multipath fades change moment by moment with motion, there is no assurance that the uncorrelated signal path can always be found. Furthermore, SM does not works well in low SNR environments where signals are weakened by distance or the noise level from the RF and channel interference is high, thereby making it more difficult for the sender and receiver to identify the uncorrelated signal paths. When the SM is not possible then the MIMO system returns to the diversity mode.

Another powerful concept for this is that of spatial diversity. In the presence of random fading caused by multipath propagation, the probability of losing the signal decreases exponentially with the number of de-correlated antenna elements being used. A key concept here is that of diversity order, which is defined by the number of correlated spatial branches available at the transmitter or the receiver. When combined together the linear array improves the coverage range versus quality trade-off offered to the wireless user.

In an MIMO link, the benefits of the conventional antennas are retained since the optimisation of multiantenna systems are carried out in a larger space, thus providing additional degrees of freedom. These MIMO systems can provide a joint transmit–receive diversity gain as well as an array gain when the antenna elements are coherently combined.

5.6 All-IP Network

For communication and networking purposes a large number of protocols are being used, but the IP (Internet protocol) provides a universal network layer protocol for wired packet networks and brings into play the same role in wireless systems. An all-IP wired and wireless network can make wireless networks more robust, scalable and cost-effective. It allows software and applications developed for wired IP networks to be employed on the wireless IP networks at present, most of the different wireless systems, such as PANs (permanent access networks), WANs and wide area cellular networks, are not compatible with each other, making it difficult for a user to roam from one radio system to another. To date, no wireless technology has emerged as a long-term universal solution for it. With IP being a

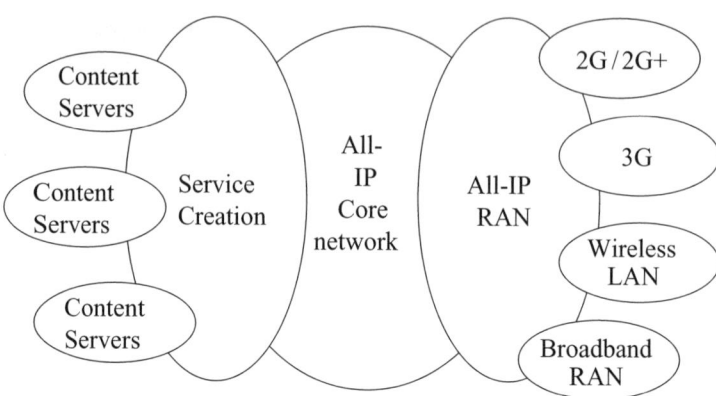

Figure 5.6 All-IP based network model

common network layer protocol, an IP based mobile device (with multiple radio interfaces) could roam between different wireless systems.

There needs to be a convergence of the core networking infrastructure on the IP protocol suite and a serious effort has been made to replace the IPv4 deployed throughout the Internet by IPv6, which simplifies the mobility support. This is important because then instead of having only one technology available there will be different technologies to assist mobility from one place to another with varying data rates.

New Network Model

The new model should be flexible enough for a user to be able to access its services independently of its locations, in a transparent way. The user terminal must be able to pick the 'best' available access technology at its current location and use it seamlessly for the provision of the desired services.

This homogeneous, high speed, secure, multiservice, multioperator network concept is required to support multiple types of services, from simple network access to complex multimedia virtual reality, and including traditional telecommunication services such as telephony in mobile environments. A type of network is therefore required that must be able to associate service agreements to network control constrains, to monitor this usage per service and user and to provide these services while the user moves (with its terminal changing access technologies). An example of an all-IP based network model is shown in Figure 5.6.

The most important point in this context is the development of a quality of service (QoS) architecture for 4G networks, which should be able to support 'any' type of user service in a secure and auditable way. Both user interfaces and interoperator interfaces have to be clearly defined, and multiple service providers should be able to interoperate under the guidelines of this architecture. The IP-enabled terminals, apart from allowing the users to access, should be able to replicate the voice service environment as in previous generations of networks.

5.6.1 Planning Model All-IP Architecture

Network Configuration

A layout for a possible 4G network based on the all-IP core network is shown in Figure 5.7. It comprises the core network, which performs the service control, and the radio access network, which performs the

All-IP Network

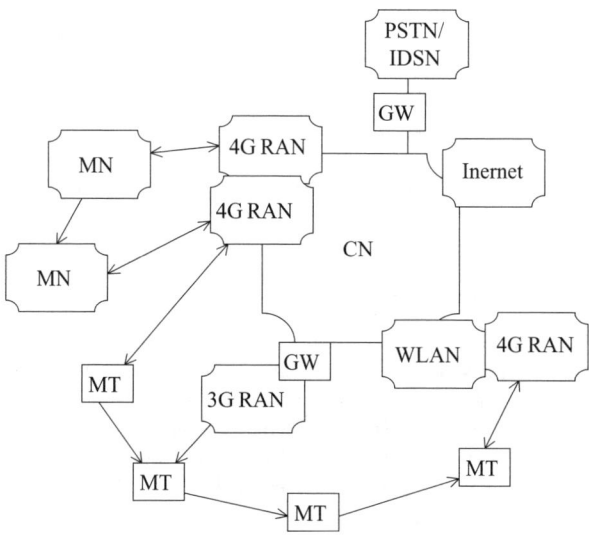

Figure 5.7 All-IP based network

radio transmission and radio resource control. Mobility control is performed by the cooperation between the CN and the RAN. Information transportation in 4G systems (CN and RAN) is based on the IPs. Each mobile terminal has its own IP address and therefore the 4G systems connect to the Internet directly but connect to the PSTN or ISDN through gateways (presently also the connection to the Internet is also through the gateways). This type of network can integrate multiple types of RANs, which are based on IPs, with each other and services can be offered through a common CN.

RAN Configuration

In order to deploy the RAN efficiently, features of distributed base station control and multihop link connections in the new configuration are included. Comparing a 3G RAN with a 4G RAN, it can be seen that a 3G RAN has a vertical tree like structure and multiple BSs are connected to a radio network controller (RNC), which intensively controls those BSs. To enhance the radio signal quality at the cell edges, a diversity handover scheme was introduced in the 3G systems, in which, whenever a mobile terminal can communicate with multiple BSs, i.e. the MT is handed over between adjacent BSs, all layer 1 signals of the uplink (MT to BS) received at each BS are transmitted to the RNC and the RNC combines them into a user data stream. Because of this, the traffic on the BS approach links is multiplied, as compared with the original user information. If the same configuration is adopted in the 4G RAN, in which the handover may occur more frequently, the load on both the approach links and the RNC signal processing equipment will be heavier, thereby causing a serious cost increase in the RAN.

The RAN structure proposed for the 4G RAN is a cluster cellular RAN, as shown in Figure 5.7. In this configuration, BSs are grouped into a 'cluster' and there is a 'cluster head' BS connected to the CN. BSs in a cluster are linked to each other by a kind of local area network (LAN). RNC functions are distributed to each BS.

Distributed Base Station Control

In order to reduce the load on both the approach links and signal processing on specific equipment, most of the layer 1 signal processing control is distributed among the BSs. The uplink signal is received by multiple BSs (slave BSs) and then the received signal is forwarded to one of the BSs that process the uplink signal as a temporal agent (master BS). As the MT moves, the role of the master BS (the temporal agent) is passed to an adjacent BS, and as a result some BSs are newly incorporated into the group and some are removed from the group. Thus, the BS communicating with an MT forms a kind of 'virtual BS' represented by the master BS. If the received layer 1 signals are transferred only within the cluster, diversity handover control is closed within the cluster, but if they are transferred by IP packets and the BSs work as a router, then the diversity handover can be carried out independently of the cluster. Downlink packet signals are multicasted from the cluster head BS to the other BSs in the cluster.

Multihopping Wireless Connections

The approach links in the present RANs are constructed both of optical fibre and radio links, and radio media are preferred as link transmission media from the viewpoint of system cost. The 4G RAN is expected to transfer 23 times more traffic than the current RANs and to cover more BSs, and so the approach link requires a greater traffic capacity and better economical performance. In order to increase the capacity, the radio link bandwidth must be broadened. Therefore, the radio frequency used for the links will become higher in order to support the wider bandwidth, making the applicable transmission distance of a radio link shorter. In order to compensate for this problem, increasing the output power and using a high directivity antenna are applicable techniques, but this is a costly option.

In order to solve these problems multihop wireless connections are used, taking advantage of the cluster-type RAN structure. In multihop connections the signal transmission distance is shorter than in a single-hop connection, but the necessary transmission capacity is larger because of the relay traffic. The multihop wireless connection is superior to the single-hop connections from the viewpoint of the total output power, although it carries relay traffic. In addition, due to the multihop configuration it is also not necessary to find a line-of-sight radio path, which directly connects a BS with the cluster-head BS.

Network Modelling

4G architecture uses technologies such as the Ethernet (802.3) for wired access, Wi-Fi (802.11b) for wireless LAN access and WCDMA (the physical layer of UMTS) for cellular access, and is IPv6 based. In such an environment seamless intertechnology handovers need to be supported. Thus mobility cannot be simply handled at the physical layer, and requires implementation at the network layer. Here, the same IPs handle the movement between different cells.

It is required that the users/terminals handover between any of these technologies without breaking their network connection and maintaining their contracted QoS levels. Also, the QoS and AAAC (authentication, authorisation, accounting and changing) subsystems are responsible of serving each user according to the SLA (service level agreement) previously negotiated, operating respectively at the network level and at the service level. On the other hand, service providers should be able to keep track of the services being used by their customers, both inside their own network and while roaming. The multiple network functions can then be listed as:

- physically supporting the mobility of the terminals and multiple technologies;
- guaranteeing planned QoS levels to specific traffic flows;
- supporting interoperator information interchange for multiple-operator service provision;
- realising appropriate monitoring functions for providing information to the service operator about network and service usage;
- implementing confidentiality both on user traffic and on network control information;
- entities involved in the architecture.

All-IP Network

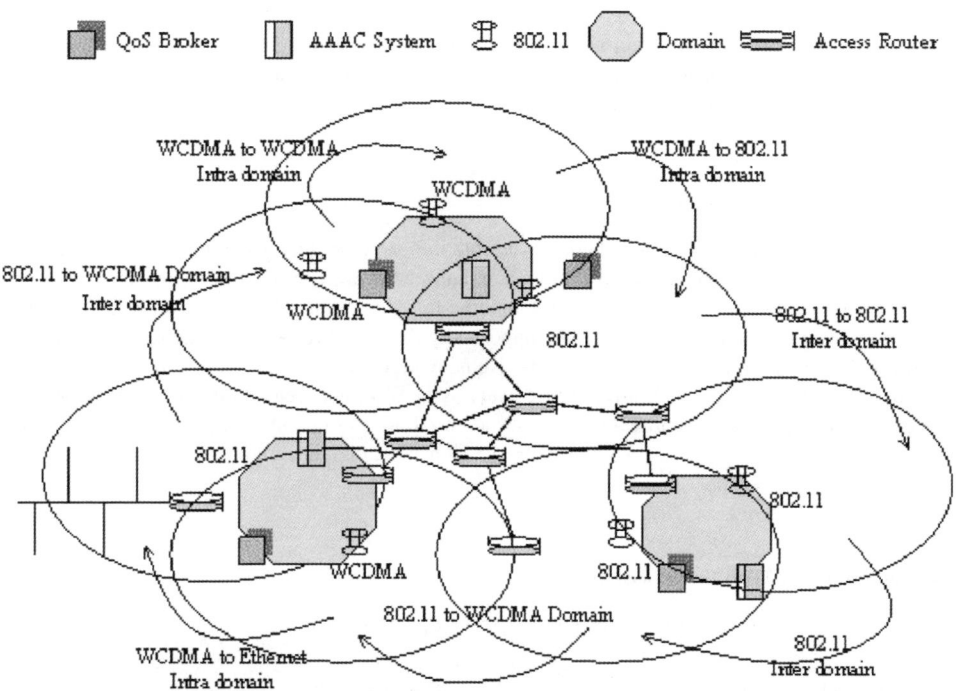

Figure 5.8 Conceptual diagram of a 4G network

The two main entities considered in this architecture are:

- the user, a person with a service level agreement (SLA) for a set of services;
- the network operator, who owns the administrative network domain, thereby providing the transfer of information across end points.

Conceptual Network Architecture

A conceptual layout of a 4G network is shown in Figure 5.8, illustrating some of the handover possibilities during user movement. There are basically three administrative domains in the figure, each using different types of access technologies. Each administrative domain contains:

- an AAAC system;
- at least one network access control entity;
- the QoS broker (QoSB).

From the diagram it is observed that the major entities involved in the architecture are:

- mobile terminal (MT): the user terminal from where the user accesses the subscribed services;
- access router (AR): the MT point of attachment to the network;
- wireless gateway (WG): when there is an AR for wireless access;

- QoS broker: the entity responsible for managing one or more ARs and controlling user access and access rights according to the information provided by the AAAC system;
- AAAC system: the authentication, authorisation, accounting and charging system, responsible for service level management (including accounting and charging).

5.6.2 Quality of Service

QoS architecture should lead to a potentially scalable infrastructure that is possible to expand to span large areas, to maintain contracted levels of QoS and ultimately to replace today's telecommunications infrastructure. No specific network services should be imposed by the QoS architecture. On the other hand, no special charging models should be imposed by the user-level AAAC system, and the overall architecture must be able to support very restrictive network resource usage. QoS architecture needs to provide different types of network-layer services, such as the Web, email, etc., and hence is a differentiated services approach. The architecture for QoS provision needs to consider multiple aspects: QoS provision both at the core and the access network, user and session signalling, integration of mobility and AAAC with a user-dependent QoS.

Management in Mobile Networks

Two layers are actually designed for the network service provision. They are:

- the service layer, with specific interoperation mechanisms across different administrative domains;
- the network transport layer, which will have its associated interoperation mechanism between network domains.

An administrative domain may be composed of one or more network domains, but a terminal crossing over the administrative domains also changes its current network service provider. Each domain has an AAAC system and a QoS broker. When a mobile user enters the network, the AAAC will have to authenticate it and on successful authentication, it sends the relevant policy information based on the SLA of the user to the QoS broker, stored in its profile. The user service profile consists of terminal/service pairs, suited for Diffserv QoS signalling.

QoS on the Core and Access Networks

The core network, like the access network, is actually managed per aggregate of network services and not by user services. Packets that need to be transmitted should cross all the domains and the various core network resources are then distributed between the various classes of service according to different traffic levels.

The access network is managed by a specific entity called the QoS broker and their numbers may vary in a domain. The QoS broker basically:

- provides an interface between the user service view and the effective QoS support at the network level;
- manages and monitors the access network resources (both the ARs and radio gateways (RGs));
- monitors the internetwork edges for incoming and outgoing resource reservations/utilisations;
- can perform SAC (service admission control) decisions and does network device configurations as instructed by the network administration entities;
- is responsible for maintaining interdomain communications with neighbour QoS brokers.

User and Session Signalling

There is a DSCP code defined for every service and so every packet being transmitted contains appropriate information required by network entities to account, authenticate, forward and differentiate each packet and each service/user. After registering, the user signals by sending a DSCP code of the service that he/she intends to use. The QoS broker recognises the code and if there are available resources for the requested service then the packets will be forwarded and accounted for; otherwise the AR informs the user/terminal.

For termination, a timeout is associated with every network service. If no packets belonging to that service are sent or received for this timeout period then the network elements will free those resources.

Services

Network Services
These are unidirectional or bidirectional services provided by the Diffserv architecture. The architecture is capable of supporting any type of network service that the network provider is willing to support.

User View
The users subscribe to SLAs in sets of network services like 'inexpensive service', 'exclusive pack', etc. This service pack in terms of network level services is called the 'network view of the user profile' (NVUP), which basically contains the subscribed network-level services and the user identification.

Entities of the QoS Architecture

QoS Entities at the Network Level
There are basically three different entities that interoperate in order to ensure QoS provision at the network level. They are the MT, AR and QoSB.

Mobile Terminal (MT)
The mobile terminal consists of:

- an enhanced IPv6 stack, able to perform marking according to the user-subscribed services;
- a networking control panel able to perform AAA (authentication, Authorisation and accounting) registration/deregistration);
- a mobile terminal networking manager who will take the decision to execute the handover and attach procedures, based on information received from the user and information received from the networking devices (WCDMA, WLAN, Ethernet drivers);
- a radio convergence function that interfaces with the radio layer, in compliance with the IP-level QoS requirements.

Access Router (AR)
The access router consists of:

- an enhanced IPv6 stack with IPSec and DiffServ filtering, implementing basic transport, authentication and security functionalities, added with QoS-related functions (dropping, shaping, scheduling);
- a fast handover (FHO) module, implementing fast handover procedures in an MIPL (mobile IP for Linux) distribution, added with proper QoS-related signalling;
- an AAA attendant, responsible for all communication with the AAAC, acting on behalf of the MT to establish initial user registration and basic security functionalities;

- a QoS attendant, able to configure QoS policies in the stack and interoperate with context transfer mechanisms:
  - for exchanging FHO metering data.

QoS Broker

The QoS broker (QoSB) comprises:

- an engine, responsible for taking all management decisions, and several different interfaces to interoperate with other entities, using COPS (common open policy service);
- an AR interface used both for letting the AR pool the broker for policies related to the services that the attached users are requesting and for configuring the AR with that specific policy;
- an AAAC interface used to exchange messages with the AAAC, to inform the QoSB of every new MT/user that attaches to its domain (including the NVUP information for that user) and to require specific user and service information;
- a QoSB interface used for exchanging information between QoS brokers.

QoS End-to-End Support

The provision of end-to-end QoS gives rise to three different aspects.

Registration

The registration process initiates after the MT acquires the care of address (CoA) via auto configuration. However, this does not entitle the user to use the resources and the MT needs to send the authentication information to the AR in order to use the resources later. This information is then forwarded to the AAAC system and on successful authentication it dumps the NVUP to the QoS broker and informs the MT via the AR.

Authorization

In the next step each network service is authorised. The MT first sends a packet with an appropriate DSCP code to request a particular service. If the requested packet does not match any policy set in the AR, the AR sends a request to the QoSB via the QoS manager. On analysing the request and based on the available NVUP and the available resources, the QoSB sends a confirmation back to the AR. The QoS manager in the AR then configures the appropriate policy for that user/MT service in the AR and after that any packet matching the configured policy will be able to cross the network.

QoS Enabled Handover

Maintaining a constant level of QoS will be the main problem involved in IP mobility scenarios. Fast handover techniques are guaranteed to the user in the context of a transfer between the network elements (ARs – old and new – and QoSBs). When an MT starts to lose the signal to the current AR (called 'old AR', or AR1), it will start a handover procedure to a neighbouring AR (now called 'new AR', or AR2), from which it receives a beacon signal with a network prefix advertisement. The MT builds its CoA and initiates the handover procedure, sending a handover request to its new AR, still through the old AR. The old AR will forward this request to both the new AR (known by the network prefix) and to the QoS broker. The old QoS broker (QoSB1) sends a handover request to the new QoS broker, indicating the user's NVUP and the list of services currently being used. This acts as a context transfer from the old QoS broker to the new QoS broker. With this information, the new QoS broker (QoSB2) will verify the availability of resources and send a message to the new AR indicating whether the MT may or may not perform the handover. If the handover is possible, the new AR will be configured with a copy of the policies that the old AR has for that user. If the handover is not possible, multiple operator policies can be supported, according to network usage and user preferences. Meanwhile, the new AR will forward the answer to the old AR, which will then send it to the MT. Upon a positive answer, the MT may perform the layer 2 handover because all the layer 3 resources are then reserved. During this phase, both ARs are bicasting, to minimise packet loss.

5.7 Challenges and Limitations of 4G Networks

No technology is complete in itself and this is also the case with 4G. Up to now the discussion has been about the 4G technologies and its advantages over the previous generations, but a careful look shows that it also faces some serious limitations and challenges. There are three major limitations/challenges in 4G networks:

- Mobile station
- Wireless network
- Quality of service

5.7.1 Mobile Station

For a large variety of services and wireless networks in 4G systems the multimode user terminals are essential for adapting to the different wireless networks, thereby eliminating the need for separate multiple terminals. The most promising approach is that of the software radio. Unfortunately, the current software radio technology is not completely feasible for all wireless networks due to the following problems:

- It is impossible to have just one antenna and one LNA to serve the wide range of frequency bands (i.e. to cover all the bands of all 4G networks). One solution could be to use multiple analogue parts to work in different frequency bands. This certainly increases the design and complexity and physical size of a terminal.
- The existing analogue-to-digital converters (ADC) used in mobile stations are not fast enough. The GSM requires at least 17-bit resolution with very high sampling rates. To provide such a bit resolution, the speed of the fastest current ADC is still two or three orders of magnitude slower than required.
- In order to allow real-time execution of software-implemented radio interface functions, such as frequency conversion, digital filtering, spreading and de-spreading, parallel DSPs have to be used, thereby increasing the circuit complexity and high power consumption and dissipation.

5.7.2 Wireless Network

To use 4G services, the multimode user terminals should be able to select the target wireless systems. The process of broadcasting messages periodically to mobile stations becomes complicated in 4G heterogeneous systems because of the differences in wireless technologies and access protocols. One of the possible solutions is to use software radio devices that can scan the available networks and, after scanning, will load the required software and configure themselves for the selected network by downloading the required software models by scanning the available wireless networks. The software can be downloaded from media such as a PC server or over the air (OTA). OTA is the most challenging way to achieve a wireless system discovery, but its availability frees users from the medium of downloading.

Wireless Network Selection

With the support of 4G user terminals, it is possible to choose any available wireless network for each particular communication session. The correct network selection can ensure the QoS required by each session. However, it is complicated to select a suitable network for each communication session, since network availability changes from time to time.

Terminal Mobility
There are two main issues in terminal mobility: location management and hand-off management. With location management, the system tracks and locates a mobile terminal for a possible connection. It involves handling all the information about the roaming terminals, such as the original and current located cells, authentication information and QoS capabilities. On the other hand, hand-off management maintains ongoing communications when the terminal roams. The mobile IPv6 is a standardised IP based mobility protocol for IPv6 wireless systems, where each terminal has an IPv6 home address. Whenever the terminal moves outside the local network, the home address becomes invalid and the terminal obtains a new IPv6 address (called the care-of-address) in the visited network. This hand-off process causes an increase in system load, high handover latency and packet losses. It is very difficult to solve these problems in 4G networks. The reason for this is that besides horizontal hand-off, vertical handover is also needed. Moreover, 4G networks are expected to support real-time multimedia services that are highly time-sensitive. It is very hard to calculate the hand-off between different wireless systems as it is very complicated.

Network Infrastructure
Existing wireless systems can be classified into two types: non-IP based and IP based. The non-IP based systems are basically optimised for voice delivery (e.g. GSM, CDMA) and the IP based systems are basically optimised for data services (e.g. 802.11, HIPERLAN). In 4G wireless networks, the problem in integrating these two systems becomes apparent when end-to-end QoS services are concerned.

Security and Privacy
The key concern in security designs for 4G networks is flexibility. Since the existing security systems are basically designed for voice services it becomes very difficult to implement them for heterogeneous environments. Moreover, the key sizes and encryption and decryption algorithms of existing schemes are also fixed. They become inflexible when applied to different technologies and devices.

Fault Tolerance and Survivability
There is presently an inadequate study on the survivability of wireless access networks, though they are more vulnerable than wired networks. A cellular wireless network is typically designed as a tree like topology. A major weakness of this topology is that when one level fails, all the other levels below it are affected. The situation becomes worse when multiple-tree topology networks work together in 4G systems. Their fault-tolerant designs should consider power consumption, user mobility, QoS management, security, system capacity and link error rates of different wireless networks. There are two different ways to achieve fault-tolerant architectures to support QoS in failures. The first is to use a hierarchical cellular network system; the second is to use collaborated or overlapping heterogeneous wireless systems.

Multiple Operators and Billing Systems
Presently, an operator usually charges customers with a simple billing and accounting scheme, where a flat rate based on subscribed services, call durations, and transferred data volume is usually enough. However, with the increase in service varieties in 4G systems, more comprehensive billing and accounting systems are needed. The customers may no longer belong to only one operator, but instead subscribe to many services from a number of service providers at the same time.

Along with this, equalisation on different charging schemes is also needed as a result of different billing schemes (e.g. charging based on data, time or information). It is basically very challenging to formulate one single billing method that covers all the billing schemes involved. Furthermore, 4G networks support multimedia communications, which consist of different media components with possibly different charging units. In order to build a structural billing system for networks, several frameworks have already been studied. The requirements on these frameworks include scalability, flexibility, stability, accuracy and usability.

Personal Mobility

Personal mobility concentrates on the movement of users instead of user terminals, and involves the provision of personal communications and personalised operating environments. A video message addressed to the mobile user, no matter where the user is located or what type of terminal is being used, will be sent to the user correctly. A personalised operating environment, on the other hand, is a service that enables other adaptable service presentations. A user also belongs to a home network that has servers with the updated user profile; when the user moves from his/her home network to a visiting network, his/her agents will migrate to the new network.

5.7.3 Quality of Service

Supporting QoS in 4G networks will be a major challenge due to varying bit rates, channel characteristics, bandwidth allocation, fault-tolerance levels and hand-off support among heterogeneous wireless networks. QoS support can occur at packet, transaction, circuit, user and network levels.

Packet-Level QoS

This applies to jitter, throughput and error rate. Network resources such as buffer space and access protocol are likely influences.

Transaction-Level QoS

This describes both the time it takes to complete a transaction and the packet loss rate. Certain transactions may be time-sensitive while others cannot tolerate any packet loss.

Circuit-Level QoS

This includes call blocking for new as well as existing calls. It depends primarily on a network's ability to establish and maintain the end-to-end circuit.

User-Level QoS

This depends on user mobility and application type. In providing end-to-end QoS support, developers need to do much more work to address end-to-end QoS. They may need to modify many existing QoS schemes, including admission control, dynamic resource reservation and QoS renegotiation to support diverse QoS requirements of 4G users.

Hand-off delay poses another important QoS-related issue in wireless networks. The delay can be problematic in internetwork hand-offs because of authentication procedures that require message exchange, multiple-database accesses and negotiation–renegotiation due to a significant difference between needed and available QoS. During the hand-off process, the user may experience a significant drop in QoS, which will affect the performance of both upper layer protocols and applications. Deploying a priority based algorithm and using location-aware adaptive applications can reduce both the hand-off delay and QoS variability.

Appendix A

Roll-Out Network Project Management

Joydeep Hazra

A.1 Project Execution

Network planning is one of the core foundations for any cellular project. Its intricacy lies in the fact that without it there cannot be any further logical execution or implementation of the project. It is either done by the network operator or purchased as a service from the equipment vendors, often supervised by a core team of experts. The remaining planning work and field teams are outsourced to subcontractors. It is priced and offered to the operators in three different sales packages:

- Radio network planning
- Transmission planning
- Core network planning

When the project begins, the planning activities are split into further stages:

- The preoptimisation service is offered to operators who carry out nominal and detailed planning. They ensure the performance of the network in order to attain the guaranteed key performance indicators, or KPIs.
- Indoor network planning is a service provided to operators who want to extend their dedicated capacity and coverage to strategic indoor areas.
- Network planning consultancy services are offered to next generation networks deploying WCDMA and EDGE technologies, requiring a high level of expertise from the vendor's side.
- A swap-planning service is offered to operators who are replacing their existing RAN or/and core network supplied by a previous vendor with the current vendor. This service offers parameterisation of the swapped sites according to the new equipment placed.

Once the initial network plan is made then, based on its grid structure, the site selection team starts to look for sites, that are to be acquired.
The next/other stages in a roll-out project will now be discussed.

Advanced Cellular Network Planning and Optimisation Edited by Ajay R Mishra
© 2007 John Wiley & Sons, Ltd

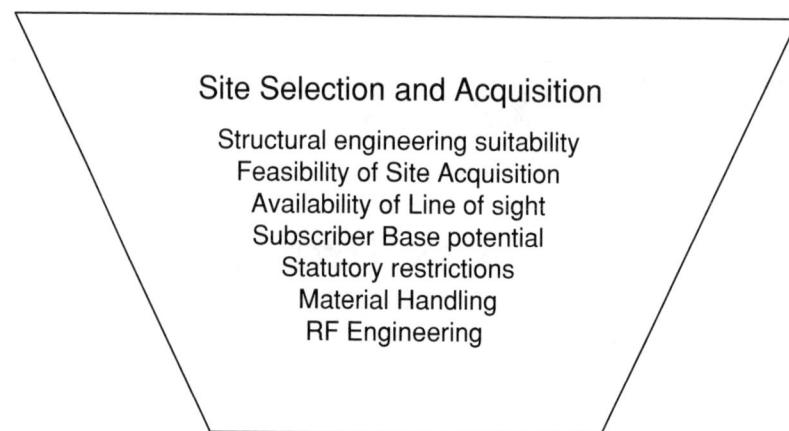

Figure A.1 Site selection criteria

A.2 Network Implementation

Network implementation is one of the most laborious parts of any network deal as it involves joint action from all the teams of the vendor as well as the operator. Aiming to achieve total customer satisfaction, it tests the effectiveness of the project teams in balancing the three main elements of time, cost and quality. Site selection, and acquisition is done after the initial planning.

A.2.1 Site Selection and Acquisition

The success of a cellular network is dependent on many factors, but one of the principal factors is the right site selection. Numerous valid arguments decide on the crucial issues of the entire cellular business, ranging from rental figures, civil engineering, RF optimisation, marketing, profitability and market share. The site selection process relies on basic experience, relating to good engineering as a result of learning from past mistakes and plain luck. Local culture, traditions and bylaws also play an important role in site selection, due to which the whole process may vary from place to place. The site selection process lays the foundation pillars of an entire network. Mediocre site selection leads to all kinds of troubles during the network launch and even more during the expansion phases. Collapse of even one site can lead to an ugly hole in a beautifully designed network. Site selection is based on some main criteria, which are classified in Figure A.1.

Subscriber Base Potential

The whole cellular business is based on subscriber and revenue generation. An adequate survey is done to identify the potential subscriber base and their needs, habits and lifestyles, which play a major role in generating average revenue per user (ARPU). Market survey departments provide inputs like the number of potential subscribers, density of users, hot spots, hubs, traffic flow, major junctions, commercial centres and offices, residential concentrations, busy roads, etc. The operator's marketing section always prefer the best coverage of all dwelling areas, including in buildings and on road coverage. Financial teams cannot permit such a generous layout of equipment. Hence the operator needs to come up with an optimal plan, deciding on the number of sites per city or town based on the subscriber base, coverage requirement and expected revenue. During the optimisation stages, they have to allocate resources in terms of finances and equipment to support in-building coverage in underground areas such as Metro stations, tunnels, subways and other basement areas in commercial areas with adequate levels of human density. Figure A.2 gives

Network Implementation

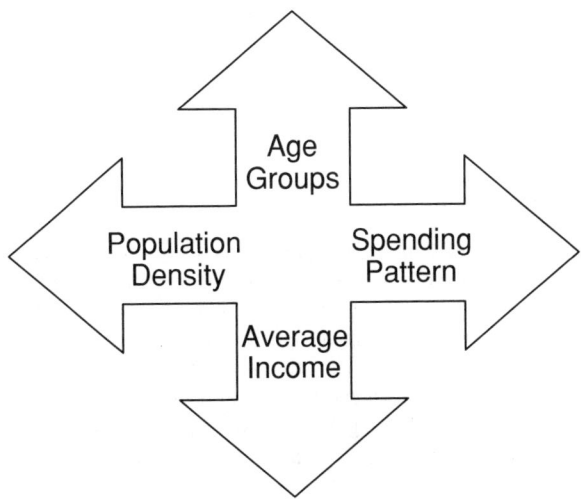

Figure A.2 Factors defining a demographic area

a brief idea of some factors that are chosen when selecting a demographic area for launching a radio network.

RF Engineering

An ideal philosophy would be to provide good coverage everywhere. That will not justify the cost of sites in relation to the subscriber base or future revenue forecasts. Signal level requirements are based on a preference order. The RF engineer designs a network through the process of site selection, satisfying both the need for good coverage and a subscriber base in the area. The initial work is done by software on digitised maps resembling actual clutter distribution. Thereafter the link budget, fading margin and penetration losses are calculated, as they are different for areas like urban high density, urban low density, semi-urban and rural areas. These determine the hexagonal (Hex) radius of the different classified areas, which finally appear on the plot obtained from the grid structure. The RF engineer goes to subscriber areas physically to choose the sites as close as possible to the grid centres. In case of practical problems, a suitable distance is taken from the grid centre, with future expansion and easy upgradability considered in subsequent phases. The operator intends to cover a maximum area under blanket coverage and provide good coverage to prime areas in order to draw subscribers. The RF engineer seeks mostly tall buildings, providing a bigger footprint with a lesser number of sites. While tall buildings provide blanket coverage in the initial stages, too high buildings can lead to interference problems. Tall buildings also create a 'shadow region' due to limitations of vertical beam width. Highways are an important source of revenue due to regular and heavy traffic. They are usually associated with high towers, providing maximum coverage without substantial penetration loss. Their heights are chosen while keeping in mind the future backbone networks spanning between cities. Finally, the surrounding obstructions are to be noted. Tall constructions like skyscrapers, towers or a dense forest cover with wide leaves should not be in the near vicinity as they might cause possible obstruction of the signal. Water bodies, solar panels and wide metallic sheets in close proximity can lead to signal losses due to reflection.

Line-of-Sight (LOS)

All sites are interlinked by microwave links, which require the placement of microwave dishes on towers, with subsequent line of sight and adjacent site(s) depending on the network architecture if the sites are daisy-chained or linked in a star configuration. If a site is totally isolated from other sites then it is usually avoided, unless the site is mandatory from an RF point of view or some hill obstruction prevents any line of sight. In such cases repeaters are used for MW connectivity. More information on the LOS is given in Chapter 3.

Structural Engineering Suitability

The tallest buildings are short-listed during site surveys, effectively reducing the height of the tower and proving more cost-effective. Another issue is that of building strength – whether it can withstand the gross weight of the equipment, room, tower, antennas and all the ancillaries and whether it has a carpet area of about 10 m^2 for setting up an equipment room and has space for the tower and running of the RF and power cables. Lastly, the building construction is checked to see that it is strong enough not to crumble under the weight of the equipment and tower load if there is a low level earthquake or gusty weather. Additionally, the overall wall loading of the building as well as the adjacent attached buildings are checked in order to ensure that the neighbouring buildings are not heavily dependent on the wall strength of the chosen building. This can cause a long-term damaging effect on the building once the equipment loads also add to the total stress on the walls. Finally, it is a trade-off between height and structural strength.

Feasibility of Site Acquisition and Material Handling

After all the technical queries comes the basic question: 'Is the site feasible?' Archaeological buildings, institutional buildings and hospitals should be avoided. Private buildings are preferred as the owner can be taken into confidence and cordial relations can be developed for smooth continuation. Permission is taken from the residing community as they might have restrictions on erecting towers and antennas on their buildings. A full and fair description of the work to be done, weight and size of equipment and ancillaries is provided to the owners. They are assured of the safety of the building to obtain permission for making minor drilling and chipping of the roof reinforcements. Site access is arranged for long working hours while the site is implemented and also at night in the case of any emergency or for routine maintenance. The site is secured from unauthorised entry or attempts of vandalism. The staircase should be wide enough to turn the equipment while lifting it to the top. In the absence of a clear space for lifting the equipment, it is hoisted up the side of the building using a chain and pulley arrangement, which requires a clear space at the side of the building. There should not be any obstructions from electrical wires or junctions and proximity to glass walls. Lastly, the building owner should be able to arrange for a commercial power supply. Below are some of the basic questions and factors to be considered about the feasibility of the site before it is acquired.

Statutory Restrictions

Local authorities, municipal or central governments, city council and development boards restrict the maximum height permissible of buildings, including roof-mounted constructions like towers and antennas. In many countries simple reasons such as the conservation of building uniformity or preservation of the beauty of buildings can be the sole reason why that building cannot be provided to the operator for commercial use. Defence organisations, airport authorities, TV and radio broadcast agencies can raise objections against putting up towers and antennas close to their installations. Sites in the vicinity of hospitals and diagnostic centres are avoided as they may cause interference with the sensitive medical equipment. Where there is a lack of choices available permission should be taken from such entities to avoid any hassles later on. Legal disputes or charges of illegal constructions should be checked before finalising the site. Antenna sites close to windows or bedroom walls are normally rejected by the owners themselves before any resident raises any objection.

Installation of Equipment Room, Tower and Ancillaries

Property prices are high in urban areas due to the abundance of commercial activity and overloaded or overbooked public street properties like pillars, balconies, hoardings on rooftops, main streets, corners, etc. Legislation concerned with the use of public places and property is also stricter compared to semi-urban or rural areas. There are no global solutions for any single type of site solutions, but various alternatives are used and customised locally. There are basically three types of installations:

- Open
- Indoor
- Outdoor

Open Installations
A simple outdoor unit (ODU) or a wall-mounted radio unit combined with a basic pole site provides one of the simplest solutions. It benefits from short cable lengths as the radio unit is mounted on a wall close to the pole, from where the antenna is radiating. Depending on the coverage needs, directional antennas are placed on the wall below the average rooftop level and oriented towards the street lengths or the next major building. In open outdoor places with a high density of subscribers, like market squares, installation of pole sites is preferred. A strong pole is chosen which houses the radio unit and a low power omni antenna can be installed at a suitable height to cater for a large number of people in the vicinity of that site. The same solution with directional antennas can be used for places like the entrances of shopping malls or narrow streets in city centres. These are commonly used to provide microcellular solutions.

Indoor Sites
The basic indoor site is the most common model deployed in any kind of cellular network. It has the same philosophy of renting a room for long-term occupancy, the only difference being that the tenant is the operator and the occupants are the network elements irrespective of what technology is being used. These sites are useful when there is a network expansion, as the same site can house more than one base station. A well-designed equipment room ultimately decides on the running life of the installed equipment. A normal equipment room designed for a BTS site provides a carpet area of roughly 10 m^2 (3.3 m × 3.3 m). Equipment rooms or shelters are of two major kinds, prefabricated or constructed. Depending on the situation, the operator chooses between erecting a prefabricated shelter and constructing a room. These models are used mainly to provide macrocellular solutions. In a basic indoor site the BTS and transmission equipment along with ancillaries are housed indoors and the cellular antenna is mounted outdoors. In such cases the tower can be mounted on the building rooftop or on the ground if it is a low building. Prefabricated equipment shelters are made of modular panels of galvanised steel sheets having an insulated separation in between. These are easy to transport and provide easy construction of equipment rooms in areas where it is not cost-effective to construct equipment rooms. These also allow quick construction of equipment sites for faster roll-out and can be dismantled if the site needs to be closed down or shifted to another building. Special indoor sites are installed in the same way bank teller machines are installed. The equipment modules are sunk inside the building and are accessible from the outside. The cables, transmission links and antenna are mounted outside on poles or a tower adjacent to the site.

Outdoor Sites
These are typical examples of microcellular solutions found in cities and big towns, where existing municipal or government property on the streets, like light poles, bus stops, free-standing structures,

power lines, etc., are used to hide the BTS and the antennas. They are integrated together depending on the regulations governing the usage of such property.

For a pole site, a BTS is mounted inside the base of the foot of an electric pole, the cables run inside the body of the pole and the antennas are attached to the top of the pole. The local city or municipal authorities generally approve these adaptations, since they do not make any major cosmetic change to the existing structure and utilise the existing framework in providing cellular services to the normal public. If there is not sufficient strength, the existing light pole needs to be modified, reinforced or replaced with a stronger frame of the same size and shape.

For free-standing outdoor structures, a BTS can be totally concealed inside a billboard, bus stop or an outdoor advertisement. Many such public installations can be used in a similar fashion, where even the antenna can be mounted inside the same structure and coloured to match the existing structure. In other cases huge commercial signs on frames, located in prime locations like business districts, highway entrance/exit, major crossings, etc., provide adequate space for hiding base stations.

A.2.2 Provision of Site Support Elements

Normal equipment sites consist of the following constituents:

- BTS equipment
- Transmission equipment
- DC power supply and power distribution board including circuit breakers or fuses
- Battery backup
- Tower or pole with required cable trays
- Antenna with feeder cable
- Heating, ventilation and air conditioning (HVAC) unit
- Accessories: fire and smoke alarm, fire extinguisher, door alarm, lighting
- Generator backup (for remote, critical or rural locations)

Site support elements comprise battery backup units, HVAC units, cable trays and also diesel generators. During any maintenance procedure battery backup takes care of the power outage, before it is restored or a diesel generator arranges a backup power. In many rural areas it is often difficult to guarantee 24 hours of electric power supply, hence a diesel generator becomes an integral part of the site. The equipment is housed in a dust-free sealed environment and generates heat during its normal operation. If the room temperature rises above the specified limit then the equipment may fail to operate normally or lead to a burnout. Beyond a certain temperature threshold, the equipment guarantee may also cease to exist, so every site is equipped with an operational and a redundant air conditioning unit. The cable trays guide the antenna feeders and other interconnecting cables. Cable trays are either hung from the ceiling of the equipment room or are mounted on thin poles of rods when they are placed outside, bringing the cables from the tower to the equipment room. All sites are fitted with mandatory fire and smoke alarms. These alarms are fed straight to the external alarm system, which transmits them to the network management centre. They are carried over the transmission or IP backbone. For security reasons, in every site there is a 'door alarm' connected to the external alarm system, notifying any kind of illegal entry into the sites.

A.2.3 Site Planning and Equipment Installation

A site survey is done to choose a technically viable solution for the whole team. The initial BTS configuration with required orientation, height and position of the antennas is given by the RF engineer, along with an estimate for future capacity expansions. The transmission engineer provides information regarding the number, type, position, height and orientation of the microwave dishes together with an indication of whether additional dishes will be required, when they might be required and their approximate orientation. The site engineering team works out the tower height and effective antenna loading with all safety factors involved. The site planners make a site layout, taking into consideration the suggestions of the RF planners, transmission planners, site engineers, etc., with all the equipment including future expansions and then an estimated bill of materials will be prepared. This technical site survey report is checked by the network and transmission planning teams and if accepted continues for further approval from the other site engineering teams, after which the legal teams can proceed with site acquisition. Before actual work begins a site design document is prepared for that particular site, which contains:

- Installation plan
- General arrangement drawings
- Detailed construction design
- HVAC and electrical design
- Bill of quantities

The installation plan defines the site configuration, installation instructions for telecom implementation, site layout for general arrangement drawings and requirements for construction design. This forms the basis of the bill of materials prepared and equipment is ordered from that for that particular site.

General arrangement drawings include a local plan drawing describing legal government maps depicting the building and the surroundings with reference to the true north direction. A rooftop site layout drawing describes the roof plan for building sites showing true north, antenna pole positions, indoor/outdoor/shelter locations, cable tray layout, lightning protection and equipment ground routing, roof paving stone layout and an approximation of an outdoor/shelter base frame system. Elevation Drawings depict the overall height of the building/tower as well as profile details.

The detailed construction drawing contains drawings from foundations (concrete, reinforcement), framework (steel, concrete), complementary structures and holes. Statistical calculations, work instructions and reports are added to the detailed design as documents. The HVAC and electrical design describes heating, piping, ventilation and air conditioning planning, electrical, grounding and lightning protection with drawings and documents. Once the site is made ready by the site engineering team the ordered equipment is delivered to the site and the installation teams are called on to start their work as given in the site plans and diagrams. The site is effectively grounded and all static electricity is eliminated before the equipment is made ready for permanent installation. All equipment is fastened tightly to the floor of the equipment room with the cables laid in a uniform fashion using an adequate colour-coding or numbering scheme. During the process of equipment installation, electrical connections are made between the mains power supply to the respective circuit breakers, leading to their nominated terminations in the power supply equipment and in turn to the batteries, finally terminating at the telecom equipment.

A.2.4 Legal Formalities and Permissions

Legal formalities are taken care of by the legal departments of the vendor and the operator themselves, though some cases are done through a local legal agent, licensed architect or the site owner. Local regulations are studied and analysed during the contractual stages to avoid bottlenecks during project implementation. The first permission comes from the land or building owner, who grants the required rights for making a new construction or modifying any existing construction as per the site requirements. All statutory requirements of health and safety hazards are taken care of as part of the written contract signed by the local authorities, before the site is launched. If there is a hospital nearby, adequate provisions have to be made so that they do not complain of any physical or psychological problems faced by the residents or patients due to their proximity to the antennas. Land and municipal authorities should give

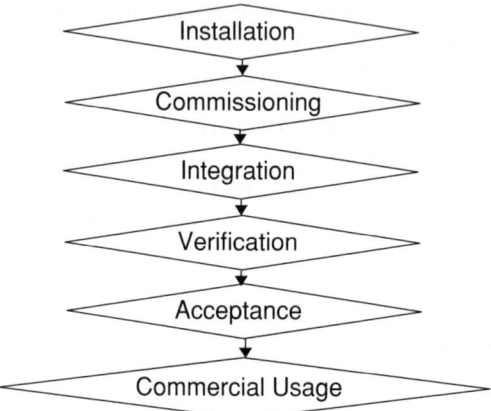

Figure A.3 Steps from site installation to site launch

clearance for running a commercial operation in a residential area and that its presence does not affect the local residents. Subsidiary permits are obtained from sanitation, fire, water, electricity and environmental authorities, depending on the country and district regulations. Aviation authorities need to give clearance that the tower height well not affect their flight paths and also that the frequency radiation will not affect their communication channels. Finally, local municipal authorities have to give clearance before any site construction can be started.

A.2.5 Statutory and Safety Requirements

The site has to follow all safety rules and regulations of the land. In case any mishap or accident occurs at the site, a first aid box should be clearly visible and readily available. There should be a fire extinguisher duly certified and checked for its operational period. The site should be equipped with fire and smoke detectors with sensors activating the external alarm mechanism. Extra harnesses should be available for the maintenance engineers who take care of the towers and fitted antennas. A battery-operated emergency light should be available in case of any power outages. Exits should be clearly marked in case of an emergency clearance procedure. The mains distribution board and the circuit breakers should be well within reach of a normal person and within sight, so that they can be turned off or reset in case any kind of electrical problem occurs. There should be antistatic mats and wristbands available on site for maintenance engineers to perform operations on the live hardware without risking any static electricity discharges. The towers should have lightning arrestors. The site, including all electrical equipment, should have a common functional grounding system. For maintenance purposes there should be a microwave communication phone available on the site; for instance, if somebody is locked in or if the cell site fails and no coverage is available then the alternate system can be used to relay a distress call to the main operations centre.

A.3 Network Commissioning and Integration

The equipment has been installed and stationed at the site but is not ready to be put into commercial use until it has undergone the steps shown in Figure A.3.

Network Commissioning and Integration

✓	Punch list
✓	Tool Availability
✓	Documentation
✓	Installation Material Delivery
✓	Rack Installation
✓	Grounding
✓	DC Power
✓	Internal Cabling
✓	External Cabling
✓	Plug in Unit Installation
✓	Alarm Panel Installation
✓	DDF Installation
✓	Construction Works Miscellaneous

Figure A.4 Categories in a checklist

A.3.1 Confirming the Checklist

The commissioning engineers take over the equipment from the installation team, after conforming to a preset checklist, which contains all data related to physical positioning and layout of the equipment as per the planned layout and also the distribution of plug-in units in the racks as per network dimensioning. Checklists are made in various ways, e.g. order of work, degree of completion of work, process followed, signing off to the next level, etc. The project manager ensures uniformity in these checklists in order to maintain effective documentation that can be referred to by site engineers and maintenance personnel. These checklists also contain minor tests done on individual items or parts, before the whole unit is assembled together. Then the cable lists and layouts are checked with the layout diagrams and colour-coded. The installation teams make sure that these minor but important points are taken care of and mentioned in the checklist, which forms the handover document between the installation and commissioning teams. The list in Figure A.4 contains some categories mentioned in a checklist.

A.3.2 Powering-Up and System Precheck

Powering-up the site equipment is a very crucial step as a simple mistake can lead to burnout of the whole equipment. The installation team connects the power supply modules to the battery backup but leaves the power termination to the equipment main power input in a switched-off condition. The commissioning engineer's prime responsibility is to check the incoming voltage and equipment earthing before proceeding further. After powering-up, all racks and subracks are rechecked for the correct voltage before any plug-in units are plugged in and then the system loading is checked.

A.3.3 Commissioning

The complexity of commissioning can depend on the type and range of equipment; a BTS can be commissioned in a few hours whereas an MSC may take days. It involves a variety of activities performed in a sequential or staggered manner as long as the whole process has its required continuity. It begins with checking all plug-in units and their interchangeability codes and versions. All cables and terminations are rechecked in accordance with the site-specific documentation. All bootprom chips, labels and then

> **Commissioning Checklist**
> Check jumper settings and EPROM versions
> Run Diagnostic tests
> External alarm test
> Customer alarm test
> Clock alarms
> Total power shutdown
> Power Supply Unit fuse test
> Power break-up in Power Distribution Unit
> Wired alarm test
> Message bus test
> Debugger test on all units
> Synchronisation test
> Set printer connection
> Load node software and hardware data
> Load customer specific parameters
> Change delivery installed
> Safe copy on hard disk & Magnetic drive

Figure A.5 Commissioning checklist for a BSC unit

jumpers (including EPROM, or erasable programmable read-only memory, versions) are checked for their correct places. Thereafter, as the equipment is powered-up, all startups are monitored. In cases of older software the commissioning engineer has to load the latest or certified software into those units. The units are brought up in their respective order of software loading process and the faulty units can be spotted by checking for faulty light emitting diode (LED) colours for replacement. Figure A.5 shows a commonly followed checklist for a BSC unit.

A.3.4 Inspection and Alarm Testing

The plug-in units are run through diagnostic tests to ascertain their individual hardware integrity and unit stability. After the units are tested, a complete power-off test is done in order to check the uniformity of the equipment parts, working in unison. Redundant units and switchovers are checked by manually separating the working units from their normal state of operation. Minor fault conditions are simulated in order to check the generation of basic alarms. When there are many subelements of the equipment, individual units, their port connection and interconnectivity are also checked. Figure A.6 shows a checklist of some common alarms.

A.3.5 Parameter Finalisation

All data and parameters are consolidated before the integration process starts. Network planners prepare the data sheets from the inputs they receive from the various vendor and operator teams. The integration process cannot be started without them, even though the site is ready and the equipment is working in a stand-alone mode. These parameters can vary depending on the data requirements; e.g. for open interfaces the data headings will be same irrespective of the equipment vendor or in other cases it can be vendor specific. Typical data can be related to information exchanged between two network elements interfaces, e.g. the number of links, size of transmission packets or pipes, IP addresses, protocols followed, naming

> **Common Alarm Inspection**
> Door Open alarm
> Smoke alarm
> Fire Alarm
> HVAC failure alarm
> Power Outage alarm
> Battery low alarm
> Unit failure alarm
> Unit State Change alarm
> Transmission link failure alarm

Figure A.6 Alarm list

conventions, delays, limits, error rates, user data content type and volumes, etc. Figure A.7 shows an example of a parameter sheet designed by core network engineers, providing data that are used for the creation of an Iu interface integrating a radio network controller with a media gateway unit.

A.3.6 Tools and Macros

After finalisation the parameter sheets can be fed into the network integration tools or composed into macros, which are then run on to the network elements either manually or downloaded via the centralised monitoring systems. Some common software programs used for running macros or individual commands are Reflection, HyperTerminal, HIT (holistic integration tool), Unix or Linux shells, ProComm or even DOS (disk operating system) windows. Other programs can have graphical user interfaces where data can be just entered and fed into the equipment via direct or remote connections. Normal laptops or workstations are capable of running these tools and macros. The DOS window is also used to connect to the network element by a simple Telnet connection.

A.3.7 Integration of Elements

Once the elements are 'up and running', they are then connected by transmission links. The field engineers are then sent out to individual equipment sites for the first stage of integration. Thereafter it may be possible to run future upgrades remotely. It is mandatory for the field engineer or network element engineers to provide adequate backups of the equipment software before they proceed with the integration process. All interfaces need to have relative consistency of data during the integration process, so that the interfaced elements do not have any lack of data compliance, which can lead to underrated performances. The interconnections between different network elements have already been discussed in earlier chapters.

In this highly competitive product business market, network elements are simply referred to as boxes. The experienced operators create their own combination of boxes by choosing multiple vendors as equipment box suppliers for their individual networks. As openness is promoted in the interfaces, current networks can have various permutations and combinations of their constituent boxes, e.g.:

- packet core network, supplier A;
- circuit core network, supplier B;
- radio access network, supplier C;
- billing solution, supplier D.

	A	B	C
4	Put_the_set_name_there	TOWARDS MGW1	TOWARDS MGW1
5	Own_Signalling_Network	NA0	NA0
6	Own_Signalling_Point_Code	1	A
7	Own_Signalling_Point_Name	RNC1	RNC5
8	Own_Signalling_Point Handling		
9	Own_SS7_Standard		
10	Own_Number_of_SPC_Subfields		
11	Own_SPC_Subfield_Lengths		
12	Signalling_Link_Number	1	1
13	External_Interface_ID_Number	2	2
14	External_VPI_VCI	1-35	1-35
15	Unit_Type	ICSU	ICSU
16	Unit_Number	0	0
17	Parameter_Set_Number	5	5
18	Link_Set_Signalling_Network	NA0	NA0
19	Link_Set_Signalling_Point_Code	100	12C
20	Signalling_Link_Set_Name	MGW1	MGW1
21	Signalling_Link_Number	1	1
22	Signalling_Link_Code	0	0
23	Signalling_Link_Priority	0	0
24	Route_Set_Signalling_Network	NA0	NA0
25	Route_Set_Signalling_Point_Code	100	12C
26	Route_Set_Signalling_Point_Name	MGW1	MGW1
27	Route_Parameter_Set_Number	0	0
28	Route_Set_Load_Sharing_Status		
29	Route_Set_Restriction_Status	N	N
30	Route_Set_Signalling_Transfer_Point_Code_Network		
31	Route_Set_Signalling_Transfer_Point_Code		
32	Route_Set_Signalling_Transfer_Point_Name		
33	RNC_ID	1	5
34	Mobile_Country_Code	240	208
35	Mobile_Network_Code	8	55
36			
37	AAL2_Service_Endpoint_Address	46708	336
38	MSC_Signalling_Point_Code	1009	C8
39	LAC	51	1000
40	List_of_RNCs	1	1
41	Default_Service_Area_Code	1	1000
42	Route_Number	1	1
43	Route_Type	ATM	ATM
44	Interface_Type	IU	IU
45	AAL2_Node_Identifier	MGW1	MGW1
46	EP_Group_Index	1	1
47	Ingress_Service_Category	C	C
48	Egress_Service_Category	C	C
49	EndPoint_Type	VC	VC
50	AAL2_Path_ID	1	1
51	TPI_Interface_ID	2	2
52	TPI_VPI	1	1
53	TPI_VCI	36	36
54	Ownership	LOCAL	LOCAL
55	Loss_Ratio	3	3
56	Mux_Delay	100	100

Figure A.7 Example of a core network parameter sheet

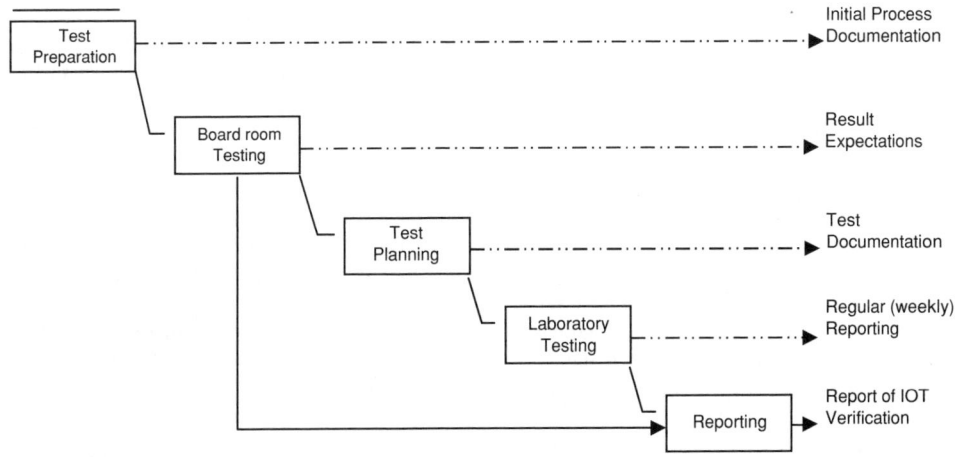

Figure A.8 Different stages of IOT

In such cases the scope of technical management is enlarged as it includes multivendor integration (MVI). Such a diversified and large-scale integration process requires its own planning, preparation, testing and reporting schedules.

A.3.8 System Verification and Feature Testing

The integration process is followed by verification. This ensures end-to-end continuity by rooting out bottlenecks and obstacles that might be present due to planning or implementation flaws. It involves all kinds of testing of all the network elements and their individual as well as combined performances, including monitoring of messages across the interfaces. Feature testing verifies that the complied features actually work in their true fashion and also that the total system performance is stable. Feature testing can involve software level testing or network element level testing or system level testing, depending on the test cases. This can be very time-consuming and laborious, generally done at the customer's pilot site. Feature testing is dependent on software and hardware levels and with every new release it has to be done again, as continuity of basic existing features need to be verified again and new features also need to be verified.

As discussed earlier, multiple vendors can also supply the network, which leads to a different field of activity known as interoperability testing (IOT) and involves verification of an interface between network elements from different manufacturers, on a functional level, in accordance with the relevant standards.

IOT is of two types:

- Network IOT
- Terminal IOT

The chart in Figure A.8 shows different stages of network interoperability testing, which are briefly explained below:

- *Test preparation.* Once the need for IOT has been identified between selected vendors, the involved vendors agree on a project plan to cover the testing. A test preparation document is prepared that contains details of the software loads to be tested, the testing schedule and any other project-specific information required by them.

- *Boardroom testing.* All suppliers exchange their interface documents with reference to the recognised standards for the interface under test. This analysis leads to a frame document charting down the expected results, which are used as a reference for test case selection. It can also be concluded that no actual testing is needed and leads straight to the reporting phase.
- *Test planning.* This phase of the IOT process is where all the participating equipment suppliers agree on the tests to be performed. Once all the tests have been agreed on, a bilateral test plan is produced, which forms the basis for laboratory testing.
- *Laboratory testing.* This is the phase where actual testing takes place. Laboratory testing can start once all the entry criteria have been met, as defined in the test plan. All tests defined in the test plan are performed and the results are documented. The testing will end only when the test exit criteria have been met, which is also defined in the test plan.
- *Reporting.* A bilateral test report is produced on completion of the testing and approved by all the participating equipment vendors. If there are outstanding issues remaining from laboratory testing, then the participants agree on an action plan in order to address these.

Terminal interoperability testing is divided into two categories, i.e. radio interface compatibility and terminal interoperability.

Radio Interface Compatibility

This testing is done to verify and maintain the compatibility between new and existing SW releases of the network elements with existing commercial mobiles. The compatibility between new network element SW releases is verified by testing with a set of mobiles against the existing verified network element SW release. The selection of mobile devices is based on available market share information, mobile feature support and user level competence.

Terminal Interoperability

This testing is done in cooperation with mobile vendors to verify and maintain the compatibility between new and existing network element SW releases with different new mobiles being launched in the market. Terminal interoperability testing is executed separately for different systems like ETSI or ANSI, which are present in different markets.

A.3.9 System Acceptance

The system acceptance procedure is conducted on all sites. The number and type of tests involved are agreed beforehand as this activity is directly related to invoice payments. The vendor's target is to raise an invoice and get the payment released as soon as possible. The operator tries to slow it down in order to hold their finances for as long as possible. The acceptance testing procedure can be a bulky document including all details regarding prerequisites, software level, hardware level, environment setup, testing results, signalling charts and diagrams. The vendors specify these tests in order to prove their equipment capability, depending on what kind of system compliance they provide to the operators' RFP (request for proposal) and system requirements. Meanwhile, the operators also create their own tests, based on what kind of system performances they expect when they order or buy the equipment. It is always a trade-off between the two, at the end of which all the acceptance criteria are defined and documented. These acceptance criterias define when the operator accepts the respective acceptance test result and the test is considered to be 'successfully completed' from the vendor's side.

Figure A.9 describes the various stages of the acceptance procedure before the final acceptance is achieved. The upper part of Figure A.9 represents the startup phase of a network and the lower side

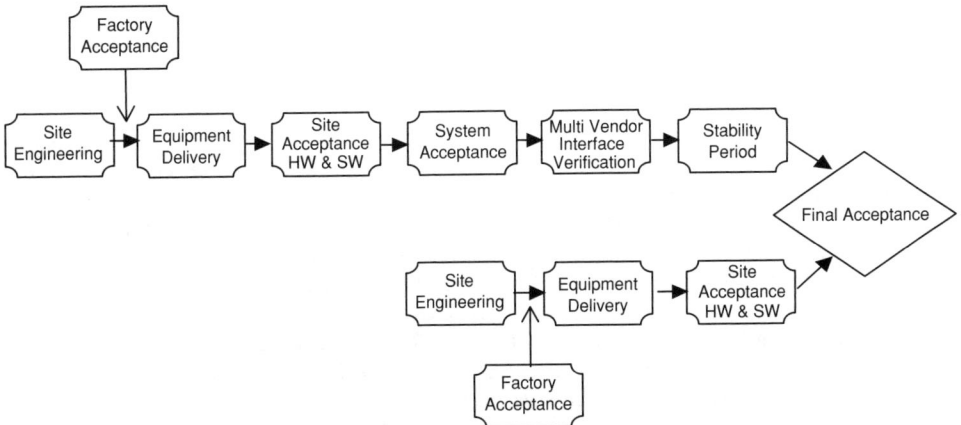

Figure A.9 Different stages of the acceptance procedure

indicates the continuation phases. Both phases contain the primary activity of site engineering or site preparation work of civil works, utility supplies and supporting infrastructure buildup. The equipment is always tested at the vendor's factory or assembly station before being dispatched for delivery to the sites. Receiving equipment is also an integral part of the acceptance procedure as it involves a sufficient inventory process, where the site in charge tallies all the delivered parts with the supplied packing list and notes down the serial numbers in accordance with the received equipment breakup. The equipment is unpacked and assembled together, installed and commissioned as per the site plans. On completion of the job the site is inspected and all hardware and software is checked before the site is deemed as accepted. Thereafter the system acceptance procedure is done for the vendor's own equipment performance, followed by multivendor interface verification with other vendors.

The whole system is kept under observation for a mutually agreed period, referred to as the 'stability period'. The stability of the system is monitored, alarms are monitored and plug-in unit state changes and resets are noted down to gauge the reliability of the whole system. Any kind of system or service outage or major failures lead to delay in the final acceptance, which remains pending till they are corrected, verified and monitored, again under the stability period. After successful performance with a minimum amount of alarms and failures and relative stability, the network operator grants the final acceptance.

A.4 Care Phase

A.4.1 Care Agreement

Once the network roll-out is completed, with network elements fully commissioned, integrated, operational and accepted by the operator or taken into commercial use, then the Care phase starts. A mature network operator can decide to do many operations on their own and restrict the vendor to supply only hardware replacements and guarantee.

A normal care agreement consists of four categories of services:

- Care management
- Hardware services (repairs and swap)

- Software maintenance services (software releases, fault management, change notes)
- Technical support service, including help desk, emergency support services and online services

Terms and Conditions

The care agreement sets out terms and conditions under which the vendor performs all or part of the care services in respect of the equipment supplied by them. An overview of a care agreement gives a description of terms and conditions under which the operator has agreed to purchase the after-sales support services. In case the operator defaults a payment by more than a set amount of days, e.g. 15 days, then the vendor has the right to issue a written notice to the Operator attention that such action can lead to cancellation of the care agreement. Conversely, if the operator has raised a high priority list of problems to be solved, which is directly affecting daily operations and customers, and the vendor is unable to provide a remedial action within a preset period of time, e.g. 30 days, then the operator can issue a written notice to the vendor, attention to the fact that such default action or lack of required support can lead to penalty clauses or liquidity damages or even cancellation of contract. However, the operator has the customer's privilege of cancelling this contract if the vendor defaults on their performance or fails to abide by the agreed terms and conditions.

Fees and Payment

Normally the operator purchases 'care services' annually in advance and the vendor invoices them quarterly in advance, unless specified otherwise. Other additional, variable fees and expenses related to the care services are invoiced on a monthly basis by the vendor and the operator pays within 30 days of the date of the vendor's invoice. In the event of any delay in payment there are penalty clauses, which apply in terms of interest charges on the amount, from the due day until the amount is credited to the vendor's bank account.

Liabilities

The care agreement limits the vendor's responsibilities to proper and timely provision of the agreed services, using due concern and a professional attitude. Both the vendor and the operator should not blame the other in case any liability arises from any injury or accident to any person or property of some third party, including those of its subcontractors involved in the care activities. The vendor is considered as the main supplier of equipment and services; hence it has to provide evidence of having insurance coverage of its own employees plus those of any third party that works on the vendor's behalf. The operator has its own liability to provide adequate security and prevent unauthorized access to the equipment sites and also to install proper warning signs outside such sites. The vendor and the operator are not to be held liable for the other's loss in profit, revenue, use or any collateral or eventful damage if it happens in connection with the care agreement.

Auxiliary Provisions

The vendor has the right to use subcontractors and third party suppliers for fulfilling its duties towards performing and providing the care services to the operator, while remaining fully responsible towards the operator for the vendor's obligations under this care agreement. All copyrights, patents, licences, design rights, trade secrets and similar proprietary rights will remain exclusively with the vendor and its licensors. The operator can neither claim on those proprietary rights and licenses, nor perform any illegal action that compromises the rights to the vendor's proprietary design and information. In case a third party is involved by either the operator or the vendor, then prior written consent has to be obtained from the other in order to assign or transfer that share or interest related to the care agreement. English is predominantly used as an official language for framing these agreements and subsequent translations are made, keeping the English version as the master version, which is legally binding in the case of a conflict. When the care agreement is signed, it brings into effect the entire understanding between the operator and the vendor. Signing of this agreement supercedes all the previous bits and pieces of offers, documents, letters, faxes, emails, memos, brochures and other documentation exchanged in the bidding period, unless some particular document is preserved as a reference or an attachment to the main agreement. A typical care agreement comes with two inherent parts: the first describes the offer of prices for the services ordered and the second describes the care services.

A.4.2 Care Services

Care services are the different kinds of services provided under the care agreement by any specific service provider to the network operator. It starts with care management itself, which is a joint effort from both sides, since both the operator and the vendor have their respective responsibilities with respect to the care agreement. The other services encompass the various kinds of support required to run the network on a planned or unplanned basis, such that there is no network outage or break in the end user level of service.

Care Management

This service refers to the care management provided by the vendor concerning the equipment sold to the operator. The vendor appoints a dedicated 'care' management team who are the operator's key contacts, to handle the contractual obligations. The whole network is logically broken into geographical or demographical partitions in order to manage all issues in a detailed way and have more control if multiple problems occur. One national level care manager is committed to monitor and provide an overall view, while regional care managers are responsible for handling all the care-related issues within their regions. The operator is also responsible for participating in the planning and scheduling of the care activities concerning the services that the vendor is expected to provide. The operator nominates its staff and management for participating in the care-related meetings and acts as an interface towards the vendor for providing relevant details of plans and activities scheduled by the operator and for reporting problems. The vendor initiates activation of required services agreed with the operator as a part of the care agreement before and during the care phases. They work in symbiosis to ensure smooth operations during transfer of the whole system or part of it into the operational phase, as well as when accepting the delivered system. The vendor organises a controlled transition of responsibilities within its teams, when the network goes from the implementation to the care stage. A master care plan is prepared and all related schedules and updates are formulated on a steady basis, with effective coordination between the operator and vendor teams. The care plan is proposed to the care management team of the operator by the care management team of the vendor. Care meetings are organised on a planned basis between different organisational levels of the two sides and reports of each meeting with agreed action points are delivered by the vendor to the operator regarding the delivery of care services, as well as various achievements and difficulties faced during the reporting periods.

Hardware Support

All hardware units come with a standard warranty period of 12 months from the date of shipment, unless some other arrangement is made. An extended hardware warranty service provides an additional warranty period of six months from the date of the return shipment or to the end of the original warranty period, whichever is longer. The vendor honours its obligation to repair or replace the defective unit within the agreed turn-around time. The vendor compiles a database of information, consisting of the serial numbers, service level and type of hardware units sold to the operator, in order to track the warranty status of all units shipped for repair or replacement. The vendor provides a repair report after a set amount of time, incorporating the type of faults found, level of service used, type and number of units found faulty for repair or replacement and a record of effective turn-around periods.

The operator is responsible for arranging, at its own cost and risk, the shipment of the defective unit to the vendor's repair centre, whereas the vendor is responsible for arranging, at its own cost and risk, the return shipment of the repaired or replaced unit to the operator's warehouse. The sender has to use electrostatic conductive enclosures and properly pack the shipped unit in order to withstand the jolts and jerks of handling and shipment.

Replacement Process

An online electronic failure report mechanism or a traditional fax based system in involved in this process. The operator personnel will request a warranty service unit replacement by filling the failure form online or by sending a failure report of the defective unit by fax to the vendor's hardware repair centre. The operator is responsible for checking with the vendor whether the hardware service centre has received the fax or not, after the recipt of which, the hardware service centre will ship a replacement unit from their ready stock by the fastest courier service available to deliver the replacement unit within the next business day, if the request is received, for example, before 03:00 pm. For later time stamps the turn-around time carries over to the next day and the response time is then two business days.

Service Level

The reliability of the turn-around response time of the vendor determines the service level; e.g. a 95 % service level is expected in most of the cases, i.e. the hardware service centre is able to send the replacement unit within the same day of getting the failure report and also ensures that the courier company delivers the replacement unit to the operator within the next business day. A drop in service levels agreed in the care agreement can lead to penalty clauses or the operator can get a rebate equal to the difference between the promised and the actual service level. Conversely, the operator must take the responsibility to send the defective unit to the hardware service centre within the next few days of submitting the failure report, as agreed in the care agreement. In case of late returns, penalty fees or late return fees can be levied as a fine by the hardware service centre. The replaced units covered under the warranty should be either new or repaired and of the latest compatible version. If the hardware unit is damaged beyond repair due to mishandling, negligence, unauthorised tampering or modification, even by the operator itself, then the vendor will not be obliged to repair the unit, but should explain the reasons for not repairing it and replace it from the operator's stock of spare parts. If no fault is found in the unit sent by the operator for repair then the hardware service centre can have the right to charge the operator for false reporting and invoice them for testing and handling.

Third Party Equipment

Third party elements are exempt from this scope of hardware service plan and are handled separately as per their own contracts with the vendor or the operator directly.

Software Maintenance Support

Software development is an ongoing process and every software release always has some scope for improvement with respect to previous shortcomings, or some additional features that can be added on top of the existing version. The software maintenance support service is categorised into three main categories:

- New software release
- Fault management
- Correction time

New Software Release
This service authorises the operator to get the new software releases free of any extra charges. The new software release will have the latest improvements incorporated into the base software as well as new optional software functions and features being offered by the vendor. The vendor presents its feature release roadmap to the operator well in advance, in order to provide relevant information about the content and delivery of the coming releases. The new software release process will include production, testing and provision of a software delivery containing storage media and required EPROMs of the new software release as well as production and delivery of one documentation set. If the operator agrees to accept the new software release, then the vendor arranges for its installation in the operator's testbed. Once satisfied with the performance, if the software installation is to be extended to other parts of the network, it is done like a software roll-out process followed by a testing schedule and a field acceptance procedure. If certain flaws or bugs are found then the vendor has to agree to bring about a change delivery with the required corrections. If the operator does not agree to accept the software on the basis of the results obtained from field-testing activities, but still goes ahead by putting it into commercial use, then automatically the software is deemed to be accepted.

Fault Management
Fault management encompasses the vendor's responsibility to perform initial fault detection procedures and subsequently issue change notes and change deliveries. The operator also has to carry out regular software fault detection procedures at various sites and identify the problem and notify the vendor by filling a failure report. If required by the vendor, the operator will have to provide additional information, like logs, symptoms, alarms, counters, etc. If the operator fails to respond to the request for additional information within a specific period of time then the vendor can close the failure case. Software corrections are prepared based on failure reports received from the operator and other users or customers. Multiple corrections are collected and compiled together as a change delivery of the software release delivered.

The main purpose of publishing a change delivery is to correct the existing faults and also bring about improvements of the existing software release. The change delivery comes in a package comprising a storage media from which the change delivery will be copied, documentation and master EPROMs. The change delivery will first be installed in the operator's testbed by the vendor, where it will be tested and verified. If accepted, then it will be taken forward to the network element and a change delivery installation fee will be levied on the operator.

Correction Time
Failure reports are generated using different priority classes, namely A, B and C, with the severity given in descending order:

$$A > B > C$$

Response Times
The failure classes in Table A.2 have an impact on the response timings but even a C class failure can have the fastest response time. Depending on the failure report, different priorities are assigned to the faults and their correction schedules. An emergency situation is referred as an A category failure priority. In such a case correction work starts immediately and a work-around has to be achieved. The B category of priority is given to almost 80 % of the cases, where an answer from the vendor takes about 4 weeks

Table A.1 Failure reports in descending order

Failure priority	Explanation	Examples
A	Major outage	System restart, all links down
		Charging lost
		Repeated restarts of different units
		Alarms or measurements not coming at all
B	Major service affecting problem	If any basic service is materially affected
		Single restart of signalling units
		Alarms from one or more network elements lost
		Major problems with statistics
		Difficulties with creating new subscribers
		Difficulties to activate supplementary services
		Problems with backup
		Problems with charging or measurement transfer
C	Minor problem that can wait until the next major release	Errors in statistics output (unless causing problems with NMS)
		Cosmetic errors in MML/statistic output
		Errors in MML syntax
		Errors in documents

Table A.2 Failure classes

Priority of failure	Supplier's answer available	Supplier's correction available	Type of correction
A	Emergency duty contacted. The correction work starts immediately and continues until a work-around solution is available		
B	4 weeks	3 months / 80 % cases	Change note or change delivery
C	2 months	Next release	New release

and correction availability is published in 3 months in the form of a change delivery or a change note. A C category priority is given to failures, which can be planned for correction in the next software release. Normally in such case about 2 months are taken for an answer to come from the vendor's side.

Technical Support Service

Technical support is a generic term applied to all kinds of hardware and software products, services and features sold by the vendor. The vendor provides free technical support to the operator in the time specified in the deal. It includes services like regular queries, remote support from product lines and any kind of competence transfer program enabling the operator personnel to run the network themselves. It provides a smooth transition of network operation and handling capabilities between the vendor and the operator's technical staff. An online 24 hour interface is offered to answer the operator's technical requests and get access to desired information. These online services are of two kinds: basic services such as online help in html format accessible by a web browser connected to the network element or a document library offered as a part of the equipment deal. The vendor continuously updates the online services with improved modernised versions of previous ones and also adds new services, which if charged have to be mutually agreed upon. The vendor holds the copyright of all the information provided on the online services. The operator holds a licence to use these online services, e.g. product catalogues, equipment manuals, technical description of units, multimedia library and electronic documentation. The operator is responsible for managing the use of online services within its organisation and hence has dedicated personnel taking care of the communication related to online services. All information provided on the online services is confidential and based on a mutual understanding that the contents of the online service are not covered under any kind of warranty for the accuracy, reliability or content of the online documentation. Information provided by the online services is not to be relied on as professional advice acting as a substitute for detailed technical advice provided by the vendor's technical experts, who still need to be contacted for technical advice. Online services are extended to four types of subservices:

- Problem management
- Preventive management
- Component management
- Training and coaching support

Problem Management
Routine helpdesk requests are submitted via the online services. The operator fills in a standard form and submits the request electronically to the vendor, who then ensures the receipt and recording of this request in an effective manner in the database. The vendor utilises the helpdesk follow-up mechanism to track the handling of their helpdesk requests, prepare relevant statistics and management reports.

Preventive Maintenance
During the course of R&D (research and development) and field testing, bugs and fault situations are often discovered by the vendor. Special instructions or technical notes are delivered to the operator through the online services in order to tackle these faulty situations. Adequate technical guidance is provided in the form of supplements of feature activation documents through the online services when special features are developed and are to be launched in the testbed or network. Generic failure reports are delivered, setting out failures in the vendor's equipment or any particular product line, for those products already ordered or employed by the operator in the network. Specific information related to key changes conducted in the delivered software releases by change deliveries and particular correction modules are provided through the online services.

Component Management
A catalogue of all components and spare parts is provided to the operator via the online services. All components are identified by vendor specific codes and a regular shopping basket feature is supported for spare parts ordering. Repair ordering comes with a failure report, order history and interactive order status information, with additional comments as required.

The component and spare part statistic reports are prepared in two formats: (a) fault codes and (b) delivery volumes and promptness. Fault codes depict the primary cause of defect in the unit. Delivery volumes report how many repaired units the vendor has sent within the current year. Promptness illustrates to what degree the vendor has adhered to the schedule for sending repaired units back to the operator.

Training and Coaching Support
This section gives a complete description of all the training programmes designed by the training department of the vendor for the purpose of coaching and competence development of the operator's personnel (see Table A.3). It is organised in different ways, giving course information related to the technology, product lines or competence levels, as well as providing information about dates and places where the training will be held.

Table A.3 Generic format classifying the different kinds of course offered by the training services

Training and coaching	Access network	Circuit core network	Packet core network	Network management
Basic				
Intermediate				
Expert				
Special modules				

Helpdesk

When the online service is not sufficient to report a problem, then technical support is accessed by means of an additional support channel known as the helpdesk service. The vendor provides a helpdesk support service covering all the software and hardware products sold to the operator as part of the complete network deal.

The vendor arranges for a central helpdesk number, where the operator can call during the working hours in working days to make a reasonable nonurgent query. The vendor ensures that its technical staff is immediately accessible for specifying queries and collecting relevant data like logs, statistics, symptoms, etc., during the call. A relevant product expert is referred to if the person receiving the call cannot provide an immediate solution. If the product expert is not available at that time, he/she must call back the operator within 30 minutes from initiation of the caller's request. A target of 2 to 5 working days is normally taken as a target for providing feedback to the customer. On investigation and further analysis, if the fault is located in the equipment, then the vendor has to request the operator to fill in a failure report, which will be taken up by the product line to bring out a software correction or improvement. If the case is not investigated and solved within a certain time period of, say, 2 weeks and no fault has been found in the equipment, then an escalation process is followed to bring it to the notice of the next level of management. The operator also has to nominate a team or a dedicated person for directing all the helpdesk queries and receiving the responses delivered by the vendor. All helpdesk calls are monitored for the type of problem raised, time taken to answer the call, time taken for investigation and providing a solution, and periodically reports are generated by the vendor to provide performance statistics to the operator.

A numerical report of the operator's requests and their status is provided by the request statistics. Other metrics and trend patterns provide facts and figures about the number of requests per network element, product type or unit type. These statistics and metrics are used in the management review meetings to measure the capability of the vendor in providing effective helpdesk services, for which the operator is charged as per the care agreement. A fall in performance levels can lead to activation of penalty or liquidity damage clauses.

A basic helpdesk form can be of the structure shown in Figure A.10, which can be expanded as given in the collective agreement between the operator and the vendor.

Emergency Support Service

An emergency support service is undertaken by the vendor to provide a round the clock, all through the year backup support for the operators should they face any critical problem causing serious damage to their networks, due to any kind of equipment-related fault. A typical emergency call-out can occur when a critical network element has failed, causing loss of revenue due to unavailability of the network service in a particular area.

Request Number:	Customer / Area / Name of requestor:	**Network Element Type:**
Request Date:	Hardware Level:	Software Level:
Abstract:		

Description: (Date: For each change)

Responses to the customer: (date:)

Figure A.10 Specimen helpdesk form

That situation can be found in the network operation centre due to the generation of alarms or complaints from customers from that affected region. The operator personnel makes an emergency call to the vendor's local office, where the first line maintenance engineer on call will receive the emergency case and within a stipulated time, e.g. 1 hour, reverts back to the customer and starts the investigation. As long as this situation is in this critical level both the operator and the vendor personnel on call will have to keep the communication lines open. If an immediate solution is not achieved in a few hours or if the situation is too complex for the local person to handle, then he/she can escalate to a higher level, which is usually the product line personnel who has a high level of product competence. During investigation, the product line personnel can ask for more data or logs from that fault situation in order to trace the fault cause and suggest the correct remedy. An emergency process can also involve a call-out situation involving an immediate site visit by the engineer. An escalation procedure has to be mutually agreed upon for initiating when a call-out is made. During an emergency situation the operator has the responsibility to initiate adequate efforts to carry out operation and maintenance procedures recommended by the vendor and to find a remedy for the faults. Basic first line maintenance is expected from the equipment operating personnel of the operator. The operator has to maintain an operational logbook in order to record the visible faults and changes tried by its staff to correct the faulty situation.

On-Site Operations Support

As a part of the deal the vendor may agree to provide on-site operation support for an initial first six months period, considering working days only and not round the clock support, which is covered separately. On-site support is provided with the aim of helping the operator rapidly ramp-up the operation and maintenance for the initial network system and also of providing initial support for routine technical operations for the first few months after commercially launching the network. This includes training the operator's personnel to whom the network operations will finally be handed over. The vendor's technical audit team as an integral part of this support service conducts a postlaunch performance audit, verifying the quality of the launched network and performance level of the end user services.

The operator has to agree with the vendor regarding the timing and fragmentation of services, which has to be supplied as part of the on-site operation support service, well before the scheduled start of the period of operation support. Whether the operation support service is agreed or not the responsibility of running normal operation and maintenance of the network lies with the operator at all times.

In order to perform their duties and obligations the on-site support personnel of the vendor have to be given free access to all locations at all times. The operator has to provide all such spare parts and consumables that are needed or might be needed by the vendor to perform the on-site support obligations.

On-site support personnel including engineers and technicians from the vendor's side can share their experience and equipment knowledge about configuration and fault management with the operator's operation and maintenance personnel, in order to help them learn faster, but they are not obliged to provide them on-job training or any formal training that substitutes for the course designed by the training centre.

Training and Coaching Services

Competence development and the ability to run the network operation and maintenance independently is considered to be one of the main objectives of this service. The overall network can be logically divided into subparts like the radio access network, circuit core network, packet core network, operation and maintenance network, transmission backbone network and radio frequency network and similarly regular network handling activities can be segregated as installation, commissioning, integration, O&M, planning, optimisation and troubleshooting.

Based on the above matrix of product lines, technologies or type of work, training needs are analysed and training solutions are designed based on these requirements. The Vendor's training department makes a solution plan and offers a concept that refers to the type and level of competence required by the operator's personnel to carry out their designated duties independently. Once approved by the operator, the training plan is designed, setting the training schedules, periods of self-study or hands-on practice with the network element. The operator's training department coordinates the training matters with the vendor's training department and sets an 'operator competence development program' schedule. The vendor's training department sends the final course programme and relevant course information regarding practical arrangements to the operator's training coordinator at least 2 weeks before the start of the course programme. In the event of the course being cancelled or postponed, the operator has to provide about 30 days notice to the vendor's training centre. If the notice period is crossed cancellation can lead to invoicing the full course fees to the operator plus possible expenses incurred for arranging the course.

A standard course day is 7 to 8 hours as given in the course instructor's plan, including an hour for a lunch break and appropriate tea/coffee breaks in the morning and afternoon. The vendor's training department arranges all the necessary training documentation and training environment. Training documentation and updates are generally in English unless translated and customised in other languages as per the customer's needs.

A.4.3 Other Optional O&M Assistance Services

The vendor provides O&M assistance services in order to assist, support and consult the operator in the regular operation and maintenance of the equipment. They can be long term or short term. Various kinds of assistance services are available to the operator as given below.

HW Expansions

Hardware expansions are recommended when the existing hardware capacity becomes staturated. Expansions can be simple processes involving the addition of extra plug-in units or memory additions without any major changes in the software. Alternatively, they can be major activities involving the addition of whole racks and subracks. Such activities are always tried in the vendor's testing laboratories or in the operator's test network, and only after running all stability and verification tests are they implemented in a live network. These activities can be free from the vendor's side due to limitations of existing hardware or they can be charged as a totally separate sale, having their own separate or combined warranty, after sales support and maintenance clauses. These activities are only done by the vendor's technical team or under their guided supervision, as it involves equipment installed in a live network with existing subscribers.

SW Upgrades

With the onset of new business models software sales are becoming some of the major business models, where feature based software modules are kept separate from the basic operating software module. Software upgrades are done by putting together totally new software packages or by installing some new change notes compiled together as 'change deliveries'. It brings in new functionalities, improved performance figures, higher success rates, new or improved additional features, improved network stability and reliability. All software upgrades are pretested in the network operator's test network and all features and functionalities are tested thoroughly to their last details before being installed on a live network. Before any software upgrade is done, there is always a team from both sides, i.e. the vendor and the operator, the former taking responsibility for doing the complete software upgrade and the latter taking the part of overall supervision and doing the final checking and testing before signing off the activity sheet with normal acceptance.

Most live upgrade activities are done in two shifts. In the first one the upgrade engineer either goes to the site physically or takes charge of the equipment remotely from the operation and maintenance centre. Before beginning at a software level, the latest technical notes are checked for any hardware modifications or additions. In typical cases memory blocks are added or replaced with bigger ones, new plug-in unit versions are installed or the interchangeability codes are updated to the latest ones.

Firstly, the alarms are checked to ascertain the equipment's stability and to find out whether any major alarms need to be taken care of. Then the current software package is backed up in the hard disk as well as in a DAT (digital audio tape) or magnetic optical disk for future reference or fallback, in case the upgrade fails for any human or technical reason. Then precheck macro is run to extract the running data of the equipment, which can be used later as a reference in case there is any variation from normal performance after the software upgrade.

The new software is downloaded into the equipment's hard disk, after which the process of new package creation is started. Firstly, files are copied that do not require conversion and the remaining system files are converted to suit the new package delivery. The new package is created but only activated at midnight, so that traffic disruption affects the minimal traffic volume and if any technical faults happen they can restore the situation before the traffic picks up in normal working hours.

The second shift starts after midnight, when the crucial cutover takes place causing a network outage for a few minutes, depending on the network element. This outage lasts for as long as it takes to load the new software into all the critical units and for the interfaces with the connected elements to come up in their respective working condition. Once everything is back in normal working order, the system is checked for any abnormal behaviour or any alarms being generated due to some unit or element failure caused by the new software.

Thereafter all normal processes are observed as they were performed before, like alarm uploading to the operation and maintenance centre, measurements being taken and reported, statistics generation, working of links and interfaces. Then normal subscriber activities, like calls, messages, downloads, etc., are done to ensure normal functioning of the network. Once all the units of the network element have come up with the new software then all the basic tests have been done in order to verify that the equipment is in correct working order and its performance is stable with respect to the rest of the network. The breakdown of a normal software upgrade process is shown in Figure A.11.

Network Maintenance

In order to maintain normal and stable running of any network, adequate equipment care needs to be taken and regular performance and health checks need to be done. This can be done either by physically checking all sites or centrally from the network operating centre. Counting on their operating lives and remaining periods it is possible to monitor battery health. New hardware units can replace old or outdated hardware versions, especially when these retrofits are necessary for enhancing the performance or launching new features. In addition, constant effort need to take place to reduce the level of alarms

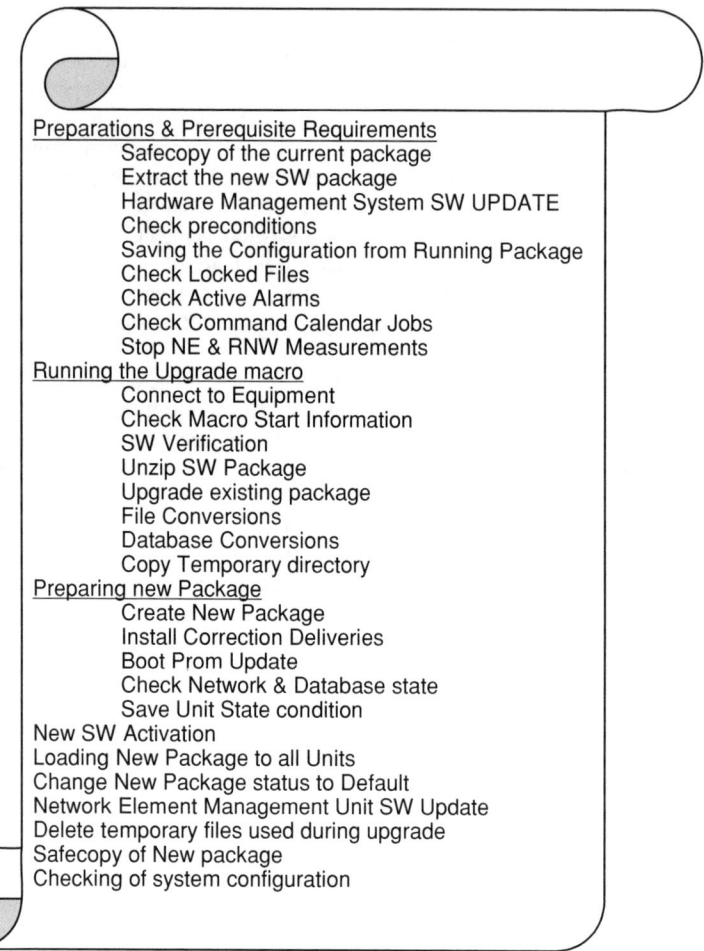

Figure A.11 Software upgrade process (NE, network element; RNW, radio network)

that are found on the system, so that the load on the system is reduced and chances of complete failure are kept to minimum levels, ensuring a high level of system stability at all times.

On-site visits can be arranged to check the sealing of equipment sites, condition of cables and orientation of antennas with regard to wind loading and twisting. All equipment sites should be checked for dust and humidity. HVAC units need to be serviced as per their service schedules and their automatic switchover and cutover mechanisms have to be checked for effective functionality. Regular maintenance of diesel generators is required for rural equipment sites, where such site support elements are used regularly. Site safety has to be ensured from time to time, to prevent any kind of vandalism or pilferages. Towers need to be painted and checked for any kind of rusting or changes in alignment due to equipment or wind loading.

Performance Monitoring and Remote Health Check

Remote logins to the different network elements provide an excellent way of doing health checks on the various network elements. Alarms can be checked from time to time, memory disks can be checked

Care Phase

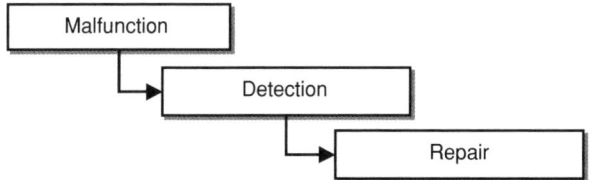

Figure A.12 Performance monitoring and remote health check-up

for remaining space, software fallbacks can be taken at regular intervals, light emitting diode (LED) indications of the hardware units can be monitored for tracing faulty units that can affect the system performance, etc. While the equipment is running many inevitable malfunctions may go unnoticed and later bundle-up into a big problem resulting in a drop in performance levels. As the network expands the more difficult it becomes to monitor it end to end. The best way is to monitor it centrally by having an automatic monitoring mechanism that searches for malfunctions by looking into the system, detects them and repairs them in a premature stage before they mature into serious issues (Figure A.12).

Network monitoring systems provide counters related to the performance of the equipment, which can be extracted online or offline from the database by performance monitoring tools used by the operator or the vendor. There are hundreds of counters covering all the network elements and not all are used in daily performance reporting. These counters are useful from a software and service point of view and help in monitoring performance. These monitoring tools use certain algorithms based on calculating key performance indicators (KPIs) in order to analyse these counter data and present them graphically in the form of tables and charts correlating performance levels with respect to time. These reports and statistics help in segregating faulty situations and faulty parts of the network, which can be separated from the rest of the network for detailed trouble shooting. For example, counters can portray a picture where a certain area of the network is having a lot of call drops. Based on this report either a drive test team can be sent to check the real performance or a BTS engineer can be summoned for checking alarms and faulty signals coming from the BTS serving that area. In this way, before the customers really start to call the customer service lines and complain, the operator can take preventive action and take advantage of the remote health checks performed manually or via some special tools.

These tools are customised, as per the key performance indicators chosen by the operator, to select the criteria used for measuring performance. On selection of the criteria, thresholds are set for the tool to function and finally corrective actions are taken that can be triggered automatically or manually. Such tools can be simple Java based programs running on typical window based laptops or they can be complex automated ones based on Unix machines. They have user friendly graphical user interfaces, easily connected to the network by means of LAN cables as they interface with the IP based backbone network running throughout the network, on top of their backbone networks.

Remote health checks are an excellent way of ensuring network stability and keeping the load of the network operation and maintenance personnel low. They help to keep high levels of network performance by proactively searching for malfunctions, bugs and alarms in the system and by timely detection, which prevents problems from turning into potential troubles. At the end of the day the checks provide a healthy system that helps to produce optimum performance from the network, thus keeping customers happy and satisfied.

Appendix B

HSDPA

Rafael Sánchez-Mejías

This appendix provides a brief introduction to high speed downlink packet access (HSDPA) as the expanded throughput solution for UMTS networks in 3GPP release 5. The objectives is to present a general overview of the main features that allow the provision of these higher throughputs, up to 14 Mbps theoretically, as well as practical limitations and resource requirements for fulfilment of the target performance. Planning and dimensioning considerations for HSDPA and the regular WCDMA network will also be pointed out.

In addition, further evolution of UMTS capabilities towards HSPA (high speed packet access) will be described. Known as HSUPA (high speed uplink packet access), this 3GPP release 6 feature promises throughputs of up to 5.4 Mbps in the uplink.

B.1 Introduction

High speed downlink packet access (HSDPA) is an extension of the capabilities of UMTS included in 3GPP release 5, with the target of providing higher bit rates and capacity. It is also called 3.5G, with transmission rates up to 14.4 Mbps and 20 Mbps (for MIMO systems) over a 5 MHz bandwidth.

HSDPA can significantly enhance downlink speeds, with average realistic throughputs of 400–700 kbps and bursts at over 1 Mbps, even in the initial stage. This dramatically improves the user experience of different applications such as web browsing, streaming or Intranet access. Also, in combination with HSUPA, it can be the driver for advance services like VoIP. HSDPA shares the spectrum and codes from WCDMA and, most of the time, only requires a software upgrade of existing UMTS R99 base stations.

HSDPA offers a lower cost per bit and is mainly intended for non-real-time (NRT) traffic, but potentially allows new application areas with higher data rates and lower delay variances. The maximum number of UEs on HSDPA does in theory depend on the number of available channellisation codes for the associated DPCHs.

The most critical parameter affecting HSDPA performance is the transmission power. Since the total power of the base station is shared with R99 DCH, a trade-off between HSDPA and R99 users needs to be considered. This trade-off is affected by the strategy chosen using HSDPA, depending on whether it is introduced as a high bit rate service for top users or as a way to improve efficiency and capacity of background NRT traffic.

B.2 HSDPA Performance

In terms of performance, in a 5 MHz channel HSDPA can provide maximum peak rates of up to 14.4 Mbps with 15 spreading codes and with no channel coding. However, this would mean that one unique subscriber will have to use all available codes on the high speed downlink share channel (HSDSCH). This is not a realistic approach, especially for initial implementations, since typically the capacity is also shared with regular UMTS DCH channels. Codes allocated for HSDPA are fixed, and not usable for DCH, so in practice implementations with 15 codes will require at least several 5 MHz carriers to be available in the system. It is expected that HSDPA realisations will follow a progressive approach, starting with five codes, and evolving to 10 or 15 as higher capacity or resources to support it become available. Additionally, it is important to consider that not all UE classes will support 10 or 15 codes, so in order to get the full benefit from maximum throughputs, terminal availability needs to be considered. A realistic data rate will be about 600–800 kbps, taking real network conditions into account, while an estimated network round trip time (RTT) would be 80–100 ms.

As a summary, the overall performance of an HSDPA network will depend on:

- the number of spreading codes (support of 5, 10 or 15 multicodes);
- the modulation mode (QSPK (quadrature phase shift keying), 16-QAM (quadrature amplitude modulation)), where 16-QAM is optional for the network and also for the UE;
- the error correction level;
- capabilities of end user devices.

On the other hand, the impact of HSDPA on WCDMA R99 will be driven by sharing resources and the allocation of fixed codes and constant transmission power to HSDPA. HSDPA will cause a drop in downlink E_c/N_0, since fixed codes are fully allocated. The higher the power for HSDPA, the bigger the drop in E_c/N_0.

Additionally, HSDPA may lead to quality problems and lower data rates for WCDMA R99 connections if the network RF planning is not designed to tolerate the extra interference caused by lack of power control in the HSDPA transmission. In some cases, it may be a need to limit the amount of power for HSDPA in order to protect WCDMA R99. In such cases, the gain in HSDPA performance coming from increasing its transmission power should be closely checked with the degradation in WCDMA R99 performance. The type of modulation (QPSK or 16-QAM) can have an important impact in this case.

B.3 Main Changes in HSDPA

Table B.1 present basic features introduced in UMTS architecture and protocols in order to support HSDPA. In addition, other capabilities like the multiple-input multiple-output (MIMO) receiver would be supported to provide further signal gain and higher throughputs.

HSDPA introduces a new radio bearer in the UMTS system, the high speed downlink share channel (HSDSCH). This channel allows several users to be time-multiplexed so that during silent periods the resources are available to other users. The HSDSCH uses 2 ms transmission time intervals (TTIs) and a fixed spreading factor of 16, which allows a maximum of 15 parallel codes for user traffic and signalling. In addition to accelerating service access for users and improving data transfers, this reduced TTI allows the system to adapt itself faster to changing conditions. The uplink data transmission of the HSDPA user initially relies on release 99 DCH with different available rates (i.e. 64, 128 or 384 kbps).

A new MAC-hs entity in added on the BTS to handle all these new features needed for HSDPA traffic, as shown in Figure B.1. Layers above MAC-hs (for the high speed downlink shared channels), such as MAC-d (for the dedicated transport channels) and RLC, are similar to those in the release 99 networks.

The adaptive modulation and coding (AMC) technique is used in order to compensate for variations in radio transmission conditions, while the transmission power remains constant. HSDPA-enabled user

Table B.1 Main HSDPA features

Feature	HSDPA
MAC layer split	Functionality moved to Node B to improve efficiency of packet scheduler and retransmissions
Downlink frame size	2 ms TTI (3 slots)
Channel feedback	Channel quality reported at 2 ms rate (500 Hz) for CQI (channel quality indication), ACK (acknowledged)/ACK, TPC (transmission power control)
Data user multiplexing	Time multiplexing/code multiplexing (optional)
Adaptive modulation and coding (AMC) scheme	QPSK and 16-QAM mandatory
HARQ	Fast layer 1 retransmission (improves RTT); chase or incremental redundancy (IR)
Spreading factor (SF)	16, using UTRA (universal terrestrial radio access) OVSF (orthogonal variable spreading factor) channellisation codes
Control channel approach	Dedicated channel pointing to the shared channel
Packet scheduling	Fast scheduling done in Node B with 2 ms time basis; types: round-robin, proportional fair, fair throughput, etc.
Shared channel transmission	Dynamically shared in the time and code domains
Multiple-input multiple-output (MIMO)	Optional; higher performance

equipment sends channel quality reports to the base station at 2 ms intervals, which are used to adapt the modulation or resources accordingly. At layer 1, the hybrid automatic repeat request (HARQ) with a Stop and Wait (SAW) Protocol is used as a retransmission mechanism. Unlike the UMTS R99, the HARQ is processed directly in Node B, which allows a faster response, instead of being handled by the RNC. The fast scheduling feature is also implemented in Node B, compared to UMTS R99, where the scheduler is located in the RNC. The scheduler determines to which terminal the transmission in HSDSCH will be directed and, depending on the AMC, at what data rate.

B.3.1 HSDPA Channels

There are five different physical channels that are used by HSDPA services. HSDPA data are carried on HSPDSCH, which is a shared channel for all HSDPA users in the cell. There are two physical control

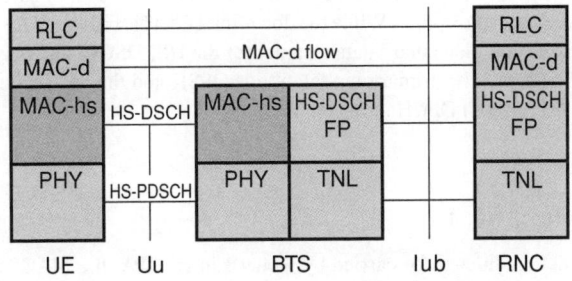

Figure B.1 HSDPA protocol layers

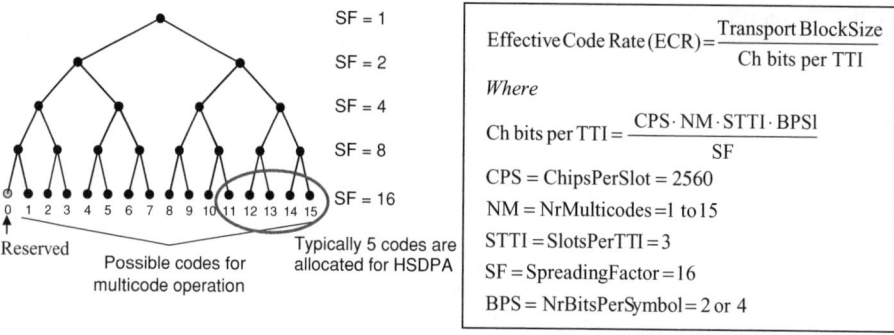

Figure B.2 Code allocation for HSPDSCH and code rate estimation

channels, one dedicated channel in uplink (HSDPCCH) and one shared channel in downlink (HSSCCH). In addition to these there are associated DPCHs for uplink and downlink.

- *HSPDSCH (high speed physical downlink shared channel)*. This transfers actual HSDPA data of the transport HSDSCH and can use 1 to 15 code channels, with a spreading factor (SF) of 16. All these associated physical channels should be adjacent, as shown in Figure B.2. QPSK or 16-QAM modulation is supported over 2 ms TTI slots. No power control is supported. In addition, the HSDSCH does not support soft handover due to the complexity of synchronising the transmission and scheduling from different cells. Instead, cell reselection through a normal DCH would be implemented; i.e. the HSDPA user is given a DCH in the SHO area, and is then moved to the new cell, where it would get an HSDPA channel again after the procedure is completed.
- *HSSCCH (high speed shared control channel)*. This includes information to tell the UE how to decode the next HSPDSCH frame. It uses QPSK modulation and a fixed SF of 128. It shares the downlink power with the HSPDSCH, but may support power control in order to maximise the available power for the data channel. More than one HSSCCHs are required when code multiplexing is used, but a maximum of four is supported by the UE. Soft handover is not supported.
- *HSDPCCH (high speed dedicated physical control channel)*. This channel carries the ACK/NACK (not acknowledged) (repetition encoded) and channel quality indicator transmitted from the UE in the uplink direction, which is needed for L1 procedures. The primary modulation is BPSK (binary phase-shift keying) with an SF of 256 (15 kbps). The transmission power used is typically the same as that used for the uplink DPCH plus additional offset to provide higher protection. The HSDPCCH may be received by two different sectors in the same Node B, but in general soft handover is not supported.
- *Associated DPCH (dedicated physical channel)*. Two DPCHs are needed for each HSDPA UE, one in the downlink and another in the uplink. While the downlink DPCH is only used for signalling purposes, the uplink DPCH is the complementary data channel for the HSPDSCH, and may be allocated a data rate of 64, 128 or 384 kbps. The primary modulation is QPSK and the SF can be from 4 to 512. Soft handover is supported for both DPCHs.

B.3.2 MAC Layer Split

As a result of new functionalities to be carried by Node B in HSDPA, the MAC layer is split into two entities. While MAC-d remains in the RNC in the same way as for R99, MAC-hs is located in Node B to allow rapid retransmission of NRT data.

MAC-d is responsible for mapping between logical channels and transport channels, selection of and appropriate transport format and handling priorities. It also has to identify UEs in the common transport channels and multiplex/demultiplex upper layer PDUs and to measure the traffic volume. Ciphering for the transparent mode RLC is also managed by MAC-d. MAC-hs is responsible for packet scheduling, link adaptation and layer 1 error correction and retransmission (HARQ).

Due to this split in the functionality of the MAC layer, the user data buffers, which used to be in the RNC, are moved to Node B. This makes the introduction necessary of a flow control mechanism in the Iub interface, in order to avoid the buffer overflow and throughput degradation due to buffers becoming empty. MAC-d schedules the number of RLC PDUs according to the credits granted by MAC-hs at each interval of 10 ms, and the aggregated rate of the HSDPA connections is controlled by the rate control implemented in MAC-hs. The MAC-d PDUs are framed into FP-HSDSCH frames, while a maximum number of 16 MAC-d flows per BTS are supported.

B.3.3 Adaptive Modulation and Coding (AMC) Scheme

Link adaptation (LA) is the key feature to the success of HSDPA, since there is no power control in HSPDSCH, and it is used to adapt HSPDSCH to different radio conditions. If LA does not work properly, cell capacity is lost and other techniques such as fast scheduling will not work.

Link adaptation is done by changing the modulation and number of codes. The UE signals information to the network about the highest data rate it can accept under the current channel conditions while still maintaining a controlled block error rate (i.e. under 10%). This CQI (channel quality indication) is signalled through the HSDPCCH. The network uses this in order to reconfigure the HSDSCH format for subsequent transmission to that UE. For example, if the CQI shows that the quality is degrading, the scheduler can choose a less aggressive coding/modulation format that will cope better with the poor conditions.

Typically the link adaptation is divided into two phases, known as the inner loop and outer loop algorithms:

- *Inner loop algorithm.* This takes the decision for the modulation and coding scheme to be used in the next TTI. This selection will be done only for new transmissions (i.e. not for retransmissions), and will be based on the received CQI, the available HSDSCH transmit power, the number of HSPDSCH codes, the RLC PDU size, input from the outer loop HSDSCH algorithm and the UE category. It is important that input parameters (CQI reports and DPCH power measurements) to the inner loop algorithm are subject to a minimum delay, because otherwise the LA would not be able to track fading in the radio channel properly.
- *Outer loop algorithm.* The primary goal is to compensate any bias introduced by the inner loop algorithm. This bias might be introduced due to offsets in relative UE performance, due to improved receiver architecture, etc. Typically, the outer loop may be based on the BLER target obtained from RLC ACK/NACK information, but also the CQI may be used directly.

B.3.4 Error Correction (HARQ)

Layer 1 retransmissions expand the system recovery capability from air transmission errors, and are subject to significantly shorter delays than RLC retransmissions, due to the closeness of UE and Node B (See the MAC layer split in Table B.1). This results in lower delay jitter, which can be very beneficial for data services based on TCP or streaming applications. Figure B.3 shows the different retransmission layers in the HSDPA architecture.

The use of HARQ adds increased robustness to the system and a spectral efficiency gain. Two retransmission strategies are supported: incremental Redundancy (IR) and chase combining. The basic idea of the chase combining scheme is to transmit an identical version of an erroneously detected data

Table B.2 Layer 1 transmission–retransmission link budget (3GPP)

Delay event	Delay
HSDSCH transmission	2.0 ms (3 TS)
Before the UE sends ACK/NACK on the uplink HSDPCCH	5.0 ms (7.5 TS)
For ACK/NACK transmission	0.67 ms (1 TS)
From reception of ACK/NACK until HSSCCH transmission	3.0 ms (4.5 TS)
For HSSCCH transmission before starting the HSDSCH transmission	1.33 ms (2 TS)
Total	12 ms

packet before the decoder combined the received copies weighted by the SNR prior to decoding. With the IR scheme, additional redundant information is incrementally transmitted if the decoding fails on the first attempt, by means of different puncturing schemes used in the coding of the retransmitted data packets. In the case of HSDPA, IR with a one-third punctured turbo code would typically be used for the retransmissions, although it has the drawback of requiring higher memory buffers in the UE than chase combining.

During the scheduling phase, the MAC-hs layer will give priority to retransmissions over new RLC packets, which will be transmitted with the same code as the original transmission. HARQ can be used in the stop-and-wait mode or in the selective repeat mode. Stop-and-wait is simpler, but waiting for the receiver's acknowledgement reduces efficiency; thus multiple stop-and-wait HARQ processes are often done in parallel in practice. When one HARQ process is waiting for an acknowledgement, another process can use the channel to send more data.

There are a few aspects to consider for the HARQ mechanism:

- If Node B receives an ACK from a UE, everything is fine.
- If Node B receives a NACK from a UE, it means that the packet was received, but could not be detected properly. In this situation, Node B should retransmit using incremental redundancy.
- If Node B never receives any ACK/NACK, it should retransmit using another self-decodable rate matching scheme.

B.3.5 Fast Packet Scheduling

The objective of the packet scheduler is to optimise the cell capacity while delivering the minimum required service experience for all active users excluding an allowed outage target. Outage is defined from blocking, dropping and QoS requirements related to a given application.

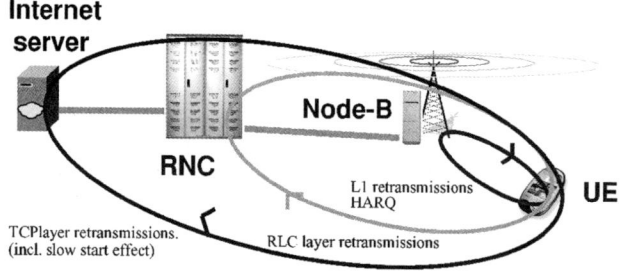

Figure B.3 HSDPA retransmission breakdown

Table B.3 Hacket scheduler strategies

Packet scheduler	Selection metric	Scheduling rate	Fairness/performance
Fair throughput (FT)	The user with least throughput is scheduled	Fast (2–20 ms basis); in R99 architecture this is slow	Same average user throughput over activity time
Fair resource (FR) or round-robin in time	Round-robin in random order	Fast (2–20 ms basis); in R99 architecture this is slow	Same average physical resources but uneven user data rates
Maximum C/I (M-C/I) or maximum throughput (M-TP)	Maximising throughput or channel quality	Fast (2–20 ms basis)	Optimum cell capacity but very uneven user data rates and limited coverage
Proportional fair resource	Ratio of instantaneous to average throughput (relative instantaneous channel quality, or RICQ)	Fast (2–20 ms basis)	Approximately same average physical resources and uneven user data rates (fairer than fair resource PS) with very high capacity

The actual packet scheduling algorithm is not specified in 3GPP and there is a large degree of freedom available to manufacturers. However, with the definition of various QoS parameters such as discard timers and guaranteed bit rates, it is expected that the packet scheduler does its best to fulfil the requirements given for any user. This is especially significant in multivendor environments (e.g. RNC and node B from different vendors) where the QoS responsibility is distributed. The packet scheduler needs to be flexible and adjustable by the parameters defined in 3GPP/release 5, including some indirect parameters such as complying with the power targets specified from the RNC.

Different approaches have been proposed for the packet scheduler, as described in Table B.3.1 Typically the most simple approach is round-robin scheduling, or best effort, which performs a 'blind' allocation of resources without using quality information. It has low complexity and allows a fair distribution of power and code resources among users. On the other hand, algorithms like proportional fair scheduling (Figure B.4), use information of user quality and fast fading behaviour to select the most appropriate transmission turn for each user. This scheduler has a higher complexity, but can provide 20–60 % gains in throughput compared with round-robin scheduling. The gain depends on the number of HSDPA users in the cell, the radio conditions and transmitted power.

UE capabilities also have an effect on scheduling. The UE's ability to receive data depends on the UE category it supports. Category 1 and 2 UEs can receive data in every third TTI. Categories 3, 4 and 11 UEs can receive data in every second TTI. The remaining UE categories can receive in every TTI.

B.3.6 Code Multiplexing (Optional)

Code multiplexing gives the possibility of sending data to several UEs in the same TTI by using different sets of codes for each UE. For example, if 10 codes are used in the HSPDSCH, five codes could be used to send data for UE1 and the remaining five codes could be used for UE2 in the same TTI. In the case of time multiplexing all 10 codes would be scheduled for one UE in a TTI.

Combining time and code multiplexing makes scheduling algorithms much more complicated. This option will therefore probably not be used in early HSDPA implementations.

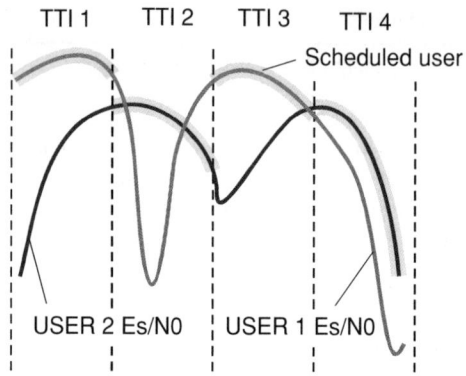

Figure B.4 Packet of proportional fair scheduling

B.3.7 Impact on the Iub Interface

With higher throughputs in HSDPA, there is a high probability of congestion between Node B and the RNC, which requires careful planning of the Iub interface. A careful flow control mechanisms needs to be introduced in order to avoid a strong reduction of data rates due to ATM discards, which would generate RLC retransmissions and TCP window reduction.

The understanding in 3GPP is that Node B is in control of the flow, so it will send capacity allocation messages to the RNC. Node B knows the status of the buffers in the RNC from the capacity request, and uses this message to modify the capacity at any time, irrespective of the reported user buffer status.

Two messages are defined for Iub flow control, as shown in Figure B.5.

- The HS-DSCH capacity request procedure allows the RNC to request HSDSCH capacity by indicating the user buffer size in the RNC for a given priority level (CmCH-PI, or common transport channel priority indicator).
- The HS-DSCH capacity allocation is used by Node B to allocate resources for a given flow. It includes a number of parameters: the number of MAC-d PDUs that the RNC is allowed to transmit for the MAC-d flow (HSDSCH credits), the associated priority level indicated by the CmCH-PI, the maximum MAC-d PDU length, the time interval during which the HSDSCH credits granted may be transmitted (HSDSCH interval) and the number of subsequent intervals that the HSDSCH credits granted may be transmitted (HSDSCH repetition period).

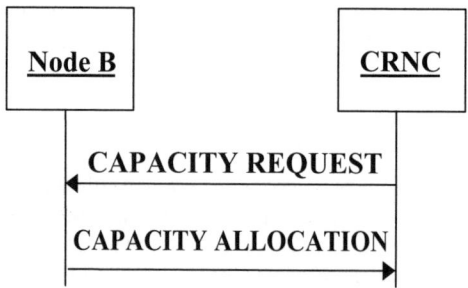

Figure B.5 Flow control messages in Iub

Handset Capabilities

Table B.4 HSDPA UE capabilities

Category	Supported modulations	Max number of HSPDSCH codes	Inter-TTI interval	Maximum TB size (bits)	Number of soft bits (memory size)	Maximum data rate (Mbps)
1	16-QAM and QPSK	5	3	7 298	19 200	1.2
2	16-QAM and QPSK	5	3	7 298	28 800	1.2
3	16-QAM and QPSK	5	2	7 298	28 800	1.8
4	16-QAM and QPSK	5	2	7 298	38 400	1.8
5	16-QAM and QPSK	5	1	7 298	57 600	3.6
6	16-QAM and QPSK	5	1	7 298	67 200	3.6
7	16-QAM and QPSK	10	1	14 411	115 200	7.2
8	16-QAM and QPSK	10	1	14 411	134 400	7.2
9	16-QAM and QPSK	15	1	20 251	172 800	10.8
10	16-QAM and QPSK	15	1	27 952	172 800	14.4
11	QPSK	5	2	3 630	14 400	0.9
12	QPSK	5	1	3 630		1.8

There are different possible approaches for the flow control algorithm, but they need to be a trade-off between performance and implementation complexity. For instance, a simple flow control implementation may consist in sending periodic capacity allocations that either follow a round-robin approach among the UEs or is based on the MAC-hs or has RLC buffer status. However, this kind of implementation can have constraints regarding the guaranteed bit rate and number of simultaneous users. On the other hand, a more a dvanced flow control may take into account the buffer status in Node B, the guaranteed bit rate, scheduling priority, the buffer status in the RNC, the air interface bit rate and the discard timer for sending the capacity allocation messages.

B.4 Handset Capabilities

HSDPA handsets are becoming more complex, due to the addition of several new features to existing 3G devices in order to achieve maximum capability, such as support of the 16-QAM modulation method, a roadmap for advanced receivers (equaliser, diversity), HARQ in layer 1, a gfaster turbo decoder, the need for increased and faster buffer memory, etc.

The UE capabilities, presented in Table B.4, are sent from the serving RNC (SRNC) to Node B when the HSDSCH MAC-d flow is established. They include among others:

- The maximum number of bits a UE can receive within one TTI;
- The maximum number of HSDSCH codes the UE can receive simultaneously;
- The minimum inter-TTI arrival;
- The total buffer size minus the RLC AM buffer size;
- Five main parameters used to define the physical layer UE capability level (3GPP TS 25.306):
  - the maximum number of HSDSCH multicodes that the UE can simultaneously receive; at least five multicodes must be supported in order to facilitate efficient multicode operation;
  - the minimum inter-TTI interval, which defines the distance from the beginning of a TTI to the beginning of the next TTI that can be assigned to the same UE; e.g. if the allowed interval is 2 ms, this means that the UE can receive HSDSCH packets every 2 ms;
  - the maximum number of HSDSCH transport channel bits that can be received within a single TTI.
- The maximum number of soft channel bits over all the HARQ processes;

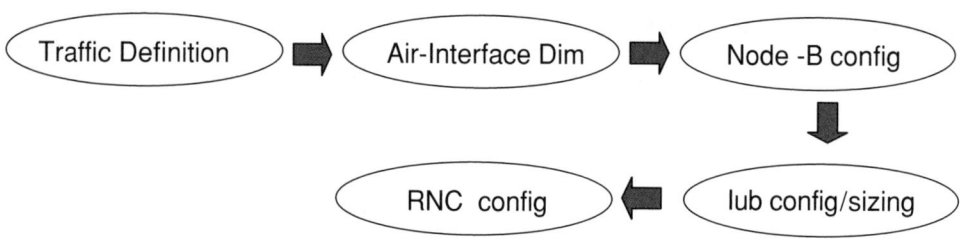

Figure B.6 Dimensioning process

- If the UE supports 16-QAM (e.g. code efficiency limitation);
- Parameters are also available for specification of the L2 buffer capability (RLC+MAC). A UE with a low number of soft channel bits will not be able to support IR for the highest peak data rates and its performance will thus be slightly lower than for a UE supporting a larger number of soft channel bits.

B.5 HSDPA Planning and Dimensioning

B.5.1 Planning Basics

Since HSDPA is a shared data channel, the end user throughput will depend on the total number of HSDPA users in the cell. Capacity planning and dimensioning is therefore different in the case of HSDPA compared with the regular WCDMA DCH. However, as power and codes are shared between all users, overall cell capacity will be considered when doing the analysis.

There are several aspects that should be considered, especially for HSDPA planning. In the first place, HSDPA throughput depends on the SINR conditions in the UE. The transmission power of the HSPDSCH is constant and reserved, so it cannot be used by the DCH (i.e. link adaptation is used instead of power control). Secondly, the uplink transmission is done with a normal UL NRT DCH bearer. In addition, the higher bit rates require a review of the Iub and RNC capacity, which need to be dimensioned accordingly.

B.5.2 HSDPA Dimensioning

There are different approaches that can be used for HSDPA dimensioning, assuming that DCH dimensioning has already been done. The first one would use the remaining power, not used for the DCH, and obtain HSDPA and cell throughput as a result. The second option would take the desired HSDPA throughput as the input and use it to determine the required transmission power.

The dimensioning process requires different inputs from the operator, who needs to provide the expected traffic distribution and expected throughputs rates. Based on this information, the radio design can be done and the number of network elements, interface capacity, etc., can be designed accordingly. Figure B.6 shows the general dimensioning process.

Input Parameters to the Dimensioning Process

Traffic Definition
Typically traffic is provided as the number of subscribers and the traffic mix per subscriber, or the total amount of traffic in the network. Voice Erlangs and MB for different data services are typically provided. The service distribution in different areas (urban, suburban, etc.) also needs to be specified. In addition, the difference between regular WCDMA and HSDPA traffic needs to be defined.

HSDPA Planning and Dimensioning

$$\boxed{\text{SF}_{\text{HSPDSCH}} \frac{C}{I} \geq \text{SINR}_{\text{target}} \quad \text{and} \quad \frac{E_s}{N_0} = \frac{\text{SINR}}{M}}$$

Figure B.7 SINR definition formulas

For HSDPA, in addition to the total amount of traffic and service distribution, the average and minimum cell throughput (in Mbps) should also be indicated. Throughput specification is typically a requirement, but in some cases it could also be estimated based on other factors, like the amount of traffic and available power. The total aggregated throughput in the cell will depend on the number of users, the scheduling algorithm and distribution of the 'G factor', which is an indicator of the cell topology and the relation between the power received from the dominant cell and the strongest interferer cells.

Air Interface Dimensioning

Firstly, the number and type of sites needs to be specified for the air interface (i.e. sectors/site, maximum transmission power, etc.). Then, the average throughput in a certain location should be estimated based on the average SINR (signal-to-interference and noise ratio), which can be used to create the HSDPA link budget. The reason for using SINR instead of E_b/N_0 or E_s/N_0, as done for the DCH, is that it does not depend on the bit rate or number of codes. In the case of HSDPA, the bit rate and code can change in every TTI for each user. Figure B.7 defines the SINR and its relationship to E_s/N_0, where M is the number of HSPDSCH codes used during the TTI.

The average throughput for a single HSDPA user can be expressed as a function of the average HSDSCH SINR, given the maximum number of HSPDSCHs. Since C is the transmission power of the HSPDSCH ($P_{\text{HSPDSCH\_tx}}$) divided by the path loss (L_p), the key factor to define the value of the SINR and throughput is the received interference:

$$C = \frac{P_{\text{HSPDSCH\_tx}}}{L_p} \quad I = I_{\text{own}}(1 - \alpha) + I_{\text{other}} + P_N$$

where the interference is defined by three factors: in the first place, the interference received from the own cell (I_{own}), which depends on the multipath propagation in the cell, typically named the orthogonally factor (α); in the second place, the interference from other cells (I_{other}); and finally the noise power from the thermal source and equipment (P_N). Additionally, the 'G factor' can also be defined as a function of the interference in the cell:

$$G = \frac{I_{\text{own}}}{I_{\text{other}} + P_N}$$

Combining these definitions, the SINR can be expressed by the following formula:

$$\text{SINR} = \text{SF}_{\text{HSPDSCH}} \frac{P_{\text{HSPDSCH\_tx}}}{P_{\text{tot\_tx}} \cdot (1 - \alpha + 1/G)}$$

However, since the total HSDPA power is shared between the HSPDSCH and the HSSCCH, the SINR can be expressed as a function of the maximum HSDPA power in the cell:

$$\text{SINR} = \text{SF}_{\text{HSPDSCH}} \frac{P_{\text{tx}} \text{ Max HSDPA} - P_{\text{HSSCCH\_tx}}}{P_{\text{tot\_tx}} (1 - \alpha + 1/G)}$$

For a given maximum transmission power, the SINR can be obtained by applying the 'G factor' given to the cell configuration (i.e. -5 in the 3GPP macrocellular configuration) and obtaining the value of

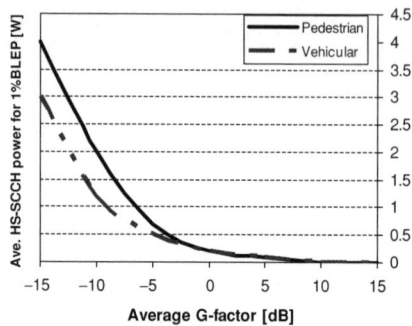

Figure B.8 HSSCCH power versus G the factor

HSSCCH power from simulated distribution tables, as shown in Figure B.8. Finally, the throughput can be obtained from throughput versus SINR tables also obtained by simulation.

Node B Configuration
The Node B configuration should be configured to enable it to carry the air interface capacity for DCH and HSDPA traffic. One Node- B will typically serve three base stations or sectors.

Iub Configuration/Sizing
The Iub configuration should be calculated based on the node B configuration, and is an estimation of the average number of simultaneous users for each of the sectors. The HSDPA throughput also needs to be taken into account.

This dimensioning will be a trade-off between the maximum air interface throughput and Iub efficiency. If Iub is dimensioned to support peak rates in the air interface, the Iub efficiency will be low, but as the Iub is dimensioned to support the average cell capacity, the peak throughput at the air interface may by limited by the Iub. Typically the average throughput criteria plus a certain margin would be defined.

It is important to note that the capacity available for HSDPA can be obtained after considering first the capacity needed for DCH, voice calls and signalling.

RNC Configuration
Typically the operator needs to define the number of RNC locations or how the Node Bs are distributed under different RNC sites. Different requirements need to be considered when deciding the number and configuration of the RNCs, i.e. the total traffic in the area, Iub connectivity and limitation of the number of sites/cells.

B.5.3 HSDPA Planning

HSDPA performance mainly depends on radio propagation and C/I conditions. Network planning needs to guarantee that the minimum criteria are fulfilled to ensure the required performance. As part of the planning, the planner should try to ensure clear cell dominance areas and minimise the SHO areas and interference from neighbour sites.

Figure B.9 shows an example of the planning process for HSDPA on a network where WCDMA has already been implemented. In the first place, the R99 network (NW) should be monitored so that current traffic can be analysed. The average BTS power used, BTS load, RNC utilisation, Iub utilisation

HSDPA Planning and Dimensioning

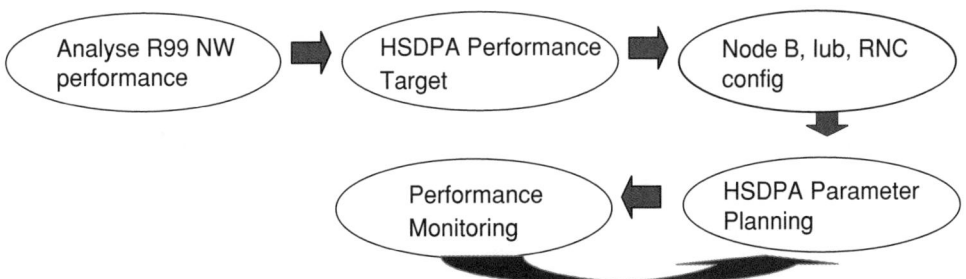

Figure B.9 HSDPA planning process

and SHO overhead should be considered. This is similar to the first phase of the dimensioning analysis. HSDPA performance target and strategy should be considered in conjunction with the load information to estimate the actual capacity needed.

From the coverage point of view, there are different strategies that the operator can consider, which will affect the planning of the cell:

- *General coverage.* This needs to include HSDPA coverage in the whole area and the impact of UL and mobility should be considered. In case SHO is not supported, additional planning of HO based on the DCH is needed.
- *Hotspot coverage.* This aims to concentrate the service only in certain areas, without the stress of service continuity.
- *Indoor solutions.* These focus on providing the service to certain customers and specific office locations. In this case throughput requirements may be more important and typically cells can manage higher bit rates. Some of the conditions that may help achieve good indoor coverage can be good isolation from the macro cell layer, a dedicated RF carrier, a high number of antennas and being able to allocate high power to the HSPDSCH.

In any case, the deployment strategy may have different phases, depending on the DCH traffic load, expected HSDPA traffic load and HSDPA throughput requirements. The objective is that in low loaded networks one single carrier can share HSDPA + DCH. However, in high load areas and hotspots dedicated HSDPA carriers should be added. In the case of a shared carrier, power allocated to HSDPA, priority of HSDPA over NRT DCH traffic, impact of higher interference on the DCH and impact of SHO areas should be considered. In the case of a dedicated carrier, the second carrier can be fully dedicated to HSDPA; however, at least some resources for the DCH may be needed in order to deal with a lack of SHO and control cell load.

For detailed Node B and Iub configurations and RNC planning, Node B and Iub configurations should be decided based on the HSDPA deployment strategy and the cell average throughput requirements. The cell capacity and Iub configuration should be done independently for each BTS, based on the assumptions. In addition, RNC planning is essential when HSDPA is introduced in the network. Since the capacity required by HSDPA users is higher than for the R99 users, the effective RNC capacity would be reduced accordingly.

HSDPA parameter planning and performance monitoring are the last two stages of the planning procedure and may be considered together, since the initial configuration of HSDPA parameters (i.e. maximum HSDP power, number codes, HO thresholds, etc.) may be modified based on the performance measured in the network and the relation between HSDPA and DCH traffic. However, performance monitoring can also be used to determine when the planning strategy needs to be changed or new dimensioning needs to be done.

B.6 Further Evolution: Release 6 HSDPA, HSUPA and HSPA

B.6.1 HSDPA Release 6 Improvements

There are a number of items in the release 6 standardisation that aim to bring further gains for HSDPA:

- fractional DPCH, where the objective is to try to avoid the need for full DCH in downlink when only PS data (SRB in HSDSCH) and a shared DL channel to provide uplink TPC commands are required;
- pre/post scheme to avoid DTX detection in base station for ACK/NACK commands, which allows lower power offsets for the ACK/NACK, and thus a better uplink range;
- RX diversity and improved receiver performance;
- L2/L3 signalling improvements on-going as well;
- mandatory HSDPA capable release 6 UE for RX diversity and an improved receiver.

B.6.2 HSUPA

High speed uplink packet access (HSUPA) is the complementary bearer for HSDPA in the uplink, which may allow very high upload throughputs (up to 5.8 Mbps). Similar to HSDPA, HSUPA is considered as 3.75G or even 4G.

The specifications for HSUPA will be included in UMTS release 6, published by the 3GPP. HSUPA will probably use an uplink-enhanced dedicated channel (E-DCH) on which link adaptation methods similar to those employed by HSDPA will be used: a shorter transmission time interval, enabling faster link adaptation, and HARQ (hybrid ARQ) with incremental redundancy, making retransmissions more effective.

A new packet scheduler will be needed, but in this case it will operate on a request–grant principle where the UEs request a permission to send data and the scheduler decides when and how many UEs will be allowed to transmit. A request for transmission will contain data about the state of the transmission buffer and the queue at the UE and its available power margin. In addition to scheduled transmissions the standards also foresee a self-initiated transmission mode from the UEs in order to support shorter delays needed for VoIP types of services.

Two basic scheduling methods are contemplated in the standard:

- *Long term grants* are issued to several terminals which can then send their data simultaneously. The grants are increased or decreased according to the current load of the cell and the requirements of the terminals.
- *Short term grants* would allow multiplexing terminals in the time domain instead of the code domain, as done for long term scheduling. In order to allow multiplexing uplink transmissions of several terminals in both the code and time domain the scrambling and channellisation codes will not be shared between different terminals.

Because in the uplink the DPDCH and DPCCH are code-multiplexed and transmitted simultaneously in time, the ratio of their transmitted powers is important for the achievable payload bit rates. The greater part of the DPDCH and DPCCH are code-multiplexed and transmitted simultaneously. This would mean that the more power of the UE that is assigned to the DPDCH, to increase the bit rate achievable on that channel, the less power is left for DPCCH and the less reliable is the signalling in the link. In UMTS R99 the ratio between the power of the DPDCH and DPCCH was set to a constant. In HSUPA this ratio can be controlled by Node B.

Table B.5 HSUPA UE categories

HSUPA category	Codes × spreading	TTI	Transport block size	Data rate (Mbps)
1	1 × SF4	10	7 110	0.71
2	2 × SF4	10	14 484	0.71
2	2 × SF4	2	2 798	1.40
3	2 × SF4	10	14 484	1.45
4	2 × SF2	10	20 000	2
4	2 × SF2	2	5 772	2.89
5	2 × SF2	10	20 000	2
6	2× SF2 + 2× SF4	10	20 000	2
6	2× SF2 + 2× SF4	2	11 484	5.74

In HSUPA, unlike in HSDPA, soft and softer handovers will be allowed for packet transmissions. The control of the transmitted power of the UE in soft/softer handovers on the E-DCH will be slightly different from that specified in R99 for the DCH; namely the main serving Node B will be able to issue both power-up and power-down commands but all other Node Bs participating in the handover will be able to issue only power-down commands. A power-down command will always take precedence over a power-up command.

Table B.51 shows the six different UE categories defined, which are mainly characterised by the number of parallel codes supported and the support of 2–10 ms TTI.

Appendix C

Digital Video Broadcasting

Lino Dalma

Digital video broadcasting, handheld (DVB-H), is a new European standard for mobile television. The standard is derived from the digital terrestrial television standard DVB-T (DVB, terrestrial) by adding enhancements for the reception in the mobile environment and integration with mobile networks. Development of mobile television is an interesting challenge from the technology point of view. There is also even a bigger challenge in predicting prospects of services based on mobile TV and their commercial aspects. This presentation gives an overview of mobile TV technology and issues of its integration with other broadband wireless systems.

C.1 Introduction

Digital television went through a full cycle of development: from basic research via standardisation to commercial introduction and building a significant market share. Today there is talk of finishing analogue TV broadcasting completely, first in some leading areas like Finland or the UK and later everywhere.

The progress of technology enabled another cycle of development to start in the area of television. This time television is being addressed to mobile devices which, judging by their numbers and variety, may open entirely new perspectives. Nowadays, however, the mobile TV has moved merely from the basic to the standardisation process so only its deployment later will judge its success. The current state of affairs offers in turn some period in which research and development can still be done.

C.2 Handheld Television: Viewing Issues

The number of picture elements and the number of colours rendered mainly describe the display quality. In fact, current high-end mobile phone displays can provide one-sixteenth and personal data assistant (PDA) devices one-quarter of the picture elements of a standard TV. Scaling the viewing distance by the same amount it is found that a handheld TV can provide home TV viewing quality. It has been also demonstrated that picture element density can be increased two times with no problems using the current technology. This would guarantee extremely good viewing quality for a handheld device.

As a result it can be seen that picture quality is not an essential problem. From this point of view, a handheld TV should rather be called a 'personal' TV since its watching experience will be solitary.

Indeed, handheld means more short-time viewing of clip material, but devices can also be set on a table top, which could also mean long-time viewing as well.

These are indications that a handheld TV is a viable concept for broadcast TV, especially with the expected progress in display technology. However, the handheld TV is not limited to the television only and can be seen as a universal system for download. This in turn puts it as one of the components in a wider landscape of emerging wireless broadband systems.

Compared to a digital terrestrial television, the handheld television is much more difficult from a technical standpoint:

- Mobile terminals have microscopic antenna size compared to standard television antennas. This requires higher signal levels with respect to television.
- Mobile reception is expected in virtually every place, especially inside buildings and in vehicles. This puts extraordinary demands on signal transmission which must be very robust.
- The power supply is limited, which may severely restrict consumption of TV content.

The mobile TV standard DVB-H addresses all of these issues with some ingenious techniques. Moreover, DVB-H is based on the digital television standard DVB-T which has capabilities for robust signal transmission. Therefore DVB-H is downward compatible with DVB-T and provides additional elements for enhancing the system robustness.

The enhancements include several aspects: signal modulation, error protection and time slicing. Signal modulation in DVB-T transmission is based on COFDM (coded orthogonal frequency division multiplex) and uses about 2000 or 8000 carrier frequencies on which transmitted streams are overlaid. This provides robust reception in vehicles and in environments with radio wave reflections.

The DVB-H standard foresees, in addition to these two types of modulation, the so-called 4k modulation mode, which uses about 4000 carriers in order to provide a good compromise between the 2k mode, which can support fast moving terminals but smaller transmitter coverage, and the 8k mode, which maximises cell size but can not support fast moving terminals, showing higher sensitivity to Doppler effects.

Mobile environment reception with small antennas is exceedingly prone to packet loss. To deal with this problem, packet error correction has been introduced in DVB-H as an option, in order to guarantee improved performances in difficult reception conditions. The final challenge is a limited battery power supply, precluding the use of TV over prolonged time periods. This is a serious problem since both the TV tuner part and the demodulator consume substantial amounts of power. To deal with this, the DVB standard uses so-called time slicing. The TV streams are sent to a receiver at higher speeds than normal and are buffered at the terminal. It is then possible to load the receiver buffer and switch the tuner and demodulator off until it is empty. For example, sending data four times faster than the nominal speed allows the receiving part to be kept switched off for 75 % of the time, extending the batter of life by the same amount. Taken together, the adoption of these techniques ensures technical viability of a handheld television.

C.3 System Issues of DVB-H and Broadband Wireless

A handheld television based on the DVB-H standard can be realised in several types of systems. In the simplest mode DVB-H will be running as part of the standard DVB-T digital television terrestrial system. This solution can be readily implemented using existing transmitter infrastructure but the problem lies in the limited transmission signal strength, which may be insufficient for large area coverage and in areas with obstacles. Another issue is limited bandwidth available within the existing television programming. It is thus inevitable that the DVB-H deployment will require a separate transmitter infrastructure to be established. Such infrastructure can be overlaid with existing cellular network, enabling planning for achieving universal coverage. Integration with cellular networks provides benefits for possible service coordination, e.g. the provision of interaction channels.

A larger task, which will come gradually in the future, is the integration of broadband wireless networks. Apart from DVB-H there are also other broadband wireless networks entering the deployment stage. A third generation UMTS system can be used for TV type services using multicast. Wireless LAN networks have significant bandwidth available for such services. It has been announced that mobile devices supporting UMTS and WLAN are soon to be available. Hence it is possible to imagine future devices supporting all three broadband wireless systems: UMTS, WLAN and DVB-H. Then the issues of coverage and access can be seen in a more general framework of network coordination, including mobile broadband handover. Here, the transmission data organisation of DVB-H can be still used in the different transport networks with mobility enhanced by handover between them.

Several types of system integration are possible. DVB-H can be integrated within the existing digital television DVB-T transmitter infrastructure. For tight integration with other broadband systems DVB-H can also be deployed by overlaying the mobile cellular network with transmitters located at base stations. There are several issues related to broadband wireless integration. The first is integration within a standard mobile network, e.g. for sending service information and provisioning the return channel. See second is integration with other emerging broadband wireless networks: 3G systems and wireless LAN. Here the integration can also be on the content level; the actual content can be sent on any of these networks and switching between them might be provided on a handover basis. Nevertheless, many issues in this area remain to be developed and tested.

C.4 Application and Services

By far the most critical aspect of DVB-H and other broadband wireless systems does not concern the technology but on providing applications and services attractive for consumers, especially the ones who would create a viable business model for operators. This is a difficult problem if consumers are not willing to pay for another TV service. On the other hand, advertising based TV for handhelds also does not seem to be attractive. Today there seem to be two main ways to solve this problem. One way is simple porting of current TV and Internet TV broadcasts to DVB-H, in the form of, for example, news clips. Incurring little costs, this would manifest the presence of content owner in another distribution channel. However, this does not seem to be universally a good idea due to limited battery power, which does not encourage watching full-length movies on mobile terminals (not asking even if this is convenient). It thus seems that mobile TV content will be more of a 'clip', 'flash', 'spot' or 'short story' type rather than full feature content. Obviously the best way to go is to devise new innovative services for which consumers would actually be enticed to pay, which is of course a very challenging issue.

These could be a form of special content tailored to handheld TV channels which might be composed of cartoons or music videos. Special attention should be made to the potential of using instant interactivity built into mobile devices. Game content and special entertainment comes to mind in this respect. It is evident, however, that applications and services for DVB-H require further studies.

C.5 Mobile Broadcasting

Today mobile streaming is a unicasting service not suitable for heavy mass communications to many receivers (broadcasting). Two promising technologies exist:

- *DVB-H (digital video broadcasting, handheld).* Specified by ETSI as an extension to DVB-T, DVB-H is based on a single-frequency network, bandwidth from 5 up to 12 Mbps depending on the modulation and protection schemes (code rates) on subcarriers and needs return channel, e.g. GPRS for interaction.
- *MBMS (multimedia broadcast/multicast service).* Specified in 3GPP release 6 (January 2005), the MBMS uses the existing EDGE/UMTS infrastructure at 64–384 kbps.

Table C.1 Technology overview

Technology	Weakness
UMTS	Not scalable for mass content delivery
DVB-T	Designed for rooftop reception
	Need for an efficient power-saving mechanism
	Inadequate impulse noise protection and mobility
DAB	Designed for devices with power constraints but too narrow a spectrum assigned for data transmission

C.5.1 DVB-H Technology Overview

To focus on significant contributions in the development of DVB-H systems and applications a substantial investment in the related research infrastructure has recently been made. The first commercially available DVB-H system in the world was ordered and installed before the end of 2004. The system is composed of a commercial DVB-H/DVB-T transmitter with output power of 50 W and a DVB-H playout system, which produces valid DVB-H streams including those for multiprogramme broadcast. Test receivers will also be available. The DVB-H test system will enable technical trials and investigation of system properties in practical conditions. It will also make it possible to investigate the impact of transmission parameters on performance and the related optimisation. In the context of research in the broader area of broadband wireless networks, the most recent WLAN (dual-band triple mode system) and, via an industrial grant, two UMTS base stations have also been installed. Due to these efforts infrastructure for more advanced research in this area has been created.

C.6 DVB-H Introduction

Convergence of digital media and communication gives users the chance to consume most digital content in the mobile environment. The emerging DVB-H standard aims to provide digital TV reception in mobile devices. Earlier known as DVB-X, DVB-H is being standardised by an ad hoc group of the DVB organisation standardised at the end of 2004. DVB-H combines traditional television broadcast standards with specific elements to handheld devices, such as mobility, smaller screens and antennas, indoor coverage and reliance on battery power.

C.6.1 Motivation for Creating DVB-H

The main reasons behind the choice of a new technology to deliver audio/video contents to many mobile users rely on some technical considerations, listed in Table C.1. This new technology is needed in order to face the following issues:

- power consumption in mobile terminals;
- performance in the mobile environment, where there is noise, the Doppler effect and impulse interference;
- network design flexibility, in order to maintain a strong enough signal for single antenna mobile reception in medium to large signle-frequency networks (SFNs).

C.6.2 Overview of Mobile TV Broadcasting Technology (DVB-H)

Enhanced User Experience

- Enables over 50 channels and interactivity TV programmes simultaneously
- Possibility for digital radio channels using DVB-H
- DRM solutions (OMA DRM2 and 18C) for access rights for pay channels

Interactivity and Connectivity

- IPDC over DVB-H allows two-way stream, enabling consumers to interact with programmes
- Two radio receivers in DVB-H handsets: one to receive TV broadcasting and another for cellular connectivity

Extended battery life

- Power saving through a packet based system; receiver only 'on' around 10 % of the time

Seamless Viewer Experience

- Soft handover; seamless switching between transmitters

Scalability

- DVB-H interleaves allow the network to be scaled
- DVB-H network scales according to the number of users

Open Standards

- Equal access to technology (ETSI)
- DVB-T compatibility (same physical layer, modulators and transmitters)

Throughput

- DVB-H has largest output capacity of broadcasting technologies up to 11 Mbps

Reception/Coverage

- Spectrum band: 470–702 MHz
- Bandwidth: 8, 7, 6 or 5 MHz
- Integrated antenna
- Indoor reception support
- Large area coverage

Flexibility for Mobile Broadcast Networks

- Single antenna in medium to large SFNs
- Many different capacity alternatives and cell sizes

The positioning IP datacasting over DVB-H and other broadcast technologies is summarised in Table C.2.

Table C.2 Technology comparison (DBM, digital multimedia broadcasting, ISDB-T, terrestrial integrated services digital broadcasting)

	Mobility	Data rate	Frequency band[a]	Comments
DVB-H	High	5–31 Mbps; typical 10 Mbps	UHF (470–702 MHz)	
DAB	High	Maximium 1.15 Mbps	VHF	VHF antenna size?
Korean terrestrial DMB	High	Maximium 1.15 Mbps	VHF	VHF antenna size?
ISDB-T	High (one segment)	Maximium 1.15 Mbps; typical 800 kbps	VHF + UHF (91–770 MHz)	Japan only
WLAN	Low	Maximum 11 Mbps	2.4 GHz	Small cell size
MBMS	High	300 kbps (multicast)	WCDMA (1920–1980 MHz + 2110–2170 MHz)	Limited capacity and medium cell size
RDS	High	~100 bps	FM–radio band ~100 MHz	Too low bit rate

[a] UHF, ultra high frequency; VHF, very high frequency

C.6.3 Overview of DVB-T

The terrestrial digital television standard is currently adopted in 36 countries worldwide, allows one-to-many broadband wireless data transport video, audio, data and, importantly, IP packets. It is scalable with a cell size up to 100 km (DVB-H cell size is smaller) with a huge capacity of 54 channels each with 5–32 Mbps. Developed for MPEG-2 (Moving Picture Experts Group 2) stream distribution, it can basically carry any data, is flexible and has many transmission modes, 4.98–31.67 Mbps at $C/N = 25$ dB.

The COFDM multicarrier modulation with 2k and 8k modes is applied and in the countries where the phase-alternating line (PAL) standard is adopted (e.g. Europe), the bandwidth of one DVB channel is ~8 MHz. Some parameters characteristics are:

- 1705 subcarriers (spacing, 4464 Hz) for the 2k mode;
- 6817 subcarriers (spacing, 1116 Hz) for the 8k mode;
- carrier modulation: QPSK, 16-QAM or 64-QAM;
- error correction: convolution code and Reed–Solomon encoder.

A classic example mode can be:

- 64-QAM, code rate = 2/3, guard interval 1/8, which gives a 22.12 Mbps capacity when $C/N = 19.2$ dB with an 8 MHz channel bandwidth.

DVB-T includes hierarchical modes where two transport streams can be sent simultaneously with low and high priorities. DVB-T can also be used for broadcasting to mobile devices, altough a suitable mode has to be selected:

- 8k 64-QAM: < 50 km/h;
- 2k QPSK: > 400 km/h tolerable.

Nevertheless, a separate network for DVB-H is desired in order to achieve optimisation of speed, coverage and capacity.

DVB-H Introduction

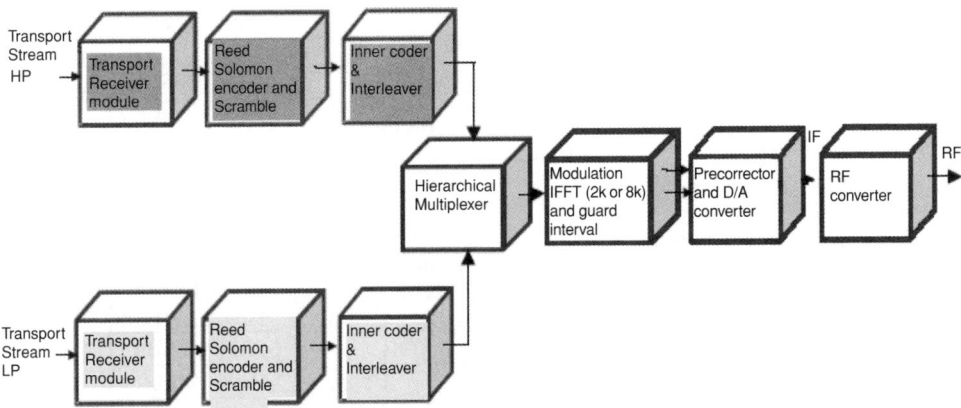

Figure C.1 Hierarchical encoding

Hierarchical Encoding

Three possible constellation offsets, $\alpha = 1, 2$ and 4, are defined, but the principle is the same. Two transport streams are fed into the modulator; one of the transport stream controls which is the quadrant where the constellation is placed (high priority stream) and the other defines the 16-QAM signal (low priority stream). The benefit of hierarchical encoding (see Figure C.1) is that the system allows one programme for a mobile or portable application together with a service supporting typically three programmes to be picked up in a stationary application.

SFN operation represents some difficulties as the low priority (LP) and high priority (HP) channels might come from different programmes providers and via different networks. Data are read from each of the megaframe information pockets (MIPs) (LP and HP).

The modulation parameters such as the guard interval, modulation parameters and time to transmit are read from the HP MIP. Only code rates are read from both the LP and HP channels.

DVB-H 4k Mode and In-Depth Interleavers

Some features are:

- interpolated solution between the 2k and 8k modes;
- directly scaled parameters;
- dedicated 4k mode symbol interleaver;
- continual pilots from the same arrangement (8k);
- easy implementation; only some control logic needed;
- 8k interleaver can be used with 4k or 2k DVB-T physical level native interleaver and works within one OFDM symbol;
- when 8k interleaver is used with 4k, interleaving happens over two symbols;
- when 8k interleaver is used with 2k, interleaving happens over four symbols.

A native inner interleaver for the '4k' mode as well as the option to use the 8k inner interleaver on the encoded bit flows produced for the 2k and 4k modes are specified in Figure C.2.

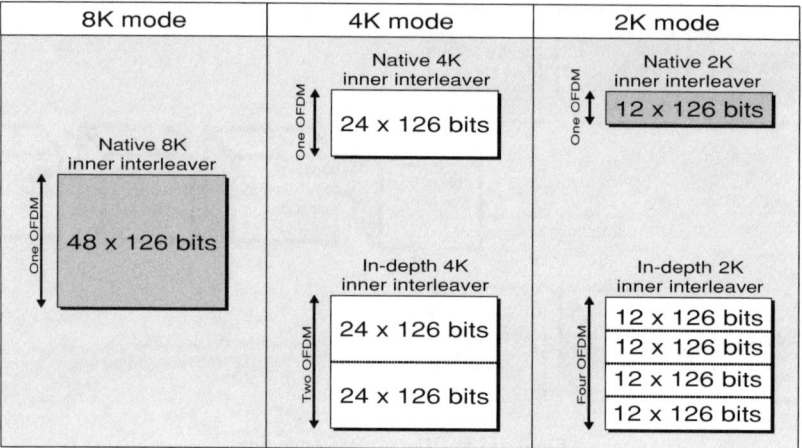

Figure C.2 Native inner interleavers

C.6.4 DVB-H Innovative Elements

Time Slicing

Time slicing (Figure C.3) is adopted in order to achieve battery saving in handsets for mobile reception applications. It is applied on top of DVB-H modulation, using the time division multiplexing (TDM) technique. Therefore, a video stream is sent in bursts on the transmission side and the DVB-H receiver, which has a buffer, is switched on only during the bursts related to the selected content; this allows approximately 90 % battery power savings and the terminal reads data from its buffer, displaying content continuously. In other words, the time between bursts gives the power saving (off-time). Moreover, time slicing offers, as an extra benefit, the possibility of using the same receiver to monitor neighbouring cells during the off-time.

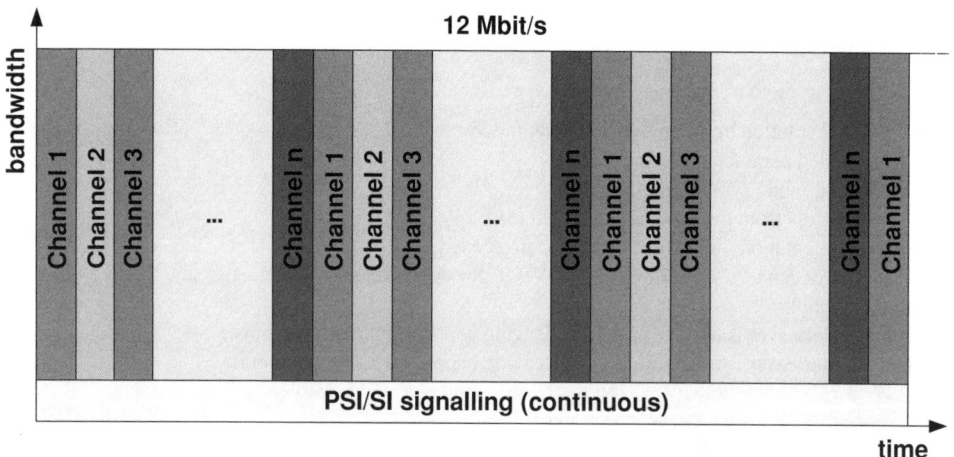

Figure C.3 Time slicing (PSI, program specific information; SI, service information)

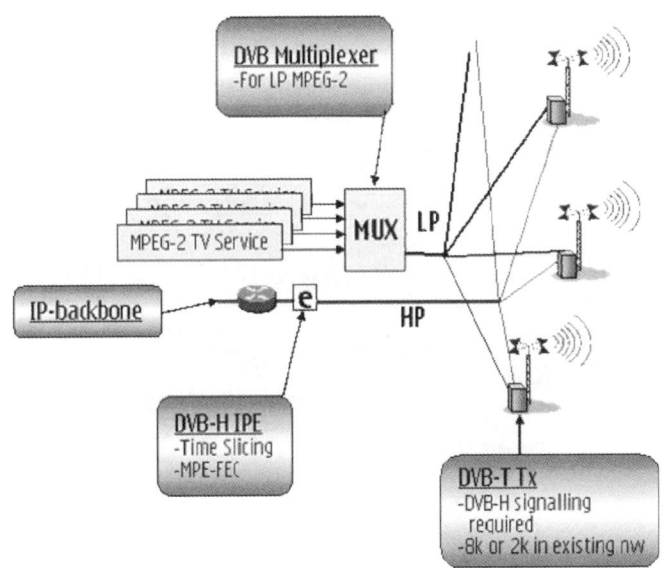

Figure C.4 DVB-H services in the existing DVB-T network with multiplexing

MPE-FEC for Performance

MPE-FEC stands for multiprotocol encapsulation forward error correction and is applied on the IPv6 packet level in order to:

- add impulse noise protection;
- enable higher speed terminal movement.

The main characteristics of this coding scheme are the following:

- additional data link layer Reed–Solomon (RS) coding for IP datagrams;
- IP data are filled in the vertical direction;
- the table is padded;
- RS code words are calculated in the horizontal direction;
- RS columns are formed in the vertical direction;
- data are transmitted in the vertical direction as MPE and FEC sections.

Regarding the frame structure, the practical MPE-FEC consists of 256, 512, 768 or 1024 rows and constant 255 columns, divided into 191 columns of IP data and the remaining 64 of parity bytes. Different sizes of MPE-FEC frames are needed for optimal planning (number of services, capacity of each service, time slicing on/off times, etc.) and the maximum buffer size for the frame is approximately 2 Mbit ($1024 \times 255 \times 8$).

C.6.5 DVB-T and DVB-H Coexistence

Since DVB-H is fully compatible with the DVB-T standard, the possible deployment scenarios foresee network sharing. The first scenario proposes the introduction of DVB-H services in the existing DVB-T network with multiplexing, with the mandatory support of portable indoor reception provided by the DVB-T network, as shown in Figure C.4. A second possibility might be to introduce DVB-H services in the existing DVB-T network with hierarchy (Figure C.5).

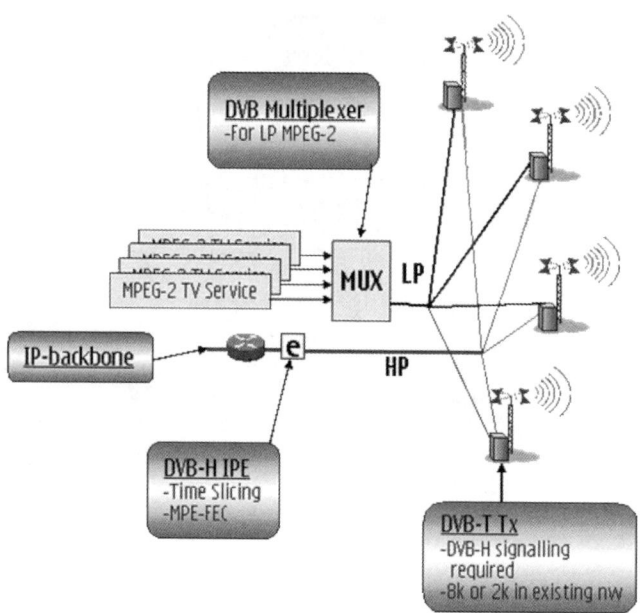

Figure C.5 DVB-H services in the existing DVB-T network with hierarchy

Conclusion

The future of handheld TV based on the DVB-H standard cannot be predicted today as ultimately it will be driven by commercial and revenue aspects. An even broader vision of an integrated broadband wireless system also including 3G and WLAN, while technically becoming possible, can only be given support by observing the ever-growing need for greater network bandwidth. Once the level of applications and services are achieved this vision requires further studies. There is no doubt that broadband wireless systems will be tested, validated, deployed and undergo ultimate verification in the market place. In preparation for this on the research side, Tampere University of Technology is assembling a unique test and development infrastructure that includes a DVB-H television transmitter and full play-out system plus 3G base stations and WLAN.

Appendix D

TETRA Network Planning

Massimiliano Mattina

D.1 TETRA Standard

The terrestrial trunked radio (TETRA) standard is an open digital trunked radio standard defined by the European Telecommunications Standardisation Institute (ETSI) to meet the needs of the most demanding private mobile radio (PMR) and public access mobile radio (PAMR) users. Usually the PMR network is owned and operated by a private company or organisation (police, ambulance and fire services, security services, utilities, military, public access, fleet management, transport services, closed user groups, factory site services, mining, etc.). The PAMR network is usually owned commercially and is technically operated by the operator, which sells the network services to companies and organisations as the TETRA facilities allow the network to be shared between the organisations.

The TETRA standardisation work was started by ETSI in 1988. The TETRA release 1 standardisation is almost complete and has reached a 100 % acceptance from all European administrations. Elsewhere in the world there has been formal adoption of the standard in China and contacts have been let throughout Europe and in the Middle East, Asia, Australia, South America and South Africa, although TETRA is not currently available in North America.

TETRA offers various features very useful for PAMR and PMR users that are not available in other standardised current technologies, such as the fast call setup time, excellent group communication support, direct mode operation between radios, high frequency efficiency and excellent security features; furthermore, it also provides packet data and circuit data transfer services.

In the current cellular radio systems TETRA is a trunking system from the channel assignment mode of operation point of view. This means that for each cell there is a pool of available traffic channels and each user, in the dedicated mode, have a traffic channel allocated on a call basis. Upon termination of the call, the previously occupied channel returns immediately to the pool of available channels. Using this channel allocation strategy, a large amount of random users can be accommodated in a relatively small number of traffic channels.

The TETRA standard has been developed with the aim of supporting international roaming by the provision of national and multinational networks. This means that the TETRA network can be connected to various different networks (e.g. PSTN, other TETRA networks, PABX, etc.) and each of them can have a large number of users. Furthermore, the use of shortened identities for intra-TETRA calls is also possible, reducing the signalling information in the call setup messages.

Advanced Cellular Network Planning and Optimisation   Edited by Ajay R Mishra
© 2007 John Wiley & Sons, Ltd

TETRA standardisation has reached a mature state. Major future developments are expected to continue with standardisation of the next generation of TETRA release 2 and maintenance of existing TETRA standards. The objective of TETRA release 2 is that EP (ETSI project) TETRA produces an additional set of ETSI deliverables (and maintenance thereafter) in order to enhance TETRA in accordance with the following requirements:

(a) Evolution of TETRA to provide packet data at much higher speeds than are available in the current standard. This is to support multimedia and other high speed data applications required by existing and future TETRA users.
(b) Selection and standardisation of additional speech codec(s) for TETRA, to enable intercommunication between TETRA and other 3G networks without transcoding and to provide enhanced voice quality for TETRA by using the latest low bit rate voice codec technology.
(c) Further enhancements of the TETRA air interface standard in order to provide increased benefits and optimisation in terms of spectrum efficiency, network capacity, system performance, quality of service, size and cost of terminals, battery life and other relevant parameters.
(d) Production and/or adoption of standard to provide improved interworking and roaming between TETRA and public mobile networks such as GSM, GPRS and UMTS.
(e) Evolution of the TETRA SIM, with the aim of convergence with USIM, to meet the needs for TETRA specific services while gaining the benefits of interworking and roaming with public mobile networks such as GSM, GPRS and UMTS.
(f) Extension of the operating range of TETRA, to provide increased coverage and low cost deployments for applications such as airborne public safety, maritime, rural telephony and 'linear utilities' (e.g. pipelines).
(g) Provision of new ETSI deliverables in order to support further user/market-driven requirements that may be identified during study work in the early stages of the work programme.
(h) Ensure full backward compatibility and integration of the new services with existing TETRA standards, in order to provide future proof of existing and future investments by TETRA users.

TETRA standards are freely available from ETSI. The following list includes the major standard series of TETRA voice plus data (V + D) and direct mode operation (DMO):

- Series 300 392   Voice plus data $(V + D)$
- Series 300 394   Conformance testing specification
- 300 395   Speech codec for a full-rate traffic channel
- 300 396   Technical requirements for a direct mode operation (DMO)
- 300 812   Security aspects; subscriber identity module to mobile equipment (SIM ME) interface
- TR 101 494   SIM review
- 01040   Security lawful interception (LI) interface
- TS 101 789 1   Trunked mode operation (TMO) repeaters. Part 1: requirements, test methods and limits
- EN 301 435   Attachment requirements for TETRA
- EN 303 035   Harmonised EN for TETRA equipment covering essential requirements under article 3.2 of the Radio and Telecommunications Terminal Equipment (R&TTE) Directive
- ETR/TR 300   Voice plus data $(V + D)$ designer's guide
- TR 102 021   User requirement Specification TETRA release 2
- ETR 300-1/2 is suggested as reading for beginners who are approaching the study of this system in order to gain an overview of the standard and the features of TETRA.

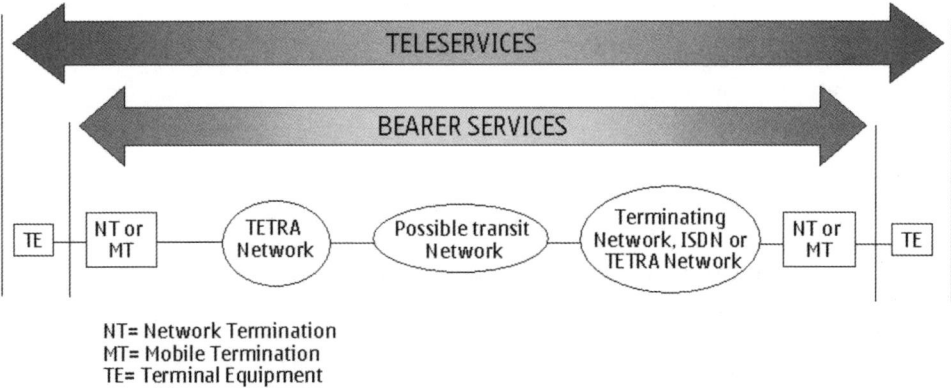

Figure D.1 TETRA services

D.2 TETRA Services

The TETRA standard defines the following services (Figure D.1):

- Bearer services
- Teleservices
- Supplementary services

A bearer service provides a communication capability between terminal network interfaces, excluding the functions of the terminal. It is characterised by some lower layer attributes (layers 1 to 3). The bearer service includes the individual call, group call, acknowledged group call and broadcast call for each of the following:

- Circuit mode unprotected data 7.2; 14.4; 21.6; 28.8 kbps
- Circuit mode protected data 4.8; 9.6; 14.4; 19.2 kbps
- Circuit mode protected data 2.4; 4.8; 7.2; 9.6 kbps
- Packet connection oriented data
- Packet connectionless data

A teleservice provides the complete capability, including terminal functions, for communication between users. Thus in addition to the lower layer attributes it also includes high layer attributes (layers 4 to 7). The teleservice includes clear speech or encrypted speech in each of the following:

- Individual call (point-to-point)
- Group call (point-to-multipoint)
- Acknowledged group call
- Broadcast call (point-to-multipoint one way)

A supplementary service modifies or supplements a bearer service or teleservice. PMR type supplementary services include:

- Access priority, preemptive priority, priority call
- Include call, transfer of control, late entry

- Call authorised by despatcher, ambience listening, discreet listening
- Area selection
- Short number addressing
- Talking party identification
- Dynamic group number assignment

Telephone type supplementary services include:

- List search call
- Call forwarding – unconditional/busy/no reply/not reachable
- Call barring – incoming/outgoing calls
- Call report
- Call waiting
- Call hold
- Calling/connected line identity presentation
- Calling/connected line identity restriction
- Call completion to busy subscriber/no reply
- Advice of charge
- Call retention

D.3 TETRA Network Elements

The TETRA standard describes the following network elements:

- Mobile station (MS)
- Switching and management infrastructure (SwMI)

In practical implementations the SwMI is composed of the radio base stations and digital switching, but their internal structure and the interfaces between them are not standardised. Furthermore, the standard does not specify details about the dispatcher workstation, but it is considered as an external network user like PSTN or PABX. Figure D.2 shows the TETRA network elements and their connections in the most common applications.

The network elements are:

- Radio terminal
- Radio base station
- Digital switch
- Dispatcher workstation

The radio terminal is used by subscribers to get services from the network. Its main task is to guarantee the best possible service performance for the user. When the MS is switched on, cell selection and reselection processes allow it to be camped on the cell with the highest received power level.

Different types of radio terminals are available depending of the scope of the use and nominal output:

- Handheld radio terminal
- Vehicular radio terminal
- Motorcycle radio terminal
- Aircraft radio terminal

Table D.1 Radio terminal power output classes

Class	Nominal output power (dBm)
1	45
1L	42.5
2	40
2L	37.5
3	35
3L	32.5
4	30
4L	27.5

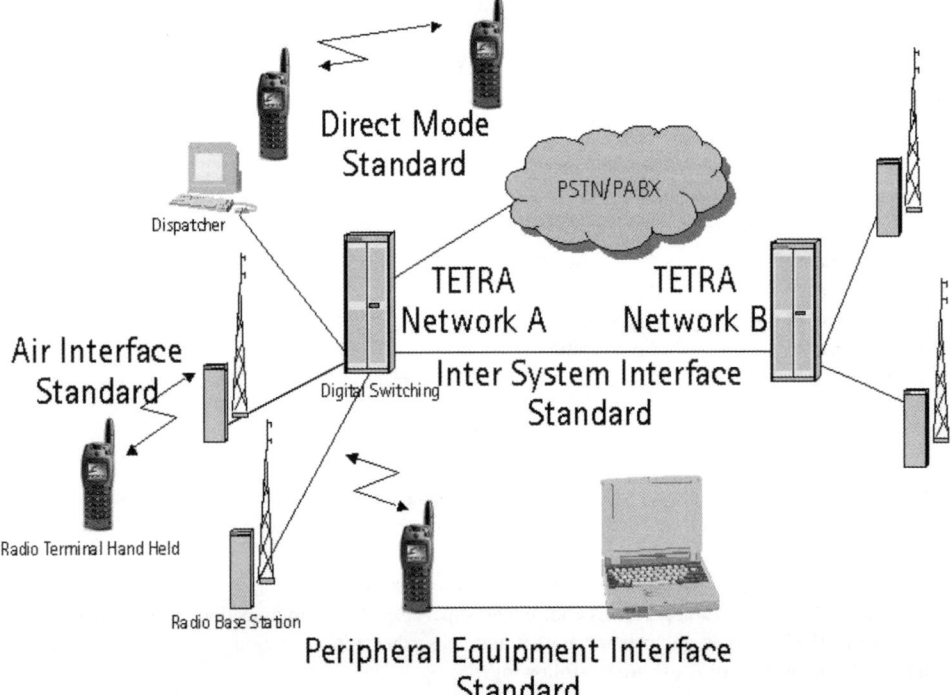

Figure D.2 TETRA basic equipment and standard interface

The radio terminal power output classes are specified by ETSI 392-2 (air interface) (Table D.1). The radio terminals have specific features for TETRA. For example, they usually have special buttons for emergency calls, for selecting the groups and push to talk and are also provided with a loudspeaker (Figure D.3).

The SwIM is the heart of the system and contains the higher hierarchical layers in the network. It is mainly dedicated to the management of the overall network resources and to the system synchronisation. The most important functions of the digital switch are the following:

Figure D.3 Handheld radio terminal

- Call control functions
- Subscriber management
- Mobility management
- Management of the radio channel
- Management of the TETRA base station
- Management of the dispatcher workstation
- Signalling

The functions of the TETRA base station are the following:

- Provides the air interface between the switching and management infrastructure and the radio terminals
- Conversion of the digital base band signal coming from the digital switch to the RF band

Table D.2 shows the output power classes of the radio base station.

The dispatcher workstation performs control and supervises the communication in the field and management functions such as group management and subscriber administration. The most common dispatcher communication function are the following:

- Receive calls and callback requests
- Make individual calls
- Make calls to and receive calls from PABX/PSTN and other external systems
- Handle emergency calls
- Communicate with one or several groups
- Send and receive status and short data messages

The most common management functions performed by the dispatcher are the following:

- Add and remove radio users from groups
- Create new groups and temporary groups

Table D.2 Base station power classes

Class	Nominal output power (dBm)
1	46
2	44
3	42
4	40
5	38
6	36
7	34
8	32
9	30
10	28

- Add new radio users to the system or modify communication rights and other attributes of existing radio users
- Manage groups and group areas
- Manage organisations
- Manage workstations and client applications
- Manage parameters, connection groups and circuit groups

Figure D.2 shows the interfaces that the TETRA standard specifies:

- Air interface (between the radio base station and the radio terminal)
- Peripheral equipment interface (between the radio terminal and peripheral equipment such as the computer)
- Intersystem interface (between the TETRA network of different manufacturers)
- Direct mode operation (between radio terminals for direct communication)

Each of these interfaces is described in detail in the relevant ETSI recommendation. The manufacturers can freely design the other interfaces (e.g. the interface between the radio base station and the digital switch).

D.4 TETRA Main Features

The TETRA standard introduces very interesting features for the emergency and safety networks and most of them are innovative for existing analogue systems and nonstandard networks. Each user, depending by his/her rights, can perform individual or group communication. Individual communication can be full duplex or half duplex. This type of call is point-to-point, i.e. between two users. On the contrary, in a group call different users can be involved in the same time (point-to-multipoint). However, only one user can speak at a time according to the priority. The creation of this group of users is not restricted to an area or number of channels; furthermore, with additional features the group calls allocate traffic channels only in the cells where the users of the specific group have a significant advantage from traffic resource use.

Thanks to digital technology, the network can be shared between multiple organisations without interfering with them. Each organisation can independently manage the right of the users or dispatchers or the group configuration. Each user or group has a specified priority to define the queing order for the channel resource. This feature allows optimisation of the use of the network, saving the investment put in for implementation of the network.

The radio base station fallback feature helps to improve network resilience. This feature allows communication within the cell when the radio base station is disconnected by digital switching, for example, due to unavailability of the communication links or unavailability of digital switching itself.

The data communication features allow the user to query databases, telemetry, remote control, email and fax, etc. Additional features are:

- Special calls type such as group calls and emergency calls
- Priorities for serving the calls
- Dispatcher workstations for the correct user management and group control
- Direct mode operation (direct communication between terminals without the support of the radio base station)
- Short data service (similar to the SMS of GSM)
- Implementation of end-to-end encryption
- Multivendor environment because the air interface is standardised (with the possibility of using radio terminals of different vendors under the same infrastructure)
- Efficient use of the frequency spectrum
- Fast call setup
- Radio base station fallback
- Data communication

D.4.1 Physical Layer

The TETRA air interface is partitioned in frequency and each frequency is partitioned in time (FDMA and TDMA). The base station and mobile station communicate in duplex; therefore one set of frequencies is allocated to the uplink path and another set to the downlink path. One pair of frequencies for each cell is reserved for the main control channel (MCCH). Each frequency is divided into four timeslots, allowing four simultaneous communications.

The carrier spacing is 25 kHz width, the modulation is $\pi/4$ DQPSK and (differential quadrature phase shift keying) the gross transmitting rate is 36 kbps. Figure D.4 shows the relevant modulation constellation. The bandwidth efficiency is very high:

$$\frac{36\,\text{kbps}}{25\,\text{kHz}} = 1.44\,\text{bps/Hz}$$

Figure D.5 compares the available number of channels in the 200 kHz bandwidth for various technologies. The TETRA standard provides 32 channels as against 16 or 8 of the GSM and traditional PMR technologies in the 200 kHz bandwidth. In fact, the TETRA standard considers 25 kHz radio channel spacing and therefore provides 8 radio channels in the 200 kHz bandwidth, since each radio channel is divided into four timeslots, in the same bandwidth the TETRA standard can provide 32 channels. On the other hand, the technologies like GSM and traditional PMR provide a smaller number of channels in the same frequency bandwidth. The GSM standard provides 8 channels, the half-rate GSM 16 channels and the traditional PMR applying only frequency division multiplexing (FDM) 8 or 16 channels depending on the radio channel spacing (12.5 or 25 kHz).

The TETRA speech codec is based on the code-excited linear predictive (CELP) coding model. This is a parametric codec using linear predictive coding with a bit rate of about 4.5 kbps. This technique is used to transmit information about the excitation sequence and the filter coefficients. The TETRA codec uses an excitation generator based on a codebook of algebraic codes (ACELP) and contains many excitation sequences of pulses with different amplitudes and positions. Figure D.6 shows the process applied to

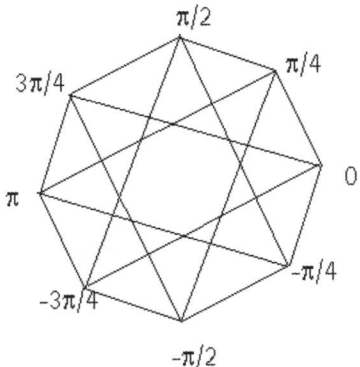

Figure D.4 TETRA modulation constellation

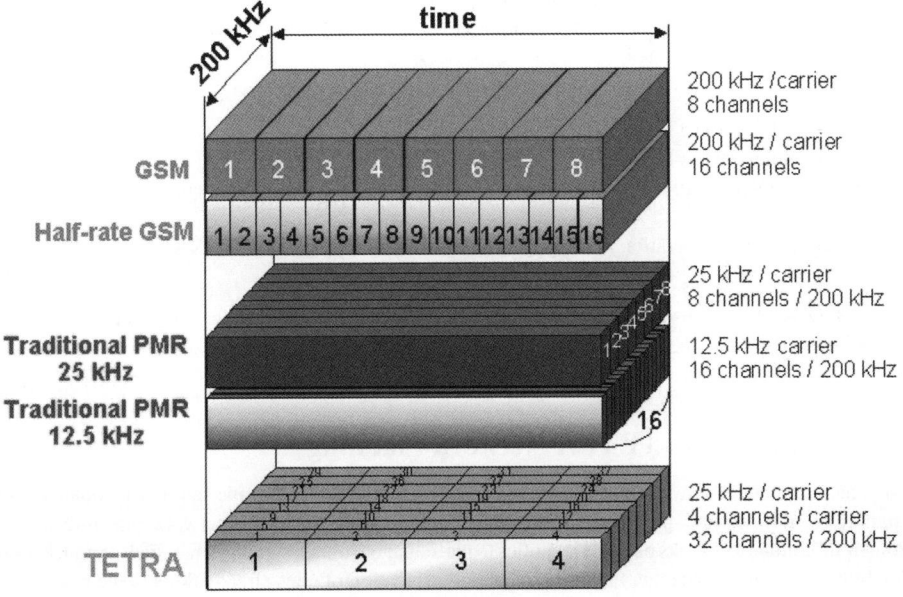

Figure D.5 TETRA spectrum usage

encode the voice and to build the frame. The physical channel is obtained by assembling the timeslots belonging to different frames. The frequency bands standardised by CEPT for TETRA are the following:

- 380–390 MHz and 390–400 MHz
- 410–420 MHz and 420–430 MHz
- 450–460 MHz and 460–470 MHz
- 806–824 MHz and 851–869 MHz

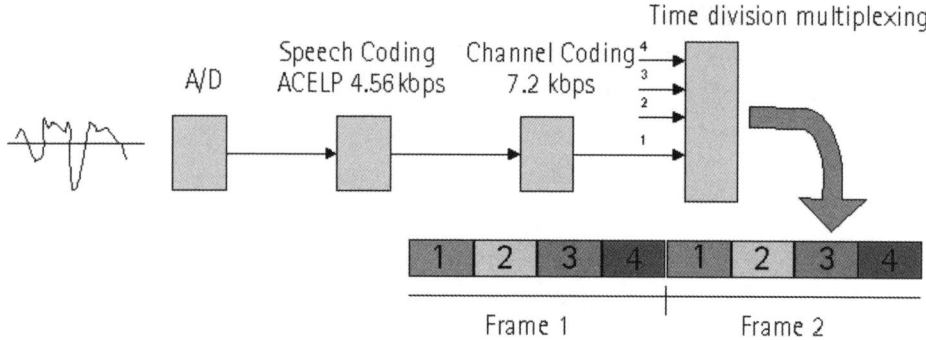

Figure D.6 Channel coding and frames

D.4.2 TETRA Memorandum of Understanding

The objective of the TETRA Memorandum of Understanding (MoU) is to support and promote the TETRA standard worldwide and to provide a forum to share and exchange information and ideas among a wide variety of individuals who share a common interest in the success of the standard. The purpose of the TETRA MoU is to:

- support the ETSI project TETRA to the full extent;
- support the initiative to obtain additional TETRA frequencies;
- develop a TETRA certification mark that will uniquely identify all equipment meeting the TETRA standard;
- ensure appropriate cooperation and support of the validation process of TETRA.

On the TETRA MoU website, the interoperability certificate for radio terminals and infrastructure can be found (www.tetramou.com).

D.5 Introduction to TETRA Network Planning

Although the general network planning process is very similar to the one applied to other cellular technologies, the network planning for TETRA usually differs from the network planning of other commercial cellular networks because it applies to particular customer requirements, such as high network availability, wide area coverage, low investment and protection of ones already done in the past.

The general network planning process is usually divided into the following steps (Figure D.7):

- Radio network planning
- Traffic capacity network planning
- Frequency planning
- Transmission and core network planning

Each of these steps could require a loopback to the beginning of the process because the network planning objectives are not met. This could mean that there is a need to go back to the previous step where, for example, new radio base station positions must be selected or the offered traffic capacity must be modified. Therefore the whole process could be iterative.

Introduction to TETRA Network Planning

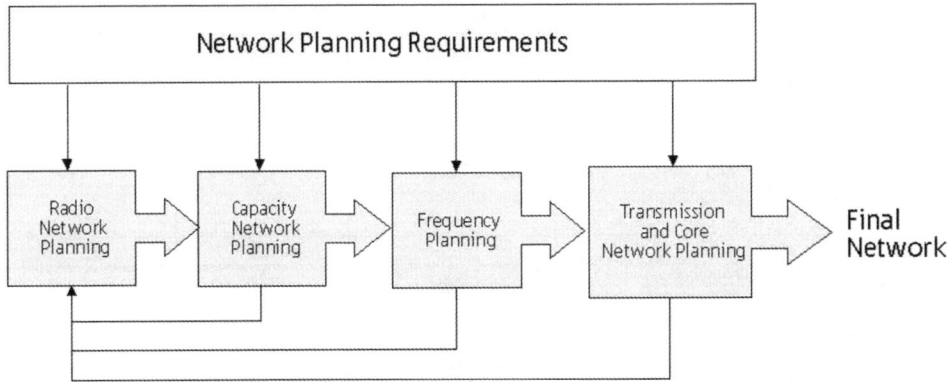

Figure D.7 Network planning process

Radio network planning consists basically of the definition of the quantity and position of the radio base stations in order to meet radio coverage requirements. This objective can be met by properly choosing the antenna system and selecting the best available sites.

The customer wants to use the minimum number of sites; therefore very often sites on the top of a mountain or hill are preferred. This choice can often produce problems in the traffic capacity and frequency planning phases. In other cases, the TETRA system is used to replace an existing old PMR system and therefore the customers want to re-use as much as possible of the existing infrastructure. This places constraints on site selection, with the result that the number of sites will probably not be optimised.

Traffic capacity planning is a very important step in the process because the cells must be dimensioned avoiding at the same time the saturation of the traffic resource and waste of frequency carriers. In a PS&S network the traffic resources must be dimensioned very carefully since in the case of an accident or other situations requiring police force intervention, the network can suffer from an anomalous concentration of users under the same cell and the peak traffic could reach unexpected values. When the traffic capacity resources are defined for each cell, frequency planning can be carried out with the scope to re-use the same frequency as much as possible while minimising the probability of co-channel and adjacent channel interference.

At the end of the process is transmission network planning. This phase is very important because the transmission network should be planned to provide the required protection against equipment failure. The network topology, transmission media and transmission equipment characteristics should be selected carefully in order to design a network having a high level of availability and resilience. Constraints about the network topology could affect the selection of radio base station sites; in these cases a trade-off is required between the selection of sites to meet the radio coverage objectives and network topology requirements.

This is a simplified view of TETRA network planning, because each of the steps shown in Figure D.7 must be exploited and many of the different details need to be explored and resolved. The following section provides some more information about radio network planning.

D.5.1 Radio Network Planning

Radio network planning can be divided into various phases, starting from preliminary radio network planning and arriving at the detailed radio network. Preliminary radio planning is very useful for an initial dimensioning of the network and in this section an overview of this first step is given.

	Downlink	Uplink	Unit	Formula
TX Power	P_{BS}	P_{MS}	dBm	A
TX Cable and Filter Loss	B_{BS}	B_{MS}	dB	B
TX Antenna Gain	C_{BS}	C_{MS}	dBi	C
Peak Effective Isotropic Radiated Power	$P_{BS} - B_{BS} + C_{BS}$	$P_{MS} - B_{MS} + C_{MS}$	dBi	$D = A - B + C$
Propagation Loss	L	L	dB	E
Signal Level at RX Antenna	$P_{BS} - B_{BS} + C_{BS} - L$	$P_{MS} - B_{MS} + C_{MS} - L$	dBm	$F = D - E$
RX Antenna Gain	G_{MS}	G_{BS}	dBi	G
RX Cable Loss	H_{MS}	H_{BS}	dB	H
Receiver Input Power	$R_{BS} = P_{BS} - B_{BS} + C_{BS} - L + G_{MS} - H_{MS}$	$R_{BS} = P_{MS} - B_{MS} + C_{MS} - L + G_{BS} - H_{BS}$	dBm	$F + G - H$
NOTE:	The BS antenna gain may be different for transmission and reception. This permits allowance to be made for techniques such as antenna diversity.			

Figure D.8 General radio link power budget from ETSI

The preliminary radio network planning process begins with the definition of the radio coverage objectives. The area to be covered and the available services must be defined in the way that allows verification of the compliant of the project. Usually the area is defined by an administrive boundary (state, region, province or city) and the services are defined by the possible use of the radio terminal, e.g. inside the buildings or outside the buildings or a mix of them.

When the objectives are clear the radio base station configuration and its antenna system must be defined. The network planner should know the various radio base station configurations (e.g. indoor base station, operational frequency band, maximum output power, diversity system, etc.) and the available antenna systems. According to the scope of the project and specific requirements, the proper radio base station configuration and antenna system will be selected.

For each selected system and each radio terminal type, a radio link power budget must be calculated. The radio link budget analyses the gains and losses between the base station transmitter and the mobile receiver (downlink) and between the mobile station transmitter and the base station receiver (uplink) in order to calculate the path loss in the uplink and downlink. The path loss in the uplink should be the same as that of the downlink in order to calculate correctly the allowable base station EIRP.

The main parameters to be taken into account in the radio link power budget are:

- Maximum output power of the radio base station and mobile terminal
- Receiver sensitivity of the radio base station and mobile terminal
- Cables losses
- Antenna gains
- Other losses, such as human body or combiner, duplexer, etc.

The ETR-300-1 shown in figure D.8 is an example of a radio link power budget. The scope is to calculate the equal propagation loss for the uplink (mobile station->radio base station) and downlink (radio base station->mobile station) by modulating the output power of the radio base station within its limits.

When the base station EIRP has been calculated, using the proper radio propagation mode the theoretical cell range can be calculated in the different environments composing the area of interest (e.g. urban,

Introduction to TETRA Network Planning 505

Figure D.9 Prediction of radio coverage offered by a radio base station

suburban areas, etc.). Knowing the cell area for the desired type of service and the area to be covered, the number of base stations can easily be estimated.

In a next step, to obtain more accurate results, a software tool for radio coverage predictions using a digital map can be used (Figure D.9). This method gives more reliable results than the previous one because the terrain height and the actual distribution of the various terrain types are taken into consideration. In the tool, the base stations for each site should be properly configured with the EIRP and antenna radiation pattern information. The software tool will calculate the radio coverage map and will show using different colour codes the different levels of the received signal level for each point on the geographical map.

D.5.2 Traffic Capacity Planning

At the end of the previous step the total number of required base stations to meet the customer requirements has been calculated. As a next step the number of carriers for each radio base station can be requested. This item can be calculated by estimating the required traffic offered by each cell and then the number of traffic channels per cell.

As already explained, the TETRA is a trunking system, which means that a large number of users can share the relatively small number of channels in a cell by providing access to each user on demand from a pool of available channels. Upon termination of a call the channel is immediately returned to the pool of available channels. When all traffic channels are busy a further channel request is blocked. The calls

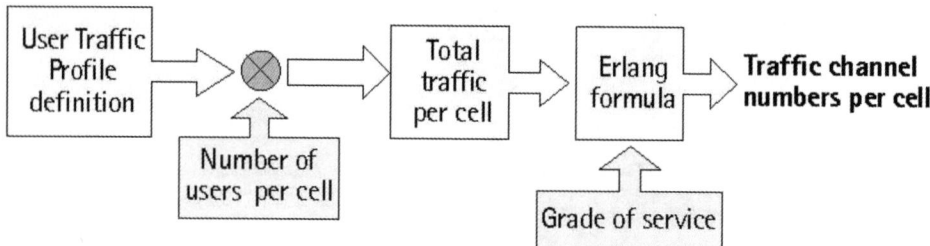

Figure D.10 Traffic channel estimation process

can be rejected or delayed. The TETRA standard, in the case of blocking at the air interface, can insert further requests of traffic channels in a queue and they will be satisfied following the priority of the users.

In order to calculate the traffic offered by each cell, the following information is usually required:

- Total number of active users
- Distribution of users in the area of interest
- Average user traffic profile

By using the above data the average number of users per cell and the traffic per cell can be calculated.

In order to estimate the number of traffic channels per cell the following objectives should be specified:

- Grade of service
- Traffic model

The grade of service (GoS) is a measure of the ability of a user to access a trunked system during the busiest hour. It is specified as the probability of a call being blocked or the probability of a call being delayed beyond a certain amount of time. The traffic model specifies the probability distribution of the calls in the time domain, the trunking system strategy, the time between successive calls, etc. The most common formula used to estimate the traffic channels was developed by Erlang. In cases where the blocked calls are delayed, the Erlang C formula will be used. Figure D.10 summarises the process to be applied for traffic channel estimations.

Suggested Reading

3GPP, 3G TR 23.923 ver.3.0, 2000–05, *Combined GSM and Mobile IP Mobility Handling in UMTS IP CN.*
3GPP, 3G TS 22.796, ver.2.0.0, 2000-06, *Study on Rel.2000 Services and Capabilities.*
3GPP, TS 23.002, *Rel.4 Network Architecture.*
3GPP, TS 23.205, *Rel.4 Bearer-Independent Circuit-Switched Core Network.*
3GPP, TS 23.228, *IP Multimedia Subsystem.*
3GPP, TS 23.907, *Services and System Aspects, QoS Concepts.*
3GPP, TS 25.414, *Rel.4 UTRAN Iu-Interface Data Transport and Transport Signalling.*
3GPP, TS 29.414, *Rel.4 Core Network Nb-Interface Data Transport and Transport Signalling.*
3G TS 25.213, *Spreading and Modulation* (FDD).
3G TS 25.223, *Spreading and Modulation* (TDD).
Agilent technologies (CDMA systems drive), http://cp.literature.agilent.com/litweb/pdf/5968_5554E.pdf.
Allen, K.C., "Observation of specific attenuation of millimetre waves by rain", *IEE A&P Conference*, 1987.
Analyser 4.1 User Manual.
Anfossi, D., Bacci, P. and Longhetti, A., "An application of Lidar technique to the study of nocturnal radiation inversion", *Atmos. Environ.*, **8**, 1960, 483–494.
Ansari and Evans, "Microwave propagation in sand and dust storms", *IEE Proc. F, Communication, Radar and Signal Processing*, 1982.
Bean *et al.*, *ESSA Monograph*, US Government Printing Office, Washington, 1966.
Beck, R. and Panzer, H., "Strategies for handover and dynamic channel allocation in micro-cellular mobile radio systems", in *IEEE Vehicular Conference*, 1989, pp.178–185.
Bekkers, R. and Smits, J., *Mobile Telecommunications : Standards, Regulation and Applications*, Artech House, 1999.
Castro, J. P., *The UMTS Network and Radio Access Technology*, John Wiley & Sons, Ltd, 2001.
Craig, K.H. and Kennedy, G.R., "Studies of microwave propagation on a microwave line-of-sight link", in *IEE A&P Conference*, 1987.
Crane, R.K., "Prediction of attenuation due to rainfall on satellite systems", *Proc. IEEE*, **65**(3), 1977.
Crane, R.K., "Fundamental limitations caused by RF propagation", *Proc. IEEE*, **69**(2), 1981.
Ctop_WI.doc, Cellular transmission network optimisation tests, Product note, work instruction/procedure.
Detail of call phases (Ned 3.1/basic protocols/call related DX causes in BSC/4 internal description/).
Doble, J., *Introduction to Radio Propagation for Fixed and Mobile Communication*, Artech House, 1996.
Eberspacher, J. and Vogel, H.-J., *GSM Switching, Services and Protocols*, John Wiley & Sons, Ltd, 1999.
Effenberger, Strickland and Joy, "The effect of rain on a radome's performance", *Microwave Journal*, May 1986.
ETSI, GSM 03.30, *Digital Cellular Telecommunication System (Phase 2+): Radio Network Planning Aspects.*
ETSI, GSM 04.60, *Digital Cellular Telecommunication System (Phase 2+): General Packet Radio Service (GPRS); Mobile Station (MS)–Base Station System (BSS) Interface; Radio Link Control/Medium Access Control Protocol.*
ETSI, GSM 05.05, *Digital Cellular Telecommunication System (Phase 2+): Radio Transmission and Reception.*
ETSI, GSM Recommendations 03.09, 1991, *Handover Procedures.*
ETSI, GSM Recommendations 04.04, 1991, *Layer 1 – General requirements.*
ETSI, GSM Recommendations 08.08, 1991, *BSS–MSC Layer 3 Specification.*
ETSI, GSM Recommendations 08.58, 1991, *BSC–BTS Layer 3 Specification.*

ETSI, GSM Recommendations 04.05, 1993, *Data Link Layer – General Aspects*.
ETSI, GSM Recommendations 04.06, 1993, *MS–BSS Interface, Data Link Layer Specification*.
ETSI, GSM Recommendations 04.07, 1993, *Mobile Radio Interface, Layer 3 – General Aspects*.
ETSI, GSM Recommendations 04.08, 1993, *Mobile Radio Interface, Layer 3 Specification*.
FER, RXQUAL, and DTX DL Rate Measurements in TEMS Investigation, GSM White Paper 2002-02-18, 2002.
Gibson, J. D., *The Mobile Communications Handbook*, 2nd edition, CRC Press, Springer, IEEE Press, 1999.
GSM Technical Specification 05.08, Section 8.2.4.
Halonen, T., Romero, J. and Melero, J., *GSM, GPRS and EDGE Performance*, John Wiley & Sons, Ltd, 2002.
Halonen, T., Romero, J. and Melero, J., *GSM, GPRS and EDGE performance, Evolution Towards 3G/UMTS*, 2nd edition, John Wiley & Sons, Ltd, 2003.
Hata, M., "Emperical formula for propagation loss in land mobile radio services," *IEEE Trans. Vehicular Technol.*, **VT-29**(3), August 1980, 317–325.
Haykin, S., *Communication System*, John Wiley & Sons, Inc., New York, 1983.
Heifiska Kari and Kangas, A, "Microcell propagation model for network planning", in *Proceedings of the IEEE International Conference on Personal Indoor and Mobile Radio Communication, PIMRC'96*, 1996, pp. 148–152.
Heine, G., *GSM Networks: Protocols, Terminology and Implementation*, Artech House, 1999.
Higginbottom, G. N., *Performance Evaluation of Communication Networks*, Artech House, 1998.
Holma, Harri, Toksala and Antti, *WCDMA for UMTS*, John Wiley & Sons, Ltd, 2000.
Hopping sequence properties of the GSM frequency hopping algorithm.
Hunag, C.Y. and Yates, R.D., "Call admission control in cellular radio system", *IEEE Trans. Vehicular Technol.*, **1**, 1992.
Ishimaru, "Introduction to wave propagation and scattering in random media", *IEEE, AP*, 1985.
ITU-R, P. 529–2, "Okumara–Hata propagation model", in *Prediction Methods for the Terrestrial Land Mobile Services in the VHF and UHF Bands*, ITU, Geneva, 1995, pp. 5–7.
ITU-R, P.530–9, *Propagation Data and Prediction Methods Required for the Design of Terrestrial Line-of-Sight Systems*.
ITU-R, P.837–3, *Characteristics of Propagation for Propagation Modelling*.
ITU-T, G.826, *Error Performance and Objectives for International, Constant Bit Rate Digital Paths at or above Primary Rate*.
ITU-T, G.827, *Availability Parameters and Objectives for Path Elements of International Constant Bit Rate Digital Paths at or above the Primary Rate*.
ITU-T, P.800, *Telecommunications Standardization Sector of ITU 08/1996, Series P: Telephone Transmission Quality, Methods for Objective and Subjective Assessment of Quality/Methods for Subjective Determination of Transmission Quality*.
ITU-T, P.862, *Telecommunication Standardisation Sector of ITU 02/2001, Series P: Telephone Transmission Quality, Telephone Installations, Local Line Networks, Methods for Objective and Subjective Assessment of Quality/Perceptual Evaluation of Speech Quality (PESQ), An Objective Method for End-to-End Speech Quality Assessment of Narrow-Band Telephone Networks and Speech Codecs*.
Jakes, Jr W.C., (ed.), *Microwave Mobile Communications*, Wiley-Interscience, 1974.
Janaswamy, R., *Radio Propagation and Smart Antennas for Wireless Communications*, Kluwner Academic Publishers, 2000.
Kaaresoja, T. and Ruutu, J., "Synchronization and cell loss in cellular ATM evaluation system", in *Proceedings of the 5th International Workshop on Mobile Communication, moMuc'98*, Berlin, 12–14 October 1998.
Kazimierz Siwiak, *Radio Propagation for Antennas for Personal Communications*, Artech House, 1998.
Knorr, J.B., "Guided EM waves with atmospheric ducts", *Microwave and RF*, May 1985.
Ko, H., "A practical guide to anomalous propagation", *Microwave and RF*, April 1985.
Lagrange, X., Godlewski, P. and Tabbane, S., *Réseaux GSM-DCS*, 4th edition, *Revue et Augment*ée, Hermes, 1999.
Lee, W.C.Y., *Mobile Cellular Telecommunication Systems*, McGraw-Hill Book Company, 1990.
Lee, W.C.Y., *Mobile Communication Design Fundamentals*, John Wiley & Sons, Ltd, 1993.
Lee, W. C. Y., *Mobile Cellular Telecommunications, Analog and Digital Systems*, 2nd edition, McGraw-Hill, Inc., 1995.
Lempiainen, J., *Radio Interface Planning for GSM/GPRS/UMTS*, Kluwer Academic Publishers, 2001.
McAllister, L.G., *et al.*, "Acoustic sounding – a new approach to the study of atmospheric structure", *Proc. IEEE*, **57**, 1969, 579.

Measurements tool (genetel-nemo technologies products).
Mehrotra, A., *Cellular Radio Performance Engineering*, Artech House, 1994.
Mehrotra, A., *GSM System Engineering*, Artech House, Boston, 1997.
Mishra, Ajay R., "Observation of the fading phenomenon on the western coast of India, on a 7 GHz terrestrial path", in *IMOC Conference*, Rio de Janeiro, 1999.
Mishra, Ajay R., "Cellular transmission network optimisation", *Advance* (Nokia Research Centre Journal), March 2001.
Mishra, Ajay R., "Transmission network planning and optimisation in third-generation access networks", *Advance* (Nokia Research Centre Journal), June 2003.
Mishra, Ajay R., "EDGE network analysis and optimisation", *WPMC*, Yokosuka, October 2003.
Mishra, Ajar R., *Fundamentals of Cellular Network Planning and Optimisation: 2G/2.5G/3G...Evolution to 4G*, John Wiley & Sons, Ltd., 2004.
Mitra, A. P., et al., "Tropospheric disturbance of 17–21 December 1974 and its effect on the microwave propagation", *Boundary Layer Meteor*, **11**, 1977, 103.
Mouley, M. and Pautet, M.-B., *The GSM System for Mobile Communications, A Comprehensive Overview of the European Digital Cellular Systems*, Cell & Sys, 1992.
Ned (alarms).
Nielsen, T. T. and Wignard, J., *Performance Enhancements in a Frequency Hopping GSM Network*, Kluwer Academic Publishers, 2000.
Oguchi, T., "Electromagnetic wave propagation and scattering in rain and other hydrometers", *Proc. IEEE*, **71**(9), 1983.
Ojanpera, T. and Prasad, R, *Wideband CDMA for Third Generation Mobile Communication*, Artech House Publishers, 1999.
Ojanpera, T., Prasad, R. and Harada, H. "Qualitative comparison of some multi-user detector algorithms for wideband CDMA", in *Proceedings of the IEEE Vehicular Technology Conference*, vol. 1, 1998, pp. 46–50.
Okumara, Y., Ohmori, E., Kawano, T. and Fukuda, K., "Field strength and its variability in VHF and UHF land mobile radio service", *Review of Electrical Communication Laboratory*, **16**(9–10), September–October 1968, 825–873.
Pajukoski, K. and Savusalo, J., "Wideband CDMA test system", in *Proceedings of the IEEE International Conference on Personal Indoor and Mobile Radio Communication, PIMRC'97*, Helsinki, Finland, 1–4 September 1997, pp. 669–672.
Parsons, D. *The Mobile Radio Propagation Channel*, Pentech Press, 1992.
Rummler W.D., "A new selective fading model: application to propagation data", *BSTJ*, 1974.
Sarkar, S.K., *Radioclimatiological Effect on Tropospheric Radiowave Propagation over the Indian Sub-continent*, PhD thesis, University of Delhi, 1978.
Schoenbeck, R. J., *Electronic Communications Modulation and Transmission*, 2nd edition, Prentice-Hall Inc., New Jersey, 1992.
Sipila, K. Laiho-Steffens, J., Wacker, A. and Jasperg, M. "Modelling the impact of the fast power control on the WCDMA uplink," in *Proceedings of the IEEE Vehicular Technology Conference, VTC'99*, 1999, pp. 1266–1270.
Soldani, D. and Abramowski, M., "An improved method for assessing the packet data transfer performance across an UMTS network", submitted.
"TCP over 2.5G and 3G wireless networks", IETF Internet draft, October 2001.
Tissal, J., *Le Radio Téléphone Cellulaire GSM*, Masson, Paris, 1995.
Tissal, J., *Le Réseau GSM, L'évolution GPRS : Une Etape vers UMTS*, 3rd edition, Dunod, Paris, 1999.
Trimble Placer 455DR Sensor Calibration Guide.
Tsurimi, H. and Suzuki, Y. "Broadband RF stage architecture for software defined radio in handheld terminal application", *IEEE Communication Mag.*, **37**(2), 1999, 90–95.
UMTS 22.01, *Service Aspects and Service Principles*.
UMTS 22.05, *Service Capabilities*.
UMTS 22.25, *Quality of Service and Network Performance*.
UMTS 23.05, *Network Principles*.
Vigants, A., "Space diversity engineering", *BSTJ*, 1974.
Vigants, A., "Microwave obstructive fading", *BSTJ*, **60**, 1981.
Walfish, J. and Bertoni, H. L., "Therotetical model of UHF propagation in urban environments", *IEEE Trans. Antenna Propagation*, **AP-36**, October 1988, 1788–1796.

Walke, B. H., *Mobile Radio Networks, Networking, Protocols and Traffic Performance*, 2nd edition, John Wiley & Sons, Ltd, 2002.

Xia, H.H., *et al.*, "Micro-cellular propagation characteristics for personal communication in urban and sub-urban environments", *IEEE Trans. Vehicular Technol.*, **43**(3), 1994, 743–752.

Yang, S.C., *CDMA RF System Engineering*, Artech House, Norwood, Massachusetts, 1998.

Zander, J., "Radio resource management – an overview", in *Proceedings of the IEEE Vehicular Technology Conference, VTC-96*, Atlanta, Georgea, 1996, pp. 661–665.

Index

8-PSK, 10, 23, 162, 163, 164, 167
(IP) Servers, 277, 294, 331–335, 341, 347–348, 350–351, 353–354, 378–384, 397–403, 427–430
2 Mbps Channel, 355
2 Mbps Planning, 286
3dB compression point, 166
3GPP, 10, 328, 329, 331, 332, 333, 335, 337, 339, 341, 342, 343, 352, 467, 472, 473
4 bit cyclic redundancy check (CRC4), 113
4G, 417, 418, 419, 420, 421, 422, 423, 425, 426, 428, 429, 430, 431, 435, 436, 437
8-PSK, 10, 23, 162, 163, 164, 167
8-PSK power envelope, 168
8-PSK Vector Diagram (Phase-State Transition), 10, 163–168, 173
A_{bis} Interface, 110, 135
A Interface Trace, 110
AAL (ATM Adaptation Layer), 11, 223–224, 228–231, 273–274, 309–311, 340–341, 349
AAL2 Signalling, 274, 340
Abis, 25, 35, 110–111, 130, 135, 140, 268–271, 286–288, 300
ABR (Available Bit Rate), 231, 233
Absorption, 256
Access Link Control Application Part (ALCAP), 274
Access Point Node (APN), 174, 381, 383, 395, 402, 413
Accessibility, 131, 299, 303–307, 408, 414–415
Accessibility, Retainability, Quality (ARQ), 162
Adaptive Mean Rate (AMR), 12, 60–61, 144, 342
Adjacency, 121, 125, 147, 150–151
Adjacency control, 85, 110, 158
Adjacency Definition, 93, 121, 147, 150, 154–155

Adjacency management, 122–123, 125
Adjacency optimization, 151
Adjacency parameters, 150
Adjacency planning, 18, 20, 122, 125
Adjacent Channel, 282, 305
Administrative automation, 122–123
Admission Control (AC), 103–106, 307–308, 441
Advection, 263
AGCH, 61, 80–87
A-GPS (Assisted-GPS), 246, 289
AICH, 90, 94–95, 102
Air-Interface, 3, 9–11, 26–27, 87, 133, 135–137, 140, 173, 419, 498–500, 506
Alarm Management, 291
Alarm monitoring, 104, 110, 369
Algorithms, 8, 20, 24, 26, 101, 109, 126, 144–145, 156, 305, 388, 390, 471, 473
ALL-IP Networks, 427–428
Amplifier, 31, 53
AMR Codec, 342–343
ANSI, 273, 295, 452
Antenna directivity, 238
Antenna Filter, 283
Antenna gain, 31–33, 36, 39–40, 49, 53, 56, 234, 238, 249, 252
Antenna Height, 53, 115, 242, 244, 252, 420
Antenna system maintenance, 160, 251
Application Manager (AM), 27, 455
Architecture, 129, 135, 324, 328, 330, 335–341, 346, 361, 399–401, 417, 428–429, 432–434
ARIB, 10
Asynchronous Transfer Mode (ATM), 11, 201, 217, 219–234, 273, 276, 294–295, 311–314, 329–331, 345, 350, 353, 355, 395, 401

Advanced Cellular Network Planning and Optimisation Edited by Ajay R Mishra
© 2007 John Wiley & Sons, Ltd

ATM Adaptation Layer (AAL), 224, 228
ATM Adaptation Layer 1 (AAL 1), 230
ATM Adaptation Layer 2 (AAL 2), 11, 273, 274, 309–311, 340, 345, 377
ATM Adaptation Layer 3 and 4 (AAL 3/4), 230–231
ATM Adaptation Layer 5 (AAL 5), 11, 231, 273, 350, 353, 355
ATM cell, 224–228, 230–231, 313
ATM Cross-Connect (AXC), 293
ATM Layer, 223–226, 228, 230–231, 312
ATM Performance, 231, 312
ATM Terminations, 293
Atmospheric Fading, 239
Attenuation, 27, 40, 235–236, 240–241, 249, 256, 424
Authentication Centre (AC), 5
Authentication Centre (AUC), 5
Automatic Repeat Request (ARQ), 162, 469
Automation tools, 122
Availability, 33, 64, 98, 113, 132, 147, 188, 247, 281–282, 285, 300–302, 347, 368, 440, 447
Availability Bit Rate (ABR), 231, 233
Average Antenna Height, 246
Average Power Decrease (APD), 163–165
Averaging, 74, 144, 371

Background Class, 11
Backoff, 36, 163–164, 166–167
Base band hopping, 75, 116
Base Station (BS), 4, 8, 11, 32, 49, 54–55, 70, 80, 294, 420, 430
Base Station Controller (BSC), 4, 106–107, 120–121, 128, 132–133, 152, 155, 160, 175, 269, 281, 284–285, 289–292, 338–339, 448
Base Station Identity Code (BSIC), 76, 121, 150, 154–155
Base Station Subsystem (BSS), 4, 74, 85–86, 110, 130, 140, 155, 270, 392, 407
Base Station Subsystem GPRS Protocol (BSSGP), 85, 391
Base Transceiver Station (BTS), 4, 18, 30–33, 36, 44, 46, 51–53, 74, 77, 126, 134, 145, 146, 155, 159–161, 268–269, 278–280, 285–287, 289–291, 444, 478
BCCH, 7, 61, 74–77, 79–80, 82–83, 88, 93–94, 116, 150, 155, 164–165, 171
BCCH strategy, 116
BCCH TRX, 74, 160, 164
Bit Error Rate (BER), 53, 109, 141, 172
Bit-Error Probability (BEP), 171, 342

Block error Rate (BLER), 54, 102
Blocking, 6, 33, 37, 58–59, 62–65, 71, 133, 174–175, 179–180, 271, 276–278, 412
Body Loss, 31, 36, 38, 56
Boosting power, 142
Border Gateway (BG), 377, 395–396
BPSK, 470
Broadcast Channel (BCCH), 7, 61, 74–77, 82–83, 88, 93–96, 116, 150, 155, 164, 171
BSC and transcoder hardware maintenance, 160
BSC Load, 120–121, 160
BSC Location, 160
BSC rehosting plan, 120
BSS parameters classification, 155
BSS Protocol, 4, 8, 25, 74, 85–86, 130, 155, 391, 429
BTS Antenna Gain, 36
BTS hardware maintenance, 159
BTS power back-off, 35
BTS Sensitivity, 30
Buffering, 173, 232, 276, 340
Burst Period, 25, 164
Burst Structure, 164
BVC flow control, 35

C/I Ratio, 282
Cable Losses, 31
Call management (CM), 136
Call performance, 185
Call phase signaling charts, 110, 130, 133, 137
Call phases, 130, 137, 139
Call Quality, 131, 149
Call setup failure (CSF), 107
Call Set-Up Success Rate, 17–18, 132, 181, 188
Call Set-Up Time, 368, 371, 493
Call success rate (CSR), 132, 344
Call-Admission Control (CAC), 275
Camping, 92, 148, 155
Capacity, 2, 3, 10–13, 17–26, 29, 33–34, 46, 61–63, 65, 67, 69, 74, 79, 82–84, 86–87, 98, 103, 109, 121, 123, 126, 141–143, 5, 160, 179, 201, 205–206, 208, 215, 217, 222, 242, 261, 269, 272, 280–281, 299, 301, 312, 331, 336, 347, 375–376, 396–397, 426, 462
Capacity Analysis, 24
Capacity enhancement, 61, 141–142, 144
Capacity Estimations, 275
Capacity Planning, 20, 57–71, 503, 505
CBR (Constant Bit Rate), 231
CCPCH, 89, 96

Index

CDMA Radio Networks, 34, 89
CDV (peak-to-peak Cell Delay Variation, 231–233
CDVT, 233
Cell Coverage, 29–30, 51–52, 73
Cell Delay Variance (CDV), 231–232
Cell Error Ratio (CER), 232
Cell Loss Priority (CLP), 312–313
Cell Loss Ration (CLR), 231–232, 312
Cell Mis-Insertion Ratio (CMR), 232
Cell Overlay, 485
Cell Range, 29, 35–36, 42, 47–48, 57, 504
Cell re-selection, 80, 93–94
Cell search, 91–92
Cell Service Area, 4, 48
Cell Structure, 227
Cell Transfer Delay (CTD), 232
Cellular troubleshooting, 126
Chain Topology, 281
Channel Allocation, 142, 200, 493
Channel Configurations, 86
Channel to Interference (C/I) Ratio, 282
Channelisation Codes, 98
Charging Gateways (CG), 385, 396
Chip Rate, 26–28, 69, 105
Chips, 26, 89–90
Circuit Group, 323–324, 357–359, 369–370, 375, 378, 399, 499
Circuit Switched (CS), 315, 335, 341, 349, 352–355, 361, 368–377
Class A Mobiles, 342, 385, 392
Class B Mobiles, 342, 386
Class C Mobiles, 342, 386
CLR (Cell Loss Ratio), 231–232, 306
Cluster, 181–182, 187, 195, 429–430
Co-Channel Interference, 20, 71–72, 76
Code Division Multiple Access (CDMA), 3, 10–11, 26, 421
Code Planning, 183
Coding Schemes, 23, 54–55, 66, 163, 367, 369
Configuration Management, 291
Conformance, 494
Congestion, 21, 133, 174, 228, 304, 311, 373, 401, 474
Connection Admission Control (CAC), 103–104
Connection establishment, 138, 274, 311, 361
Connectivity, 278, 329, 399, 403, 406, 487
Connector Losses, 31, 36, 241
Constant Bit Rate (CBR), 231
Control Channels (CCH), 60, 88, 133, 275
Control Unit (RRM), 26, 29, 66, 286

Convergence (CL), 272, 309, 486
Conversational Class, 11
Core Network (CN), 11, 333, 421–422, 429
Core Network Optimisation, 368–377, 404–410
Core Network Optimisation Plan, 373, 373–377, 378–402, 404–413, 414
Core Network Planning, 315–415
Counters, 128–131, 141, 189, 309, 370, 407, 465
Coverage, 17, 20, 34, 46–47, 51, 54–55, 57, 141, 144, 337, 487
Coverage and capacity enhancement methods, 17, 34, 53, 141
Coverage Enhancements, 53, 141–142
Coverage Planning, 20, 45–47, 50–55, 57
Coverage Plans, 20, 52
Coverage Threshold, 21, 48–49, 52
CPICH, 89, 91–92, 94, 105
CS Territory, 63
CS-1, 23, 36, 54–56
CS-2, 23, 36, 55–56, 66–67, 350
CS-3, 23, 36, 55, 57, 67, 271
CS-4, 23, 36, 54–55, 67
CSW, 66, 168

DAT Tapes, 463
Data Collection and Analysis, 370
Data Collection: Signalling, 128, 140, 369
DCR, 107, 131–133, 191
Delay, 232, 242
Demodulation, 236
Destination, 7, 378
Detailed Network Planning, 323, 325, 352, 353, 355, 357
Diffraction, 40, 44, 254–255
Diffraction Fading, 254
Dimensioning, 19, 29, 33–34, 37, 51, 57–59, 62, 64, 68–69, 84, 128, 272, 275, 278, 315, 317, 319–320, 335–336, 339, 341, 345, 347, 377, 390–398, 412, 476–480
Direct Sequencing, 26
Directed retry, 23
Direct-Sequence WCDMA Frequency-Division Duplex (DS-WCDMA-FDD), 97
Direct-Sequence WCDMA Time-Division Duplex (DS-WCDMA-TDD), 11, 97
Discontinuous transmission, 74, 91, 143, 342
Diversity Effects, 31
Domain Name Server (DNS), 331, 395, 402
Double frame mode (DBLF), 113
Downlink Modulation, 70,
Downlink Power Budget, 70, 143, 146

Downlink Shared Channel (DSCH), 468–477
Downlink Spreading, 27
Drive Test, 20–21, 45, 106–108, 114, 121, 127, 140, 154, 181–182, 185, 187, 465
Drive testing (EGPRS), 23–25, 34–36, 54, 62, 66, 68, 69, 78, 79, 82, 162, 163, 164, 165, 169–171, 172–175, 179–181, 269, 271
DRNC (Drifting RNC), 100
Drop Call Rate (DCR), 107
DSCH, 469, 474
Dual band, 23, 143
Dual-Tone Multi-Frequency (DTMF), 349
Ducting, 351, 361–362
Dynamic Abis, 269–270, 300
Dynamic Abis Pool (DAP), 300

E1, 35, 314, 384
Eb/N_o, 94
Effective Radius, 245
EGPRS, 23–25, 34–36, 54, 62, 67–69, 78–80, 83, 162–165, 169–172, 174, 175, 279
EGPRS Abis transmission, 35
EGPRS Coverage Planning, 54–57
EGPRS Dimensioning, 46
EGPRS Throughput per Timeslot vs C/I, 169
EGPRS Throughput, Latency, Link Adaptation, 171–172
EIR, 5–6
Emergency Numbers, 323
Enablability, 149, 156–157
End-to-End QoS, 434, 436, 437
Enhanced Data Rates In GSM Environment (EDGE), 10, 271–272, 287, 293
Equipment Enhancements, 483, 484
Equipment Identity Registers (EIR), 352
Equipment Location, 5, 7–8, 16, 18–21, 45–46, 51, 53, 58, 74–75, 77, 86, 92–93, 96, 100, 103, 106, 115, 125, 160, 161, 182, 246, 250, 256, 280, 281, 287, 291–292, 303, 314, 316–317, 320, 336–337, 339, 347, 350–352, 378–381, 383, 384, 436
Erlang B, 58–59, 62, 71, 278
Erlang C, 58, 506
Error Performance, 251–252, 300–301
Error Rate, 102, 113, 141
Error Seconds (ES), 301
Estimated Traffic, 33, 37
ET Port, 323
ET Port Allocation Plans, 323
ETSI, 23, 30, 48, 342, 493–494

Exchange Terminals, 160, 294
Extended cell smart radio concept, 142
External Interference, 182, 187

FACH, 88–89, 103
Fade Margin, 247, 250, 266–267
Fading, 31, 34, 38, 41, 47–49, 142, 162, 169, 171–172, 174, 234–236, 239, 251, 254, 256, 261, 264–265, 281, 421, 423–424, 441, 465
Failure Analysis, 106, 107, 121, 128, 130, 132, 137, 141, 152, 154–155, 160, 172, 174, 177–179, 183, 187, 189, 191, 195, 205, 250, 284, 291, 301, 307, 310–311, 314, 318, 321, 324, 354, 373, 375, 377, 401, 403, 408–409, 421, 439, 449, 453, 456–460
Fast Associated Control Channel (FACCH), 61, 133, 135, 136
Fast moving threshold, 151
Fault analysis and improvement, 128
FDD, 94, 97
Feeder loss, 240–241, 252
Filter, 31, 33, 36, 110, 114, 140, 283, 435
Final Acknowledge Indicator (FAI), 180, 181
Firewall (FW), 396
Firewall Servers, 403, 410
First-Generation, 1, 2, 3
Flat Fading, 252
Forward Error Correction (FEC) – codec rate, 162, 171–173, 236, 491
Fourth-Generation Networks, 417–436
Frame Error Rate (FER), 106, 109, 110, 141
Frame-Relay, 220, 231, 286, 403
Free-Space Loss (FSL), 249, 252
Frequency Allocation, 115, 143, 283
Frequency Band, 2–4, 11, 18, 33, 40, 59, 71, 115–116, 125, 143, 187, 234–236, 241, 282, 419–420, 435, 488
Frequency Band Division, 59
Frequency Bands in WCDMA-FDD, 97
Frequency Correction Channel (FCCH), 61, 82–83
Frequency Diversity, 3, 265, 267
Frequency Hopping, 54–55, 61, 74–75, 116, 120, 125, 143
Frequency Plan, 20, 24, 76, 78, 108, 115, 126, 154, 283–284, 503
Frequency planning, 18, 20, 22, 26, 71–78, 115–116, 124–126, 282–284, 302, 503
Frequency Re-use, 71–72, 74, 144, 145
Frequency Usage, 26, 112, 183, 419–420, 500

Frequency-Division Multiple Access (FDMA), 2, 59, 500
Frequency-Selective Fading, 261, 265, 423
Fresnel zone, 245–246, 255, 264
Frontal System, 263
Front-to-Back ratio, 239
FTP, 166, 180–181, 183
Full-Rate (FR), 54, 60, 144, 399, 494

Gain of the Antenna, 4, 29, 31–33, 36, 39, 40, 49, 234, 238–239, 249, 252, 504
Gate Tunnelling Protocol (GTP), 35, 273, 277, 382, 384, 390–391, 401
Gateway (GW), 348, 429
Gateway GPRS Support Node (GGSN), 11, 273, 328–329, 332, 377–385, 391, 394–395, 399–400, 403, 413–414
Gateway Mobile Switching Centre (GMSC), 5, 7, 330
Gaussian Minimum Phase-Shift Keying (GMSK), 10, 163–168, 171–173, 175, 179
G_b Interface, 377–378, 384, 391–392, 407
Gb Link, 25, 35
Gb NSVC CIR, 35
G_d Interface, 407
General Packet Radio Services (GPRS), 9, 10, 11, 22–23, 54–55, 57, 62–69, 78–80, 82–86, 174, 285–286, 327–331, 377–384, 390–392, 407
Generic Flow Control (GFC), 227
G_f Interface, 407
G_i Interface, 381, 396, 402–403, 411
G-Interfaces, 401–403
Global System for Mobile Communication, 3
GMSK Modulation, 164, 166
G_n, 273, 381, 390, 394, 401–402
G_n Interface, 381–382, 390, 401–402
G_p Interface, 390–391
GPRS and EGPRS Quality of Service, 67–69
GPRS and EGPRS Rate Reduction, 67–68
GPRS Attach Request, 78–80
GPRS Coding Scheme Coverage, 23–24, 54–56, 68–69, 173, 174
GPRS Detach Request, 78, 80, 82, 400
GPRS Territory, 64–67, 79
GPRS Throughput per Timeslot vs C/I, 66–67
GPRS Tunnelling Protocol (GTP), 35, 273, 277, 382, 396
GPRS/EGPRS Cell re-selection, 90–91
G_r Interface, 381, 407
G_s Interface, 371, 381
GSN, 381, 382, 384, 390–391, 395

GTP Signalling, 277
GTPU, 273, 376, 390, 391, 401
Guard Band, 250

Half rate, 60, 143–144, 500
Handover, 145–152, 154–159, 186, 274–275, 277–279, 303, 308, 418, 430, 433–434, 436, 470, 482, 485–487
Handover algorithm, 147, 149
Handover and adjacency difference, 168–169
Handover causes, 147, 154
Handover control, 20, 97, 133, 156, 157, 183, 430
Handover distribution, 107, 133
Handover failures, 154, 155
Handover features, 149
Handover parameters, 126, 149–150
Handover performance, 145, 154
Handover phases, 160
Handover success rate, 22, 121, 132, 154, 368
Hard handover, 97, 99–100
Hardware configuration, 339
Hardware monitoring, 160
Header Error Control (HEC), 225, 228, 313
High-Speed Circuit-Switched Data (HSCSD), 377
Home Location Register (HLR), 5, 76, 318, 328–330, 333–335, 348, 352, 368, 371, 375, 377–378, 378–382, 395, 400, 403, 413
HSN planning, 118
Hyperframe, 60

IDLE, 84, 85, 103, 166, 383
Idle State, 84–85, 383
IMEI, 5–6, 136, 376–377
IMSI, 5–7, 85, 98, 106, 136, 322, 382–383, 400
IMSI Analysis, 85, 96, 136, 327, 376, 383, 400
IMSI Numbers, 318, 326–327
IMT, 10–11, 417
In Application Protocol (INAP), 330, 352–354, 397
Incremental Redundancy (IR), 469, 471–472, 480
Indoor coverage, 29, 48–49, 52, 108, 142, 479, 486
Indoor Coverage, 142
Information Theory, 28
Intelligent Network, 317
Interactive Class, 11, 12
Interface Unit (IFU), 401
Interfaces and Signalling in GSM, 22–23, 56, 59–60, 79, 133–135, 175, 327–328, 341
Interference, 22, 31, 38, 69–76, 104, 113, 143, 147–148, 154, 187, 282–283

Interference Degradation, 31–32, 48
Interference Diversity, 116
Interference Margin, 38–39, 69
Interference Plans, 282–283
Inter-Hop Interference, 282
Inter-Mode Handover, 8
International Telecommunication Union (ITU), 10, 11, 204
Internet Protocol (IP), 367, 377–392, 394–403, 407, 412, 422, 427–433, 436, 445, 448, 465
Inter-Switch Traffic, 318, 321, 324, 338, 367
Intra-System Interference, 97, 101, 183
IP Addressing, 295, 385–388, 399, 401, 411
IP Routers, 395
IP Routing, 331, 335, 351, 384, 388–389
IP-BTS, 465
ISDN (Integrated Services Digital Network), 6–8, 110, 134, 201, 205–206, 211, 217–220, 223, 224, 226, 233, 322, 326–329, 332, 335, 348, 356, 358, 429
I_u, 189, 267, 272–273, 275–277, 279–280
IU (Indoor Unit), 236–237
I_{ub}, 267, 272–276, 278–279, 295
I_{u-cs}, 272–274, 276–280
I_{u-ps}, 272–273, 276–278, 280, 295
I_{ur}, 99–100, 272, 274, 276–277, 278, 280, 295
IVR, 318

Key Performance Indicators (KPIs), 20–21, 106, 127, 128, 131, 175, 309, 368, 405, 408, 439, 465
Key Performance Indicators in WCDMA Network, 439
k-factor, 11, 243–244, 248, 253–254, 264
k-Fading, 253–254, 264
Knife-Edge Diffraction, 254–255

LAN, 378, 397, 401–402, 417–418, 425, 428–429, 433, 465, 485–486, 488, 492
LAPD, 86, 134, 135, 140, 286
Layer 1: Physical Layer, 88–89, 133–134, 135, 205, 217, 223–226, 272–273, 275, 293, 307, 309, 313–314, 345, 481, 487, 498
Layer 2: Data Link Layer, 134, 217, 223–224, 491
Layer 3 messages, 110, 170, 179
Layer 3: Network Layer, 135–136, 319, 384, 427–428, 110, 170, 179

Layer 4: Transport Layer, 136, 275, 293, 331, 335, 341, 351, 361, 432
Layer 7: Application Layer, 169, 173, 293
LCB, 289
Legal Interception Gateway (LIG), 378
Level, 18–21, 23, 29, 36, 40, 47, 53–54, 57–58, 63–66, 71, 73–77, 84–86, 91, 94, 101, 106, 108, 125, 128, 131–134, 140–154, 156–160, 163–164, 171–174, 200–202, 251–253, 283, 355, 397, 409–410, 418, 425–430, 432–433, 451–452, 461
Line of Sight Survey, 246
Linear Power Amplifier (PA), 166
Line-of-Sight (LOS), 43, 246, 441
Link (or Power) Budget, 32, 49, 504
Link Adaptation, 171–175, 179, 471, 476, 480
Link Budget, 162–163, 247, 253, 261, 254, 282–283, 302, 441, 472, 477, 504
Link Budget Calculation, 29, 31, 34, 46, 54, 57, 247, 261, 264, 282–283, 302
Link Performance, 70, 133, 163, 171, 266, 301
Linking layer, 134–135
LLC packet routing, 171
LLC/GTP buffers, 35
LMDS, 421
Load Balancing, 399
Load control, 29, 93, 103–104
Load sharing, 143, 320, 366, 375, 395–397
Loading Effect, 67
Loading Factor, 396
Location area planning, 121, 122, 126
Location Probability, 17, 37, 47–50
Location update, 7–8, 76–77, 103, 106, 121, 127, 133–134, 136–137, 140, 322, 371, 376–378
Location-Based Services (LBS), 12
Logical channels, 60, 80, 87–88, 133
Logical Link Control (LLC), 35, 86, 169–173, 273
Loop protection, 284
Loop Topology, 281
LOS (Line-of-Sight), 40, 41, 43, 44, 197, 246, 251, 430, 441
Low-Noise Amplifiers (LNA), 237, 435

Macro-Cells, 4, 23, 29, 43, 51, 61, 80, 150
MAIO planning, 118
Maximum Burst Size (MBS), 231, 233
Maximum CTD (Maximum Cell Transfer Delay), 231, 233
MCB, 289

MC-CDMA, 421
MCR (Minimum Cell Rate), 231
MCS, 23, 25, 35, 36, 54, 68, 162, 163, 171–176, 269, 270, 271, 302
Mean CTD (Mean Cell Transfer Delay), 231, 232, 233
Mean Opinion Score (MOS), 106, 109, 144, 145
Measurement Triggering, 98, 100, 101
Media Gateway (MGW), 329–332
Medium Access Control (MAC), 25, 87, 88, 173, 273, 401, 469–478, 474
Microwave link, 234–237, 246, 249, 250, 253, 272, 300–302
Microwave Link Performance, 301
Microwave Link Planning, 234
MIMO, 13, 420, 425, 426, 427, 467–475
Minimum Cell Rate (MCR), 231
Mixed Networks, 288
MMS, 381, 408, 409
Mobile allocation list (MAList), 106, 116–119
Mobile Station (MS), 4, 6, 381, 496
Mobile Switching Centre (MSC), 4, 5, 7, 33, 113, 126, 136, 138, 149, 152, 289, 328, 329, 330, 335, 348, 350, 352, 359, 367, 378, 397
Mobile Terminal, 277, 413, 420, 429, 431, 433, 436, 484, 485, 486, 504
Mobility, 23, 77, 80, 84, 85, 136, 145, 150, 154, 155, 274, 307, 329, 330, 335, 347, 349, 381, 392, 400, 405, 408, 412, 417, 428, 429, 430, 436, 437, 485, 486
Mobility Management (MM), 84, 136, 400
Modulation, 3, 10, 23, 24, 35, 54, 66, 171, 172, 173, 175, 199, 236, 282, 420, 423, 425, 468–477, 484, 485, 488, 489, 500
Modulation and Coding Scheme, 23, 24, 163, 471
Modulation Coding Scheme (MCS) distribution, 35, 54, 66, 171, 172
Modulation Coding Scheme allocation, 35
Modulation Types, 282
MS Antenna Gain, 31, 33
MS Sensitivity, 30
Multi-BCF, 121, 155, 156, 269, 287, 292, 293
Multi-Carrier Code-Division Duplex (MC-CDMA), 421
Multiframe, 23, 25, 54, 60, 82, 83, 84, 109, 113, 163, 166, 169
Multimode device, 421
Multipath, 29, 40, 41, 44, 142, 234, 235, 239, 256, 261, 263, 265, 266, 268, 420, 424, 425, 426, 427, 477
Multipath Propagation, 41, 44, 420, 425, 427, 477

Multiple-Access Techniques, 11, 423
Multiplexing, 134, 135, 199, 200–204, 206, 207, 213, 215, 216, 217, 218, 220, 223, 225, 230, 236, 309, 328, 423, 480, 490
Multivendor interoperability, 161

NBAP, 274, 275, 294, 304
Neighbour Cells, 89, 93, 94, 155
Neighbour list, 94, 98, 182, 183, 185
NETACT Planner, 50
Network Analysis, 375, 376, 378
Network assessment, 128, 315, 316, 318, 319, 320, 323, 336, 339, 397
Network Assisted Cell Change (NACC), 167, 169
Network care activity, 159
Network Coverage, 17, 50, 141, 403, 414, 419
Network dimensioning, 19, 33, 48, 58, 64, 128, 315, 316, 317, 319, 322, 323, 336, 390, 391, 447
Network elements restructuring, 121
Network Interface Unit, 407, 410, 495
Network layer, 135, 136, 319, 384, 427, 430, 432
Network Management, 11, 221, 228
Network Management Planning, 197
Network Management System (NMS), 78, 106, 124, 128, 159, 291, 300, 301
Network Optimisation, 20, 106
Network Performance Monitoring, 377
Network Switching Subsystem (NSS), 106, 113, 140, 155, 161
Network Topology, 116, 205, 373, 406, 503
Network-to-Network Interface (NNI), 227, 350, 401
NMS maintenance, 161
Node B Application Protocol (NBAC), 274, 275, 296, 304
Nominal Planning, 19
Non-Real-Time Variable Bit Rate (NRT-VBR), 231
Nordic Mobile Telephone (NMT), 1, 2
NRT, 37, 467
nrt-VBR (Non-Real-Time Variable Bit Rate), 231
NSS faults, 161
NSS performance, 106, 113

Octagonal Phase-Shift Keying (8-PSK), 10, 23, 35, 163, 167, 173, 179
OFDM, 420, 423, 424
Offline Observation, 110
Okumara-Hata Model, 57

OMC (operation & maintenance center), 5, 127, 140, 160
One-phase access, 35, 174
Optimisation, 17, 19, 20, 21, 106, 107, 110, 114, 115, 122, 125
Orthogonal Signals, 423
Orthogonal Variable Spreading Factor (OVSF), 27, 469
OSI Reference Model, 223, 224
Othogonality, 27, 420, 423
Othogonality Factor, 38, 70, 105
OU (Outdoor Unit), 237, 443
Outage, 233, 260, 267, 406, 463
Outer-Loop Power Control, 97, 102, 106
Overheads, 212, 276, 277, 390, 391
Overlay Network, 421

Packet Access Grant Channel (PAGCH), 80
Packet Associated Control Channel (PACCH), 163, 166
Packet BCCH (PBCCH), 171
Packet Call Quality, 179
Packet Common Control Channel (PCCCH), 80
Packet Control Unit (PCU), 25, 286
Packet Data Channel (PDCH), 23
Packet data connection, 102
Packet Data Scheduling, 26
Packet Downlink/Uplink Ack/Nack (PDAN, PUAN), 180
Packet Paging Channel (PPCH), 80
Packet Random Access Channel (PRACH), 90, 94, 95
Packet Scheduling Optimisation, 295, 471, 279
Packet Switched (PS), 11, 96, 270, 276, 331, 377, 421
Paging, 85, 88, 126
Paging and Access Grant Channel (PAGCH), 80
Paging procedure, 85, 96
Parameter extraction, 159
Parameter implementation, 140, 159
Parameter management, 155
Parameter Planning, 19, 20, 78
Parameter Setting, 17, 20, 77, 78, 301
Parameter Tuning, 146, 151
Parameters, 6, 17–24, 29
Path balance, 133, 141
Path Loss, 30, 33, 40, 43, 44
Path loss improvement, 141
Payload Type (PT), 228, 231, 343
PBGT, 147, 148
PCH, 61, 84, 89, 96, 103

PCM, 113, 200, 285, 299, 355
PDCH Multiframe Structure, 163, 166
PDCP, 272, 273
PDH (Plesiochronous Digital Hierarchy), 201–207, 217, 272, 288
PDH Interface, 288, 293
PDP Context, 85, 378, 382, 383, 384, 400, 407, 413
Peak Cell Rate (PCR), 231, 232
Peak-Hour Multiplier, 69, 405
Peak-to-average, 166, 167
Performance and Fault Management, 127, 160, 184, 457
Performance Enhancing Proxy (PEP), 174
Performance indicators, 20, 127, 128, 175, 187, 188, 368, 369
Periodic Routing Area Update (PRAU), 175
Periodicity, 149, 157
Physical Channels, 60, 89, 96, 102, 470
Physical Layer, 88, 133, 134, 223, 224, 225, 272, 309, 430
Physical Medium (PM), 89, 224
PICH, 89, 92, 96
Pico-Cells, 29, 51
Pilots Bits, 89, 91, 92
PLMN selection, 91, 94
Point-to-Point Topology, 280, 281
Polarisation, 235, 236, 239, 282, 283
Power Amplifier, 163
Power Boosters, 31, 53, 142
Power Budget, 28, 32, 148, 149, 504
Power budget margin, 151
Power control, 11, 38, 73, 90, 98, 101, 102, 146, 156
Power Control Headroom, 38
Power Control Parameter, 23, 156
Primary Reference Clock (PRC), 289
Priority and load, 158
Probe, 410, 414
Processing Gain, 28
Propagation, 17, 20, 22, 29, 41, 43–48, 115, 304, 420, 425
Propagation Model, 20, 22, 40, 41, 45, 48
Propagation Over Vegetation, 253, 255
Propagation Over Water, 262
Protection, 23, 54, 144, 264, 342, 445
Protocol analysers, 136, 140, 410
Protocol Layer, 349, 393, 404
Protocol Overhead, 276, 390, 391
Protocol Stacks, 272, 341, 390
Protocol trace, 136, 140

PSK, 10, 23, 163, 166, 167, 470
PSTN, 8, 330, 338, 360
Public Land Mobile Network (PLMN), 5, 91, 376, 377, 381, 383, 403
Puncturing, 472
PVCs, 377

QoS Class, 11
QoSB, 431, 434
Quadrature Phase-Shift Keying (QPSK), 468, 475
Quality, 4, 12, 19–22
Quality of Service (QoS), 11, 68, 104, 220, 412, 421, 428
Quality Optimisation, 34
Queuing, 50, 143, 222, 413

RACH, 35, 61, 80, 83, 88, 90, 170, 182
Radiation Night, 263
Radio Access Network (RAN), 11, 275, 412, 462
Radio channel allocation technics, 164
Radio Link Control (RLC), 23, 163, 169, 174
Radio Link Management (RLM), 110
Radio Network Controller (RNC), 11, 37, 97, 98–101, 191, 275, 429
Radio Network Optimisation, 106
Radio Network Planning Process, 18, 504
Radio Parameter, 115, 183
Radio performance increase, 144
Radio Refractivity, 243, 253
Radio Resource Control (RRC), 11, 189, 429
Radio Resource Management (RRM), 26, 29
Radius of the Fresnel Zone, 245
Rain Attenuation, 256
RAKE Receivers, 29
RANAP, 275, 349, 356, 361, 401
Random Access Channel (RACH), 61, 80
Random access procedure, 94, 95
Rate Matching, 472
Rayleigh Fading, 41, 172
Real-Time Variable Bit Rate (RT-VBR), 231
Receive Block Bitmap (RBB), 180
Received Power, 38, 40, 249, 250
Receiver Diversity, 426, 427
Redundancy, 24, 162, 376, 396, 397
Reflections, 41, 44, 45, 53, 297, 484
Refractivity, 242, 243, 244, 253, 263
Reliability, 123, 191, 221, 456
Reports generation, 128
Retainability, 178, 307, 308, 409
Re-Transmissions, 173, 174, 308, 471, 472
RLC Block Transmission, 25

RNSAP, 275
Round Trip Time (RTT), 468
Routers, 386–387, 395, 403
Routing, 4, 7, 80, 84, 86, 292, 320, 323, 358, 366, 376, 384, 390
rt-VBR (Real-Time Variable Bit Rate), 231
Rural, 2, 31, 48, 117

SACCH Frame, 151
Scattering, 40
Scope of Radio Network Planning, 15, 18, 21, 107, 439, 503, 504
SCR (Sustainable Cell Rate), 209, 231, 342
Scrambling Codes, 27, 92, 186, 187
SDCCH, 57, 59, 61, 110, 133, 138
SDCCH drop, 107, 130, 133
SDH (Synchronous Digital Hierarchy), 201, 203, 204, 224
SDH Equipment Clock (SEC), 289
SDH Interface, 205, 226
SDH Network, 205–207, 288
SECBR, 232
Second-Generation, 2, 3, 5, 7, 9, 10
Second-Generation System, 3, 10
Security, 2, 3, 5, 8, 291, 410–411, 433, 436, 493, 494
Security Management, 5, 291
Segmentation and Reassembly (SAR), 173, 228–230, 345
Servers, 334, 369, 410, 428
Serving GPRS Support Node (SGSN), 11, 81, 84, 85, 328, 377, 378, 391–395, 400–408
Serving RNC, 97, 100, 275, 276, 475
Setup success, 18, 132, 175, 181, 188
Severe Error Seconds (SES), 301
Severely Errored Cell Block Ratio (SECBR), 232
Shadowing, 41
SHO process, 99
Short Message Service Centre (SMSC), 326, 352
Signaling Payload, 174
Signalling Links (SL), 296, 304, 307, 353–356, 399, 407
Signalling Network Dimensioning, 347
Signalling Overhead, 173
Signalling Plans, 399
Signalling Point Codes (SPC), 295, 318, 323, 378
Signalling Points (SP), 324, 346, 354, 355, 356, 378
Signalling Transfer Point (STP), 331, 335, 376
Signal-to-Interference Ratio (SIR), 90, 101, 102, 106

SIM Card, 26, 136
Site follow-up, 122
Site re-engineering, 114, 115, 126, 160
Site Selection, 440
Site survey, 18
Slave Clock – Synchronisation Supply Unit (SSU), 289
Slow Associated Control Channels (SAACH), 34, 61, 133
Slow Fading, 41, 47–49, 256
Smart antenna, 143, 508
Smart Radio Concept, 142
Smooth-Sphere Diffraction, 254, 255
SMS, 3, 4, 77, 82, 352, 408
SMS establishment phases, 138
Soft Capacity, 70
Soft Handover (SHO), 97–102, 479
Softer handover, 98
Spatial Multiplexing, 426, 427
Spectrum Efficiency, 73, 74, 494
Spreading, 26–28, 95, 435
Spreading and De-spreading, 27, 435
SRNC, 100, 276, 475
SS7, 134, 285, 321, 347, 352, 355, 356, 401
STANDBY, 85, 385, 395
Statistics, 106, 107, 121, 124, 127, 128, 154, 371
Statistics' extraction, 127, 128
STM (Synchronous Transport Module), 207, 216
Streaming Class, 11, 12
Sub-Destination, 316, 366, 375, 378
Subscriber Identity Module (SIM), 5, 6
Subsidence, 263
Suburban, 29, 30, 48, 57, 405
Superframe, 60
Sustainable Cell Rate (SCR), 231
Switches, 319–321, 401
Switching, 2, 4, 7, 11, 22, 219, 222, 320, 377, 397
Switching and Multiplexing, 217
Symbols, 29, 90, 165, 425, 489
Synchronisation, 5, 61, 89, 91, 92, 162, 201, 225, 288, 289, 370, 375
Synchronisation Channel (SCH), 61
Synchronisation Planning, 288, 289, 323, 324, 370
Synthesised frequency hopping, 116, 120, 125
System information, 61, 86–88, 92–94, 171
System Information Messages, 170, 171, 180

T1, 35, 201, 393, 394
Target cell evaluation, 148
TBF, 35, 66, 172–175, 177–181
TBF Blocking, 174, 175, 179, 180
TBF Performance, 157, 180
TBF Release Delay, 35, 174, 179
TCH and SDCCH availability, 132
TCH strategy, 117
TDD, 11, 97
Technical automation, 122, 124
Temporary Block Flow (TBF), 35, 66, 173, 175, 177, 178, 179, 181
TETRA, 493–506
The A Interface, 328, 329, 349, 384
The I_u Interface, 273, 276, 328, 329, 333
The I_{ub} Interface, 273, 274, 276, 279, 471, 474
The I_{ur} Interface, 100, 274, 277
Third-Generation, 10, 11, 13, 485
Third-Generation Networks, 1
Thresholds, 21, 35, 48, 104, 157
Time Slots (TS), 9, 23, 24, 35, 60, 63, 65–67, 78, 79, 163, 269–271, 355, 384, 500
Time-Division Multiple Access (TDMA), 3, 10, 25, 54, 59, 60, 73, 74, 115, 134
Timeslot allocation, 286
Time-Slot Allocation Planning, 286
Timing advance, 77, 107, 114, 133
Topology, 22, 116, 152, 155, 280, 281, 340, 351, 364, 373, 402, 406, 503
Tracing, 44, 43, 106, 110, 407, 409
Traffic, 4, 5, 9, 11, 26, 33, 34, 37, 58, 59, 60, 62–67, 69, 70, 107, 113, 121, 126, 133, 149, 227, 231, 269, 276, 278, 317–321, 337–341, 346, 369, 370, 375, 391–401, 406–415, 422, 499
Traffic Calculation, 33
Traffic Channels (TCH), 54, 60, 61, 74, 76, 116, 134, 137, 269
Traffic Distribution, 114, 319, 321, 476
Traffic Estimates, 58, 402
Traffic Management, 23, 143, 397
Traffic Mix, 306, 413, 476
Traffic Parameters, 227, 233
Traffic Routing Tool, 323, 338, 370, 375, 398
Transceiver (TRX), 19, 20, 33, 59, 74–76, 107, 113, 116, 118, 133, 142, 163
Transceiver Signalling (TRXSIG), 287
Transcoder Sub-Multiplexers (TSCM), 269
Transit Layer, 321 324, 330, 340
Transit Switch, 320
Transmission, 1–3, 11, 19, 25, 26, 29, 31, 35, 53, 54, 62, 66, 67, 73, 74, 77, 88–91, 94, 96–98, 101, 102, 104, 113, 161, 199, 204, 219, 224–226, 296

Index

Transmission Convergence (TC), 224
Transmission Network, 19, 113, 161, 189, 195, 307, 308
Transmission network auditing, 106, 113
Transmission Network Optimisation, 296, 299
Transmission Network Planning Process, 197, 198
Transmission Unit (TRU), 293
Transport channels, 88, 89, 102, 471
Transport Control Protocol (TCP), 351, 353, 295, 471, 474
TRAU, 286
Triggering, 98–101, 130, 131
Troubleshooting follow-up, 122, 124
TRX quality, 107, 133

UBR (Unspecified Bit Rate), 231
Umbrella, 4, 116, 147, 148, 149, 157
Umbrella level, 151
Unspecified Bit Rate (UBR), 231, 294
Uplink and downlink power control, 143, 156
Urban, 2, 17, 21, 39, 31, 40, 41, 42, 44, 46, 48, 50, 107, 115, 116, 117, 143, 274, 275, 405, 441, 443
User Data Protocol (UDP), 272, 273, 277, 343, 344, 345, 351, 390, 391
User Equipment (UE), 11, 37, 39, 88, 89, 91, 92, 93, 96, 97, 100, 186, 400, 403, 471, 472, 473, 475, 476, 480, 481
User-to-Network Interface (UNI), 227
USIM, 11, 494
Utilisation, 270, 271, 368, 371, 389, 422, 478
UTRAN, 11, 93, 96, 276, 329, 333, 349

Value Added Services (VAS), 317, 398
VC (Virtual Container), 205, 207, 212, 213
VCC (Virtual Channel Connection), 220, 221, 222, 226, 228, 231
VCI, 221, 226, 227, 228, 294, 295, 355
Video Call Quality, 16
Virtual Channel (VCI), 220, 221, 226, 227, 228, 294, 295, 355
Virtual Channel link, 226
Virtual Path (VPI), 226, 227, 228, 294, 295, 355
Virtual Path link, 226
Visitor Location Register (VLR), 5, 11, 76, 328
Voice Activity Factor (VAF), 278, 279
Voice Call Quality, 62, 132, 355, 346, 399, 478
Voice Mail System (VMS), 318
Voice-Activity Detection, 74, 399
VP, 220, 226, 227, 294
VPC (Virtual Path Connection), 220, 221, 222, 226
VPI (Virtual Path Identifier), 226, 227, 228, 294, 295, 355

Walfish-Ikegami Model, 40, 43, 44, 57
WAN, 205, 399, 402, 427
WAP, 381
Wave Propagation, 29, 40–45, 247, 253–267
WCDMA, 12, 26, 28–29, 34, 38–39, 57, 69–70, 72, 89, 92–93, 97, 101, 181, 183, 184, 272, 273, 274, 329, 332, 467
WCDMA carrier, 26
WCDMA Technology, 57
Wideband CDMA Mobile Switching Centre (WMSC), 350
Wireless Local-Area Network (WLAN), 333, 417, 418, 425, 485, 486, 492